Sensor Networks: Physical and Social Sensing in the IoT

Sensor Networks: Physical and Social Sensing in the IoT

Editors

Suparna De
Klaus Moessner

MDPI • Basel • Beijing • Wuhan • Barcelona • Belgrade • Manchester • Tokyo • Cluj • Tianjin

Editors
Suparna De
University of Surrey
UK

Klaus Moessner
Technische Universität Chemnitz
Germany

Editorial Office
MDPI
St. Alban-Anlage 66
4052 Basel, Switzerland

This is a reprint of articles from the Special Issue published online in the open access journal *Sensors* (ISSN 1424-8220) (available at: https://www.mdpi.com/journal/sensors/special_issues/sn_sensors).

For citation purposes, cite each article independently as indicated on the article page online and as indicated below:

LastName, A.A.; LastName, B.B.; LastName, C.C. Article Title. *Journal Name* **Year**, *Volume Number*, Page Range.

ISBN 978-3-0365-7468-4 (Hbk)
ISBN 978-3-0365-7469-1 (PDF)

Contents

About the Editors

Suparna De

Dr. Suparna De (Lecturer) is a Lecturer in Computer Science at the University of Surrey. She is a Surrey AI Fellow in the Surrey Institute for People-Centred AI and also holds an Honorary Senior Research Fellow post in the Social Research Institute at UCL. Her research interests include Internet of Things; data analytics; social computing and semantics.

Klaus Mößner

Prof. Dr. Klaus Mößner (Professor) holds the Professorship Communications Engineering at the Technische Universität Chemnitz. He also is a Professsor in Cognitive Networking at the University of Surrey and a Senior Member of IEEE. His research interests include situation awareness in shared spaces, connected autonomous mobility, and reliable wireless communication for mission critical services.

sensors

MDPI

Editorial

Sensor Networks: Physical and Social Sensing in the IoT

Suparna De [1,*] and Klaus Moessner [2]

[1] Department of Computer Science, University of Surrey, Guildford GU2 7XH, UK
[2] Faculty of Electronics and Information Technology, Chemnitz University of Technology, Str. der Nationen 62, 09111 Chemnitz, Germany
* Correspondence: s.de@surrey.ac.uk

Advances made in the Internet of Things (IoT) and other disruptive technological trends, including big data analytics and edge computing methods, have contributed enabling solutions to the numerous challenges affecting modern communities. With Gartner reporting 14.2 billion IoT devices in 2019 [1] and, according to some reports [2], a projected 30.9 billion devices to be deployed by 2025 in areas like environment monitoring [3], smart agriculture [4], smart healthcare [5] or smart cities [6], one could be tempted to think that most related issues are already resolved.

However, there remain practical challenges in large-scale and rapid deployment of sensors for diverse applications, such as problems affecting siting optimization methods and participant recruitment and incentive mechanisms. On a higher level, the deluge of data sources that drive the IoT phenomenon grows every day. With the rise of smartphone-enabled citizen sensing data via social networks or personal health devices, as well as with increasing connectedness in transport, logistics, utilities, or manufacturing domains, this range and complexity of available data calls for even more advanced data processing, mining and fusion methods than those already applied.

The goal of this Special Issue was to solicit high-quality original papers aimed at demonstrating effective and efficient deployment of sensor networks in the IoT, encompassing both physical and virtual sensor networks (through modelling) as well as social networks. Related issues of data processing, in addition to challenges of the fusion and visualisation of the resultant IoT big data, are also reflected in the published papers. This Special Issue consists of 11 papers. All submissions were strictly and thoroughly peer-reviewed by experts. These submissions cover many of the relevant research issues. In the following sections, we summarize these articles and highlight their major contributions.

In the article entitled "egoDetect: Visual Detection and Exploration of Anomaly in Social Communication Network", Pu et al [7]. present a visualisation system for the analysis of anomalies in social graphs. Moreover, the authors design a novel glyph to explore an ego's topology and the relationship between egos and alters. The proposed unsupervised method addresses the lack of labelled data in social networks, and the functionality of the developed system is demonstrated on a real-world operator's call record dataset.

The article entitled "Recognizing Context-Aware Human Sociability Patterns Using Pervasive Monitoring for Supporting Mental Health Professionals", contributed by Moura et al. [8], presents a proposal to detect context-aware sociability patterns. This would enable the identification of patterns in the periods of day in which users socialize, while also supporting the detection of abnormal behaviour and changes in daily routine. The solution presented does not detect, classify or predict mental health problems, but aims to identify situations of interest that can be further explored by mental health professionals. The work is evaluated using a well-known dataset of StudentLife, with interesting and logical results, as the identified sociability patterns show a strong positive correlation with individuals' social routine. The detection of behavioural changes is important to mental health professionals and may indicate the occurrence of mental disorders.

Citation: De, S.; Moessner, K. Sensor Networks: Physical and Social Sensing in the IoT. *Sensors* **2023**, *23*, 1451. https://doi.org/10.3390/s23031451

Received: 11 January 2023
Accepted: 11 January 2023
Published: 28 January 2023

Zurbuchen et al. [9], in their article "A Machine Learning Multi-Class Approach for Fall Detection Systems Based on Wearable Sensors with a Study on Sampling Rates Selection", present a fall detection system (FDS) using an inertial measurement unit worn at the waist and evaluated with a public dataset. Their work extends the application of machine learning classification algorithms to a multi-class problem and an investigation into the effect of the sensors' sampling rate on the performance of the FDS is performed, finding that the sampling rate of 50 Hz is generally sufficient for an accurate detection.

The article entitled "IoTCrawler: Challenges and Solutions for Searching the Internet of Things" by Iggena et al. [10] addresses the issues of the interoperability of IoT solutions and data fragmentation. The proposed IoT search framework, IoTCrawler, connects existing solutions and provides solutions for crawling, indexing and searching IoT data sources. A comprehensive evaluation combined with real-world case studies showcase the validation of the developed framework.

The paper entitled "Energy Management Expert Assistant, a New Concept" by Linan-Reyes et al. [11] presents a detailed report on a real-world deployed (2 years of testing, before full deployment) home energy management system that integrates the emerging technologies of IoT, AI, big data and expert systems for a home assistant, through a multi-objective optimization (MOP) problem. The resultant deployed system presents an interactive platform for optimized energy consumption in the home, which is both efficient and comfortable, while also improving security.

The topical theme of user device location privacy is explored in the paper "Perturbed-Location Mechanism for Increased User-Location Privacy in Proximity Detection and Digital Contact-Tracing Applications" by Lohan et al. [12], which presents perturbation-based location privacy protection, applied to location-based and proximity-based services (e.g., COVID-19 contact tracing). The approach is validated with simulation-based results in multi-floor building scenarios, enabling devices to adjust the accuracy level for location sharing with service providers.

The predictive maintenance of sensors is addressed in the article entitled "Providing Fault Detection from Sensor Data in Complex Machines That Build the Smart City" by Gascón et al. [13] through a case study of the application of a scheme of sensor data pre-processing and feature extraction to fault identification and classification in a bill-counting machine. Feature extraction is performed using the Kullback–Liebler divergence measure, enabling the visualization of the differences between normal and failure operating states, followed by fault classification using a neural network.

The application of IoT-enabled services to smart tourism scenarios is explored in the article "An Architecture for Service Integration to Fully Support Novel Personalized Smart Tourism Offerings" by Sabbioni et al. [14]. The article presents an innovative architecture for smart tourism services by integrating event and place information with transport ticketing. It successfully blends a technology-assisted experience with human-centric interaction and personalization, applying these to the Italian part of a historical pilgrim path, the "Francigena way".

The use of short texts from social networks as a data source in the IoT is the focus of the article entitled "Investigating the Efficient Use of Word Embedding with Neural-Topic Models for Interpretable Topics from Short Texts", submitted by Murakami and Chakraborty [15]. The authors study eight neural topic models, using simulation experiments with several benchmark data sets to assess the effectiveness of fine tuning and pretrained word embedding in generating interpretable topics. The paper concludes that the additional fine tuning step improves the performance of the neural topic models, a measure assessed through the topic coherence and topic diversity metrics, with GloVE [16] as the pre-trained word embedding.

"A Novel Privacy Preserving Scheme for Smart Grid-Based Home Area Networks", submitted by Ali et al. [17], presents a privacy-preserving method for smart grid-based home area networks (HAN). Using homomorphic Paillier encryption, Chinese remainder theorem, and a one-way hash function, the proposed approach enables the detection of replay and false data injection attacks.

A review article, entitled "Applications of Wireless Sensor Networks and Internet of Things Frameworks in the Industry Revolution 4.0: A Systematic Literature Review" by Majid et al. [18], completes this SI. The systematic survey focuses on research solutions and new techniques to automate industry 4.0. Among the research questions explored in this paper, prominent are those related to network intruders and network security attacks on the IoT and wireless sensor network (WSN) layers, with the authors proposing a taxonomy for these issues. Challenges related to adaptation to 6G, the environment, supply chain management and limited resources are analysed.

Acknowledgments: Finally: we, the Guest Editors would like to thank the external reviewers for volunteering their time to review and discuss the submissions. Most importantly, we want to thank all the authors for their contributions. We hope that this special issue will provide useful insights for researchers, scientists, engineers, practitioners, and academics in the field of sensor networks for the Internet of Things.

Conflicts of Interest: The authors declare no conflict of interest.

References

1. Zhou, Y.; De, S.; Wang, W.; Moessner, K.; Palaniswami, M.S. Spatial Indexing for Data Searching in Mobile Sensing Environments. *Sensors* **2017**, *17*, 1427. [CrossRef]
2. Vailshery, L.S.; IoT and non-IoT connections worldwide 2010–2025. Statista. Available online: https://www.statista.com/statistics/1101442/iot-number-of-connected-devices-worldwide/ (accessed on 3 January 2023).
3. Zhou, Y.; De, S.; Ewa, G.; Perera, C.; Moessner, K. Data-Driven Air Quality Characterization for Urban Environments: A Case Study. *IEEE Access* **2018**, *6*, 77996–78006. [CrossRef]
4. Gligoric, N.; Popovic, T.; Drajic, D.; Gajinov, S.; Krco, S. Qualitative parameter analysis for Botrytis cinerea forecast modelling using IoT sensor networks. *J. Netw. Netw. Appl.* **2022**, *3*, 129–135.
5. Catarinucci, L.; De Donno, D.; Mainetti, L.; Palano, L.; Patrono, L.; Stefanizzi, M.L.; Tarricone, L. An IoT-Aware Architecture for Smart Healthcare Systems. *IEEE Internet Things J.* **2015**, *2*, 515–526. [CrossRef]
6. De, S.; Zhou, Y.; Larizgoitia Abad, I.; Moessner, K. Cyber–Physical–Social Frameworks for Urban Big Data Systems: A Survey. *Appl. Sci.* **2017**, *7*, 1017. [CrossRef]
7. Pu, J.; Zhang, J.; Shao, H.; Zhang, T.; Rao, Y. egoDetect: Visual Detection and Exploration of Anomaly in Social Communication Network. *Sensors* **2020**, *20*, 5895. Available online: https://www.mdpi.com/1424-8220/20/20/5895 (accessed on 3 January 2023). [CrossRef] [PubMed]
8. de Moura, I.R.; Teles, A.S.; Endler, M.; Coutinho, L.R.; da Silva E Silva, F.J. Recognizing Context-Aware Human Sociability Patterns Using Pervasive Monitoring for Supporting Mental Health Professionals. *Sensors* **2021**, *21*, 86. Available online: https://www.mdpi.com/1424-8220/21/1/86 (accessed on 3 January 2023). [CrossRef] [PubMed]
9. Zurbuchen, N.; Wilde, A.; Bruegger, P. A Machine Learning Multi-Class Approach for Fall Detection Systems Based on Wearable Sensors with a Study on Sampling Rates Selection. *Sensors* **2021**, *21*, 938. Available online: https://www.mdpi.com/1424-8220/21/3/938 (accessed on 3 January 2023). [CrossRef] [PubMed]
10. Iggena, T.; Bin Ilyas, E.; Fischer, M.; Tönjes, R.; Elsaleh, T.; Rezvani, R.; Pourshahrokhi, N.; Bischof, S.; Fernbach, A.; Parreira, J.X.; et al. IoTCrawler: Challenges and Solutions for Searching the Internet of Things. *Sensors* **2021**, *21*, 1559. Available online: https://www.mdpi.com/1424-8220/21/5/1559 (accessed on 3 January 2023). [PubMed]
11. Linan-Reyes, M.; Garrido-Zafra, J.; Gil-de-Castro, A.; Moreno-Munoz, A. Energy Management Expert Assistant, a New Concept. *Sensors* **2021**, *21*, 5915. Available online: https://www.mdpi.com/1424-8220/21/17/5915 (accessed on 3 January 2023). [CrossRef] [PubMed]
12. Lohan, E.S.; Shubina, V.; Niculescu, D. Perturbed-Location Mechanism for Increased User-Location Privacy in Proximity Detection and Digital Contact-Tracing Applications. *Sensors* **2022**, *22*, 687. Available online: https://www.mdpi.com/1424-8220/22/2/687 (accessed on 3 January 2023). [CrossRef]
13. Gascón, A.; Casas, R.; Buldain, D.; Marco, Á. Providing Fault Detection from Sensor Data in Complex Machines That Build the Smart City. *Sensors* **2022**, *22*, 586. Available online: https://www.mdpi.com/1424-8220/22/2/586 (accessed on 3 January 2023). [CrossRef]
14. Sabbioni, A.; Villano, T.; Corradi, A. An Architecture for Service Integration to Fully Support Novel Personalized Smart Tourism Offerings. *Sensors* **2022**, *22*, 1619. Available online: https://www.mdpi.com/1424-8220/22/4/1619 (accessed on 3 January 2023). [CrossRef]
15. Murakami, R.; Chakraborty, B. Investigating the Efficient Use of Word Embedding with Neural-Topic Models for Interpretable Topics from Short Texts. *Sensors* **2022**, *22*, 852. Available online: https://www.mdpi.com/1424-8220/22/3/852 (accessed on 3 January 2023). [CrossRef] [PubMed]

16. Pennington, J.; Socher, R.; Manning, C.D. GloVe: Global Vectors for Word Representation. In Proceedings of the 2014 Conference on Empirical Methods in Natural Language Processing (EMNLP), Doha, Qatar, 25–29 October 2014; Association for Computational Linguistics: Doha, Qatar, 2014; pp. 1532–1543.
17. Ali, W.; Din, I.U.; Almogren, A.; Kim, B.S. A Novel Privacy Preserving Scheme for Smart Grid-Based Home Area Networks. *Sensors* **2022**, *22*, 2269. Available online: https://www.mdpi.com/1424-8220/22/6/2269 (accessed on 3 January 2023). [CrossRef]
18. Majid, M.; Habib, S.; Javed, A.R.; Rizwan, M.; Srivastava, G.; Gadekallu, T.R.; Lin, J.C.W. Applications of Wireless Sensor Networks and Internet of Things Frameworks in the Industry Revolution 4.0: A Systematic Literature Review. *Sensors* **2022**, *22*, 2087. Available online: https://www.mdpi.com/1424-8220/22/6/2087 (accessed on 3 January 2023). [CrossRef]

sensors

MDPI

Article

egoDetect: Visual Detection and Exploration of Anomaly in Social Communication Network

Jiansu Pu [1,*], Jingwen Zhang [1], Hui Shao [1], Tingting Zhang [1] and Yunbo Rao [2]

1 Visual Analytic of Big Data Lab, School of Computer Science and Engineering, University of Electronic Science and Technology of China, Chengdu 610054, China; 15928603008@163.com (J.Z.); sophyond@163.com (H.S.); tingting_uestc@163.com (T.Z.)
2 School of Information and Software Engineering, University of Electronic Science and Technology of China, Chengdu 610054, China; uestc2008@126.com
* Correspondence: jiansu.pu@uestc.edu.cn

Received: 1 September 2020; Accepted: 14 October 2020; Published: 18 October 2020

Abstract: The development of the Internet has made social communication increasingly important for maintaining relationships between people. However, advertising and fraud are also growing incredibly fast and seriously affect our daily life, e.g., leading to money and time losses, trash information, and privacy problems. Therefore, it is very important to detect anomalies in social networks. However, existing anomaly detection methods cannot guarantee the correct rate. Besides, due to the lack of labeled data, we also cannot use the detection results directly. In other words, we still need human analysts in the loop to provide enough judgment for decision making. To help experts analyze and explore the results of anomaly detection in social networks more objectively and effectively, we propose a novel visualization system, egoDetect, which can detect the anomalies in social communication networks efficiently. Based on the unsupervised anomaly detection method, the system can detect the anomaly without training and get the overview quickly. Then we explore an ego's topology and the relationship between egos and alters by designing a novel glyph based on the egocentric network. Besides, it also provides rich interactions for experts to quickly navigate to the interested users for further exploration. We use an actual call dataset provided by an operator to evaluate our system. The result proves that our proposed system is effective in the anomaly detection of social networks.

Keywords: anomaly detection; visualization; social communication; egocentric network; internet of things

---

1. Introduction

Social communication is a necessary part of people's daily life. However, it seems that we suffer from all kinds of harassment every day, like sales phone calls, robots, harassment on social platform and so on. These seriously affect our daily life. Therefore, this has led to the development of anomaly detection. To best of our knowledge, the majority of the anomaly detection methods can be divided into two categories: unsupervised [1] and supervised [2] methods. However, there are still many challenges in anomaly detection. Firstly, the valid and objective tag data is hard to gain [3]. If we collect them all manually, it will take a long time and make the data more subjective and likely to lose their meaning. The number of anomalies is usually small and anomalies come in a variety of forms [3]. Besides, in practical use, we can only infer from the behavior or some characteristics of a user in the network without knowing exactly whether he or she is an abnormal user or not, which poses a greater challenge to the accuracy of anomaly detection.

Social network is a durable research topic, and theories for analyzing social networks are also emerging, such as the structural hole theory [4], egocentric network and Dunbar's number [5]. Dunbar and Zhou discovered through research that an ordinary person's social network is hierarchical [6,7], which means that if a person lacks this hierarchical structure, it is very likely that he or she is an anomalous user. In addition, the egocentric network allows experts to start by the topology and have an intuitive understanding of the user's network, which is very helpful in identifying anomalies in social networks [8].

Visualization has a huge impact on evaluating data analysis results and mining data, and can help provide additional evidence to support ideas and conclusions. It also allows users to access the information and discover hidden connections in the data quickly by mapping the data into recognizable graphics. Therefore, it has been widely used in various fields, such as anomaly detection [9–11], social analysis [12–14] and so on. However, to design a general and effective visualization is a difficult problem, especially for anomaly detection in social networks. As the types of social network data are diverse, containing text, audio files, and video files, etc, it is hard to design a suitable model to cover all of them.

Combining the above questions and thoughts, we design a novel visualization system, egoDetect, which can explore anomalies from both global and local perspectives, and then, combine the time series to analyze users' anomalies from multiple perspectives. It can detect anomalies in the data of social networks without tags. egoDetect based on ego central network provides three views for exploring and analyzing suspicious users. (1) A macroscopic view using the features of egos, analyzes all the egos' degree of anomaly and displays from a group level. (2) Inspired by the solar, we propose a mesoscopic view to explore the nodes we are interested in. We can learn the topology of egos with alters and the characteristics of them from an ego central network perspective. (3) We also provide a microscopic view to reveal the behavior patterns and hobbies of the ego and the detail of the ego with a specific alter. In summary, our system can analyze users from three levels from multi-perspective. We also add friendly and intuitive interactions to help experts quickly get the information they want.

Our contributions in this paper as follows:

- We provide a novel visualization system for anomaly detection of social communication data, especially the unlabeled data. It combines anomaly detection algorithm with sociological theory, and then uses time sequence together for validation.
- Inspired by the solar system and the social brain hypothesis [5], we design a novel glyph to explore an ego's topology and the relationship between egos and alters.
- We use a call record data provided by an operator to demonstrate the effectiveness of our system.

2. Related Work

2.1. Anomaly Detection

Anomaly detection is to identify points which are significantly different from other data [3,15]. For example, in the field of social analysis [16–18], anomalies refer to users with anomalous behaviors compared to the general public. They may be robots or highly active anomalous users [19]. Its core is how to identify the real anomalous points and avoid the wrong partition. With the input of data, we can get the results, like scores or labels. It is a very important issue in various fields, and many methods and tools are proposed. One of them is supervised machine learning methods [2,20], using tag data training model to classify the data. Another one is unsupervised machine learning methods [1,21,22], with no training data and thus has been widely used. H. Shao et al. use multi-modal microblog content features with analysis of propagation patterns to determine veracity of microblog observations [22]. However, because of the lack of an objective evaluation system, both of them is hard to evaluate. Therefore, an increasing number of experts and scholars tend to apply visualization analysis to anomaly detection. Histogram visualization is the most mature and popular method [23] due to its easy and intuitive to use. Especially in fraud detection and denning, their behavior is easy to capture and model into histogram [24,25]. Thom et al. design a visualization system for detecting anomalies based on label cloud [10]. N. Cao et al. propose a visualization system detecting anomalous users via Twitter [8]. Our system compares to them, based on ego central network to analyze the relationship between egos and alters, using a novelty design to explore users' behavior effectively. The LOF algorithm we use can quantify the user's anomalies into scores rather than just a single label, which is very helpful for the follow-up work. Besides, our system can be used in all social communication records.

2.2. Social Network Visualization and Analysis

With the analysis of social network, we can know individuals, groups or the whole network in a more effective way [12,26]. There are many kinds of topics in social network research, such as character recognition [27], information diffusion research [22,28,29], group detection [30], etc. Z. Qin et al. use homophily to increase the diffusion accuracy in social network [22]. J. Gao et al. find that the volatility of weak ties is very important for a person to make decisions and information diffusion [29]. Visualization methods are used extensively in social networks to enable intuitive research on abstract networks [31]. Zbigniew Tarapata et al. consider applying multicriteria weighted graphs similarity (MWGSP) method to examine some properties of social networks [32]. Vincent D Blondel et al. survey the contributions made so far on the social networks and explore large-scale anonymized datasets [33]. Jian Zhao et al. incorporate machine learning algorithms to detect anomalies and present an interactive visual analysis system called FluxFlow, which also offers visualization designs for presenting the detected threads for deeper analysis [34]. Based on node link network, J.Heer et al. design a system to explore the large graph structures using visualization [35]. Nardi et al. utilize colors to distinguish the communities that exist in users' email contacts [36]. Mutton's PieSpy gives us the opportunity to research the real-time dynamic community visualization in Internet-based chat systems [37].

However, all of the above studies are focused on a specific social network, which causes to a result that one research can only be used for one purpose, and does not have good scalability and generality. As is stated above, with the development of the Internet, there have been various social communication platform, which also leads to the complexity and heterogeneity of social data. N. Cao et al. proposed an initiator-centric model and a responder-centric model to tackle this problem [38]. Based on their study, we design an ego centric based model to crush this challenge.

2.3. Ego Centric Network Visualization and Analysis

Ego centric network analysis has been widely applied in anthropology and sociology. In the method, it assumes that one node called ego is in the center, and some nodes around it called alters. From the ego centric network, we can have a deeper understanding of interested nodes, obtaining the ego's behavior and the structure with alters [39,40]. Mesoscopic and microscopic perspectives are two common starting points in it [4]. On the one hand, many researchers study the network structure and attributes of one or part of egos from a mesoscopic perspective. It is found that network size has a large effect on ego's features and the composition of the network [41]. In Lubbers et al. research, they summarized that how long a relationship can be maintained depends mainly on the strength of the relationship, the density of the network [42]. On the other hand, the microscopic level focus on network properties, alters and their behaviors' effect to the ego. L. Backstrom et al. found that the intimacy between couples can be indirectly determined based on the relationship between their mutual friends [19].

Nowadays, more and more researchers use advanced visualization methods to reveal more patterns in ego central network [43]. Shi et al. show the time dimension and the ego network's structure by a 1.5D form [44]. Node-link model is the most commonly used visualization method in ego centered networks [4,45,46], where each vertex represents a person and each edge represents the strength of the relationship between two vertices. However, this method does not directly reflect the relationship between each alter and ego, and as the number of vertices increases, visual clutter will become serious. Different from the above studies, our system uses a novel glyph based on the solar, and is more intuitive to study the topology of the network and the relationship between egos and alters.

3. Problem Description

Our goal is to design a visual analytic system that can help detect, analyze and reason about why this person in the system is considered to be the anomalous user. In this section, we will define the problems that we need to solve.

Problem Description

Generally, the data of social communication is noisy and huge. Typically we usually need to analyze on a real-time. Besides, experts with different backgrounds may have different ideas of whether a user in the network is anomalous or not. For example, Sociologists may judge whether a user is anomalous by the number of contacts he/she has; researchers of the algorithm may focus on calculating the similarity between the user and others in the network. Thus, we need to design a novel visualization system to help experts from different domains. Our goal is to find anomalies in unlabeled social data, which is difficult to analyze through only one perspective. For the follow-up study, we define the following questions.

- P1 Rate Ego's Anomaly. Traditional anomaly detection only draws a conclusion of whether the point is an anomaly or not. Then here come problems: Are the two anomalous users having the same degree of anomaly? Is there something in common or in a difference of anomalous users. Therefore, we need to quantify a user's degree of an anomaly and rank the users. Then experts can determine whether the user needs deep analysis.
- P2 Multi-perspectives Analysis. Since we know little about the unlabeled data, it easily leads to different people owning different results. In order to eliminate the influence of subjective factors, we need to design some indicators that can help us dissolve from multiple perspectives, such as the anomaly in the global social network and local topology, the time series of the user and alters anomalies.
- P3 Identify Anomalies. After we quantify the ego's anomaly from multi-perspectives, the next thing we need to do is to identify the anomalies in the social networks. In our work, we mainly focus on those who have different behaviors from the general public. e.g., the number of contacts exceeds the Dunbar's number, the topology of the user is strange, or the time series and behavior patterns are different from other people. We need to propose a novel visualization system to reveal the features and patterns of users and validate them, because they are the abstract of all types of social networks and play an important role in depicting the egos' portraits.
- P4 Generality and Scalability. With the development of social communication, there are many kinds of social ways, such as Email, Twitter, Weibo and so on, which makes the data of them more and more complex and noisy. It is time-consuming and laborious if we design different systems for each type of social communication. Therefore, we want to design a more general model, which can be used in each type of social communication.
- P5 Rich Interaction. The main goal of our system is to help experts find out the anomaly in the social network full of unlabeled data, so it is necessary to have various interactions to help experts exploit the network by a custom way. For example, we need filtering and zooming for finding the ego we are interested in and looking for the detail about the network.

4. Detection in Social Networks

Anomaly usually refers to the part of users that behave differently from other users throughout the social networks. Thus, only by identifying those weird users can we make follow-up analysis and validation. In this section, we will introduce the egocentral based data model and the metrics used in detecting anomaly in our system. Then, we combine them into anomaly detection to find out the egos we need to explore deeply.

4.1. Data and Model

Communication data records the behaviors of how people communicate with each other and how they organize their social networks. For example, the Call Detail Records (CDRs) can be used not only to study the human communication behaviors but also to analyze Ego Networks (ENs). The call detail records are collected by mobile operators for billing and network traffic monitoring. The basic information of such data contains the anonymous IDs of callers and callees, time stamps, call durations, and so on.

In order to design a system using in all social networks, the first thing we need to do is to build a general model. However, this is not an easy task, because it not only needs to summarize from a wide variety of data, but also requires to make the features meet the requirements for anomaly detection. On the basis of existing research, we build a general model based on the egocentric network. The egocentric network can reveal the topology and features of egos and is useful in understanding the egos and validating them.

An egocentric network usually consists of a central node and several other nodes surrounding it, and there is a bond that allows egos and alters to connect with each other. For social networks, it is called contact. Contact is an important measure of the intimacy between egos and alters, as well as the structure of the egos' network. Different social communication has different contact methods. For example, in the telecommunication or e-mail, it means I have a phone call or email with you, while in the Tweeter or Weibo, it means I retweet, comment or like under your tweet or vice versa.

It is a common way to use graphs to represent social networks [33]. Both directed and undirected graphs are used in the research [4,42]. General speaking, bidirectional contact usually shows stronger intimacy than unidirectional [33] and is full of research value. Thus, in order to preserve the difference and information between bidirectional contacts and unidirectional contacts, we abstract the social network into a directed graph $G(V, E)$, where V and E respectively represent the number of nodes and the number of links in the graph. $l^t_{i,j}$ is a link from i to j at start time t, and the weight $w^t_{i,j}$ means the value of contact between i and j from the start time t.While the methods to quantify contact are different in different social networks, it can be concluded that the contact between two people is measured by the total counts, the strength of each contact and the direction. Figure 1 shows our data model using the above concepts, and the longer the arrows are, the less intimate the relationship is. The dotted border indicates the alter is a bidirectional alter. The square box represents the ego and the color of the box indicates which group the ego belongs to. The color of nodes means that whether they have something in common. For telecommunication or email, it means whether they are from the same operator or service provider, while for Twitter and Weibo, it means whether they have joined in the same topic. Without special explanation, we will all represent the contact from ego to alter with contact-out, and from alter to ego with contact-in. The same to alters, we use alter-in and alter-out to represent alters who have contact-in or contact-out behavior, and we use local alter and alien alter to show whether the alter and the ego have something in common, such as interests.

4.2. Metrics Abstract

In the last section, we build a general model based on the ego central network for anomaly detection. In addition to it, abstracting metrics is another important thing. However, since the different egos may have different behaviors in different social communication methods, it requires us to deeply abstract the data, and extract the common metrics of the egos. After investigating various social networks, we designed the following metrics for analysis. For the ego i in the social network, we can divide its alters into two parts: S^i_{in} and S^i_{out}, representing the alter-in set and the alter-out set respectively. $k^i_{in} = |S^i_{in}|$ and $k^i_{out} = |S^i_{out}|$ mean the in-degree and out-degree. They can characterize the influence of the ego, which means the role of ego in the whole network. For example, the ego with more out-degree indicates that he is more willing to maintain relationships, while the one with more in-degree shows that he is more attractive to alters [47], thus suggesting the ego's size of the network.

From the previous subsection, we can infer that the $w_{i,j}^t$ is important for us to measure the importance of alter j to ego i. Therefore, p

$$W_i = \sum_{j \in S_{out}^i} \sum_t w_{i,j}^t \tag{1}$$

can quantify how ego i pays attention to his/her community. For normal egos, their total weight should not in an anomaly range. Contact-in and contact-out can help us judge whether the ego has a strong attraction and the importance of alters to egos, respectively. So we use the balance of attraction τ_i:

$$\tau_i = \frac{k_{in}^i}{k_{out}^i} \tag{2}$$

to measure the relationship between alters and egos. Among them, the closer the ratio is to 1, the more balanced the ego is, and the more stable the network structure is. The closer to 0, the greater the attraction of the ego, while the larger than 1, the weaker the ego's attraction. In the last 2 cases, they all mean anomalous. From the previous section we know that the direction of contact is of great research value. Bidirectional alters are more likely to show intimacy than unidirectional alters, and if the ego's alters are basically unidirectional, then he is very likely to be an anomalous user. We introduce relationship balance δ_i:

$$\delta_i = \frac{|S_{in}^i \cap S_{out}^i|}{|S_{in}^i \cup S_{out}^i|} \tag{3}$$

to measure the abnormal degree of the ego's relationship. $\delta = 1$ means all the alters in the network are bidirectional, while $\delta = 0$ means all the alters are unidirectional. The proportion of bidirectional and unidirectional alters in the normal user's network should be maintained within a normal range. In other words, too big or too small are both anomalous.

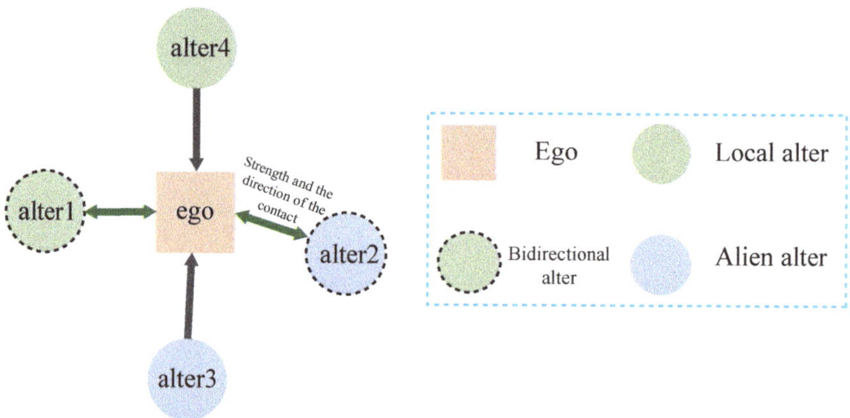

Figure 1. Data structure for social model based on egocentral network. The square box represents the ego and the color of the box indicates which group the ego belongs to. The color of the circle indicates whether they have something in common. The length of arrow tells us the relationship between ego and the alter. The color of an arrow represent the direction of contact. The dotted border indicates the alter is a bidirectional alter.

The temporal features of egos, such as posing/calling interval /frequency, can characterize their behavior habits, patterns and properties. Therefore, we use the time sequence vector T_i:

$$T_i = \{t_k^i, k = 0, 1 \cdots, 23\} \tag{4}$$

to show the behavior of the ego i. t_k^i means at time k the ego i's features. As normal egos' energy is limited, the reflect on the T_i is that the time sequence of normal egos should be regular and have or resemble a shape of hump (having meals or break) and there should have none-active place (sleeping), which means there is no behavior during this period.

Above all, we propose the following metrics:

M1. Ego network's in-degree k_{in}^i and out-degree k_{out}^i.

M2. Ego network's weight W_i.

M3. Attractiveness Balance τ_i.

M4. Relationship Balance δ_i.

M5. Time sequence vector T_i

These features can help us to make a preliminary classification of egos from a group perspective. The egos who unlike others can be found out. In order to analyze and judge these suspicious egos more deeply, we also need sociological methods to help make decisions. We will introduce them in next section.

4.3. Anomaly Detection

In anomaly detection, there is a dilemma that we have difficulty obtaining or only having a very small portion of the tag data. This makes it difficult for us to make a clear distinction between anomalies and normality. Besides, anomaly detection should be quick, but in social networks, the network shape is changing and growing at any time. These are the problems we need to solve. Finally, after careful consideration in our mind, we choose the local outlier factor (LOF) [48] model to solve the challenges. The reason why we choose it is beacuse: (1) Compared with supervised learning methods, it can be used directly without label dataset training and thus has a wider application range. (2) It can quantify the anomalies into scores instead of the labels, and let us find users who need to be explored in depth intuitively and efficiently. (3) It is a density-based detection method. Its detection results can be directly responded by dimension reduction.

An ego p's anomalous score $LOF_k(p)$ based on his k-distance neighborhood is defined as followed:

$$LOF_k(p) = \frac{\sum_{o \in N_k(p)} \frac{lrd_k(o)}{lrd_k(p)}}{|N_k(p)|} = \frac{\sum_{o \in N_k(p)} lrd_k(o)}{|N_k(p)| \cdot lrd_k(p)} \tag{5}$$

$$reach - distance_k(p, o) = max\{d_k(o), d(p, o)\} \tag{6}$$

$$d_k(p) = d(p, o) \tag{7}$$

$$lrd_k(p) = 1 / \frac{\sum_{o \in N_k(p)} reach - dist_k(p, o)}{|N_k(p)|} \tag{8}$$

$d_k(p)$ is the k-distance of p. $reach - distance_k(p, o)$ is the k-reach-distance from node o to p. It represents the maximum value between the k-distance of o and the real distance between p and o. $N_k(p)$ is k-distance neighborhood of node p, which means it is a set include all the nodes that less than or equal to the k-distance of p. $lrd_k(p)$ is the local reach-ability density of node p based on $N_k(p)$. It means the local density of the current point and its surroundings. From the equation we can know that when $lrd_k(p)$ is higher, the p is more likely to be a normal node. Above all, we can conclude that if a node's $LOF_k(p)$ is higher, it indicates the node is more different with its local k-neighbor. Generally, an ego i's feature vector is shown below:

$$f_i = \left[k_{in}^i, k_{out}^i, W_i, \tau_i, \delta_i, T_i\right] \tag{9}$$

It can describe the global properties of the ego, which is important in anomaly detection. We use $F = \{f_i, i \in G\}$, the collect of f_i and the LOF to find out a preliminary detection results, and then do more in-depth research.

5. System Design and Overview

5.1. System Design

With the development of social communication, the data in it is getting larger and larger, which is very painful for experts to analyze and thus they want to have a convenient tool to do research. After our study, we find that multi-level analysis can help them make decisions, especially in unlabeled data. In addition to these, although there have been many studies aiming at detecting anomalies in social networks, they only focus on specific types of social networks, such as blogs, e-mail or telecommunications, and fail to make full use of the common patterns. Based on these requirements, we design a novel and general visualization system, egoDetect, which can be explored from macroscopic, mesoscopic and microscopic levels. The design requirements for these three levels are as follows:

Through the model in the previous chapter, we can select some suspicious users. While displaying these users, we should also show those users who are suspected to be normal, because the results of the algorithm cannot guarantee that they are all correct and we need a comparison to find out the difference between them. Therefore, the macroscopic level view's aims are to display the whole picture of the network and help us make a preliminary classification of data.

- T1 Revealing Egos' Features and Patterns. Since the model we use is multidimensional, we need a descending dimension to display the whole egos in the network. We should ensure that the relationship between egos is not lost in dimensionality reduction, which can help us to determine which points of the network are suspected egos.
- T2 Simple and Intuitive. The view should allow researchers to intuitively understand the inside of the network by using the anomaly detection's results and though it, they can determine the node they want to study in depth is where.

Through the mesoscopic level view, the experts can select the ego they want to study deeply. Then, they need a more detailed view to help them make decisions. It is found that the relationship between two persons is important and a person's relationship with others should be stratified [5–7]. In other words, everyone's energy is limited, so people choose to spend a lot of time socializing with a small number of people and a small part of time communicating with others, but the hierarchical structure of abnormal users is generally vague. For advertisers, they are more likely to show an outward-spreading structure, that is, they like to contact other people but the relationship is not strong, while robot accounts show a high degree of intimacy with many people. Therefore, we need to reveal the relationship between the egos and alters in a mesoscopic level view.

- T3 Showing the Relationship Between Egos and Alters. As discussed above, network topology is very useful for exploring the ego. Therefore, we need to reveal the connection between egos and alters. In this way, experts can make a more in-depth analysis from a sociological perspective.
- T4 Drawing Egos' Portrait. In addition to showing the relationship between egos and alters, we also need to build a user portrait based on these data to help experts understand the whole network.

The mesoscopic view is to analyze egos from a holistic perspective, so it does not provide more detailed information about egos. Anomalous egos not only differ in topology from others, but also in many other aspects, such as the behavior, patterns, active time and so on. Sometimes abnormal egos' topology cannot be judged directly. Above all, the microscopic level view needs to provide the behavior, patterns and other detailed information.

- T5 Exploring Time Sequence. Normal egos should have a specific active time, which may vary according to their occupations, but people's energy is limited, so it should have a hump shape and 0 valued region. Robot accounts and anomalous accounts will show long-term or even full-time behavior, while others may display local and random behaviors.

- T6 Analyzing Alters. The alters are those who egos contact-in or contact-out with. They form the networks of egos. Through the abnormal score of alters, we can judge egos from the other side. Besides, when we find an interesting alter from a mesoscopic perspective and want to dig deeper, or when we want to have a deeper understanding of the ego's behavior with each alter, it can give us more information.

5.2. Solar Ego Network Model

In order to use the egocentric network to display the relationship between users and alters, we did a lot of studies and find that the traditional egocentric network view adopts node-link mode, where each edge represents the strength of the relationship between the ego and alter. However, this method does not intuitive enough, and with the increase of data, visual clutter is very serious. Therefore, we believe that a new view should be designed to clearly show the network structure and relationship with each contact in any situation.

Previous research indicates that all alters of an ego should have a hierarchy [6,7]. However, we find that anomalous users do not have this feature. Therefore, we decided to use this model to display the egocentric network and analyze egos by observing its network structure. However, here comes another question: What method should we use to determine the number of layers? There are already many researches and mature algorithms, such as Jenks Natural Breaks Classification [49] and Head/Tail method [50]. The main problem of them is that different people may have different network structures. Especially, anomaly egos have various structures, so it is hard to use algorithms to define uniform measurement indicators. Thus, we have come up with a compromise method to quantify each alter of ego according to the formula, and then map it to different layers in the graph according to the value. This ensures that different users can be properly displayed. Besides, previous studies have shown that most users' alters can be divided into five categories, so our model uses a five-layer network structure.

Through the above research, and inspired by the solar system, we design a novel view to solve it. The pipeline of it is shown in Figure 2. It consists of a central node and five layers of tracks, the closer to the central node, the more intimate with egos. The distance $\theta_{i,j}$ represents the relationship between the ego i and the alter j, The equation of $\theta_{i,j}$ is as follows:

$$k_{i,j} = \frac{max(C_{i,j}, C_{j,i})}{min(C_{i,j}, C_{j,i}))} \tag{10}$$

$$\theta_{i,j} = k_{i,j} \cdot \frac{1}{(C_{i,j} + C_{j,i})} \tag{11}$$

$C_{i,j}$ means the number of contact-out from i to j. From the equation, we can get that it is determined by the number of bidirectional contacts, so if the alter is unidirectional, the $k_{i,j} = -1$, which means the relationship between them is weak. With this glyph, experts can analyze more efficiently. With $k_{i,j}$, each alter can be placed on the corresponding layer.

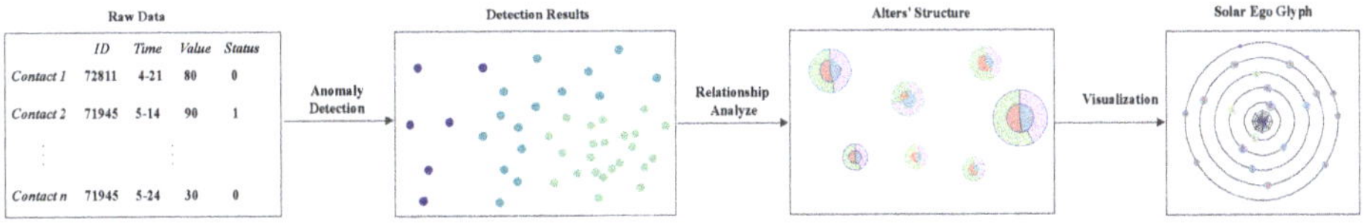

Figure 2. Workflow of the solar ego model.

5.3. System Overview

Motivated by the above requirements, we design the egoDetect to detect and analyze the anomalous users at three different scales: A group view to show the scatter of all egos through their features and anomaly scores, the topology and features in ego network with an ego view, and the more detail of the ego and between egos and alters showing in the detail view.

The data pipeline of the system is shown in Figure 3. The raw data storage in HDFS (Hadoop Distributed File System). We use Spark to model graph and compute the metrics of it, and use the results to build egos' features. With the features, we can get the anomaly detection score of each ego in the network. Then, we use multidimensional scaling (MDS) [45] to reveal the scatter of them and use a novel glyph to show the ego network of them. We design our view through the D3 and Echarts, using Flask as our framework.

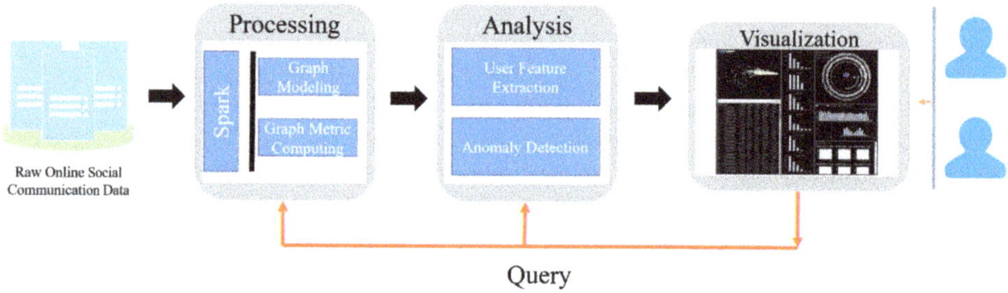

Figure 3. The data processing pipeline.

6. Visualization

The goal of our system is to assist experts in identify and verify anomalies of users in social networks which are lack of unlabelled data, and provide multiple views to validate. We also offer a lot of interactions to help researchers. In this section, based on the design tasks and requirements outlined in the previous sections, we will give a detailed description of the system and each view in the system.

6.1. User Interface

Guided by the above requirements, our system is designed as Figure 4. The interface consists of six major UI components: (a) a view mapping user's multi-dimensional features into two-dimensional space and showing in the feature space; (b) a list of whole users' detailed features sorted based on LOF anomaly scores; (c) some statistical information for each segment; (d) a novel ego network glyph inspired by the solar system for visualizing the structure of the ego's network and the relationship between alters and egos; (e) a statistical view representing the active time and habit of the ego; (f) a detail view to display the contact between ego and each alter and the anomaly of each alters. In summary, (a), (b), (c) compose our group view, (d) is our ego view, and (e) and (f) are our detail view. We will introduce the design of them as follows.

6.2. Group View

The group view is aimed at displaying the whole topology of the network and helping us make a preliminary classification of data. Multidimensional scaling (MDS) reducts based on the distance between points, so if the points are more similar, the closer they are after dimension reduction. With the using of it, we can reveal the distribution of each ego by their features and find out whether the algorithm results are effective (T1). The result is shown in Figure 4a, and in order to distinguish anomaly scores of each ego, we color code their score into five colors, representing their anomaly degree (T2). Scatter plot can help experts to have a detailed understanding of the internal situation of the social network. However, due to dimensionality reduction, on the one hand, it is impossible for experts to have an understanding of egos through x and y coordinates, so we need to provide the specific characteristics of each ego to help them make decisions. As shown in Figure 4b, we design a list based on the ranking of LOF scores to show the detail of each ego. On the other hand, the MDS dimension reduction brings similar points together. With the increase of data, visual clutter becomes more serious and we cannot understand the overall situation of the network, so we provide statistical data based on the detection results. After experiments, we find that the score of users below 1.5 are more than 90%, so we take 1.5 as the boundary and classify the users into 6 groups. We select the maximum, minimum and average values of alters and contact times for each segment. These allow us to have an intuitive understanding of each fraction. Above all, we design a statistical view to help experts analyze. (T2). The design is shown in Figure 4c.

6.3. Ego View

Through the analysis of the group view, experts can pick out the points they want to continue analyze. In this view, we need to display the relationship between the ego and all his/her alters, so that experts can explore through a sociological perspective. Figure 4d is the ego view of our system. It consists of two parts: An ego network glyph based on solar system and an ego's portrait.

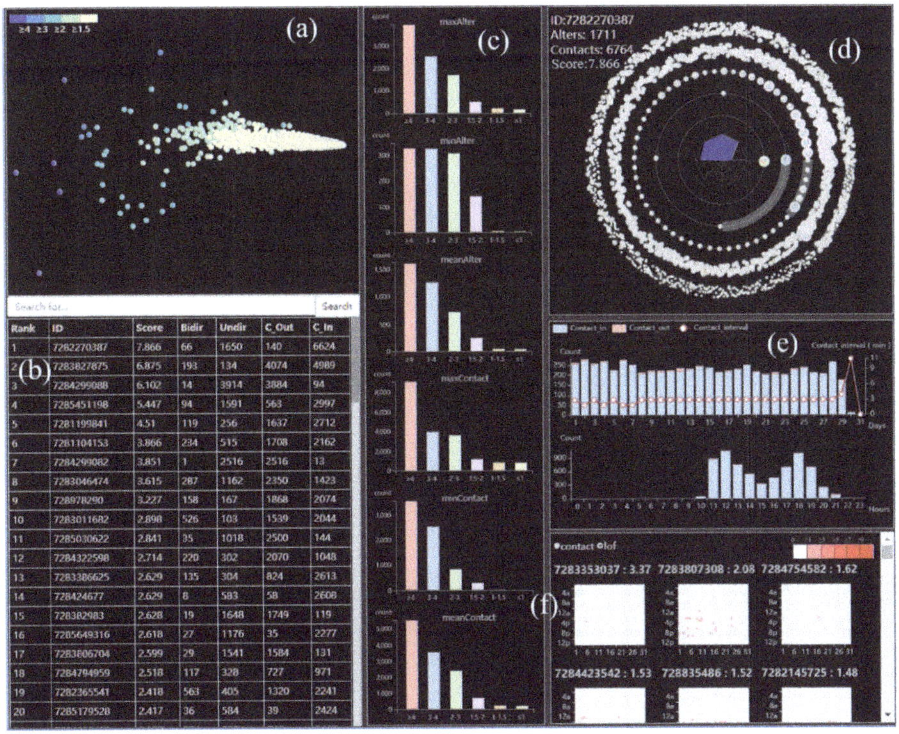

Figure 4. The overview of egoDetect based on the call record data. The user interface consists of six parts: (**a**) The distribution of users with their features, (**b**) a list sorted by users' anomaly scores, (**c**) statistical information for each segment, (**d**) the ego network glyph inspired from solar system, (**e**) the statistical view of ego's active time and behavior, (**f**) the detail view about the contact between the ego with each alter.

Ego Network Glyph. As mentioned above, we design a novel glyph to reveal the patterns between egos and alters. Figure 4d is our solar network glyph. Each alter is a node surrounding by the ego. The distance between the ego and an alter and the radius of the node represents the relationship between them. The grey transparent ring in each track tells us the proportion of local alters' number to the total number in this track, and the nodes in it mean they are the local alters, while the nodes not in it mean the alien alters. Above all, it can give us an intuitive and detailed understanding of the internal structure and circumstances of an ego network (T3). It is worth noting that how to code alters, as an important part of the network, is a very important issue. We consider three alternatives to visual encodings of each alter, showing in Figure 5. In the first design, the inner ring shows the count of contact-in and the contact-out, like the count of contacts in Twitter, while the outer ring is the value of the inner ring, such as time spending on calls. We also use the line color to show the anomaly score of the alter. However, when the circle is too small to see, it is hard to get the color of the line. The second design uses the inner and outer circle radii to encode the information of alters, and the center circle represents the anomaly score of them. When the alters become large, we find it is sometimes confused about the color coding. The third design is based on the bar chart, where each column represents one information of the alter. It is limited by shape. When the amount of data is large, the data clutter is serious. While all three designs have some drawbacks, we finally used the first design after careful consideration. Because there are egos contact with many alters in social networks, we need to make sure that they are still as clear as possible when they are presented (T3).

Portrait Glyph. Through the contact data of ego, we can conclude the various characteristics of the ego and they can help us judge whether the ego is abnormal. In addition to some common indicators, like total count of contact-in and contact-out, we introduce average relationship strength with alters, average anomaly score of alters ,local alters and alien alters, unidirectional and bidirectional alters features into the system. The average relationship strength with alters helps us analyze ego's behavior patterns. Many anomalous egos are machine users generated by software, so there will be many similar behavior patterns and like to contact with each other. Besides, for advertisers, they tend to contact-out more than contact-in and have little bidirectional alters, leading to a low level of average relationship strength with alters, which can respond from these attributes. The local alters , alien alters and unidirectional and bidirectional alters properties help us to understand the egos' alters' structure. Normal egos are more likely to contact people who have similar interests and hobbies with them and the proportion of unidirectional and bidirectional alters should be balanced. The average anomaly score of alters can tell us how the average anomaly of alters is. As mentioned above, machine users will contact other machine users. If an ego's alters' anomaly is high, the ego can also be considered an anomalous user. In the solar network layout, the ego's information is located in the middle of the whole network (T4). In order to better display ego's data from multiple dimensions, we choose the radar map as the center of the whole layout. All features of the user are mapped to each direction of the radar map, and the center of each direction means the minimum while the outside means the maximum. The color of the area represents the anomaly score of the ego. As shown in Figure 6, radar maps can simply and directly highlight important information.

6.4. Detail View

It is often not enough to only provide the overall information of an ego. Sometimes anomalous egos may have no difference with others in a holistic perspective, or we want to dig deeper about egos' behavior with alters, and thus we need more detail. Based on these, the design of a detail view is shown as Figure 4e,f. Figure 4e is a statistical view showing the time sequence information about the ego and Figure 4f represents the detail contact between the ego and each alters.

Statistical View. In addition to exploring the topology of alters and egos, time series information about the ego is also one of the most important means of analysis. It can reflect egos' social habits and behavior patterns, and is also a part of the common criterion in anomalous analysis. The active time of the ego can reflect his habit, and tell us his general lifestyle. The value of his contact-in and contact-out represents the structure of his contact. Normal users should have a regular or relatively regular active time, as well as a normal structure and number of contact. In addition, the ego's daily contact interval is also very important to measure whether he/she is normal. Experience shows that there is no regularity in the contact interval of normal people. Thus, the statistical glyph is designed as Figure 4e. We use a polygon and a histogram to display time sequence data, making it easier for us to know the ego's active time and habit (T5).

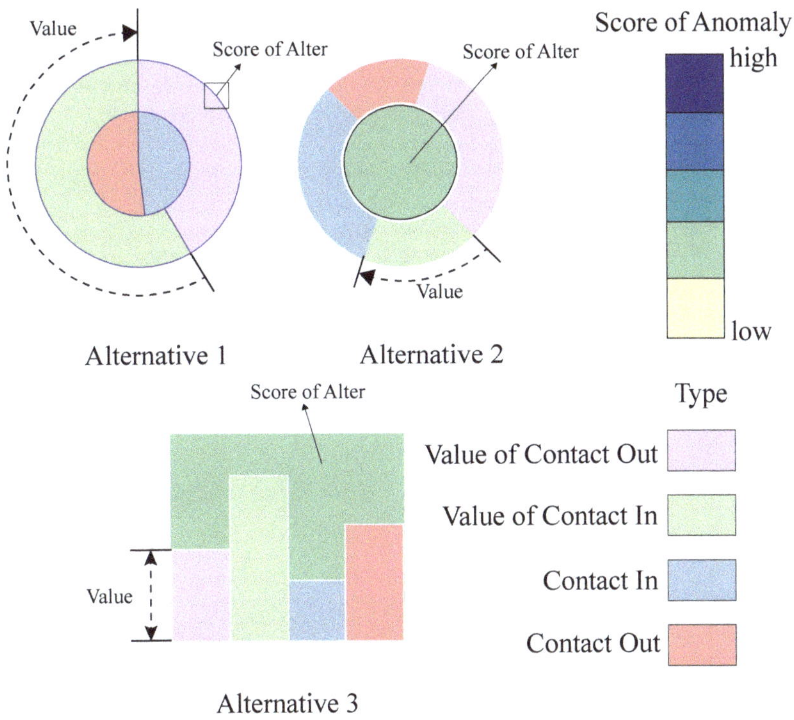

Figure 5. Design alternatives of alters.

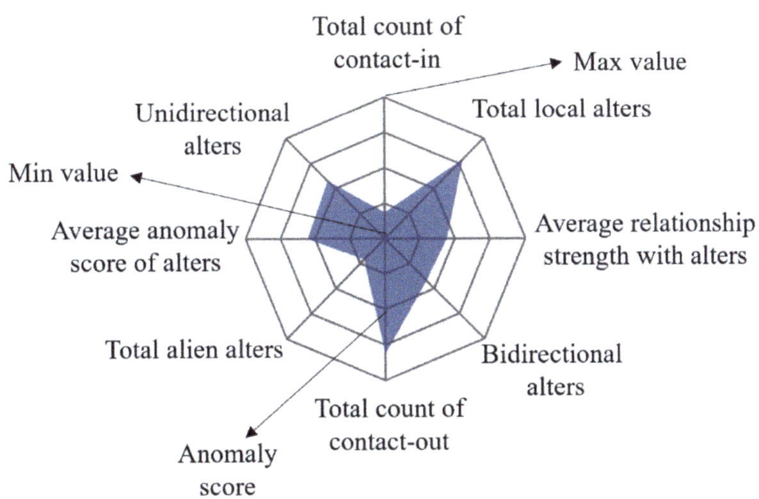

Figure 6. Ego features radar view with six dimensions.

Detail View. The contact between the egos and alters may occur at any time of the day, if we display the full time series of every day, it is very difficult to visualize in a small space, so we make a compromise, using the heat map to display the contact between the ego and each alter. The heat map shows the size of the data values in a highlighted way and can intuitively show local anomalies. In the heat map, the horizontal axis represents the time period, the vertical axis represents the hours during a day, and the color of each grid represents the number of times that the ego and the alter communicate with each other during that period. The title of each view consists of alter ID and its anomaly score. We combine the heat map with the timing information to show the ego's contact timing diagram with each alter (T6). Besides, in order to explore the anomalies of the ego from the alters, the detail view can be sort by the anomaly score from high to low or by the strength with ego and visualize them (T6). It is shown in Figure 4f. We also experiment with another design, using a circular layout. In Figure 7, each sector represents a day, and a lattice in the sector shows a period of time. However, we do not use this because the utilization of circular layout space is not high, and it is difficult to make horizontal and vertical comparisons.

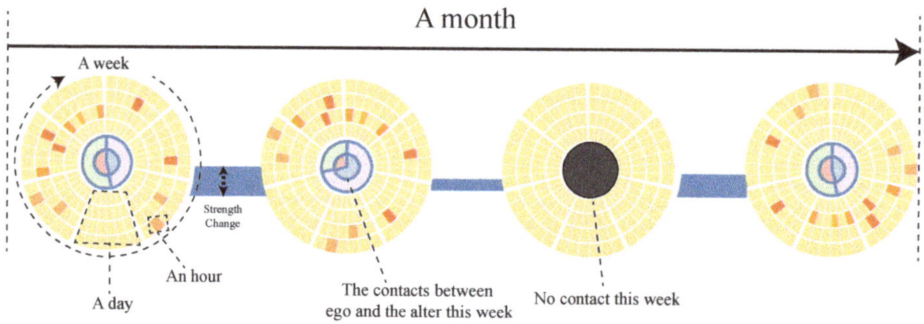

Figure 7. An alternative design for detail view based on circular layout.

6.5. User Interactions

We add some interactions in the system to help the researchers to use the system more smoothly.

Filter. In order to help the experts have an understanding of the users of different score segments in the network, we provide filtering interaction through the color bar in the Figure 4a.

Search. The search interaction is provided by the search bar in Figure 4b. The researchers can search for the ego they find interesting in the Figure 4a.

Highlight. Most of the elements in the system can be highlighted. For example, when the mouse is over one line in the Figure 4b, the node of this line can be highlighted in the Figure 4a synchronously. When the mouse is over the alters in the Figure 4d, the location of it can be found and highlighted in the Figure 4f.

Zoom In and Out. The view in the Figure 4a can be zoomed in and out for the detail.

Click. If we find an ego in the Figure 4a,b or an alter in the Figure 4d we want to know more, we can click on the line or the circle, then the ego view or the detail view will be drawn.

Tips. Our system can give you some tips when you are confused or forget something.

7. Case Study

In this section, we will apply our system in the task of anomaly detection with a call record data provided by an operator to demonstrate the effectiveness of our system.

In this study, the dataset is provided by one of the largest mobile operators in China. It covers 7 million people of a Chinese provincial capital city for half a year spanning from January to June 2014. According to the operator, all the users can be divided into two categories the local users (customers of the mobile operator who provide this dataset) and the alien users (customers from the other operators). The reason for such distinction is that the communication behaviors of alien users are not recorded and cannot be collected by our data provider based on policies. Therefore, we have to put our focus on the local users whose entire calling behaviors are recorded within the dataset. We won't show the alien alters' details, such as the score of them in the detail view and ego view. In order to protect the user's personal information, we encrypt all telephone numbers and embody the characteristics of local and alien directly on the user ID. So 728 indicates local user, and 719 indicates alien user. Each user has his own unique ID. The basic statistics of the mobile communication data are summarized in Table 1.

Table 1. Basic statistics of the mobile communication networks.

Time	N_t (*Total* Users)	N_l (*Local* Users)	L_t (*Total* Links)
Jan.	6520121	751643	32521180
Feb.	6234877	742504	27600221
Mar.	6481767	783751	32720452
Apr.	6526250	777486	32383231
May	6561107	787614	34119390
Jun.	6531076	787156	33461297

Parameter Selection. As described in Chapter 4, for the LOF algorithm, the most important parameter is its neighbor number n. We find that the n has a great influence on the score of egos in sparse and dense boundaries. In order to ensure that the points in these areas can be classified more accurately, we need to compare them with enough points, so we select a larger n, which can ensure that the points in dense or sparse areas do not have an impact, but can give more full consideration to the points in the boundary. In feature selection, we find that if we do not consider the time sequence, some points with anomalous behavior cannot be detected. We found that because their attributes, such as the number of alters, the number of calls and so on, are not different from normal people, leading to the failure to detect. However, when we introduce the time sequence, the problem is improved.

Exploratory Analysis. First of all, we have made a preliminary exploration to get the whole picture of the entire dataset. From the group view's MDS map, shown in Figure 4a and containing all the ego information, we can find that most of the points inside the network are concentrated together, and only a small number of points are distributed in the periphery with high anomaly scores. This shows that most users are regarded as normal users when they gather together, while a few users are regarded as outliers when they are distributed at the edge. When we zoom in on the view for observation, we find that egos with higher scores also appear in places with lower scores, where is a place worth studying.

From the list in Figure 4b, we notice that nine egos with more than three points, and from the Figure 4c, the data of each segment shows a downward trend. We also find that the number of alters with anomaly scores less than 1 does not exceed 150, but they make up only about 6% of the population, while the percentage of egos whose anomaly scores are less than 1.5 is about 95%, and the alters' number of them is no more than 216, which is larger than the Dunbar's Number. We think the reason for this is that communication is bidirectional firstly, which means that you may receive calls you don't want to answer, leading to an increase in the number of contacts. Secondly, it may be affected by the algorithm, and there is the possibility of misclassification. We need a deep analysis to validate.

Results. In order to further verify the effectiveness of the system, we proceed from the actual case and demonstrate the system. We have drawn ego views of all users with ratings greater than 3. As shown in Figure 8, their alters and calls are particularly numerous. They mainly show two kinds of structure, either focusing on the outer layer, showing the characteristics of advertisement users, or mainly focusing on the inner layer, showing the characteristics of robots. From their central radar maps, we can see that they have distinct convex shapes. Figure 4f is the first ego's detail view. We find that this one contacts many users with higher anomaly scores and contacts in more than contacts out, through understanding, we find that this is a customer service of the scam group. For the second ego, shown in Figure 9, we find that he is a highly active user with average contact interval, so we think he is a robot account. Those users who scores more than three points can initially identify anomalous users from the group view, which is further confirmed by the analysis of ego view.

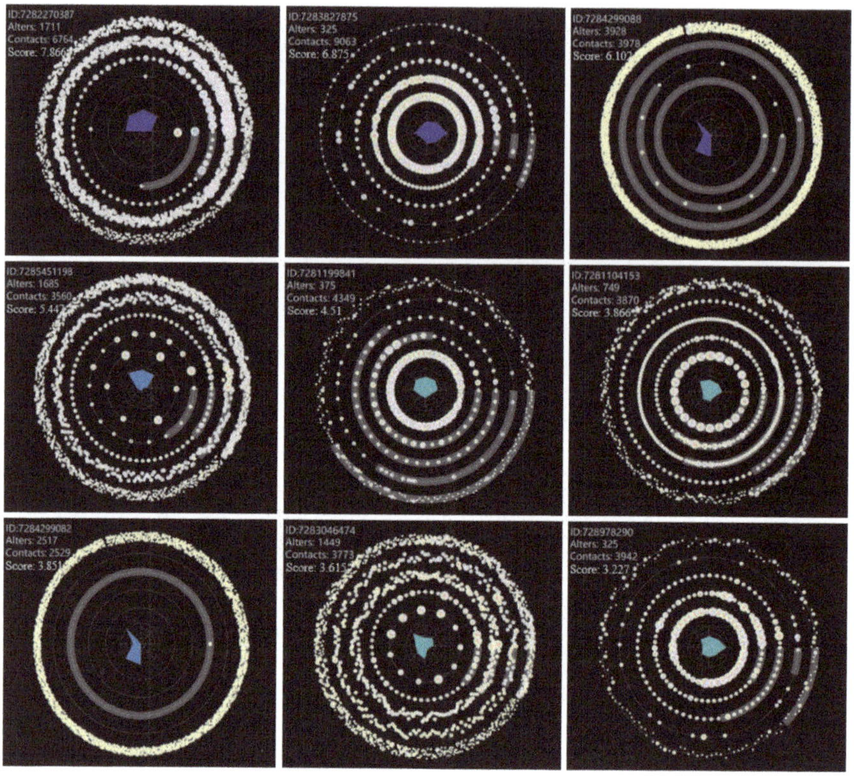

Figure 8. Top nine users' ego views. Decrease from left to right and from top to bottom.

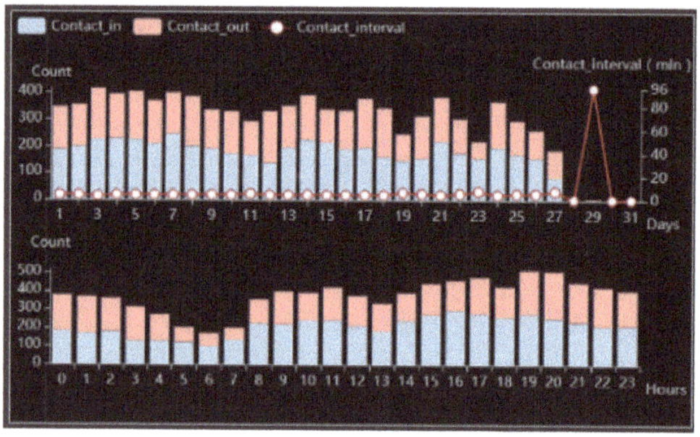

Figure 9. Second ego's statistical view.

At the same time, some egos cannot directly judge whether they are anomalous users through group view, such as ego: 7285322362, shown in Figure 10a, whose alters and calls are not high, but from his solar ego glyph we can find that his connection with alters is weak, and from his detail view we can also find that his active time is different from normal people, so he is an abnormal ego. As mentioned above, there are some nodes with high anomaly scores in an area full of low scores. We speculate that they may be normal egos with abnormal behavior or abnormal egos disguised as normal egos. So we have a detailed analysis of these points. Like ego: 7281468187, shown in Figure 10b, his score is 1.738, and can be considered as an anomaly ego. However, after our exploration, we find that although the behavior pattern is slightly different from that of ordinary people, the network structure of his contacts is not abnormal, so we think that this is a normal ego who shows abnormal behavior. While the above ego has high anomaly scores, further analysis shows that he is not really abnormal, just because he behaves differently from normal egos. In the dense areas, we believe that there are also really anomaly egos. As shown in Figure 10c, it is an anomalous ego that we have found. While we can't tell if he's abnormal from his solar ego glyph, when we go deep into his behavior patterns, we find that he has signs of full-time activity, so we conclude that he's an abnormal user. We think that this kind of nodes is anomalous egos who want to mix up with normal egos and disguise as normal egos, but they can still be detected by our system. After communicating with experts, they confirm that our speculate is correct and show that our research is very helpful for them to mine potential abnormal users.

We have fully investigated the anomaly detection results by our system and many interesting patterns are found. We also invite several experts from our data provider and telecom data analysis field to help us understand and check our findings. First, we show them the abnormal users with high scores, as shown in Figure 8. They share the same idea with us. For example, the first ego, ID:7282270387, is a customer service of the scam group, while the second user, ID:7283827875, is a robot account. Next, we verify some controversial nodes, like nodes with high scores in an area full of low scores, with them. They all feel that our discovery is valuable and prove that our discovery is useful. They verify that, for example, ego:7285321419, shown in Figure 10c, is indeed a potential abnormal user. However, the results involve the privacy of users, so they do not disclose more detail to us.

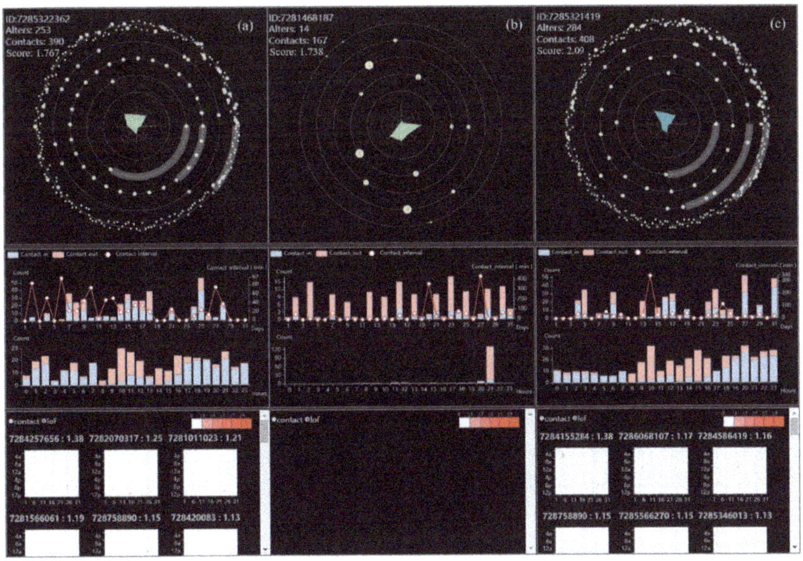

Figure 10. Some examples; (**a**) is the ego who cannot be judged directly by ego view, (**b**) is the ego who is misclassified, (**c**) is the ego who intends to disguise as a noraml one, but actually is an anomlous ego.

8. Discussion

The egoDetect, a method proposed in this paper, mainly uses visualization to capture and analyze the anomalies in online social network from the perspective of ego network. The analysis of communication data has been verified by experts, which proves the effectiveness and usefulness of the method. When designing the corresponding visual coding and layout algorithm, scalability is our special concern. Our method starts from the macroscopic level, looks for suspicious anomalies by comparing the status of all nodes in the network. Then, it analyzes suspicious objects by ego network from the mesoscopic level. Finally, from the microscopic level, it analyzes suspicious objects using time series data. In other words, through the high-level abstraction and generalization of the objects, we ensure the scalability of the method. Based on this, we can focus more on the objects themselves, rather than how many attributes there are and what each attribute means. Therefore, this method has good scalability and generality, and can perform well in different fields of time series data. For example, in the field of IOT and cyber-physical social systems, sensors, the basis for data collection, are the most critical part. If the sensors function abnormally or are attacked maliciously, the validity of the data cannot be guaranteed, and the processing and analysis cannot be carried out. Therefore, it is very important to detect abnormal sensors. Just as people can be compared to sensors, and vice versa. Besides, the data collected by sensors can be compared to the call data of people. Through anomaly detection and visual analysis of all sensors, we can quickly target dubious sensors. Then we can analyze dubious sensors based on the visualization of ego network and time series data, and finally we can find out abnormal sensors to ensure the security of the whole network and systems. Therefore, our method has strong scalability, and has great value for sensor networks, cyber-physical social systems and IOT.

9. ConclusionS

In this paper, we propose a novel visualization system, which has novel visual glyphs and uses multi-view to explore, detect and analyze the anomaly in the social network. The system analyzes from macroscopic, mesoscopic and microscopic perspectives. We show the abnormal situation in the online social communication network after anomaly detection from a macroscopic point of view; in the mesoscopic view, we introduce galaxy maps, combined with the ego central network analysis method, to display the interested users in multi-dimensional, from the network structure, active time, alters intimacy and other aspects to judge the abnormal degree of users; and through the microscopic view, combined with timing, we can evaluate the abnormal degree of users from the point of view of alters. We also add friendly and intuitive interactions to help researchers quickly get the information they want. We use a call record data to demonstrate the system is beneficial for detecting abnormal behavior in online social communication. We also discuss the feasibility of applying the method to other fields, like IOT and cyber-physical social systems.

However, limited by time and energy, our work still has a lot of room to improve. Through the case study, we find that although the LOF algorithm can help us to mine latent anomalous egos by combining time series, it also incorrectly classifies some normal errors. Restricted by datasets, it is difficult for us to analyze alien alters and egos, which is disadvantageous to our analysis.

In the future, we plan to design better anomaly detection algorithms. This can make our detection accuracy higher. Besides, the dataset used in this experiment is only provided by a certain operator, so there are limitations in the analysis of specific contacts. In the follow-up experiments, we hope to deepen cooperation with other operators, obtain more and more communication data from the external network, and conduct more in-depth research.

Author Contributions: Conceptualization, J.P.; methodology, J.P. and J.Z.; software, J.Z. and H.S.; validation, J.P. and J.Z.; formal analysis, J.P. and J.Z. and H.S. and T.Z.; resources, J.P.; data curation, J.Z. and T.Z.; writing—original draft preparation, J.P. and J.Z.; writing—review and editing, J.P. and J.Z. and Y.R.; visualization, J.Z. and H.S.; supervision, J.P. and Y.R.; project administration, J.P. All authors have read and agreed to the published version of the manuscript.

Funding: This research was funded by the National Natural Science Foundation of China, grant number 61872066 and U19A2078 and the Science and Technology project of Sichuan, number 2020YFG0056, 2019YFG0504 and 2020YFG0459 and the Science and Technology Service Industry Demonstration project of Sichuan, number 2019GFW126 and the Aeronautic Science Foundation of China, number 20160580004.

Conflicts of Interest: The authors declare no conflict of interest.

References

1. Eskin, E.; Arnold, A.; Prerau, M.; Portnoy, L.; Stolfo, S. *A Geometric Framework for Unsupervised Anomaly Detection*; Springer: Boston, MA, USA, 2002.
2. Steinwart, I.; Hush, D.; Scovel, C. A Classification Framework for Anomaly Detection. *J. Mach. Learn. Research.* **2005**, *6*, 211–232.
3. Chandola, V.; Banerjee, A.; Kumar, V. Anomaly detection: A survey. *ACM Comput. Surv.* **2009**, *41*, 1–58. [CrossRef]
4. Yanhong, W.; Naveen, P.; Jian, Z.; Sixiao, Y.; Guowei, H.; Huamin, Q. egoSlider: Visual Analysis of Egocentric Network Evolution. *IEEE Trans. Visual Comput. Graph.* **2015**, *22*, 260–269.
5. Dunbar, R.I.M. The Social Brain Hypothesis. *Evol. Anthropol. Issues News Rev.* **1998**, *6*, 178–190. [CrossRef]
6. Hill, R.A.; Dunbar, R.I.M. Social network size in humans. *Hum. Nat.* **2003**, *14*, 53–72. [CrossRef] [PubMed]
7. Zhou, W.X.; Sornette, D.; Hill, R.A.; Dunbar, R.I. Discrete hierarchical organization of social group sizes. *Proc. Biol. Sci.* **2005**, *272*, 439–444. [CrossRef] [PubMed]
8. Nan, C.; Conglei, S.; Sabrina, L.; Jie, L.; Yu-Ru, L.; Ching-Yung, L. TargetVue: Visual Analysis of Anomalous User Behaviors in Online Communication Systems. *IEEE Trans. Vis. Comput. Graph.* **2015**, *22*, 1.
9. Miao, X.; Liu, K.; He, Y.; Liu, Y.; Papadias, D. Agnostic Diagnosis: Discovering Silent Failures in Wireless Sensor Networks. *IEEE Trans. Wirel. Commun.* **2013**, *12*, 6067–6075. [CrossRef]
10. Thom, D.; Bosch, H.; Koch, S.; W'Orner, M.; Ertl, T. Spatiotemporal Anomaly Detection through Visual Analysis of Geolocated Twitter Messages. In Proceedings of the 2012 IEEE Pacific Visualization Symposium, Songdo, Korea, 28 February–2 March 2012.
11. Tao, J.; Lei, S.; Zhou, Z.; Huang, C.; Yu, R.; Su, P.; Wang, C.; Yang, C. Visual Analysis of Collective Anomalies Through High-Order Correlation Graph. In Proceedings of the 2018 IEEE Pacific Visualization Symposium (PacificVis), Kobe, Japan, 10–13 April 2018.
12. Abbasi, A.; Chung, K.S.K.; Hossain, L. Egocentric analysis of co-authorship network structure, position and performance. *Inf. Process. Manag.* **2012**, *48*, 671–679. [CrossRef]
13. Halgin, D.S.; Borgatti, S.P. An introduction to personal network analysis and tie churn statistics using E-NET. *Connections* **2012**, *32*, 37–48.
14. Jarvenpaa, S.L.; Majchrzak, A. Knowledge Collaboration Among Professionals Protecting National Security: Role of Transactive Memories in Ego-Centered Knowledge Networks. *Organ. Sci.* **2014**, *19*, 260–276. [CrossRef]
15. Jie, Z.; Li, Y.; Liu, R. Social Network Group Identification based on Local Attribute Community Detection. In Proceedings of the 2019 IEEE 3rd Information Technology, Networking, Electronic and Automation Control Conference (ITNEC), Chengdu, China, 15–17 March 2019; pp. 443–447.
16. Weixin, L.; Vijay, M.; Nuno, V. Anomaly detection and localization in crowded scenes. *IEEE Trans. Pattern Anal. Mach. Intell.* **2013**, *36*, 18–32. [CrossRef]
17. Smith, M.A. NodeXL: Simple Network Analysis for Social Media. In Proceedings of the International Conference on Collaboration Technologies and System, San Diego, CA, USA, 20–24 May 2013.
18. Ghani, N.A.; Hamid, S.; Hashem, I.A.T.; Ahmed, E. Social media big data analytics: A survey. *Comput. Hum. Behav.* **2019**, *101*, 417–428. [CrossRef]
19. Backstrom, L.; Kleinberg, J. Romantic Partnerships and the Dispersion of Social Ties: A Network Analysis of Relationship Status on Facebook. In *Proceedings of the 17th ACM Conference on Computer supported Cooperative Work Social Computing*; Association for Computing Machinery: New York, NY, USA, 2014.
20. Phua, C.; Alahakoon, D.; Lee, V. Minority Report in Fraud Detection: Classification of Skewed Data. *ACM Sigkdd Explor. Newsl.* **2004**, *6*, 50–59. [CrossRef]
21. Zhang, K.; Shi, S.; Hong, G.; Li, J. *Unsupervised Outlier Detection in Sensor Networks Using Aggregation Tree*; Springer: Berlin/Heidelberg, Germany, 2007.
22. Qin, Z.; You, Z.; Jin, H.; Gan, X.; Wang, J. Homophily-Driven Evolution Increases the Diffusion Accuracy in Social Networks. *IEEE Trans. Netw. Sci. Eng.* **2020**, 1. [CrossRef]
23. Portnoy, L.; Eskin, E.; Stolfo, S. Intrusion Detection With Unlabeled Data Using Clustering. In Proceedings of the ACM CSS Workshop on Data Mining Applied to Security (DMSA-2001), Philadelphia, PA, USA, 5–8 November 2001.

24. Fawcett, T.; Provost, F. Activity Monitoring: Noticing interesting changes in behavior. In Proceedings of the Fifth ACM SIGKDD International Conference on Knowledge Discovery and Data Mining, San Diego, CA, USA, 15–18 August 1999.
25. Eskin, E. Modeling system calls for intrusion detection with dynamic window sizes. In Proceedings of the Proceedings DARPA Information Survivability Conference and Exposition II. DISCEX'01, Anaheim, CA, USA, 1–14 June 2001.
26. Lorrain, F.; White, H.C. Structural Equivalence of Individuals in Social Networks †. *Social Netw.* **1977**, *1*, 67–98.
27. Yan, Q.; Wu, L.; Zheng, L. Social network based microblog user behavior analysis. *Phys. Stat. Mech. Appl.* **2013**, *392*, 1712–1723. [CrossRef]
28. Wei, J.; Bing, B.; Liang, L. Estimating the diffusion models of crisis information in micro blog. *J. Inf.* **2012**, *6*, 600–610. [CrossRef]
29. Gao, J.; Schoenebeck, G.; Yu, F.Y. The Volatility of Weak Ties: Co-Evolution of Selection and Influence in Social Networks. In Proceedings of the 18th International Conference on Autonomous Agents and MultiAgent Systems, Montreal, QC, Canada, 13–17 May 2019.
30. Papadopoulos, S.; Kompatsiaris, Y.; Vakali, A.; Spyridonos, P. Community detection in Social Media. *Data Min. Knowl. Discov.* **2012**, *24*, 515–554. [CrossRef]
31. Bakshy, E.; Eckles, D.; Yan, R.; Rosenn, I. Social Influence in Social Advertising: Evidence from Field Experiments. In Proceedings of the 13th ACM Conference on Electronic Commerce, Valencia, Spain, 4–8 June 2012.
32. Jie, T.; Sun, J.; Chi, W.; Zi, Y. Social influence analysis in large-scale networks. In Proceedings of the 15th ACM SIGKDD international conference on Knowledge discovery and data mining, Paris, France, 28 June–1 July 2009.
33. Blondel, V.D.; Decuyper, A.; Krings, G. A survey of results on mobile phone datasets analysis. *Epj Data Sci.* **2015**, *4*, 1–55. [CrossRef]
34. Pizzuti, C. Evolutionary Computation for Community Detection in Networks: A Review. *IEEE Trans. Tran. Comput.* **2018**, *22*, 464–483. [CrossRef]
35. Heer, J.; Boyd, D. Vizster: Visualizing Online Social Networks. In Proceedings of the IEEE Symposium on Information Visualization, Minneapolis, MN, USA, 23–25 October 2005.
36. Nardi, B.A.; Whittaker, S.; Isaacs, E.; Creech, M.; Johnson, J.; Hainsworth, J. Integrating communication and information through ContactMap. *Commun. ACM* **2002**, *45*, 89–95. [CrossRef]
37. Mutton, P. Inferring and visualizing social networks on Internet chat. In Proceedings of the Eighth International Conference on Information Visualisation, London, UK, 14–16 July 2004.
38. Xing, H.; Liu, Y.; Gao, J.; Chen, S. Episogram: Visual Summarization of Egocentric Social Interactions. *IEEE Comput. Graph. Appl.* **2016**, *36*, 72–81.
39. Xiong, R.; Donath, J. PeopleGarden: Creating Data Portraits for Users. In Proceedings of the Acm Symposium on User Interface Software & Technology, San Antonio, TX, USA, 28 February–2 March 1999.
40. Lei, S.; Tong, H.; Jie, T.; Lin, C. Flow-Based Influence Graph Visual Summarization. In Proceedings of the IEEE International Conference on Data Mining, Shenzhen, China, 14–17 December 2014 .
41. Roberts, S.G.B.; Dunbar, R.I.M.; Pollet, T.V.; Kuppens, T. Exploring variation in active network size: Constraints and ego characteristics. *Soc. Netw.* **2009**, *31*, 138–146. [CrossRef]
42. Lubbers, M.J.; Molina, J.L.; Lerner, J.; Brandes, U.; Ávila, J.; Mccarty, C. Longitudinal analysis of personal networks. The case of Argentinean migrants in Spain. *Soc. Netw.* **2010**, *32*, 91–104. [CrossRef]
43. Chen, S.; Chen, S.; Wang, Z.; Liang, J.; Yuan, X.; Cao, N.; Wu, Y. D-Map: Visual analysis of ego-centric information diffusion patterns in social media. In Proceedings of the IEEE Conference on Visual Analytics Science and Technology, Baltimore, MD, USA, 23–28 October 2016.
44. Lei, S.; Chen, W.; Zhen, W.; Huamin, Q.; Chuang, L.; Qi, L. 1.5D Egocentric Dynamic Network Visualization. *IEEE Trans. Vis. Comput. Graph.* **2015**, *21*, 624–637.
45. Liu, D.; Guo, F.; Deng, B.; Qu, H.; Wu, Y. egoComp: A node-link-based technique for visual comparison of ego-networks. *Inf. Vis.* **2016**, *16*, 179–189. [CrossRef]
46. Liu, Q.; Hu, Y.; Lei, S.; Mu, X.; Jie, T. EgoNetCloud: Event-based egocentric dynamic network visualization. In Proceedings of the IEEE Conference on Visual Analytics Science and Technology, Chicago, IL, USA, 25–30 October 2015.

47. Wang, Q.; Pu, J.; Guo, Y.; Hu, Z.; Tian, H. *egoPortray: Visual Exploration of Mobile Communication Signature from Egocentric Network Perspective*. In Proceedings of the International Conference on Multimedia Modeling, Reykjavik, Iceland, 4–6 January 2017.
48. Breunig, M.M. LOF: identifying density-based local outliers. In Proceedings of the Acm Sigmod International Conference on Management of Data, Dallas, TX, USA, 16–18 May 2000.
49. Jenks, G. The Data Model Concept in Statistical Mapping. *Int. Yearb. Cartogr.* **1967**, 7, 186–190.
50. Jiang, B. Head/Tail Breaks: A New Classification Scheme for Data with a Heavy-Tailed Distribution. *Prof. Geogr.* **2013**, 65, 482–494. [CrossRef]

sensors

MDPI

Article

Recognizing Context-Aware Human Sociability Patterns Using Pervasive Monitoring for Supporting Mental Health Professionals

Ivan Rodrigues de Moura [1,†,*], Ariel Soares Teles [1,2], Markus Endler [3], Luciano Reis Coutinho [1] and Francisco José da Silva e Silva [1]

1 Laboratory of Intelligent Distributed Systems (LSDi), Federal University of Maranhão, 65080-805 São Luís, Brazil; ariel.teles@ifma.edu.br (A.S.T.); luciano.rc@ufma.br (L.R.C.); fssilva@lsdi.ufma.br (F.J.d.S.eS.)
2 Federal Institute of Maranhão, 65570-000 Araioses, Brazil
3 Department of Informatics, Pontifical Catholic University of Rio de Janeiro, 22453-900 Rio de Janeiro, Brazil; endler@inf.puc-rio.br
* Correspondence: ivan.rodrigues@lsdi.ufma.br

† This paper is an extended version of our paper published in I. Rodrigues de Moura, F. José da Silva e Silva, L. Reis Coutinho and A. Soares Teles, "Mental Health Ubiquitous Monitoring: Detecting Context-Enriched Sociability Patterns Through Complex Event Processing", in Proceedings of the 2020 IEEE 33rd International Symposium on Computer-Based Medical Systems (CBMS), Rochester, MN, USA, 28–30 July 2020; pp. 239–244, doi:10.1109/CBMS49503.2020.00052.

Abstract: Traditionally, mental health specialists monitor their patients' social behavior by applying subjective self-report questionnaires in face-to-face meetings. Usually, the application of the self-report questionnaire is limited by cognitive biases (e.g., memory bias and social desirability). As an alternative, we present a solution to detect context-aware sociability patterns and behavioral changes based on social situations inferred from ubiquitous device data. This solution does not focus on the diagnosis of mental states, but works on identifying situations of interest to specialized professionals. The proposed solution consists of an algorithm based on frequent pattern mining and complex event processing to detect periods of the day in which the individual usually socializes. Social routine recognition is performed under different context conditions to differentiate abnormal social behaviors from the variation of usual social habits. The proposed solution also can detect abnormal behavior and routine changes. This solution uses fuzzy logic to model the knowledge of the mental health specialist necessary to identify the occurrence of behavioral change. Evaluation results show that the prediction performance of the identified context-aware sociability patterns has strong positive relation (Pearson's correlation coefficient >70%) with individuals' social routine. Finally, the evaluation conducted recognized that the proposed solution leading to the identification of abnormal social behaviors and social routine changes consistently.

Keywords: mental health; pervasive computing; context awareness; sociability; social behavior; sociability pattern

Citation: Moura, I.; Teles, A.; Endler, M.; Coutinho, L.; Silva, F. Recognizing Context-Aware Human Sociability Patterns Using Pervasive Monitoring for Supporting Mental Health Professionals. *Sensors* **2021**, *21*, 86. https://dx.doi.org/10.3390/s21010086

Received: 27 September 2020
Accepted: 15 December 2020
Published: 25 December 2020

1. Introduction

Mental health refers to the psychological, social, and emotional well-being, so influencing our behaviors, feelings, and thoughts. Mental well-being contributes to individuals perceive their skills, work productively, contribute to their community, interact with other people, and recover from their daily routine stresses [1]. Mental disorder is a term used to describe mental health problems, such as depression, schizophrenia, and social anxiety. These disorders are responsible for affecting aspects such as mood, sleep, personality, thoughts, and social relationships [2]. Mental disorders are a health problem prevalent in a large part of the world population, affecting about 700 million people worldwide [3].

Depression is a mental disorder that affects more than 300 million people worldwide, while around 800,000 people commit suicide each year [4]. Therefore, it is possible to recognize that the prevalence of mental health problems has reached a significant part of the world population.

Mental disorders cause behavioral change that represents a relevant indicator of their onset, presence, or development. These behavioral changes are situations of interest to mental health professionals since they are used as a basis for performing assessments and interventions. In particular, social behavior changes can represent relevant indicators of mental disorders as individuals' sociability has significant implications for their state of well-being [5]. Social relationships' characteristics can represent aspects capable of protecting or contributing to the development of mental disorders. For example, there is evidence that social support is a relevant factor for mental health [6,7], since there is a higher likelihood of depression among people who do not have social support. There is also evidence that social isolation is associated with mental health problems, such as depression, anxiety, and suicidal ideation [8]. Moreover, social isolation imposed to reduce the rate of contagion by the COVID-19 coronavirus may further impact global mental health [9]. Therefore, symptoms of mental disorders can be externalized through changes in social behaviors, so characterizing a situation of interest for monitoring mental health.

Traditional methods of evaluating social behavior performed by mental health professionals are based on clinical evidence and information self-reported by the patient [10]. These approaches generally use retrospective reports of social experiences lived by individuals in their daily lives, in which memory time can be days, weeks, and even months. As a result, cognitive biases limit these methods, hence contributing to an incoherent exposure of the lived experience [11,12]. For example, memory bias can prevent patients from reporting their feelings and behaviors accurately [11]. Social desirability bias encourages patients to hide or modify the truth of their reports to achieve socially desirable results [12]. The clinical context in which mental health assessments occur is also a limitation since it does not represent the patients' natural environment, implying a low ecological validity of traditional mental health assessment methods.

Currently, ubiquitous devices (e.g., smartphones, smartwatches, smart bands, and fitness bracelets) represent a promising means of mitigating those limitations [13]. The pervasive nature of these devices combined with a large amount of behavioral data from their sensors make ubiquitous computing a natural option to incorporate new system proposals for monitoring social behaviors related to mental health. Among the methodologies in this research area, the approach called Digital Phenotyping stands out. The term Digital Phenotyping refers to "moment-by-moment quantification of the individual-level human phenotype in-situ using data from smartphones and other personal digital devices" [14]. The goal of digital phenotyping is to learn and monitor patterns overtime that characterize behaviors of individuals (e.g., physical activities performed, their social interactions and mobility), based on context data derived from mobile, wearable, and Internet of Things (IoT) computing devices [13]. By using this concept, it is possible to create computational mechanisms able to perform continuous and discrete detection of individuals' social behaviors [15]. These mechanisms can integrate computerized systems of Ecological Momentary Assessment (EMA) and Ecological Momentary Intervention (EMI), which allow mental health professionals to collect daily behavioral information from their patients and perform interventions in their natural environment. These solutions contribute to the effectiveness of the treatment and provide real-time support to the patients' daily life.

The current literature presents solutions that use pervasive devices to recognize social behaviors related to mental health [13,16,17]. These solutions usually aim to make association, detection, classification, and prediction of mental states through features of the identified social situations [17]. However, there is still a need to develop solutions capable of recognizing sociability patterns representing the patients' social routine, so providing a valuable tool for assessing social behavior. Consequently, it is also essential to develop solutions to monitor changes in a patient's sociability pattern because these behavioral changes can mean the manifestation of mental disorders.

Given the need to objectively monitor social behavior, this study proposes a solution for processing social activity derived from pervasive devices to detect context-aware sociability patterns and social behavior changes. The proposed approach is able to perform incremental learning of context-aware sociability patterns through the combination of Frequent Pattern Mining (FPM) [18] and Complex Event Processing (CEP) [19]. Specifically, our proposed solution detects the time intervals in which social activities habitually occur. The recognition of sociability patterns is performed for specific contexts (e.g., weekdays, rainy days, and weekends), which enables the identification of behavior variability in different context conditions. The proposed solution is also able to identify changes in sociability patterns that reflect abnormal social behaviors and variations in social routines.

This article is an extended version of [20], where we outlined our approach to detect sociability patterns, but we did not present the solution for identifying changes in social behaviors. This paper has the following contributions: (i) we present an update of the formalization of the algorithm to detect context-aware sociability patterns; (ii) we introduce a solution for recognizing abnormal social behaviors and social routine changes; (iii) we use fuzzy logic to model knowledge of the mental health specialist needed to recognize social behavior changes; (iv) we evaluate the ability of the sociability patterns identified by the proposed solution to explain and predict users' social behaviors; and (v) we present an extensive analysis to evaluate the social behavior change detection solution.

The remaining of the paper is organized as follows. Section 2 discuss the related works. Section 3 presents the proposed solution to detect context-aware sociability patterns and changes in social behavior. Section 4 exposes an experimental evaluation of the proposed solution using a real-world data stream. In the end, we drive our conclusions and future works in Section 5.

2. Related Work

Several studies have proposed solutions to identify social situations through mobile and wearable devices to support mental health professionals [17]. In particular, studies have developed solutions to transform contextual data into sociability information. We categorize these studies according to their primary objectives [17]: solutions that aim to classify, predict, or associate social features to mental state, and solutions that focus on detecting and quantifying sociability levels. Table 1 presents the related works categorized by their primary objectives.

Table 1. Categorization of related works.

Type of Study	Goal	References
Detection	Detecting and quantifying sociability.	[21–27]
Classification	Classifying a mental state through social features.	[26,28–33]
Prediction	Predicting a mental state through social features.	[22,30,32,34]
Association	Associating sociability with a mental state.	[30,33,35–40]

2.1. Detecting and Quantifying Sociability

Some studies aim to develop solutions to identify social situations through passive detection to derive high-level information such as behavioral patterns and sociability levels. Exler et al. [21] designed a classification model capable of recognizing whether a person is alone or accompanied with an accuracy of 91.1%. This model uses location data, time of day, and activity information to perform this task. Barnett et al. [22] present a statistical approach to detect changes in sociability patterns by using phone calls and text messages, which were used to predict schizophrenic relapses. Harari et al. [23] identify patterns of stability and changes in social behavior (i.e., daily duration of conversations) of a student group over ten weeks. Bonilla et al. [24] found a set of patterns related to intensity functions of all interactions in which patients were involved by analyzing data from the use of their phone calls and social applications.

Studies also aimed to explore the passive detection of social situations to quantify sociability. Eskes et al. [25] propose the use of context data produced by smartphones (e.g., call logs, GPS locations, and Bluetooth encounters) to capture social communication and social exploration (e.g., mobility patterns and social density). These social behaviors were used to develop a statistical approach able to generate a sociability score, in which higher scores represent greater social engagement. Wahle et al. [26] monitors participants' involvement in device-mediated communication (i.e., call logs and text messages) to quantify their sociability. Additionally, this solution recommends social exercises based on information on the intensity of social activities, time and location. Lane et al. [27] present a mobile application called *BeWell*, which has a classifier able to infer the human voice through microphone audio. This application calculates a sociability score by applying a linear regression on the total duration of conversations. In addition, *BeWell* provides feedback on the social engagement level of its users.

2.2. Associations between Mental State and Social Features

Researchers also work on correlating social features (e.g., duration of phone calls, frequency of using social applications) with mental states (e.g., bipolar disorder, anxiety, depression). Mental states are typically identified using clinically validated self-report questionnaires. In general, researchers use correlation coefficients, as the Pearson and Spearman correlation coefficients [41,42], to calculate the degree of association between these variables.

In a study involving university students, Wang et al. [35] recognize that social routines (i.e., conversations and Bluetooth co-locations) of students presented correlations with their depression symptoms and stress levels. Results indicate that students who had lower frequencies and shorter duration periods in their daily social interactions also had higher levels of depression and stress. Chow et al. [36] identify temporal associations between depression, affective state and social anxiety with social isolation (i.e., home stay duration measured by GPS data). Boukhechba et al. [37] demonstrate that the social roles of visited places and communication patterns (i.e., phone calls and text messages) of the students show consistent associations with depressive states and social anxiety.

Some studies focus specifically on investigating association of social features with participants' stress levels. Wu et al. [30] found significant correlations between student stress levels with social features extracted from their social encounters measured through Bluetooth co-locations. Ono et al. [38] use a wearable device equipped with an infrared sensor to identify face-to-face interactions. This study found relationships between participants' stress levels with frequency, duration, and number of people involved in the social interactions. To do so, evaluation and validation methods were used to measure the performance of social models.

Some studies have as the main topic of social anxiety. Gong et al. [39] performed an association between participants' social anxiety levels and their physical behaviors, which were based on accelerometer-tracked body movement during device-mediated social interactions such as phone calls and text messages. Also, this study investigates whether the location where users performed technology-mediated communication influenced the social anxiety levels. Results indicate that people with higher levels of social anxiety exhibit more movement variations when making phone calls, especially in unfamiliar environments.

Other studies aim to correlate participants' social activities with their mood status. Servia-Rodriguez et al. [33] found associations between sociability patterns measured by analyzing phone calls and text messages of a large number of participants with their self-reported mood assessments. Matic et al. [40] found associations between time spent on speech activities (i.e., participation in verbal social interactions) and changes in positive affects.

2.3. Classifying and Predicting Mental State

In this study category, researchers design solutions capable of classifying and predicting mental states. Specifically, these approaches train machine learning algorithms from social features.

Different social models have developed to classify mental states of individuals. Gu et al. [28] developed a wearable device equipped with a microphone capable of automatically identifying and analyzing paralinguistic features (e.g., Brightness_sp and MFCC5_sp) contained in the human voice during social interactions. These features were used to train the K-Means algorithm to classify the participants' anxiety level, which obtained an accuracy of 72.73%. Chen et al. [29] use the transfer learning technique [43] to identify autism symptoms through the analysis of speech features extracted from microphone data of the wearable device.

Solutions presented by the studies also developed social models to predict mental states of individuals. For this, Wu et al. [30] use features extracted from physical social interactions identified by smartphone Bluetooth encounters to train the Random Forest algorithm, which was able to predict participants' stress levels. Barnett et al. [22] developed and applied a statistical approach to recognize changes in patients' communication patterns. The proposed method can predict schizophrenic relapses at two weeks in advance.

Some solutions recognize other human behaviors (e.g., sleep, mood, mobility) combined with sociability to design solutions that can identify mental states. Researchers identified several types of behaviors that have implications for mental health, allowing them to design more appropriate features to develop mental state classification models and predictions. For example, these multimodal features are used to design machine learning models to identify and predict patients with depression [26,31], bipolar disorder [32,34], and mood states [33]. Therefore, these solutions represent potential tools for supporting digital phenotyping of mental health as they recognize and utilize various patient behaviors to perform this task.

2.4. Discussion

Related works aim to make association, detection, classification, and prediction about mental health. From the analysis of these studies, we identified some open issues about applying passive detection of social situations to support mental health professionals. First, it is necessary to develop solutions able to identify sociability patterns that represent information about social routine of individuals. Second, analyses of such solutions should consider contextual information to identify social situations. Finally, there is a need for solutions that can detect social behavior changes to allow specialized professionals to investigate whether there is a relationship between the identified change and the patient's mental state.

Although the related works aim to identify social behaviors to support mental health monitoring, these studies differ from our solution, so it is a challenge to use objective comparison metrics. For example, some studies develop machine learning models to classify and predict mental states, while our solution aims to extract sociability patterns and detect behavior changes. The works [22–24] also propose solutions capable of detecting sociability patterns, but they differ from our solution. These works design sociability patterns to quantify the duration and frequency of social interactions, while our solution recognizes periods of the day representing individuals' social routine. Besides, these works do not recognize sociability patterns based on contextual data (e.g., weekends, holidays, and rainy days) and do not perform incremental learning, then requiring to execute batch processing.

In comparison to related works, our research has the following contributions. First, this study does not focus on the diagnosis of a specific state or mental disorder but works on identifying situations of interest (i.e., the sociability routine) for mental health professionals. Second, the proposed solution recognizes context-sensitive sociability patterns, so enabling to distinguish normal behavioral variation from behaviors that are considered anomalies.

Third, besides identifying the periods of the day when the individual generally socializes, the proposed solution can recognize unusual social routines and significant changes in the patient's social habit. Finally, the proposed solution uses a data stream mining approach to learn from social observations continuously.

3. Proposed Solution

In this section, we present the solution proposed to detect context-aware sociability patterns and behavioral changes. It performs incremental learning of context-sensitive sociability patterns through the combination of FPM and CEP. FPM is a computational technique that aims to discover patterns that occur with significant frequency in different data collection types, such as relational and non-relational databases, text files, and data streams [18].

The algorithm used in our solution was proposed by Lago et al. [44], which aims to learn activity patterns in smart homes (e.g., activity sequence). We applied this algorithm to digital phenotyping of mental health through the recognition of sociability patterns. First, we present a formalization update of this algorithm through unit step functions to represent the appropriate logic to identify time intervals in which social events routinely occur. We used the formalized algorithm to implement an event processing network capable of incrementally identifying context-sensitive sociability patterns. For this purpose, we used CEP concepts [19], which provide a set of tools to process data streams efficiently, so performing tasks such as data aggregation and filtering, context partitions, data window, high-level information derivation, and pattern recognition.

The proposed solution can also detect abnormal social behaviors and changes in social routines through the application of concepts of drift identification techniques. Additionally, we use fuzzy logic to model the knowledge of the mental health specialist to detect behavior change. Finally, the developed solution provides an Application Programming Interface (API) to enable the rapid implementation of strategies to identify context-aware sociability patterns and configure behavioral changes.

Figure 1 presents an overview of the processing flow to identify context-aware sociability patterns and social behavior change. The first layer represents the generation of social events from data of online social networks and physical and virtual sensors embedded in ubiquitous devices. As examples of social events, it is possible to cite conversations identified from microphone data and interactions mediated by technology (e.g., phone calls, text messages, and social media posts). Next, the layer responsible for detecting context-aware sociability patterns supports a set of CEP rules designed to implement the algorithm to identify sociability patterns. The next layer performs tasks of detecting abnormal behaviors and routine changes. This layer also contains a Fuzzy Inference System (FIS) that models specialist knowledge needed to recognize social behavior changes. The last layer refers to client applications that receive notifications of new patterns and behavior changes emitted by the proposed solution's components.

Ubiquitous devices (e.g., smartphones, wearable sensors, IoT devices, social networks) represent valuable social data sources. Computational methods (e.g., data mining, machine learning) can process context data from physical and virtual sensors embedded in these devices to identify social situations, such as face-to-face interactions (i.e., socialization in physical environments) and device-mediated interaction (i.e, socialization in virtual environments). For example, computational methods can process microphone and wireless communication interfaces data (e.g., WiFi, Bluetooth, NFC) to identify conversations and physical proximity [23,35]. Call logs and text messages can represent device-mediated interactions [31,33].

We emphasize that our proposed solution does not include the generation of events, but focuses on processing high-level sociability events inferred by other solutions to identify patterns and behavior changes. Next, we present in detail its components.

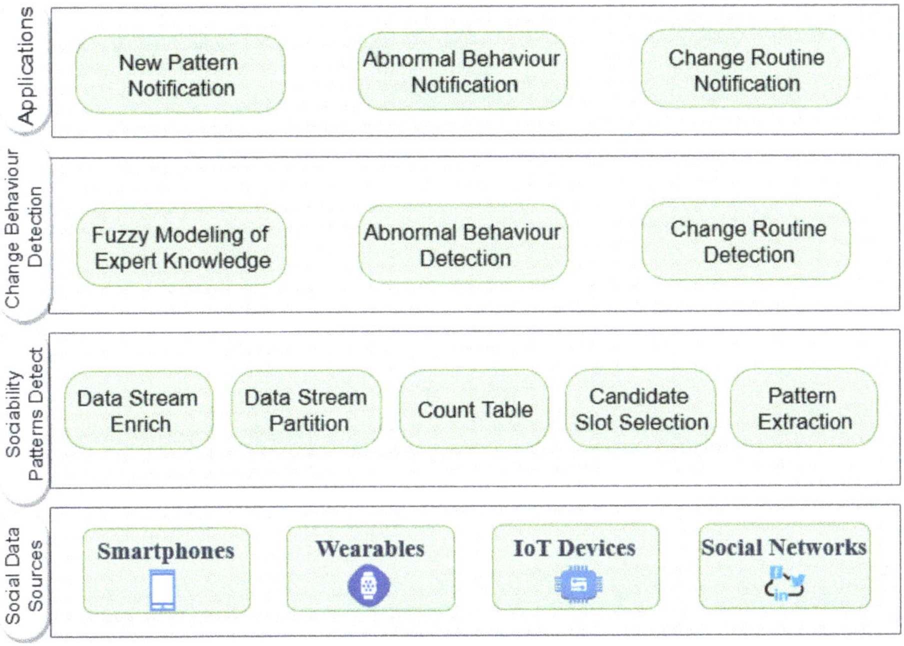

Figure 1. Components of the proposed solution.

3.1. *Learning Context-Aware Sociability Patterns*

In this section, we present the algorithm for identifying context-sensitive sociability patterns and its implementation using CEP.

3.1.1. Algorithm for Identifying Sociability Patterns

We consider that, if the social activities are detected frequently at a specific time interval, this interval composes the sociability pattern of monitored individuals. Thus, we define sociability patterns as *periods of the day in which the individual usually socializes, that is, the set of time intervals in which social activities habitually occur*. The algorithm processes the data stream to recognize time intervals $[T_{start}, T_{end}]$ in which the number of occurrences of social activities is higher than $\phi * |n|$. In this regard, $|n|$ is considered as the total number of processed observations in a defined time window to model social behavior, and ϕ is a parameter to be manually set, which is responsible for indicating the sensitivity of the algorithm.

The algorithm input is a social event stream that has the start time of each social activity. The first step of the algorithm is to determine, based on the timestamp, which time frame each social event belongs. For this, the algorithm segments the time in slots with equal sizes. Each slot represents a slice of the day and has a sequential identifier. To define the size of the slot (i.e., in how many periods should divide the day), the programmer is required to specify a value for the t parameter, which is responsible for creating an array of counters with the total number of slots. For example, if the programmer decides to divide the day into periods of 30 min, the parameter t is equal to 0.5, since $\frac{24}{t} = 48$ slots. This equation is responsible for creating the storage structure for counting occurrences of social activities in each slot.

After defining the size of the slots, we now describe the counting phase of the algorithm. At this stage, the algorithm uses the timestamp of each event to define its slot. By identifying the slot of social event, the algorithm increments the counter value that represents this slot in the structure responsible for storing these statistics. Therefore, when processing the flow of events, the frequency of social activities in each slot is updated. This approach of saving only the summary (i.e., the count) allows reducing the data volume since it is not necessary to memorize the full content of the events.

The next phase of the algorithm is the sociability pattern discovery, which uses the summary of the counting phase to identify frequent intervals of sociability. At this stage, it is necessary to define which slots have a sufficient number of observations, that is, a quantity that enables them to be candidate slots to form a frequent period. For this, the number counting of social observations of the analyzed slot must be greater than or equal to S_{th}. The algorithm uses Equation (1) to define the value of S_{th}. The θ parameter is entered by the programmer to set up the sensitivity of the equation.

$$S_{th} = |n| * \theta * \frac{1}{\frac{24}{t}} \tag{1}$$

Equation (2) is responsible for iterating the slot array C_s and verifying which slots are candidates to form a sociability pattern, so assigning zero to the count of non-candidate slots. For this, this equation defines the multiplication between the count of each slot (*slot[i]*) and the unit step function, which returns zero value in cases of negative arguments ($slot[i] - slot\_th < 0$) and one for non-negative arguments ($slot[i] - slot\_th > 0$). In the end, the slot array C_s is sent to the process of identifying frequent sociability intervals.

$$C_s[i] := slot[i] * \textbf{unit\_step}(slot[i] - slot\_th) \tag{2}$$

Finally, after defining the requirement for a slot to be candidate, the next step is to identify which sets of slots compose an interval at which social activities are routine for the monitored individual. Equation (3) groups the adjacent non-zero candidate slots in the array C_s into a sociability pattern. The unit step function verifies whether the sum of the event counts for the grouped slots (i.e., time intervals) subtracted from $\varphi * |n|$ results in a positive value. If this condition is satisfied, the time interval formed by these sets of adjacent slots represents a sociability pattern. In the end, the array P_s will contain sociability patterns, that is, the time intervals in which social activities routinely occur for the monitored individual.

$$P_s[i : i+n] := \textbf{unit\_step}\Big(\big(\sum_{j=i}^{i+n-1} C_s[j]\big) - \varphi * |n|\Big) \tag{3}$$

3.1.2. Context-Aware Sociability Patterns

So far, the algorithm allows identifying the individual's sociability routine, so mapping the frequent start time of social activities. However, this context-free analysis may result in inefficiency when outlining the social habit, since the individual's behavior may vary due to specific contexts, such as workdays, weekends, rainy days, among others. For this, we use a strategy with Context Attributes (CAs), in which several scales can be used to represent them. For example, a temporal feature may have several scales, as a broad scale, so differentiating days of week and weekends, or a more specific, so distinguishing each day of the week (e.g., Monday, Tuesday, Wednesday). We inject these CAs into the stream of social observations, which can be derived directly from event properties (e.g., timestamp) or retrieved from external sources (e.g., climate APIs). By enabling this setting, mental health professionals can define which contexts are considered more suitable for each patient and treatment.

Each CA is used as a data segmentation dimension to identify behavior change due to specific context situations. Therefore, the identification of sociability patterns is performed from a subset of data that has a particular CA. For example, all social events that occurred over the weekend (i.e., CA = Weekend) are used to identify the individual's social routine in this context condition. The algorithm needs to create a structure (e.g., a matrix) to store slot counters for each context dimension. During the counting phase, each social event increments, in the index of its respective slot, values in the structures that store statistics for each CA of the processed social observation. In summary, we partitioned data flow based on CAs and performed incremental learning of sociability patterns for each derived data stream.

3.1.3. CEP Implementation

We have used CEP concepts to implement the algorithm. CEP enables to react in real time to data stream through a continuous query language [19]. This method processes data as a sequence of events, in which each event models an observation in a specific domain. Atomic events, or only events, are immutable records of something in the real world or a software system. For example, in this study, an event represents a social activity at a specific time. These social events are generated by ubiquitous devices through the processing of context data. In the implementation of the algorithm, we use *Esper* (http://www.espertech.com/esper/), which is an engine developed for CEP and continuous stream analysis. This mechanism provides an Event Processing Language (EPL) that implements and extends the Structured Query Language (SQL) standard. The processing workflow (Figure 2) consists of the following steps:

- **(a)** ***Enrich Social Event*:** it injects the slot (extracted from its timestamp) and CAs in social events. The result of this process is the emission of an enriched event named *SocialUpdate*;
- **(b)** ***Context Partition EPA*:** it segments *SocialUpdate* streams based on CAs. A derived event called *SocialContext* that has the slot and its context label is triggered for each context record of the observation;
- **(c)** ***Count Table*:** it is responsible for maintaining the count of events that occurred in specific contexts in each slot. This count is updated with each *ContextEvent* emitted;
- **(d)** ***Candidate Slot EPA*:** it is responsible for verifying which slots have reached an adequate number of observations to become candidate, so sending these to the pattern extraction phase;
- **(e)** ***Extract Pattern*:** it identifies which sets of adjacent candidate slots compose a time interval that the individual habitually socializes.

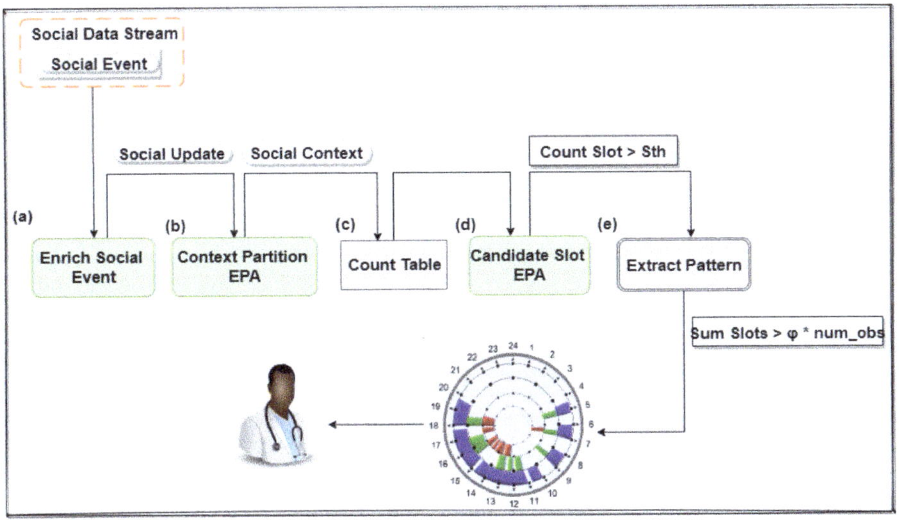

Figure 2. Processing workflow.

3.2. Social Behavior Change Detection

Another contribution of this study is an approach to recognize social routine changes and abnormal social behaviors. This approach aims to continuously detect information about changes in social habits of the monitored individual, since this information is of vital importance for treatment and monitoring, so enabling to increase the chances of effective interventions performed by mental health professionals.

3.2.1. Behavioral Change Detection Strategy

We developed a strategy for detecting social behavior changes by applying the concept of drift identification techniques [45], which allows recognizing socialization flow changes of the monitored individual. Specifically, we explore unsupervised techniques to detect changes in the data stream distribution, since there is no ground truth available in social event streams to monitor performance indicators (e.g., accuracy, sensitivity, specificity, recall). Therefore, we combine the processing of data windows with a similarity metric to verify pattern change of an instant from *t1* to *t2*. It is important to note that the proposed approach detects social behavior changes for each CA, so differentiating common social routine variation from abnormal social habits.

We explored data window processing strategy provided by the Esper engine to partition data flow, so extracting sociability patterns at different time intervals. Therefore, it is possible to detect sociability patterns through current data and, subsequently, compare them with the pattern identified in a future data window to recognize occurrences of social routine changes of the monitored individual. We used the Jaccard similarity coefficient (Equation (4)) to quantify overlaps between intervals of identified sociability patterns (i.e., proportion between intersection and union of the compared patterns).

$$J(A, B) = \frac{A \cap B}{A \cup B} \tag{4}$$

Based on the strategy described above, we developed a solution to detect two types of situations of interest for specialized mental health professionals: abnormal social behaviors and social routine changes. Abnormal social behaviors reflect a low similarity between the current sociability pattern and the individual's social habit on a specific day. When abnormal behaviors occur with substantial frequency, it is necessary to assess whether the individual's sociability routine has changed. Therefore, routine change is recognized when a sociability pattern extracted from a data window at time t has low similarity to a pattern identified from the data window $t + n$. The proposed solution maintains the first identified sociability pattern (i.e., a reference pattern) until it finds a new pattern significantly different from the reference pattern, so allowing to identify changes that happen slowly, over a long period.

Figure 3 shows the strategy for detecting abnormal social behaviors. To perform this task, the specialist (e.g., a mental health professional) is required to define the time window size that models an observation of the individual's social behavior in a given context (e.g., Mondays, rainy days, weekends). For example, in Figure 3, an observation could be composed of a one-week data window, so requiring two observations to extract a consistent sociability pattern, i.e., to conceive a predictive model of intervals of the day in which the individual habitually socializes. Therefore, to recognize abnormal behaviors, our proposed solution uses the Jaccard similarity coefficient (Equation (4)) to compare the current sociability pattern with the social behavior of the next observations. Hence, if the similarity between the current pattern and the recognized social habit (i.e., an observation) has value below a threshold defined by the specialist, the solution sends an event to notify interested parties about the identification of abnormal social behavior.

Figure 4 presents the strategy for detecting social routine changes. In this scenario, sociability patterns are extracted through a determined number of observations (e.g., two observations), which represent the current social routine of the monitored individual. In principle, the solution persists in memory the first sociability pattern identified, which is used as a comparison model with the next patterns to recognize significant social routine changes. Specifically, when the similarity is below the defined threshold, two tasks are triggered to: (i) update the current sociability pattern with the most recent routine and; (ii) notify interested parties by sending a social routine change event.

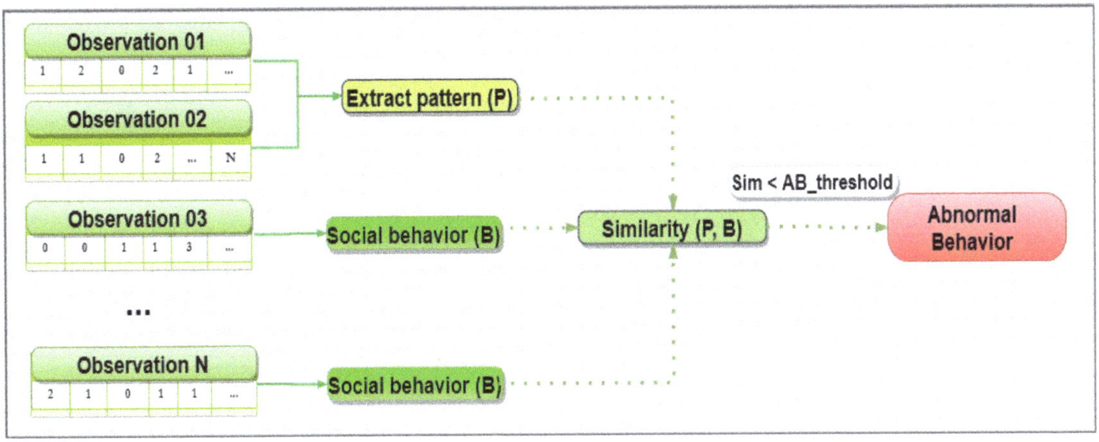

Figure 3. Strategy for detecting abnormal social behavior.

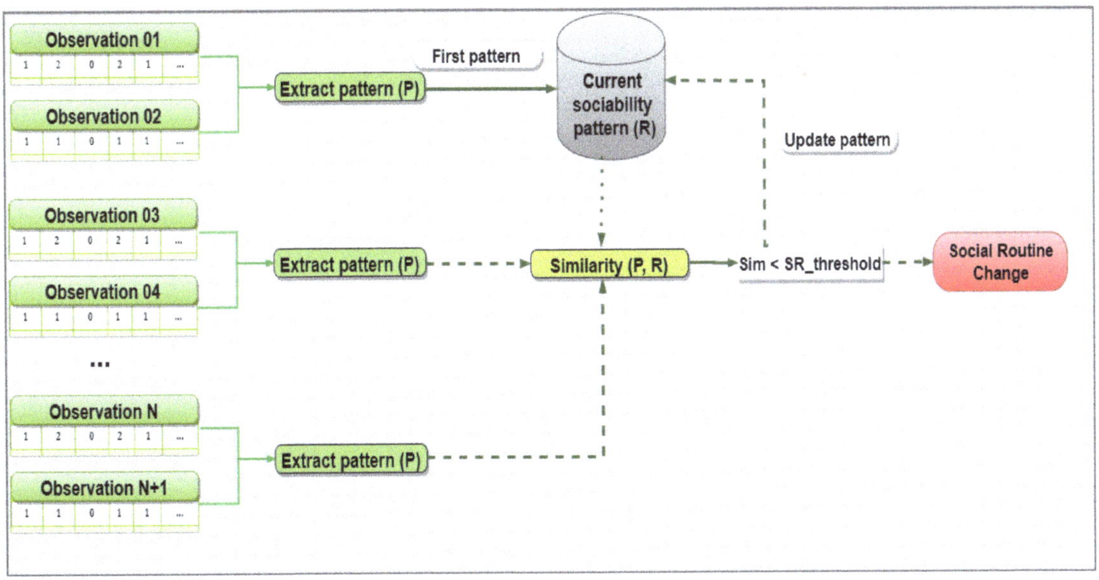

Figure 4. Strategy for detecting social routine changes.

3.2.2. Expert Knowledge Modeling

As we could see, it is possible to identify the specialist's need to define a similarity threshold between patterns. The specification of the change threshold is subjective, as it depends on the specialist's knowledge, and has an imprecise essence, since a rigid limit may not adequately model the change that has occurred. For example, consider that an specialist has configured the behavior change detection solution with a similarity threshold of 60%. In this scenario, if the similarity between sociability patterns is 59.9%, abnormal social behavior is detected even though it is very close to the specified threshold. Therefore, it is necessary to define strategies that allow modeling the specialist's knowledge automatically, so enabling to send behavior change alerts with a judgment of the belief degrees.

We implemented expert knowledge modeling through fuzzy logic concepts to consider the imprecise nature of this task. Specifically, we used the open-source library *jFuzzyLogic* (http://jfuzzylogic.sourceforge.net/) [46,47] to develop fuzzy controllers. This library uses Fuzzy Control Language (FCL), which is a domain-specific language to facilitate the development of fuzzy systems. *jFuzzyLogic* enabled the integration of FIS into the proposed solution to allow the specialist to specify linguistic variables, fuzzy sets, and rules.

The motivation for using fuzzy logic in this task is the possibility of representing imprecise and qualitative human knowledge through fuzzy sets instead of using crisp sets [48]. Fuzzy logic allows computational systems to model human reasoning so that output variables can vary from false to true gradually, then making it possible to express partially true or partially false conclusions. Another reason is the notation of defining fuzzy rules through linguistic variables (i.e., natural language) and logical connectors [49], which presents semantics that is easy to understand for users. Therefore, we used fuzzy logic due to its ability to model expert knowledge through fuzzy sets and fuzzy rules and its notation that is easy to understand. Moreover, fuzzy logic has already been used to model situations of interest to mental health professionals [50].

First, the specialist is required to determine three fuzzy sets: sensitivity, similarity, and drift. Sensitivity sets define discrepancy levels between patterns that represent a change, i.e., it models the sensitivity knowledge to detect changes. Similarity sets define correspondence levels between evaluated patterns. Finally, drift sets represent the FIS output (i.e., defuzzifier) responsible for modeling change levels in social behaviors.

After defining fuzzy sets, the next step is to subdivide them to represent specialist's knowledge to assign linguistic terms to the intervals of each set. Therefore, the specialist should use FCL to perform this task. Code 1 shows a configuration example of the sensitivity sets using FCL, while Figure 5 shows their visual representation. The specialist divided the sensitivity input variable into three levels: low, moderate, and high. In this scenario, a certain level of intersection between intervals is observed, so representing a gradual pertinence transition.

Code 1: Configuration example of the sensitivity sets using FCL.

```
FUZZIFY sensibility
TERM low:= (0, 1) (25, 1) (50, 0);
TERM moderate:= (25, 0) (50,1) (75, 0);
TERM high:= (50, 0) (75, 1) (100, 1);
END_FUZZIFY
```

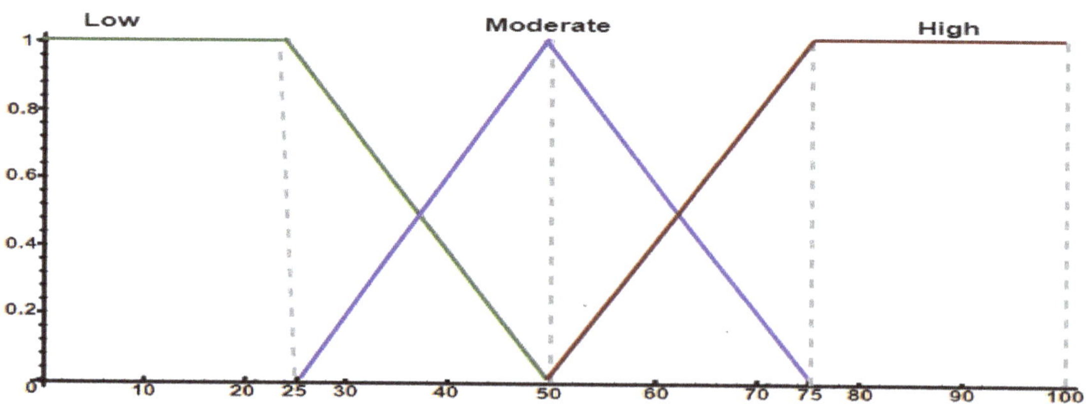

Figure 5. Fuzzy sets to model behavior change detection sensitivity.

The specialist is also required to represent his/her knowledge through FCL to define similarity set partitions and their respective linguistic terms. Code 2 shows a configuration example of the similarity sets using FCL, while Figure 6 shows their visual representation. The specialist subdivided the similarity input variable into three levels: low, moderate, and high. In this scenario, intervals have an intersection level to model uncertainty.

Code 2: Configuration example of the similarity sets using FCL.

```
FUZZIFY similarity
TERM low:= (0, 1) (40, 1) (60, 0);
TERM moderate:= (50, 0) (60,1) (70, 0);
TERM high:= (60, 0) (75, 1) (100, 1);
END_FUZZIFY
```

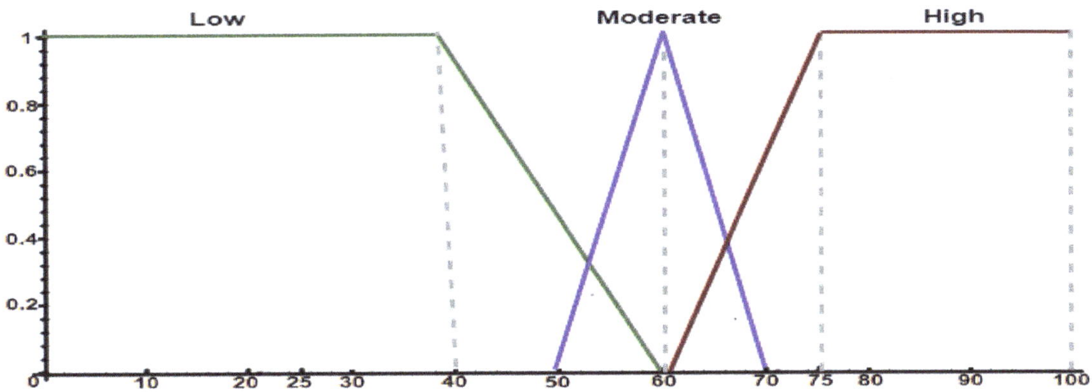

Figure 6. Fuzzy sets to model the similarity levels between patterns and behaviors.

The third set to be specified is the drift, which is representing the FIS output, i.e., it models the occurrence of social behavior changes of the monitored individual. Code 3 presents a configuration example of the drift sets, while Figure 7 shows their visual representation. The specialist described three linguistic terms: no_change, moderate_change, and change. In this example, we used the Center of Gravity (COG) defuzzification method to determine the inference's final value from activated rules. *jFuzzyLogic* supports several fuzzification and defuzzification methods.

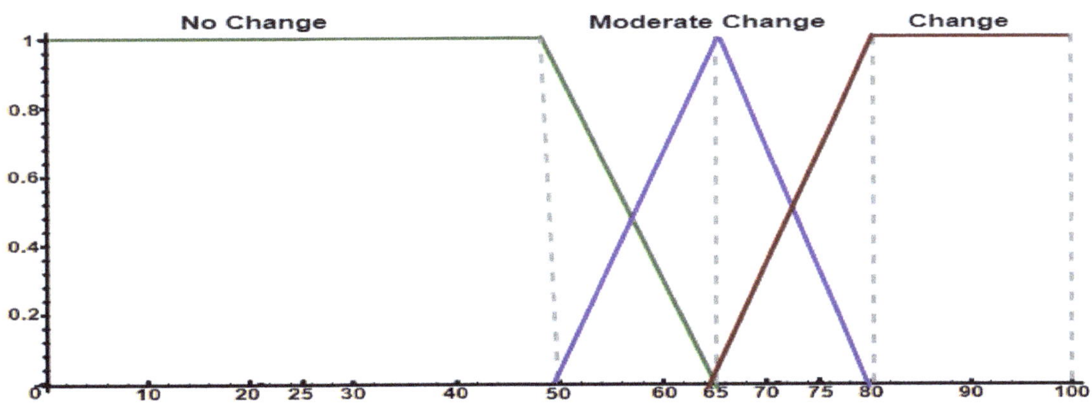

Figure 7. Output fuzzy sets defined to assess behavior change.

Code 3: Configuration example of the drift sets using FCL.

```
DEFUZZIFY drift
TERM no_change:= (0, 1) (50, 1) (65, 0);
TERM moderate_chabge:= (50, 0) (65, 1) (80, 0);
TERM change:= (65, 0) (80, 1) (100,1);
METHOD: COG;
ENDDEFUZZIFY
```

Next, the specialist is required to specify fuzzy rules that guide the solution's decision. These fuzzy rules have the primary structure <condition> *AND* <condition> *THEN* <consequence>. The specified fuzzy rules form the knowledge used by the FIS to determine the final value of the behavior change inference. Therefore, based on fuzzy sets presented above, the fuzzy rules contained in Code 4 can be specified using FCL.

Code 4: Fuzzy rules for behavior change.

```
RULEBLOCK No1
    RULE 1 : IF sensibility IS low AND similarity IS low THEN drift IS change
    ;
  RULE 2 : IF sensibility IS low AND similarity IS moderate THEN drift IS
    no_change;
  RULE 3 : IF sensibility IS low AND similarity IS high THEN drift IS
    no_change;
    RULE 4 : IF sensibility IS moderate AND similarity IS low THEN drift IS
    change;
  RULE 5 : IF sensibility IS moderate AND similarity IS moderate THEN drift
    IS
  moderate_change;
  RULE 6 : IF sensibility IS moderate AND similarity IS high THEN drift IS
    no_change;
    RULE 7 : IF sensibility IS high AND similarity IS low THEN drift IS
    change;
  RULE 8 : IF sensibility IS high AND similarity IS moderate THEN drift IS
    change;
  RULE 9 : IF sensibility IS high AND similarity IS high THEN drift IS
    no_change;
END_RULEBLOCK
```

At the end of the expert knowledge modeling process, the solution can detect social behavior changes considering uncertainty in this task. Specifically, the solution sends a JSON object that contains information such as date, context, similarity index, and the pertinence degree to each interval of the fuzzy output set. Code 5 presents an example of social behavior change notification considering the expert knowledge modeled in the previous cases. In this example, the output variable (i.e., defuzzification value) is ≈74.86, which is ≈65% contained in the interval that represents change and ≈34% in the partition that reflects moderate change.

Code 5: Notification of social behavior change.

```
{
    "date": "Sun 1, 2020, 9:17:48 PM",
    "context":"Sunday",
    "similarity":56.00000000000001,
    "defuzzification value":74.86688093051647,
    "change":0.6577920620344315,
    "no_change":0.0,
    "moderate_change":0.3422079379655685,
    "message":"Social routine change detected"
}
```

3.3. *Application Programming Interface*

The solution designed by this study provides an API to its users, which allows them to quickly implement strategies for detecting sociability patterns and configuring the recognition of behavioral changes. Development of this API used the Builder design pattern (Constructor) [51], which enables separation of the construction of a complex object from its representation, so dividing this process into parts (i.e., steps). We mitigate the complexity inherent in the creation and instantiation of the EPLs and solution configuration using this design pattern.

Next, the user specifies the RootTopic attribute (setRootTopic) with the value "com/ls-di/sociability". The solution uses this attribute to publish new sociability patterns and behavior change notifications in MQTT broker, allowing interested client applications to subscribe to this topic to receive updates. Finally, the user set true value to the parameters AbnormalBehavior and ChangeBehavior (setAbnormalBehavior and setChangeBehavior) for the solution to activate the detection of abnormal behaviors and social routine changes.

Code 6 presents an example of use of the designed API. Specifically, the developer create an object called *SociabilityPattern*, which configures and enables the entire solution operation. In this example, the developer initializes the constructor with two parameters: "MONDAY_" and 50.0, so representing the CA considered and the sensitivity level to detect behavior changes, respectively. Next, the developer specifies the *RootTopic* attribute (`setRootTopic`) with the value "com/lsdi/sociability". The solution uses this attribute to publish new sociability patterns and behavior change notifications in an MQTT broker [52], so allowing interested client applications to subscribe to this topic to receive updates. Finally, the developer set true value to the parameters *AbnormalBehavior* and *ChangeBehavior* (`setAbnormalBehavior` and `setChangeBehavior`) for the solution to activate detection of abnormal behaviors and social routine changes.

Code 6: Application Programming Interface (API) usage example.

```
SociabilityPattern sociabilityPattern = new SociabilityPattern
            .Builder("MONDAY_", 50.0)
            .setRootTopic("com/lsdi/sociability")
            .setAbnormalBehavior(true)
            .setChangeBehavior(true)
            .build();
```

4. Experimental Evaluation

The proposed solution identifies context-aware sociability patterns using incremental unsupervised learning. Consequently, applying metrics commonly used to evaluate learning models is challenging because there is no ground truth available to compare results. By considering this, we evaluate the proposed solution's feature to consistently recognize sociability patterns for modeling social routine and behavior changes.

In [20], we performed the following: (i) a comparison of the similarity between the sociability patterns identified with the intervals recognized by Gaussian Mixture Models (GMMs); (ii) an analysis of the contribution of context-aware sociability patterns to understand the monitored individual's social routine. Here, we performed a more in-depth evaluation to recognize the proposed solution ability to identify context-sensitive sociability patterns that model social behavior and detect behavioral changes.

4.1. Data Description

We evaluated the solution using a public dataset [35], which is referred to as *StudentLife*. Wang et al. [35] performed during 66 days a passive sensing of social activities (i.e., conversations) derived from microphone data gathered from smartphones of 48 Dartmouth College undergraduate and graduate students. Conversation samples are composed of two fields: start and end timestamps of conversations experienced by participants. All collected data was anonymized to preserve privacy of the monitored individuals.

The used dataset contains conversation samples composed of two features: start and end timestamps of social interactions. Figure 8 shows the first lines of the dataset of an individual. For example, the second line in this file records that he/she experienced a conversation that started at Unix timestamp 1364359600 and ended at Unix timestamp 1364359812.

```
start_timestamp, end_timestamp
1364359600,1364359812
1364382621,1364383065
1364383516,1364384993
```

Figure 8. Dataset features used in the solution.

Firstly, we performed a data cleaning process to remove users who had insufficient data to conduct experiments. Only users who contained at least 52 days of collected data (≈80% of the study days) are in this experiment. We used data from 24 individuals who had sufficient data.

We represent each record as a social event to design a proper data flow for the proposed processing network. For this, we derived the following information from social activity records: social activity type (i.e., conversation), start time, and a set of CAs. The first step was to convert the Unix timestamp to a Date Java object to represent the event start time attribute and extract CAs. We identified the weekday when events occurred using their timestamps, so enabling to specify temporal context scales. We defined two context scales: a fine-grained scale composed of weekdays (e.g., Monday, Tuesday, and Wednesday) and a broad one to distinguish weekends (i.e., Saturdays and Sundays) from midweeks (i.e., Monday to Friday). In the end, the structure of the generated social event flows was as follows:

```
event = (activity: Conversation, start: 12:20, contexts: [Saturday, Weekend])
event = (activity: Conversation, start: 14:12, contexts: [Saturday, Weekend])
event = (activity: Conversation, start: 8:35, contexts: [Monday, Week]))
event = (activity: Conversation, start: 16:20, contexts: [Tuesday, Week])
```

4.2. Experimental Design

The first evaluation consisted of verifying the ability of the context-aware sociability patterns to model and explain social behaviors, i.e., we verified whether sociability patterns could explain and predict stable social routine (i.e., individuals repeating their social behaviors over the days) and less able to explain unstable social routine. For this, we used Pearson correlation coefficient [41] to assess the association between the ability of sociability patterns to explain social routine and stability of the individual's social routine. This coefficient measures linear correlation between two variables, which assumes values between -1 (perfect negative correlation) and 1 (perfect positive correlation). Therefore, higher levels of positive associations indicate that the proposed solution recognizes consistent sociability patterns and capable of modeling monitored people's social behaviors.

In the end, we evaluated the proposed solution to detect social behavior changes. For this, we joined data of two users who had different social routine, hence enabling to identify the moment when the change occurs. We verified whether the proposed solution can accurately detect this change and its ability to adapt to the new sociability pattern.

4.3. Ability to Model Social Routines

This experiment aimed to assess the ability of the sociability patterns to model social routine. Specifically, a sociability pattern should explain the social behavior of the monitored individual, which should be correlated with his/her social habit, because the more stable the individual's social behavior, the greater the pattern's ability to predict it. We analyzed association between the identified sociability patterns' prediction level and the social routine's stability. For this reason, we used Pearson correlation coefficient for quantifying this association. By using this correlation coefficient, patterns' ability to explain and predict social routine of the monitored individuals were identified.

We defined that a social observation consisted of a data window of one week (i.e., seven days). For example, for the *MONDAY* context, an observation consisted of data from 1 day, since a week has only 1 day with that context. Therefore, the first step of this experiment was to define the number of observations necessary to extract patterns consistent with the monitored individual's social behavior. For this, we identified sociability patterns considering each day of the week as a CA (i.e., *MONDAY*, *TUESDAY*, *WEDNESDAY*, *THURSDAY*, *FRIDAY*, *SATURDAY*, and *SUNDAY*), and we checked the predictive performance of sociability patterns when using different numbers of social observations to design them. We defined prediction performance as the ability of sociability patterns to explain and predict individuals' social routines.

Figure 9 presents an example of scenario for assessing the prediction performance of sociability patterns using two observations. We performed this evaluation using one, two, three, and four observations to project sociability patterns, so allowing us to compare the predictive performance of each configuration. Each execution consisted of the following steps: (i) recognizing the sociability pattern with the number of specified observations (i.e., one, two, three, or four); (ii) measuring the Jaccard similarity index between the pattern extracted with the next social observations; and (iii) identifying a new social pattern from the evaluated observations to represent a new reference pattern.

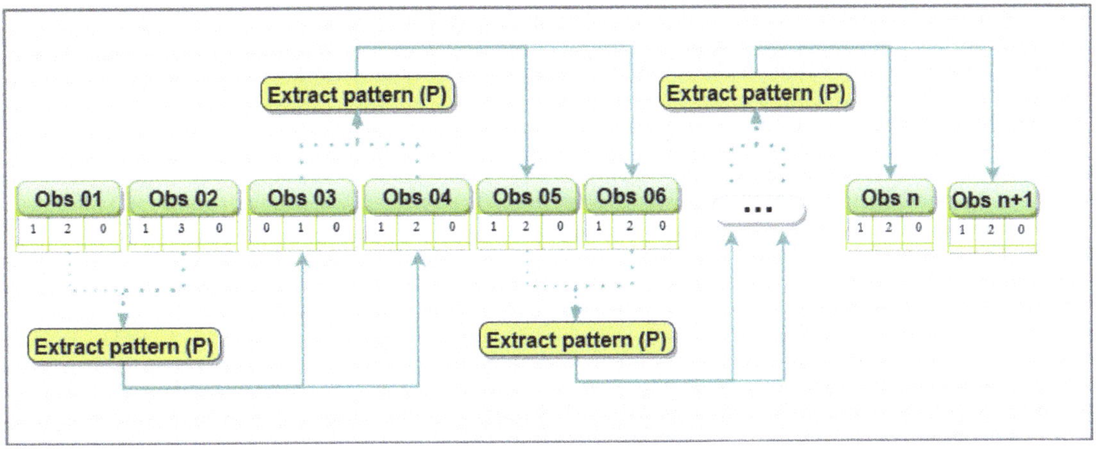

Figure 9. Evaluation of the prediction performance of sociability patterns using two observations.

We performed the experiment for all users considering CAs specified previously (e.g., *MONDAY*, *TUESDAY*, and *SUNDAY*), so making it possible to recognize the predictive performance (i.e., the similarity between sociability patterns and observations) of sociability patterns for these CAs. We calculated the average prediction performance to identify the number of observations required to extract sociability patterns to explain social routine.

Figure 10 shows the average predictive performance of the sociability patterns identified using one, two, three, and four observations. From results, we identified that extracting the sociability pattern using only one observation resulted in poor predictive performance compared to other configurations. Extraction of social patterns with two, three, or four observations showed similar predictions. Therefore, we concluded that the most appropriate approach was to use two observations to extract sociability patterns since we could identify them in less time with predictive performance similar to other configurations.

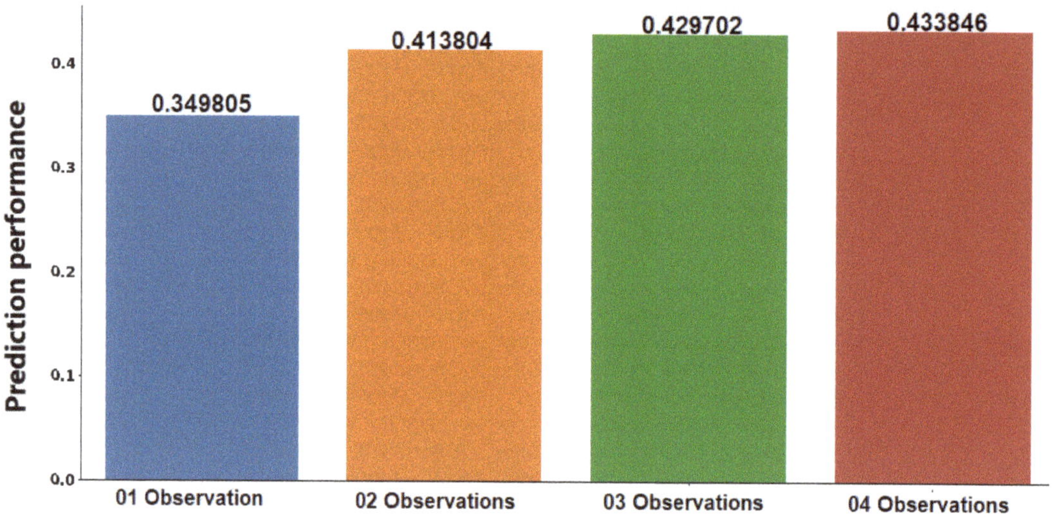

Figure 10. Prediction performance of sociability patterns.

After quantifying prediction levels of the extracted sociability patterns, we measured stability of the social routine of the individuals who participated in this study. We calculated average similarity between one day and the subsequent day (Figure 11), i.e., similarity of the individual's social behavior between consecutive days. CAs considered each day of the week, similar to the previous experiment. In the end, we calculated average stability of the individuals' social routine, so allowing us to correlate this variable with prediction levels of sociability patterns.

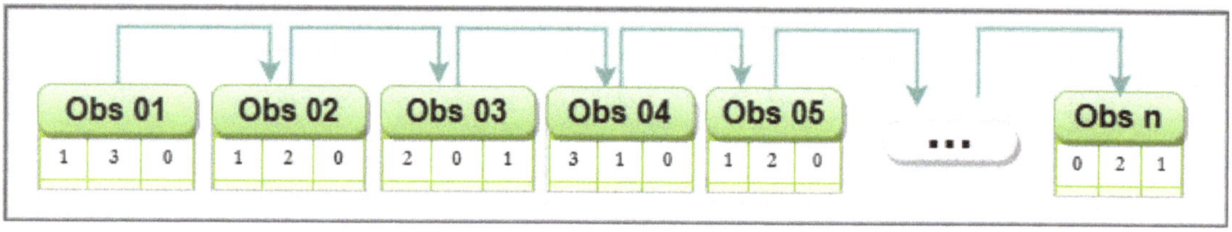

Figure 11. Configuration used to evaluate stability of individuals' social routines.

We performed a stability analysis of the individuals' social routines to recognize essential information to understand their social behaviors. Figure 12 shows the stability of the individuals' social routines (i.e., the similarity of social behaviors between days), so making it possible to identify that most users had social habits with stability below 40%. However, some users had more stable routines, such as *u04* and *u27*. From this analysis, we expect that sociability patterns could explain and predict more consistently social behaviors of the more stable users and that present lower levels of predictions when applied to individuals with more unstable social routines.

So far, we quantified the prediction performance of the extracted sociability patterns and the stability of the individuals' social routine. Therefore, we can perform association analysis between these two variables using Pearson correlation coefficient. Figure 13 shows this association, in which y-axis represents the average of prediction performance of the social patterns, and x-axis represents the average of the individuals' social routine stability. When analyzing Figure 13, we can identify a clear correlation between these two variables, so representing a linear relationship. Pearson correlation coefficient resulted in +0.86, which represents a strong positive association between these variables.

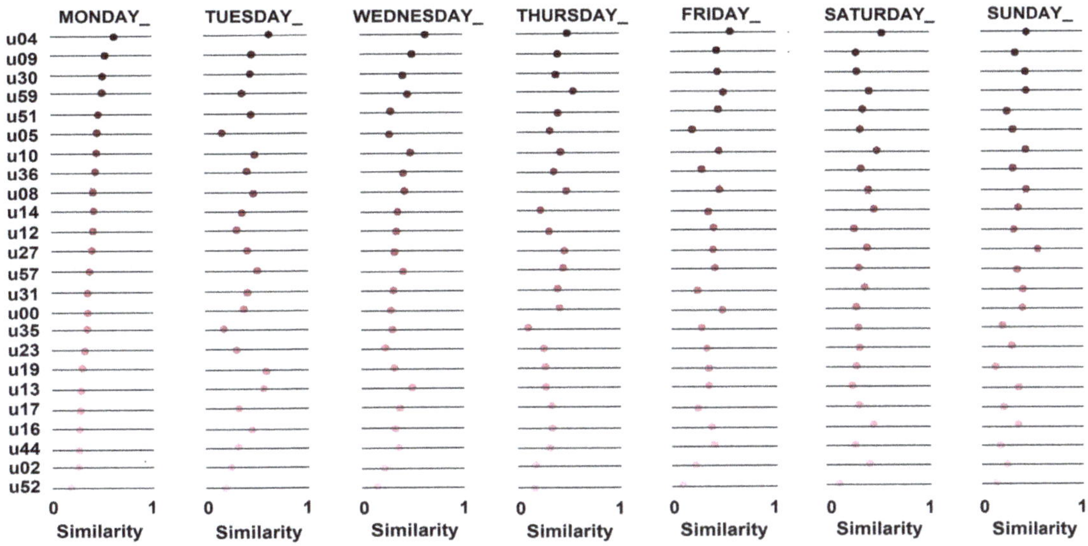

Figure 12. Stability of the individuals' social routines.

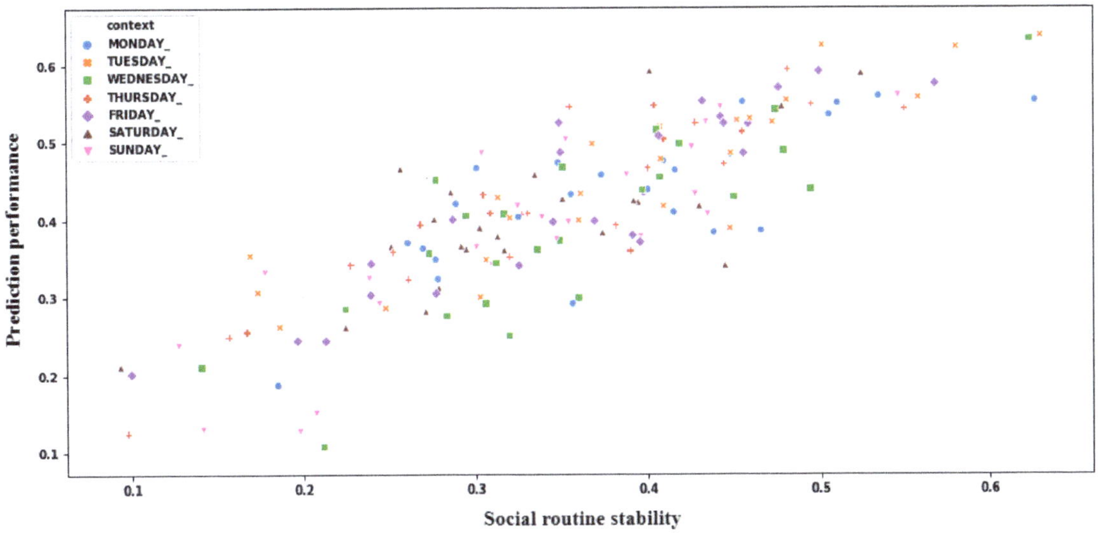

Figure 13. Correlation between prediction performance of sociability pattern and social routine stability.

Figure 14 shows the relationship between prediction performance of the sociability patterns and stability of the individuals' social routines for each CA, so making possible to identify a linear relationship between these variables. Figure 15 shows the result of applying the Pearson correlation coefficient between these variables for each specified CA, so indicating that association levels were higher than 0.7, which represents strong positive correlations. When analyzing these results, we can recognize that prediction performance of the extracted sociability patterns remains related to stability of the social routine in all evaluated CAs.

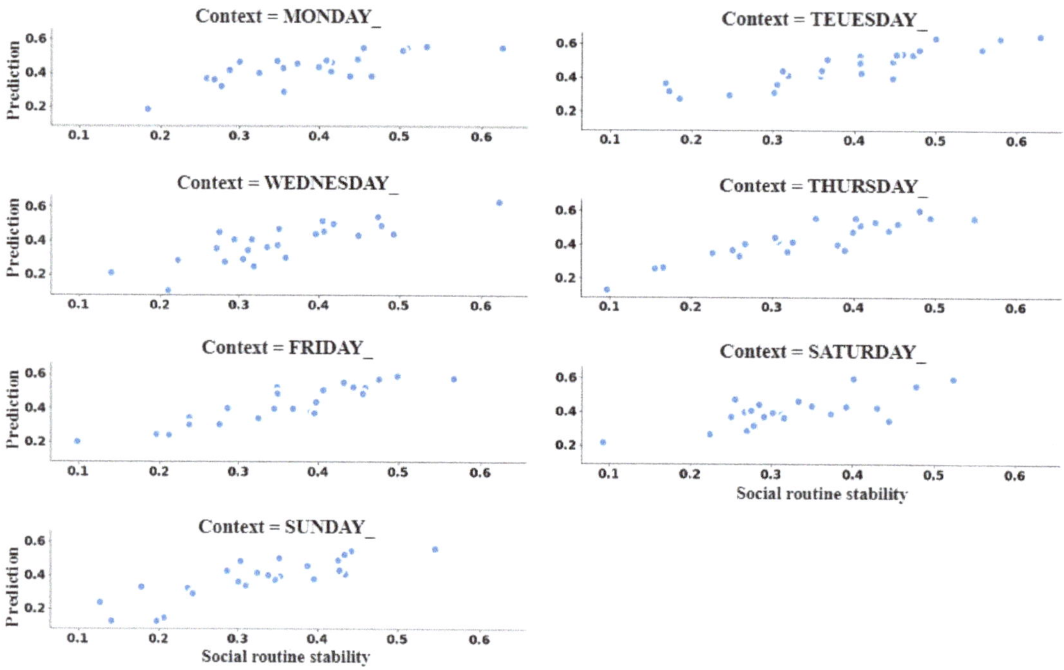

aria

Figure 14. Correlation between prediction performance of sociability pattern and social routine for each Context Attribute (CA).

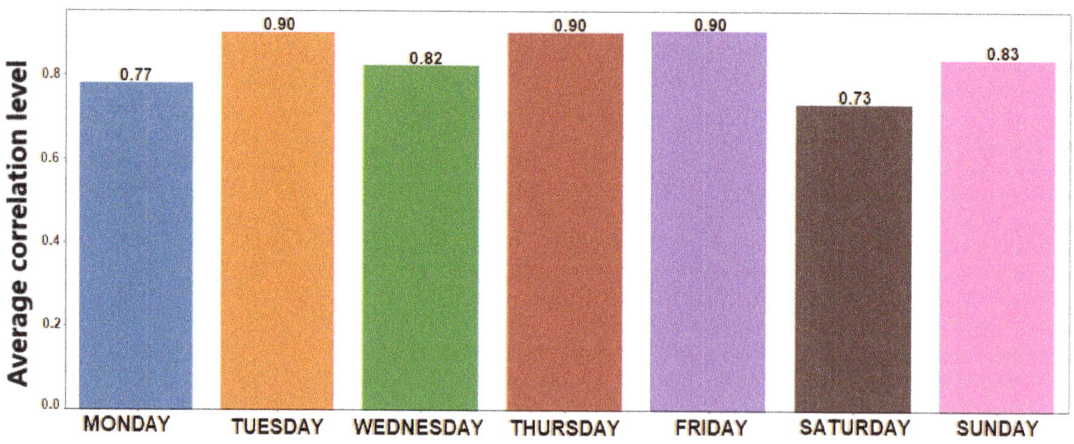

Figure 15. Average correlation level for each CA.

From this experiment, we can recognize that sociability patterns of the proposed solution satisfactorily model social routines of individuals. Therefore, sociability patterns can be used to reliably understand and predict social behavior since they have strong correlations with users' social habits.

4.4. Evaluation of Social Behavior Change Detection

This experiment aimed to verify the performance of the social behavior change detection solution. In particular, we evaluate its ability to detect social behavior changes and adaption to the new social behavior of the monitored individual. Therefore, we expect that the solution to accurately identify observations that represent abnormal social behaviors and social routine changes.

Firstly, we defined a similarity threshold that represents a change in social behavior. For this reason, we calculated mean and standard deviation of the stability of social routines for all users, so defining this threshold as $[\mu + \sigma]$. Similar to the configuration of the previous experiment (Figure 11), we calculated mean and standard deviation of the similarity of social behaviors of individuals between consecutive days considering each day of the week as CAs. In the end, we identified that users had an average of 35.4% of stability in their social routines and standard deviation of 10.7%. Therefore, we specified the threshold for social behavior changes at 46.1%.

We combined data from two users with significantly different social routines to simulate a change in social behavior. For this, we selected users who had more stable social habits. We identified that users u27 and u04 presented social routines with satisfactory stability for this experiment by considering results from the previous experiment.

Figure 16 shows social routines of the individuals u27 and u04, in which each cell contains the number of social events in a given time slot of 30 min (t = $\frac{24}{0.5}$). From this visualization, we can recognize an evident change in social behavior at the limit that separates the two users' data. In this scenario, we expect that the proposed solution would detect abnormal behaviors when starting to process data of the user u04 and recognize the social routine change. Additionally, the solution adapt to the new pattern, then providing a new sociability pattern capable of explaining and predicting the new social routine.

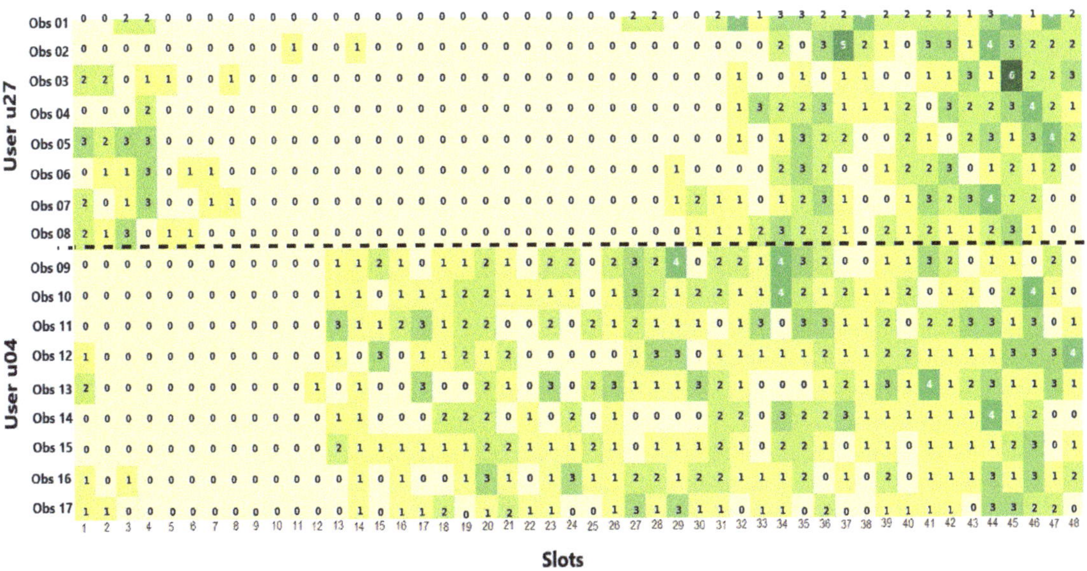

Figure 16. Merge of the social routines of the users u27 and u04.

Social event streams for the proposed solution were created from data of the selected users. After processing data stream, our solution detected social behavior changes presented in Table 2. These results demonstrate that it identified abnormal behaviors and social routine changes precisely. Specifically, our solution identified a high similarity between social patterns and observations (i.e., similarity > 46.1%) while processing user data u27, so not representing behavioral changes. From the first observation of the user u04, our solution started to detect abnormal social behaviors, so recognizing the social routine change by extracting the first pattern using data from this user (i.e., 5th pattern). This new reference pattern remained consistent with the next processed social observations, i.e., the solution can efficiently adapt to data stream changes.

Table 2. The flow of social behavior change detection. Red lines represent the identification of abnormal behaviors and social routine change.

Observation	Uid	Similarity	Change
03	*u27*	0.520000	**Normal social behavior**
04	*u27*	0.727273	**Normal social behavior**
2nd Pattern	*u27*	0.680000	**Maintained the reference pattern**
05	*u27*	0.652174	**Normal social behavior**
06	*u27*	0.520000	**Normal social behavior**
3rd Pattern	*u27*	0.640000	**Maintained the reference pattern**
07	*u27*	0.576923	**Normal social behavior**
08	*u27*	0.555556	**Normal social behavior**
4th Pattern	u27	0.666667	**Maintained the reference pattern**
09	*u04*	0.411765	**Abnormal social behavior**
10	*u04*	0.444444	**Abnormal social behavior**
5th Pattern	*u04*	0.428571	**Social routine change detected**
11	*u04*	0.657143	**Normal social behavior**
12	*u04*	0.647143	**Normal social behavior**
6th Pattern	*u04*	0.647059	**Maintained the reference pattern**
13	*u04*	0.540541	**Normal social behavior**
14	*u04*	0.575758	**Normal social behavior**
7th Pattern	*u04*	0.600000	**Maintained the reference pattern**
15	*u04*	0.666667	**Normal social behavior**
16	*u04*	0.685714	**Normal social behavior**
8th Pattern	*u04*	0.666667	**Maintained the reference pattern**
17	*u04*	0.647059	**Normal social behavior**

4.5. Discussion and Limitations

This section presented an experimental evaluation of the proposed solution that significantly extended experimental evaluations reported in [20]. We performed an in-depth evaluation of the components designed to detect context-aware sociability patterns and behavior changes. From these experiments, we found that patterns recognized by the solution can model and predict social routines of individuals considering context information. Moreover, it can detect significant changes in social habits consistently.

In [20], we compared the similarity between social intervals detected by the proposed solution with those recognized by a batch processing algorithm. From the evaluation in [20], we recognized that the compared solutions identified sociability patterns with 86.33% similarity. We also analyzed sociability patterns based on CAs to investigate their contribution to understand social habits. We identified that context-based recognition provides insights into sociability patterns hidden in the context-free analysis. Therefore, the detection of sociability patterns based on CAs improves the understanding of social habits because it enables to distinguish abnormal behaviors from expected changes due to the context.

The aims of our experiments differ from the evaluations performed by the related works, hence it is a challenge to compare their results using specific metrics. For example, studies that design machine learning models to classify and predict mental states use metrics such as accuracy, precision, recall, whereas in our unsupervised sociability pattern learning approach, we use methods to assess the ability to model social routine and detect behavior changes. The works [22–24] also assess sociability patterns, but they differ from our experiments. Harari et al. [23] computed test-retest correlations between the observed behavior durations for adjacent weeks. Barnett et al. [22] analyzed the rate of anomalies in behavioral patterns based on a statistical test inspired by Filzmoser's approach to predict schizophrenic relapse. Bonilla et al. [24] analyzed the applying of the Poisson mixture model to obtain the intensity functions of all calls in which patients were involved.

Our experiments evaluated the ability of the proposed solution to identify context-aware sociability patterns capable of explaining social routine and detecting social behavior changes. From this analysis, we identified that the prediction performance of social patterns has a strong positive correlation with the stability of the social routine (i.e., Pearson correlation coefficient greater than +0.7) in all CAs considered, so enabling to recognize that the proposed solution detects patterns consistent with the social behaviors of the monitored individuals. The evaluation of the social behavior change detection solution analyzed results of the detection processes performed by the solution when processing data containing changes in social routines. This experiment recognizes that our proposed solution can accurately detect and report abnormal social behaviors and social routine changes. Therefore, this is a promising tool for monitoring mental health, since reports of behavioral changes may indicate the onset, presence, or development of mental disorders.

This study has some limitations. First, the algorithm requires to manually enter two parameters: φ and θ. Therefore, the recognized sociability pattern depends on the predefined values chosen empirically rather than automatically setting the best values based on processed data. Second, the experimental evaluation was based on only one type of social activity (i.e., conversations). Other sources of social interactions should be considered, such as interactions on mobile social networks and telephone call communications. Third, social routine change does not necessarily imply change in sociability, as individuals can change their routine and maintain the same sociability level. However, the solution can identify social routine changes (i.e., whether there is a change in sociability), so allowing mental health professionals to interpret and investigate changes in the user's social aspect. Finally, another aspect is the homogeneous essence of the study participants (i.e., university students). Our solution should be validated with a more heterogeneous population, especially mental health professionals and their patients.

5. Conclusions and Future Work

This work presented an approach for monitoring mental health through awareness of the social situation. Specifically, we introduced a solution based on FPM and CEP concepts to recognize intervals of the day when a monitored individual habitually socializes for each contextual condition. We also presented the solution developed to recognize abnormal behaviors and routine changes. Additionally, we introduced specialized knowledge modeling through fuzzy logic to allow the solution to send notifications of behavioral changes considering the imprecision of this task. From the evaluation conducted, we demonstrate that the predictive performance of context-aware sociability patterns has a strong correlation with social routine stability and that the solution can detect social behavior changes. We conclude that our proposed solution can be integrated into mental health monitoring tools to objectively collect patterns and changes in social behaviors, then providing support to mental health professionals and contributing to the effectiveness of the treatment proposed for patients.

As future work, we plan to address some open issues. The first one is to detect sociability patterns previously specified by mental health professionals, so enabling to identify situations of interest. We would also like to create dashboards for sociability patterns in an appropriate way for professionals, so facilitating analysis of social behavior. Another task is to update the solution to add information about the sociability level (e.g., social interaction intensity) to extracted patterns and identified behavioral changes. We also intend to develop an approach capable of automatically defining the best values for the parameters of the algorithm. Moreover, we plan to extend our solution to detect patterns related to other behaviors, such as physical activity and mobility. Plans also include validating our solution with professionals and their patients.

Author Contributions: I.R.d.M. implemented the solution and wrote this manuscript. A.S.T. and I.R.d.M. conducted experimental evaluations. F.J.d.S.eS., L.R.C. and M.E. supervised the research. F.J.d.S.eS., L.R.C., A.S.T. and M.E. participated in the revision of the manuscript, so supporting the writing process. All authors have read and agreed to the published version of the manuscript.

Funding: This research was funded by Coordination for the Improvement of Higher Education Personnel—CAPES (grant 88887.200532/2018-00); INCT of the Future Internet for Smart Cities (CNPq 465446/ 2014-0, CAPES 88887.136422/2017-00, and FAPESP 14/50937-1 and 15/24485-9); National Council for Scientific and Technological Development—CNPq (311608/2017-5, 420907/2016-5, 312324/2015-4); and the State of Maranhão Research Funding Agency (FAPEMA).

Conflicts of Interest: The authors declare no conflict of interest.

Abbreviations

The following abbreviations are used in this manuscript:

IoT	Internet of Things
EMA	Ecological Momentary Assessment
EMI	Ecological Momentary Intervention
FPM	Frequent Pattern Mining
CEP	Complex Event Processing
API	Application Programming Interface
FIS	Fuzzy Inference System
CA	Context Attribute
EPL	Event Processing Language
SQL	Structured Query Language
FCL	Fuzzy Control Language
COG	Center of Gravity
GMM	Gaussian Mixture Model

References

1. World Health Organization. *Mental Health Action Plan 2013–2020*; Technical Report; World Health Organization, WHO Document Production Services: Geneva, Switzerland, 2013. Available online: https://www.who.int/publications/i/item/9789241506021 (accessed on 29 March 2019).
2. Clinic, M. *Menatl Illness*; Technical Report; Mayo Clinic: Rochester, NY, USA, 2018. Available online: https://www.mayoclinic.org/diseases-conditions/mental-illness/symptoms-causes/syc-20374968 (accessed on 29 March 2019).
3. Patel, V.; Saxena, S. Transforming lives, enhancing communities—Innovations in global mental health. *N. Engl. J. Med.* **2014**, *370*, 498–501. [CrossRef]
4. Depression—Key Facts. Available online: https://www.who.int/news-room/fact-sheets/detail/depression (accessed on 29 March 2019).
5. Umberson, D.; Montez, J.K. Social Relationships and Health: A Flashpoint for Health Policy. *J. Health Soc. Behav.* **2010**, *51*, S54–S66. [CrossRef] [PubMed]
6. Grav, S.; Hellzèn, O.; Romild, U.; Stordal, E. Association between social support and depression in the general population: The HUNT study, a cross-sectional survey. *J. Clin. Nurs.* **2012**, *21*, 111–120. [CrossRef] [PubMed]
7. Fauth, E.B.; Gerstorf, D.; Ram, N.; Malmberg, B. Changes in Depressive Symptoms in the Context of Disablement Processes: Role of Demographic Characteristics, Cognitive Function, Health, and Social Support. *J. Gerontol. Ser. B* **2011**, *67B*, 167–177. [CrossRef] [PubMed]
8. Beutel, M.E.; Klein, E.M.; Brähler, E.; Reiner, I.; Jünger, C.; Michal, M.; Wiltink, J.; Wild, P.S.; Münzel, T.; Lackner, K.J.; et al. Loneliness in the general population: Prevalence, determinants and relations to mental health. *BMC Psychiatry* **2017**, *17*, 97. [CrossRef]
9. Torales, J.; O'Higgins, M.; Castaldelli-Maia, J.M.; Ventriglio, A. The outbreak of COVID-19 coronavirus and its impact on global mental health. *Int. J. Soc. Psychiatry* **2020**. [CrossRef]
10. Morrison-Valfre, M. *Foundations of Mental Health Care-E-Book*; Elsevier Health Sciences: St. Louis, MO, USA, 2016.
11. Schacter, D.L. The seven sins of memory: Insights from psychology and cognitive neuroscience. *Am. Psychol.* **1999**, *54*, 182. [CrossRef]
12. Van de Mortel, T.F. Faking it: Social desirability response bias in self-report research. *Aust. J. Adv. Nursing* **2008**, *25*, 40.
13. Liang, Y.; Zheng, X.; Zeng, D.D. A survey on big data-driven digital phenotyping of mental health. *Inf. Fusion* **2019**, *52*, 290–307. [CrossRef]
14. Torous, J.; Kiang, M.V.; Lorme, J.; Onnela, J.P. New Tools for New Research in Psychiatry: A Scalable and Customizable Platform to Empower Data Driven Smartphone Research. *JMIR Ment. Health* **2016**, *3*, e16. [CrossRef]

15. Garcia-Ceja, E.; Riegler, M.; Nordgreen, T.; Jakobsen, P.; Oedegaard, K.J.; Tørresen, J. Mental health monitoring with multimodal sensing and machine learning: A survey. *Pervasive Mob. Comput.* **2018**, *51*, 1–26. [CrossRef]
16. Mohr, D.C.; Zhang, M.; Schueller, S.M. Personal Sensing: Understanding Mental Health Using Ubiquitous Sensors and Machine Learning. *Annu. Rev. Clin. Psychol.* **2017**, *13*, 23–47. [CrossRef] [PubMed]
17. Moura, I.; Teles, A.; Silva, F.; Viana, D.; Coutinho, L.; Barros, F.; Endler, M. Mental health ubiquitous monitoring supported by social situation awareness: A systematic review. *J. Biomed. Inform.* **2020**, *107*, 103454, doi:10.1016/j.jbi.2020.103454. [CrossRef] [PubMed]
18. Aggarwal, C.C.; Bhuiyan, M.A.; Hasan, M.A. Frequent Pattern Mining Algorithms: A Survey. In *Frequent Pattern Mining*; Aggarwal, C.C., Han, J., Eds.; Springer International Publishing: Cham, Switzerland, 2014; pp. 19–64.
19. Etzion, O.; Niblett, P.; Luckham, D.C. *Event Processing in Action*; Manning Greenwich: Stamford, CT, USA, 2011.
20. Rodrigues de Moura, I.; da Silva e Silva, F.J.; Reis Coutinho, L.; Soares Teles, A. Mental Health Ubiquitous Monitoring: Detecting Context-Enriched Sociability Patterns Through Complex Event Processing. In Proceedings of the 2020 IEEE 33rd International Symposium on Computer-Based Medical Systems (CBMS), Rochester, MN, USA, 28–30 July 2020; pp. 239–244. [CrossRef]
21. Exler, A.; Braith, M.; Mincheva, K.; Schankin, A.; Beigl, M. Smartphone-Based Estimation of a User Being in Company or Alone Based on Place, Time, and Activity. In *Mobile Computing, Applications, and Services*; Murao, K., Ohmura, R., Inoue, S., Gotoh, Y., Eds.; Springer International Publishing: Cham, Switzerland, 2018; pp. 74–89.
22. Barnett, I.; Torous, J.; Staples, P.; Sandoval, L.; Keshavan, M.; Onnela, J.P. Relapse prediction in schizophrenia through digital phenotyping: A pilot study. *Neuropsychopharmacology* **2018**, *43*, 1660. [CrossRef] [PubMed]
23. Harari, G.M.; Gosling, S.D.; Wang, R.; Chen, F.; Chen, Z.; Campbell, A.T. Patterns of behavior change in students over an academic term: A preliminary study of activity and sociability behaviors using smartphone sensing methods. *Comput. Hum. Behav.* **2017**, *67*, 129–138. [CrossRef]
24. Bonilla-Escribano, P.; Ramírez, D.; Sedano-Capdevila, A.; Campaña-Montes, J.J.; Baca-García, E.; Courtet, P.; Artés-Rodríguez, A. Assessment of e-Social Activity in Psychiatric Patients. *IEEE J. Biomed. Health Inform.* **2019**, *23*, 2247–2256. [CrossRef] [PubMed]
25. Eskes, P.; Spruit, M.; Brinkkemper, S.; Vorstman, J.; Kas, M.J. The sociability score: App-based social profiling from a healthcare perspective. *Comput. Hum. Behav.* **2016**, *59*, 39–48. [CrossRef]
26. Wahle, F.; Kowatsch, T.; Fleisch, E.; Rufer, M.; Weidt, S. Mobile sensing and support for people with depression: A pilot trial in the wild. *mHealth uHealth* **2016**, *4*. [CrossRef]
27. Lane, N.D.; Lin, M.; Mohammod, M.; Yang, X.; Lu, H.; Cardone, G.; Ali, S.; Doryab, A.; Berke, E.; Campbell, A.T.; et al. BeWell: Sensing Sleep, Physical Activities and Social Interactions to Promote Wellbeing. *Mob. Netw. Appl.* **2014**, *19*, 345–359. [CrossRef]
28. Gu, J.; Gao, B.; Chen, Y.; Jiang, L.; Gao, Z.; Ma, X.; Ma, Y.; Woo, W.L.; Jin, J. Wearable Social Sensing: Content-Based Processing Methodology and Implementation. *IEEE Sensors J.* **2017**, *17*, 7167–7176. [CrossRef]
29. Chen, Y.; Gao, B.; Jiang, L.; Yin, K.; Gu, J.; Woo, W.L. Transfer Learning for Wearable Long-Term Social Speech Evaluations. *IEEE Access* **2018**, *6*, 61305–61316. [CrossRef]
30. Wu, C.; Boukhechba, M.; Cai, L.; Barnes, L.E.; Gerber, M.S. Improving momentary stress measurement and prediction with bluetooth encounter networks. *Smart Health* **2018**, *9–10*, 219–231. [CrossRef]
31. Sarda, A.; Munuswamy, S.; Sarda, S.; Subramanian, V. Using Passive Smartphone Sensing for Improved Risk Stratification of Patients With Depression and Diabetes: Cross-Sectional Observational Study. *mHealth uHealth* **2019**, *7*, e11041. [CrossRef]
32. Abdullah, S.; Matthews, M.; Frank, E.; Doherty, G.; Gay, G.; Choudhury, T. Automatic detection of social rhythms in bipolar disorder. *J. Am. Med. Inform. Assoc.* **2016**, *23*, 538–543. [CrossRef] [PubMed]
33. Servia-Rodríguez, S.; Rachuri, K.K.; Mascolo, C.; Rentfrow, P.J.; Lathia, N.; Sandstrom, G.M. Mobile Sensing at the Service of Mental Well-being: A Large-scale Longitudinal Study. In Proceedings of the WWW '17, 26th International Conference on World Wide Web, Perth, Australia, 3 April 2017; pp. 103–112. [CrossRef]
34. Beiwinkel, T.; Kindermann, S.; Maier, A.; Kerl, C.; Moock, J.; Barbian, G.; Rössler, W. Using smartphones to monitor bipolar disorder symptoms: A pilot study. *JMIR Mental Health* **2016**, *3*. [CrossRef] [PubMed]
35. Wang, R.; Chen, F.; Chen, Z.; Li, T.; Harari, G.; Tignor, S.; Zhou, X.; Ben-Zeev, D.; Campbell, A.T., StudentLife: Using Smartphones to Assess Mental Health and Academic Performance of College Students. In *Mobile Health: Sensors, Analytic Methods, and Applications*; Springer International Publishing: Cham, Switzerland, 2017; pp. 7–33.
36. Chow, P.I.; Fua, K.; Huang, Y.; Bonelli, W.; Xiong, H.; Barnes, L.E.; Teachman, B.A. Using Mobile Sensing to Test Clinical Models of Depression, Social Anxiety, State Affect, and Social Isolation Among College Students. *J. Med. Internet Res.* **2017**, *19*, e62. [CrossRef] [PubMed]
37. Boukhechba, M.; Daros, A.R.; Fua, K.; Chow, P.I.; Teachman, B.A.; Barnes, L.E. DemonicSalmon: Monitoring mental health and social interactions of college students using smartphones. *Smart Health* **2018**, *9–10*, 192–203. [CrossRef]
38. Ono, E.; Nozawa, T.; Ogata, T.; Motohashi, M.; Higo, N.; Kobayashi, T.; Ishikawa, K.; Ara, K.; Yano, K.; Miyake, Y. Fundamental deliberation on exploring mental health through social interaction pattern. In Proceedings of the 2012 ICME International Conference on Complex Medical Engineering (CME), Kobe, Japan, 1–4 July 2012; pp. 321–326. [CrossRef]
39. Gong, J.; Huang, Y.; Chow, P.I.; Fua, K.; Gerber, M.S.; Teachman, B.A.; Barnes, L.E. Understanding behavioral dynamics of social anxiety among college students through smartphone sensors. *Inf. Fusion* **2019**, *49*, 57–68. [CrossRef]

40. Matic, A.; Osmani, V.; Mayora, O., Automatic Sensing of Speech Activity and Correlation with Mood Changes. In *Pervasive and Mobile Sensing and Computing for Healthcare: Technological and Social Issues*; Springer: Berlin/Heidelberg, Germany, 2013; Chapter 9; pp. 195–205; [CrossRef]
41. Benesty, J.; Chen, J.; Huang, Y.; Cohen, I. Pearson correlation coefficient. In *Noise Reduction in Speech Processing*; Springer: Berlin/Heidelberg, Germany, 2009; pp. 1–4. [CrossRef]
42. Myers, L.; Sirois, M.J. Spearman correlation coefficients, differences between. *Encycl. Stat. Sci.* **2004**, *12*. [CrossRef]
43. Weiss, K.; Khoshgoftaar, T.M.; Wang, D. A survey of transfer learning. *J. Big Data* **2016**, *3*, 9. [CrossRef]
44. Lago, P.; Roncancio, C.; Jiménez-Guarín, C. Learning and managing context enriched behavior patterns in smart homes. *Future Gener. Comput. Syst.* **2019**, *91*, 191–205. [CrossRef]
45. Žliobaitė, I. Learning under concept drift: An overview. *arXiv* **2010**, arXiv:1010.4784.
46. Cingolani, P.; Alcalá-Fdez, J. jFuzzyLogic: A robust and flexible Fuzzy-Logic inference system language implementation. In Proceedings of the 2012 IEEE International Conference on Fuzzy Systems, Brisbane, QLD, Australia, 13 August 2012; pp. 1–8.
47. Cingolani, P.; Alcalá-Fdez, J. jFuzzyLogic: A Java Library to Design Fuzzy Logic Controllers According to the Standard for Fuzzy Control Programming. *Int. J. Comput. Intell. Syst.* **2013**, *6*, 61–75. [CrossRef]
48. McNeill, F.M.; Thro, E. *Fuzzy logic: A Practical Approach*; Academic Press: Cambridge, MA, USA, 2014.
49. Ross, T.J. *Fuzzy Logic with Engineering Applications*; Wiley Online Library: Chichester, UK, 2004; Volume 2.
50. Soares Teles, A.; Rocha, A.; José da Silva e Silva, F.; Correia Lopes, J.; O'Sullivan, D.; Van de Ven, P.; Endler, M. Enriching Mental Health Mobile Assessment and Intervention with Situation Awareness. *Sensors* **2017**, *17*, 127. [CrossRef]
51. Fowler, M. *Refactoring: Improving the Design of Existing Code*; Addison-Wesley Professional: Boston, MA, USA 2018.
52. Shinde, S.A.; Nimkar, P.A.; Singh, S.P.; Salpe, V.D.; Jadhav, Y.R. MQTT-message queuing telemetry transport protocol. *Int. J. Res.* **2016**, *3*, 240–244.

sensors

Article

A Machine Learning Multi-Class Approach for Fall Detection Systems Based on Wearable Sensors with a Study on Sampling Rates Selection †

Nicolas Zurbuchen [1,*], Adriana Wilde [2,3,*] and Pascal Bruegger [1]

1 Institute of Complex Systems (iCoSys), School of Engineering and Architecture of Fribourg Switzerland, HES-SO University of Applied Sciences and Arts Western Switzerland, 1700 Fribourg, Switzerland; pascal.bruegger@hes-so.ch

2 Centre for Health Technologies (CHT), School of Electronics and Computer Science, University of Southampton, Southampton SO17 1BJ, UK

3 Department of Digital Technologies, Faculty of Business, Law and Digital Technologies, University of Winchester, Winchester SO22 4NR, UK

* Correspondence: nicolas.zurbuchen@hes-so.ch (N.Z.); agw106@ecs.soton.ac.uk or adriana.wilde@winchester.ac.uk (A.W.)

† This paper is an extended version of the conference paper: Zurbuchen, N.; Bruegger, P. and Wilde, A. A Comparison of Machine Learning Algorithms for Fall Detection using Wearable Sensors. In Proceedings of the 2020 International Conference on Artificial Intelligence in Information and Communication (ICAIIC), Fukuoka, Japan, 19–21 February 2020; pp. 427–431, doi: 10.1109/ICAIIC48513.2020.9065205.

Abstract: Falls are dangerous for the elderly, often causing serious injuries especially when the fallen person stays on the ground for a long time without assistance. This paper extends our previous work on the development of a Fall Detection System (FDS) using an inertial measurement unit worn at the waist. Data come from *SisFall*, a publicly available dataset containing records of Activities of Daily Living and falls. We first applied a preprocessing and a feature extraction stage before using five Machine Learning algorithms, allowing us to compare them. Ensemble learning algorithms such as Random Forest and Gradient Boosting have the best performance, with a Sensitivity and Specificity both close to 99%. Our contribution is: a multi-class classification approach for fall detection combined with a study of the effect of the sensors' sampling rate on the performance of the FDS. Our multi-class classification approach splits the fall into three phases: pre-fall, impact, post-fall. The extension to a multi-class problem is not trivial and we present a well-performing solution. We experimented sampling rates between 1 and 200 Hz. The results show that, while high sampling rates tend to improve performance, a sampling rate of 50 Hz is generally sufficient for an accurate detection.

Keywords: fall detection; wearable sensors; sampling rate; data preprocessing; feature extraction; Machine Learning

Citation: Zurbuchen, N.; Wilde, A.; Bruegger, P. A Machine Learning Multi-Class Approach for Fall Detection Systems Based on Wearable Sensors with a Study on Sampling Rates Selection. *Sensors* **2021**, *21*, 938. https://doi.org/10.3390/s21030938

Academic Editor: Klaus Moessner
Received: 22 December 2020
Accepted: 26 January 2021
Published: 30 January 2021

1. Introduction

Falls are one of the leading causes of death among the elderly [1]. Every year, 28% to 35% of the elderly fall at least once and this rate increases with age [2]. Falls can have severe physical, psychological and even social consequences. They can also heavily affect the independent quality of living. They can result in bruises and swellings, as well as fractures and traumas [3]. A significant risk is the *long-lie*. This happens when an elderly person remains on the ground for a long duration without being able to call for help. It is associated with death within the next few months following the accident [4]. It also affects the elderly's self-confidence who may develop the *fear of falling*' syndrome. It leads to anxiety when performing Activities of Daily Living (ADLs) and can lead to subsequent falls [1].

Therefore, the elderly must continuously be monitored to ensure their safety. Families organize visits but these can be inconvenient and even insufficient. Hiring caregivers or moving into nursing homes are sometimes not affordable options. Recent progresses in technology have enabled the development of Assisted-Living Systems (ALSs) [5]. They can assist the elderly and provide a safer environment through constant monitoring while relieving caregivers' workload. However, ALSs create other challenges such as privacy concerns and acceptability issues that need to be addressed [6].

Fall Detection Systems (FDSs) are part of ALSs. Their goals are to identify falls and notify caregivers so that they can intervene as fast as possible. However, fall recognition is challenging from a computational perspective. Falls can be defined as "the rapid changes from the upright/sitting position to the reclining or almost lengthened position, but it is not a controlled movement" [7]. There is a higher acceleration during falls. Another challenge is that falls can happen in innumerable scenarios. They may occur anywhere at any time [3]. Their starting and ending body posture as well as their direction (e.g., forward, backward) may vary [1]. Hence, FDSs must cover the whole living area. Their reliability must be high while minimizing false alarms, all the while respecting the elderly's privacy.

This paper is an extension of our work accepted at the ICAIIC 2020 [8]. This paper has three research questions:

RQ1: *What is the difference in performance across various types of Machine Learning (ML) algorithms in a FDS?*
To answer this, we developed a reliable FDS by the mean of wearable sensors (accelerometer and gyroscope) and ML algorithms. The goal is to compare lazy, eager and ensemble learning algorithms and assess their results. We implemented five algorithms and tested them in the same setup.

RQ2: *What is the effect of the sensors' sampling rate on the fall detection?*
To study this, we analyzed the influence of the sensors' sampling rate on the detection. We filtered the data in order to reduce the number of samples measured per second. We then experimented on the filtered data with five ML algorithms. This research question extends our previous work [8].

RQ3: *What is the difference in performance across various types of ML algorithms by adopting a multi-class approach for identifying phases of a fall?*
We experimented a different fall detection approach where falls are split into three phases. These are: the period before the fall happens (pre-fall), the fall itself (impact) and after the fall happened (post-fall). This research question extends our previous work [8].

The rest of this paper is organized as follows. In Section 2, we discuss existing FDSs and highlight their distinctive features. Section 3 covers the employed methodology. Section 4 presents and discusses the obtained results. Finally, we conclude with a comment on future work in Section 5.

2. Related Work

Scientists have employed various approaches to implement FDSs over the past years. They have been classified as presented in Figure 1. Each of them has its strengths and weaknesses. We focus on wearable technologies since we use this approach. Nevertheless, several survey studies [9,10] reported the other methods in more depth.

2.1. Choice of Sensors and Sampling Rate

Several types of sensors including accelerometers, gyroscopes, magnetometers and tilt sensors have been used to detect falls. Based on the fall characteristics, most studies, such as [11–15], employed only acceleration measurements. From our literature review, very few studies use a single gyroscope. For example, Bourke and Lyons [16] used a single biaxial gyroscope and measured changes in angular velocity, angular acceleration and body angle. Tang and Ou [17] also reported promising results, using a single six-axis gyroscope. The separate use of these sensors already produced promising results but their combination is even better [18]. Wang et al. [19] employed a heart rate monitor and discovered that the heart rate increases by 22% after a fall in people over 40 years old.

This demonstrates that physiological data can be used in such a system. Across the papers reviewed (summarized in Table 1), the sensors' sampling rate varied within a range from 10 to 1000 Hz. This variation is not small, one having 100 times more samples than the other, seemingly arbitrarily. Fudickar et al. [20] compared the detection results when varying the sampling rates from 50 to 800 Hz. The results obtained with a sampling rate of 50 Hz were as good as the ones with 800 Hz. Other studies show that low sampling rates can offer reasonable results, for example Medrano et al. [21] used data sampled up to 52 Hz. We further investigate this issue in this paper with similarly low sampling rates.

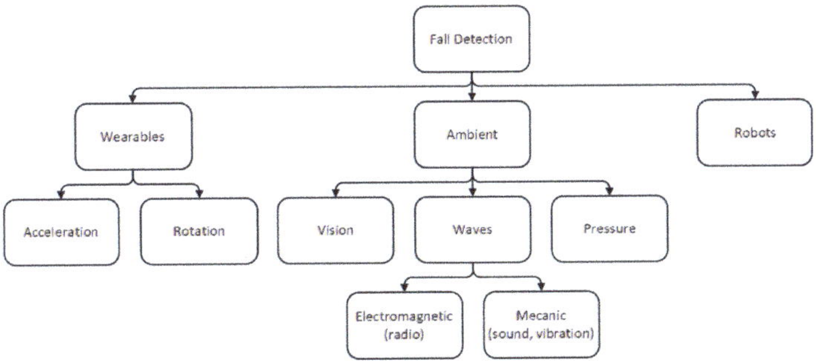

Figure 1. Classification of Fall Detection System approaches.

Table 1. Reviewed studies that used wearable sensors for fall detection (including acronyms at the end of the table).

Research Authors (Year)	Sensors	Freq.	Algorithm	Reported Outcomes
Hwang et al. [22] (2004)	Accelerometer, gyroscope and tilt sensor placed at the chest.	Not reported	Threshold on each sensor compared sequentially.	Accuracy 96.7%
Bourke et al. [12] (2007)	Accelerometer placed at the thigh and chest.	1 kHz	Double acceleration thresholds applied on both sensors.	SP 100%
Bourke et al. [16] (2008)	Bi-axial gyroscope placed at the chest.	1 kHz	Treble angular thresholds.	SE 100% SP 100%
Kangas et al. [14] (2008)	Accelerometer placed at the waist, head and wrist.	400 Hz	Several simple algorithms including thresholds and posture recognition.	SE 97.5% SP 100%
Dinh et al. [18] (2009)	Accelerometer and gyroscope placed at the chest.	40 Hz	Supervised ML algorithms (SVM, Naïve Bayes, C4.5, Ripple-down rules and RBF.	Accuracy 97%
Choi et al. [23] (2011)	Accelerometer and gyroscope placed at the belt.	10–18 Hz	Naive Bayesian Algorithm to identify specific falls and ADLs.	Accuracy 99.4%
Gjoreski et al. [24] (2011)	Four accelerometers placed at the chest, waist, thigh and ankle	6 Hz	Several simple algorithms including thresholds and posture recognition.	Accuracy 99%
Aziz et al. [25] (2011)	Three accelerometers placed at the sternum, right ankle and left ankle	120 Hz	Linear discriminant analysis to identify three causes of fall.	SE 96% SP 98%
Yuwono et al. [15] (2012)	Accelerometer placed at the waist.	20 Hz	Unsupervised ML algorithms (clustering, MLP and augmented RBF neural network) with WT.	SE 100% SP 99.33%
Bagalà et al. [26] (2012)	Accelerometer placed at the lower back.	100 Hz	Comparison of several threshold-based algorithms with posture recognition.	SE 83% SP 94%

Table 1. *Cont.*

Research Authors (Year)	Sensors	Freq.	Algorithm	Reported Outcomes
Abbate et al. [11] (2012)	Accelerometer from a belt worn smartphone.	50 Hz	Neural network with 8 features extracted as input and a 4 classes classification.	SE 100% SP 100%
Chan et al. [13] (2013)	Three accelerometers placed at the chest.	62.5 Hz	Combination of thresholds, posture measurements and posture recognition.	SE 95.2% SP 100%
Fudickar et al. [20] (2014)	Accelerometer from a smartphone worn at the hip.	50–800 Hz	Threshold-based with sequential posture recognition.	SE 99%
Wang et al. [19] (2014)	Accelerometer and cardiotachometer placed at the chest.	Not reported	Treble thresholds including impact magnitude, trunk angle and heart rate.	SE 96.8% SP 97.5%
Medrano et al. [21] (2014)	Smartphone accelerometer in a pocket (for 95% of ADL data), a hand bag (5%), or two smartphones in separate hand bags (for falls).	unstable, 16.7–52 Hz	One-class SVM, kNN (k = 1), kNN-sum (k = 2) and K-means + 1 NN (k = 800)	SE > 89% SP > 88%
Özdemir et al. [27] (2014)	Accelerometer, gyroscope and magnetometer placed at the head, chest, back, wrist, ankle and thigh.	25 Hz	Features extraction at the total peak acceleration and use of ML algorithms (KNN, LSM, SVM, BDM, DTW and ANN).	SE 100% SP > 99%
Vilarinho et al. [28] (2015)	Accelerometer and gyroscope from the smartphone and smartwatch respectively placed at the thigh and wrist.	Not reported	Acceleration threshold and pattern recognition from both devices	SE 63% SP 78%
Casilari et al. [29] (2015)	Accelerometer and gyroscope from the smartphone and smartwatch respectively placed at the thigh and wrist.	Not reported	Several thresholds compared to each other with every combination of sensors.	SE 96.7% SP 100%
Gibson et al. [30] (2016)	Accelerometer placed at the chest.	50 Hz	Combination of several algorithms (ANN, KNN, RBF, PPCA, LDA)	SE > 90% SP > 90%
Sucerquia et al. [31] (2017)	Accelerometer placed at the waist.	200 Hz	Threshold-based classifier with feature extraction.	Accuracy 96%
Hsieh et al. [32] (2017)	Accelerometer placed at the waist.	128 Hz	Threshold-based Classifier followed by SVM.	Accuracy > 98.74%
Krupitzer et al. [33,34] (2018, 2019)	Accelerometers placed at the chest, waist and thigh.	20–200 Hz	Self-adaptive pervasive fall detection system combining multiple datasets.	SE 75%
Tang et al. [17] (2018)	Six-axis gyroscope inside a bracelet worn at the wrist.	Not reported	Three-feature vector fed into SVM.	Accuracy 100%
Casilari et al. [35] (2020)	Accelerometry signals from several datasets mainly placed at the waist.	10–200 Hz	Deep Learning with Convolutional Neural Networks	SE > 98% SP > 98%

BDM: Bayesian Decision Making; DTW: Dynamic Time Warping; LDA: Linear Discriminant Analysis; LSM: Least Squares Method; PPCA: Probabilistic Principal Component Analysis; WT: Wavelet Transform; ANN: Artificial Neural Network; MLP: Multilayer Perceptron; RBF: Radial Basis Function.

2.2. Sensing Position

The sensor placement highly affects the detection performance. Previous studies [14,33,34] demonstrated that better results are achieved when sensors are placed along the longitudinal axis of the body (e.g., head, chest, waist) when compared to other placements (e.g., thigh, wrist). The movement of this axis during a fall is more consistent and steady. However, this requires to wear a dedicated device on uncommon body parts which consequently creates inconveniences. For this reason, other studies [11,28,29] used commodities (e.g., smartphones carried by the thigh, smartwatches worn on the wrist). These usually do not disturb the users since they already wear them. However, people tend to take these devices off when they are at home which makes the FDS useless. Another method is to combine various sensing positions. Özdemir et al. [27] developed a system consisting of six wearable devices that are all used together. The problem is that the elderly already have acceptability issues with one device, let alone six.

2.3. Algorithms

There are two categories of algorithms: *threshold-based* and *ML-based*. Threshold algorithms simply define limit values, outside of which, a fall is detected. They have often been sufficient but they tend to produce false alarms especially with fall-like activities such as sitting abruptly [16]. To compensate, these studies [13,22] added simple posture and pattern recognition algorithms that detect changes in body posture and level of activity. This improves the detection's robustness while keeping a low computational complexity. However, it may still fail during specific falls and ADLs. For example, Sucerquia et al. [31] used a manual threshold-based classification over their dataset *SisFall*, achieving 96% accuracy.

ML algorithms automatically learn patterns based on data, and very commonly include feature extraction. They require more computational power and are complex to optimize but produce improved results. Most of the studies such as [11,13] employed a supervised learning technique. Common algorithms are k-Nearest Neighbor [27], Support Vector Machine [18,27] and Artificial Neural Network [11,27]. Yuwono et al. [15] used unsupervised learning which works with clusters. This is a compelling solution because it does not require labeled data. The state-of-the-art Deep Learning algorithms are increasing in popularity, achieving promising results in various fields. Musci et al. [36] employed Recurrent Neural Networks to detect falls. They used a publicly available dataset (*SisFall*) [37] and reported outperforming the results of the original paper [31]. Casilari et al. [35] employed a Convolutional Neural Networks on several datasets, including *SisFall* [31]. They reported promising results with a Sensitivity and Specificity over 98%.

2.4. Classification Strategies

The objective of FDSs is to identify whether a fall happened or not, hence a binary decision. FDSs previously reported in the literature typically follow a binary classification approach, following the intuition that the event of interest is whether the participant has fallen or not. A notable exception to this trend [25] extends this common approach by aiming to differentiate amongst various causes of falls. The study differentiates three causes of falls which are trips, slips and others. Another study [23] used a different approach where the type of fall is identified (amongst the types forward, backward, lateral) as well as various ADLs. In a different context, which is Fall Prevention System [38], the goal is to detect if a fall will definitively happen in order to deploy a protection mechanism such as airbags. In such systems, it is not the fall that needs to be detected but what we could call the pre-fall, meaning what happens before the actual fall. More recent approaches combine these two ideas, for example [32] used a multi-phase model. They differentiate phases of a fall and then classify them into three classes: free fall, impact and rest phases. We further investigate this, using several ML algorithms as detailed in Section 3.4.

2.5. Strengths and Weaknesses

Wearable technologies have several advantages. They are relatively inexpensive and can operate anywhere all of it with minimal intrusion compared to other approaches, such as environmental monitoring [33,34]. In addition, their somewhat limited computational power can be easily overcome with the use of their telecommunication capabilities, which allow the transfer of data for processing outside the device. Wearables can also identify the wearer and get precise measurements. However, they may create discomfort due to their size and intrusiveness. The main disadvantage is their human dependency. These sensors must have enough battery and be worn to work properly. Furthermore, the elderly may have a cognitive impairment and thus, may forget to wear the sensor.

3. Materials and Methods

Our FDS is based on a common pipeline (Figure 2) which has been seen in the literature [27]. This pipeline is a common practice when working with ML algorithms. We first acquire raw data using various sensors and convert them into discrete values. We then preprocess the raw data to remove measuring errors which can badly affect the performance. Afterwards, we construct and extract meaningful information in a vector. Finally, we train and evaluate our ML algorithm to distinguish falls from ADLs.

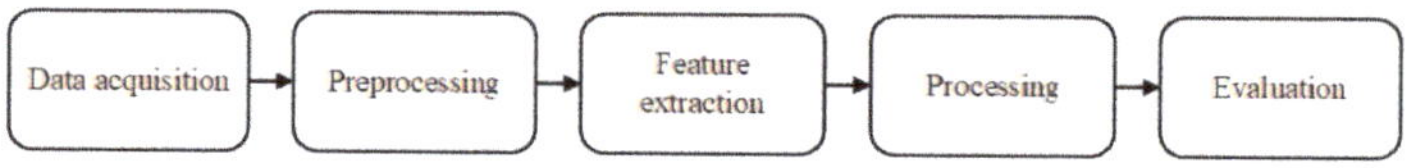

Figure 2. General architecture of Fall Detection Systems.

The steps presented above for our FDS pipeline are common to most of our research questions. However, in order to address research question 3, we have a few changes that will be highlighted. Thus, this section is organised as follows: Sections 3.1 to 3.5 details each step of the pipeline which are common to all research questions. These five subsections answers entirely the first two research questions. However, the third research question requires additional data preparation which is described in Section 3.6.

3.1. Dataset

We decided to use a publicly available dataset rather than creating our own experiment with diverse subjects, for reproducibilty purposes. Therefore, in order to select such dataset, we considered those evaluated in a recent meta-review [39]. From those, we pre-selected those who were freely available, as listed in Table 2 which describes each dataset characteristics. We decided to use acceleration measures because studies have shown that interesting performances can be achieved with it. Ultimately, we selected the dataset named *SisFall* [31] over others [40,41] because of its high quality. We assessed this quality with various criteria, namely the size of the dataset and the diversity of subjects in terms of age, gender, weight and height, as detailed in Table 2.

Table 2. Main characteristics of the considered datasets. Adapted from [39].

Characteristics	Casilari et al. [40] (2016)	Sucerquia et al. [31] (2017)	Micucci et al. [41] (2017)
Dataset name	*UMAFall*	*SisFall*	*UniMiB SHAR*
No. of sensing points	5	1	1
No. of sensors per point	3	3	1
Type of sensors	*A* | *G* | *M*	*A* | *A* | *G*	*A*
Positions of the points	*Ch* | *Wa* | *Wr* | *Th* | *An*	*Wa*	*Th*
Sampling rates per sensor [Hz]	20 | 20 | 20 | 100 | 20	200 | 200 | 200	50
No. of types of ADL/Falls	12/3	19/15	9/8
No. of samples ADL/Falls)	746 (538/208)	4505 (2707/1798)	7013 (5314/1699)
No. of subjects (Female/Male)	(8/11)	38 (19/19)	30 (24/6)
Subjects' age range	18–68	19–75	18–60
Subjects' weight range [kg]	50–93	41.5–102	50–82
Subjects' height range [cm]	155–195	149–183	160–190

A: Accelerometer; G: Gyroscope; M: Magnetometer; An: Ankle; Ch: Chest; Th: Thigh; Wa: Waist; Wr: Wrist.

We also took into account the number of falls and ADLs performed by each subject. An additional factor was the sensors' sampling rate which needed to be high in order to experiment using various sampling rates. In the *SisFall* dataset, two tri-axial accelerometers (ADXL345 and MMA8451Q) and a tri-axial gyroscope (ITG3200) were used at a sampling rate of 200 Hz. These sensors were attached to the waist, following the longitudinal axis, of the subjects in the data collection phase [31]. This location has been proven to be a reliable one from the literature, as discussed in Section 2.2.

We decided not to use the data of the second accelerometer (MMA8451Q) because usual setups only have a single accelerometer. Having decided to use only data from one accelerometer, we chose that with the highest sensing range and the lowest power consumption which seems adequate for the application. Future work could explore whether there is a significant difference between these sensors.

Twenty-three young people (19 to 30 years old) performed 15 types of falls and 19 types of ADLs including fall-like activities. Fifteen elderly people (60 to 75 years old) also performed the same ADLs for more authenticity. There were five trials per activity except for the walking and jogging activities, each of which had only one trial (See Table 3). Hence, *SisFall* contains a total of 4505 records including 2707 ADLs and 1798 falls, making it unbalanced. A total of 38 people including 19 women and 19 men participated. Table 3 lists the falls and ADLs and their duration.

Table 3. Details of the Activities of Daily Living and falls contained in the *SisFall* dataset [31].

Activity	Duration [s]
Walking slowly	100
Walking quickly	100
Jogging slowly	100
Jogging quickly	100
Walking upstairs and downstairs slowly	25
Walking upstairs and downstairs quickly	25
Slowly sit and get up in a half-height chair	12
Quickly sit and get up in a half-height chair	12
Slowly sit and get up in a low-height chair	12
Quickly sit and get up in a low-height chair	12
Sitting, trying to get up, and collapse into a chair	12
Sitting, lying slowly, wait a moment, and sit again	12
Sitting, lying quickly, wait a moment, and sit again	12
Changing position while lying (back-lateral-back)	12
Standing, slowly bending at knees, and getting up	12
Standing, slowly bending w/o knees, and getting up	12
Standing, get into and get out of a car	25
Stumble while walking	12
Gently jump without falling (to reach a high object)	12
Fall forward while walking, caused by a slip	15
Fall backward while walking, caused by a slip	15
Lateral fall while walking, caused by a slip	15
Fall forward while walking, caused by a trip	15
Fall forward while jogging, caused by a trip	15
Vertical fall while walking, caused by fainting	15
Fall while walking with damping, caused by fainting	15
Fall forward when trying to get up	15
Lateral fall when trying to get up	15
Fall forward when trying to sit down	15
Fall backward when trying to sit down	15
Lateral fall when trying to sit down	15
Fall forward while sitting, caused by fainting	15

Table 3. *Cont.*

Activity	Duration [s]
Fall backward while sitting, caused by fainting	15
Lateral fall while sitting, caused by fainting	15

3.2. Data Preprocessing

The *SisFall* dataset required minimal preprocessing. We started by equalizing the duration of each record, by equally cutting (*top and tail* in equal measure) reducing the length to 10 s. We chose 10 s to remove any outliers induced by the fall experiment, whilst preserving the fall within each record. To generate various sensors' sampling rates, we reduced the number of samples in each record. Thus, for a sampling rate of 100 Hz, we removed 50% of the sample along the record.

Regarding the two walking and two jogging activities, which only have one trial (Table 3), we extracted 5 times 10 s for each record. We did this to have the same number of trials per activity. We selected 5 windows with no overlap along each record as follows:

1. From 5 to 15 s.
2. From 25 to 35 s.
3. From 45 to 55 s.
4. From 65 to 75 s.
5. From 85 to 95 s.

The additional data preprocessing required for research question 3 is described separately, in Section 3.6.

3.3. Feature Extraction

We then extracted meaningful information from the preprocessed data. This process helps extracting information that better characterize each activity. A common practice, when working with time series, is to extract time and frequency domain features [11,27,28]. In addition to the axes' features, we calculated the *magnitude* of acceleration and rotation measures, to improve the robustness of the fall detection (e.g., in case of fall-like activities involving fast movements). Thus, we also extracted time-domain features such as the *variance, standard deviation, mean, median, maximum, minimum, delta, 25th centile* and *75th centile*. Additionally, we extracted frequency-domain features, using a Fast Fourier Transform and we extracted two features: the *power spectral density* and the *power spectral entropy*.

The feature extraction process is as follows. Firstly, various formulae are applied to each record. In our case, each record has a length of 10 s with a number of samples varying from 10 to 2000 depending on the sampling rate. We then selected a sensor axis and used all samples to extract the wanted feature class.

This process was repeated for each of the other sensor axes (3 axes, 2 sensors). We appended each calculation to a vector to characterize the record (Table 4). We applied this process also for each sensor *magnitude*, resulting in a feature vector of 88 features per record (11 feature classes $\times$ 8 axes). The resulting vector uniquely defines each activity. The algorithm compares and tries to find patterns using these features in order to correctly classify each activity. For example, a fall would most likely have a large delta on its vertical sensing axis, since a fall is usually defined by a high vertical acceleration.

Finally, we normalized the extracted features to rescale the data to a common scale. This gives more influence to data with small values which can be neglected depending on the employed algorithm. In this work, we used the common *min-max* normalization which scales the values between 0 and 1 included.

Table 4. List of extracted time and frequency domain features.

Feature Classes	Domain
Variance	Time
Standard deviation	Time
Mean	Time
Median	Time
Maximum	Time
Minimum	Time
Delta (peak-to-peak)	Time
25th Centile	Time
75th Centile	Time
Power Spectral Density	Frequency
Power Spectral Entropy	Frequency

3.4. Classification Algorithms

We selected 5 different ML algorithms: k-Nearest Neighbor (KNN), Support Vector Machine (SVM), Decision Tree (DT), Random Forest (RF) and Gradient Boosting (GB). These are described in Sections 3.4.1–3.4.5, and implemented in Python using *Scikit-Learn*. We used the default parameters value of the different classifiers from the version 0.23.2 of Scikit-Learn. It is a tool with a simple interface, built on scientific libraries such as *NumPy*, *SciPy*, and *matplotlib*. The library code is open source and is under the BSD license. Moreover, its documentation is very complete and includes many sample codes. We used the default parameters of each classifier of the version

Pedregosa et al. [42] introduced *Scikit-Learn* and presented its features, comparing the efficiency of its algorithms to other similar libraries. The results show that it is often faster and has the advantage of supporting many available algorithms. This led to its wide adoption in the ML community. In particular, *Scikit-Learn* provides several classification and regression algorithms for supervised learning. Moreover, it implements model selection and evaluation functions that allow to perform cross-validations, searches and comparisons with various metrics.

3.4.1. k-Nearest Neighbor (KNN)

KNN is a well-known algorithm with a very simple operating principle. Data are classified, by a majority vote, with the class most represented among its k-closest neighbors. This algorithm belongs to the lazy learning class because it defers the work as long as possible. During the training, it simply organizes data. However, during a prediction, it browses the recorded data to count the classes of its k-nearest neighbors. Therefore, all calculation costs are during a prediction [43].

This algorithm has two main parameters. The first one is the number of neighbors to consider. A big value allows to have a probabilistic information but the estimation locality may be destroyed. Therefore, compromises have to be made and the value of 5 is used typically in the literature. The optimal number of neighbors depends strongly on the type of data. The second parameter is the method to calculate the distance between two data and defines their closeness. The choice of this metric is complicated and the notion of distance depends on the data characteristics [43]. There are several distance formulae but the most commonly used ones are Euclidean, Manhattan and Minkowski.

Throughout our experiments, we confirmed the following characteristics. The advantages of this algorithm are simplicity, efficacy and ease of tuning to find the best hyper-parameters. In addition, the greater the number of training data, the better the performance, which is however still sensitive to noise. Data normalization can then solve this problem. As a disadvantage, this algorithm is sensitive to the curse of dimensionality. The increase in the number of features tends to improve the results but only up to a certain threshold. When this one is reached, the addition of new features degrades the results. This is because irrelevant features influence negatively in the calculation of the distance. Finally, KNN is expensive in memory and in computation in comparison to other algorithms, as corroborated by the literature [27,30].

3.4.2. Support Vector Machines (SVM)

SVM is also a well-known algorithm. It can be employed in supervised and unsupervised learning. It tries to find the best hyperplane which maximizes the margins between each class. When a linear classification is not feasible, SVM can use a technique named *kernel trick* that maps inputs into a higher dimension [43]. It is an eager learning algorithm because it creates a classification model based on the data during the training. When a prediction is asked, it uses the model to determine the class.

SVM has several hyper-parameters affecting the classification results. The most relevant ones are:

- *C* makes a compromise between the number of misclassified instances and the margins width of the hyperplane. The lower the value, the larger the margins but potentially increasing the number of errors. When the margins are thin, the number of misclassified samples is low, but this can lead to overfitting.
- *Kernel* changes the employed mathematical function which creates the hyperplane. A typically used kernel is the Radial Basis Function.
- *Gamma* defines the influence that one data has compared to the other ones. The higher the value, the bigger its influence range, but this can lead to overfitting. With a low value, the model is at risk of underfitting.

SVM has the advantage of being able to find a unique and global solution which comes from the fact that the optimization problem is convex [43]. Thanks to the *kernel trick*, it can produce good results even with a high features space. However, SVM requires greater processing power during the training to find the best hyperplane and also during the predictions to calculate the support vector for each new data, as corroborated by the literature [18,24,27,30].

3.4.3. Decision Tree (DT)

Trees are well known data structures which are used in many different problems. They are applicable in ML and their objective is to create a DT based on the features of each data. Every node of the tree is divided to satisfy the most data until there are only leaves at the end. Therefore, it is an eager learning algorithm because it tries to build the best DT during the training phase [43].

Most of the hyper-parameters are useful to decide when a node must be divided and when the DT must stop. The most relevant ones are:

- *Criterion* is the function allowing to measure the quality of the split of a node. A commonly used criterion is *Gini impurity*.
- *Splitter* is the split selection method of each node because there may be several split solutions. A commonly used splitter is the *best split*.
- *Max depth* defines the maximum depth that a tree can reach during its creation. A big depth complicates the structure and tends to create overfitting on the data. But on the contrary, a low depth tends to create underfitting.
- *Min samples split* is the minimum number of data required to enable the split of a node. This value is usually low because the higher it is, the more constrained the model becomes, which creates underfitting.
- *Min samples leaf* is the minimum number of data required to consider a node as a leaf. Its effect is similar to the previous parameter because a high value would create underfitting.
- *Max features* corresponds to the maximum number of features to take into account when the algorithm searches for the best split. This hyper-parameter depends on the employed data but also tends to produce overfitting when its value is high.

Advantages of DTs are their ease of understanding and interpretation for humans, as it can be visualized [43]. It also requires few data preparations and has a low cost during a prediction because its complexity is logarithmic. However, a tree can become very complex and not generalize enough the data which then produces overfitting. In the same way, an unbalanced dataset will create biased trees. Despite this shortcoming, DTs (J48 in particular) are commonly used in the literature [24].

3.4.4. Random Forest (RF)

RF is an improvement to DTs because it includes many of them as its name *forest* suggests. Its principle is to create multiple trees and train them on random subsets of data. During a prediction, every tree processes the data and the obtained results are then merged to determine the most likely class by a vote [43]. This method is called *bagging*. This algorithm allows to remove the overfitting problem created by DTs. It is part of ensemble learning algorithms whose concept is to combine several ML algorithms to achieve better performance.

The available hyper-parameters are the same as the ones in DTs in addition to one which allows to define the number of trees to use in the forest. A value of 1 is equivalent to the DT algorithm. A high value will usually give better results. However, this creates a high cost in computational power and memory because each tree has to be stored.

One of the strongest advantages of RF is that it can automatically create a list with the most discriminative features. It has also the ability to create confidence intervals which indicate the certainty rate of a predicted class for each data. Its disadvantage is that the ease of interpretation of DTs is lost. This algorithm has also been used in the fall detection literature [24].

3.4.5. Gradient Boosting (GB)

GB is very similar to RF because it also employs multiple trees but in a different manner. The trees do not work in parallel as in RF but sequentially. The output of each tree is used as input of the following one. The idea is that each tree learns iteratively on the errors made by its predecessor. This is called *boosting* [43]. Because GB is composed of DTs, most of the parameters are the same. However, it has additional ones which are:

- *Loss* defines the loss function which must be optimized.
- *Learning rate* slows the learning speed of the algorithm by reducing the contribution that each tree produces. This avoids to rapidly create overfitting.
- *Estimator* corresponds to the number of sequential DTs. A high number would produce good results but a number too high may create an overfitting issue and use more computational power and memory. The idea is to make a compromise between the number of estimators and the learning rate.
- *Subsample* defines the data fraction used to train each tree. When the fraction is smaller than 1, the model becomes a Stochastic GB algorithm which reduces the variance but increases the bias.

An advantage of this algorithm is that it can produce better results than RF but it potentially has overfitting issues. It also allows to reduce the variance and the bias. However, the model is more complex to create and as a result the training phase is much longer than in other algorithms. Despite this shortcoming, GB is commonly used in the literature [33,34].

3.5. Evaluation

The performance evaluation of our FDS under the selected classifiers was done using *k-fold cross-validation*. This required splitting the dataset into k sets. $k-1$ sets are used as training and 1 as testing. The process is repeated k times with a different set as the test one. Given that FDSs must be able to detect falls for new people (e.g., unseen data), the test set should not contain people data that the algorithm has been trained on.

We chose a value of $k = 5$. This creates a training set of 80% and a test set of 20%. We filtered the *SisFall* to only keep subjects that performed all activities. Thus, despite being our motivation to develop a FDS for the elderly, we found it necessary to remove the data related to the elderly subjects, as these had not performed simulated falls. Similarly, we removed three young people's data due to missing records. This leaves us with data from 20 subjects. This number turned out to be ideal as it allowed us to guarantee that no data from a given subject is used for both training and testing (in an 80/20 split). In other words, the trained models would always be tested with data from new subjects. Consequently, we have 1900 ADLs (19 ADLs × 5 trials × 20 subjects) and 1500 falls (15 falls × 5 trials × 20 subjects), resulting in a more balanced dataset of 3400 records.

During the evaluation of ML algorithms, each prediction falls in one of the following categories:

- *True negative (TN)*: Correct classification of a negative condition, meaning a reject.
- *False positive (FP)*: Incorrect classification of a negative condition, meaning a false alarm.
- *False negative (FN)*: Incorrect classification of a positive condition, meaning a missed.
- *True positive (TP)*: Correct classification of a positive condition, meaning a hit.

Each prediction is added to the count of its category which allows then to calculate various metrics such as the accuracy. A usual representation of these categories is a confusion matrix.

In fall detection, two metrics are especially important: Sensitivity (SE) (Equation (1)) and the Specificity (SP) (Equation (2)) [7]. The SE (or *recall*) corresponds to how many relevant elements are actually selected. This is basically the detection probability meaning how many falls have actually been detected. The SP corresponds to how many non-relevant elements are selected, i.e., how many events classified as non-falls are actually non-falls.

$$Sensitivity = \frac{TP}{TP + FN} \tag{1}$$

$$Specificity = \frac{TN}{TN + FP} \tag{2}$$

We also calculated the accuracy (Equation (3)) and the F1-score (Equation (4)). Additionally, we calculated the Area Under the Receiver Operating Characteristics Curve (AUROC) as provided in *scikit-learn*. The AUROC is used to evaluate classifiers' performance which is used in pattern recognition and ML [44]. In simple terms, an AUROC close to the value of one is indicative of a well-performing algorithm, with high true-positive and true-negative rates consistently.

$$Accuracy = \frac{TP + TN}{TP + TN + FP + FN} \tag{3}$$

$$F1\text{-}score = \frac{2 \times TP}{2 \times TP + FP + FN} \tag{4}$$

3.6. Multi-Class Approach Considerations

To answer our third research question, i.e., "What is the difference in performance across various types of ML algorithms by adopting a multi-class approach for identifying phases of a fall?", we needed to do one more step to prepare the data for the ML algorithms. The goal of this additional step was to divide the fall sample into three parts which are: pre-fall, impact and post-fall. In doing so, two related questions arise: Where should we split the fall sample and what duration should each part have. Given that a fall has been defined as an uncontrolled, high acceleration [7], especially around the impact point, we defined that the latter would be our reference point to split the fall data sample. Based on this definition, we calculated the magnitude of each accelerometer axis along the sample and selected the highest magnitude as the impact point for each sample. The average time between the moment of loss of balance and the impact point is 0.715 s with a standard deviation of 0.1 s [25]. Consequently, we defined the impact part of the fall as a 2 s interval in the sample which includes the impact point, with 1.5 s leading to it, and the remaining 0.5 s after it. This interval is labeled as *impact*. The remaining part of the sample before the impact interval is labeled as *pre-fall* and the remaining, final part is labeled as *post-fall* (note that based on the result of RQ2, we selected a sample frequency of 50 Hz.). Thus, each 10 s fall sample creates three features vector, one for each phase. The impact phase always represents a 2 s window. The remaining 8 s represents the pre- and post-fall phases. Since the magnitude of the fall is not always at the same timestamp, the pre- and post-fall phase duration varies. If the fall happens early in the sample, the pre-fall phase will be much shorter than the post-fall phase. The opposite if the fall happens late in the sample.

To illustrate the above process, we present Figure 3a, a fall sample of the *SisFall* dataset [31]. Each line represents one of the accelerometer's axis. In it, a peak in the middle is highlighted which is the impact point (shown as a dotted line) (Figure 3b). The dashed lines limit the three parts of the fall, including the 2 s window of the impact interval. The left-hand part is the pre-fall and the right-hand part is the post-fall. The feature extraction step is applied to each phase of the fall as well as the ADLs.

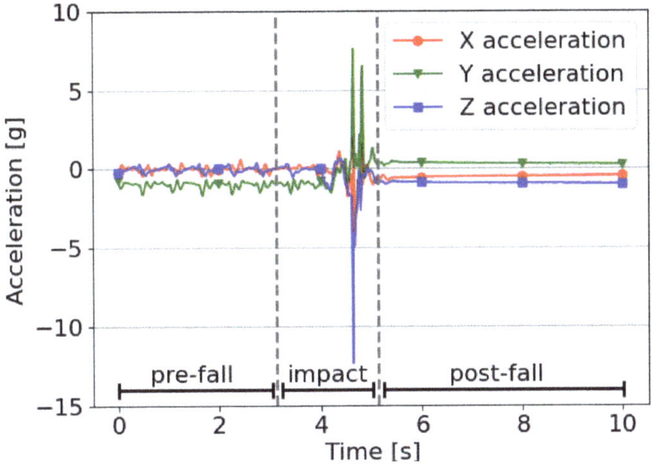

(**a**) Complete fall sample.

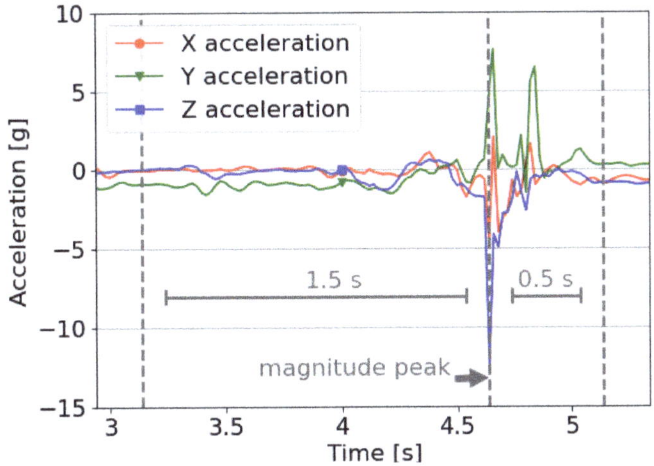

(**b**) Zoom on the impact phase.

Figure 3. Division of a fall sample into pre-fall, impact and post-fall phases.

By identifying the three different phases of the fall in the manner described above, the FDS becomes a multi-class problem. More specifically, when ADLs are taken into consideration, it becomes a four-class classification problem. The motivation behind it lies on the importance in differentiating between ADLs and any phases of a fall, as labeled in the *SisFall* dataset. In order to do that, we apply the same ML algorithms as in Section 3.4. As the SVM classifier is a binary classifier, we extended it by choosing a one-vs.-one scheme.

In order to evaluate the performance of such classification, it is possible to use metrics such as SE, SP F1-score and AUROC, presented in Section 3.5. However, these are typically defined for two-class classification and it is important to show how we have extended them for multi-class problems. We evaluated the performance with the same metrics, using the calculation of the *macro* score for SE, SP, F1-score and AUROC. This is the average metric per class which gives the same importance for each class. The other solution is the *micro* score which average the metric by giving more importance to the amount of data per class. As falls happen rarely, it creates unbalanced dataset but it is crucial to detect them correctly, thus the need to give importance to this class. In our multi-class problem, we calculated the SE for a specific class against all the others together as if they were one class. Matches for this specific class represent the positive cases and matches for the combined class represent the negative cases. Applying this step for each class offers four different Sensitivities, which then are averaged using the previously explained macro score, as per Equation (5). A similar process is applied for SP and F1-score, as shown in Equations (6) and (7).

$$SE_{macro} = \frac{1}{|Class|} \times \sum_{i=1}^{|Class|} \frac{TP_i}{TP_i + FN_i} \tag{5}$$

$$SP_{macro} = \frac{1}{|Class|} \times \sum_{i=1}^{|Class|} \frac{TN_i}{TN_i + FP_i} \tag{6}$$

$$F1\text{-}score_{macro} = \frac{1}{|Class|} \times \sum_{i=1}^{|Class|} \frac{2 \times TP_i}{2 \times TP_i + FP_i + FN_i} \tag{7}$$

4. Results and Discussion

This section presents and discusses the results for each of the research questions listed in Section 1, namely: Section 4.1 presents the comparison of various Machine Learning (ML) algorithms; Section 4.2 talks about the effect of the sensors' sampling rates on the detection performance, and Section 4.3 presents the results by splitting each fall into its phases.

4.1. Fall Detection System (FDS) Performance

Tables 5–9 present the results of the evaluation of our FDS under the selected five ML algorithms, showing that we successfully developed a reliable FDS. The Sensitivity (SE) reached 98.4% and the Specificity (SP), 99.68%, respectively with Gradient Boosting (GB) and k-Nearest Neighbor (KNN). These results outperformed those reported by Sucerquia et al. [31]. From our review of classification algorithms (Section 3.4), we expected ensemble learning algorithms to achieve better performance than the others. In practice, this trend has been confirmed even though there are some exceptions (see Table 6). This is because they use multiple ML algorithms, though the improvement in performance is at the expense of more resources. Support Vector Machine (SVM) had more difficulties to distinguish the activities. However, by tuning some hyper-parameters, its results may improve.

The high quality of these results was unexpected especially without any optimization such as hyper-parameters tuning. We infer that Activities of Daily Living and falls in the *SisFall* dataset are discriminating by default, similar to [16]. Thus, any algorithm can perform very well. However, in real-life conditions, the SE and SP would very likely drop because of the falls heterogeneity as highlighted by Krupitzer et al. [33,34]. The difficulty of obtaining real falls data is the main shortcoming in FDS studies, given that it is challenging to capture them in realistic settings with the elderly, as noted by Bagalà et al. [26], who compiled a database of only 29 real-world falls.

Table 5. Comparison of the Sensitivity across the ML algorithms, with the highest values in bold.

Frequency [Hz]	KNN [%]	SVM [%]	DT [%]	RF [%]	GB [%]
1	85.66	74.13	91.26	93.60	95.33
2	91.46	77.93	91.33	95.53	96.86
5	95.33	84.86	94.73	95.66	98.06
10	96.53	88.80	94.73	97.26	**98.40**
20	97.20	91.80	95.80	97.66	98.26
50	**97.40**	91.80	96.26	**98.20**	98.13
100	**97.40**	**93.89**	96.40	97.73	98.20
200	97.26	93.78	**96.60**	98.00	98.06

Table 6. Comparison of the Specificity across the ML algorithms, with the highest values in bold.

Frequency [Hz]	KNN [%]	SVM [%]	DT [%]	RF [%]	GB [%]
1	94.68	80.21	93.00	96.21	96.21
2	97.05	83.73	94.26	97.42	97.57
5	98.78	88.68	96.32	98.68	98.47
10	99.57	90.89	96.73	99.42	98.47
20	**99.68**	92.15	**97.52**	99.21	99.10
50	99.42	93.26	96.63	99.15	98.73
100	99.42	**93.89**	97.26	99.15	98.94
200	99.31	93.78	97.26	98.94	**99.21**

Table 7. Comparison of the accuracy across the ML algorithms, with the highest values in bold.

Frequency [Hz]	KNN [%]	SVM [%]	DT [%]	RF [%]	GB [%]
1	90.70	77.52	92.23	95.05	95.82
2	94.58	81.17	92.97	96.58	97.26
5	97.26	87.00	95.61	97.35	98.29
10	98.23	89.97	95.85	98.47	98.44
20	**98.58**	92.00	96.76	98.52	**98.73**
50	98.52	**92.61**	96.47	**98.73**	98.47
100	98.52	92.05	96.88	98.52	98.61
200	98.41	91.20	**96.97**	98.52	98.70

Table 8. Comparison of the F1-score across the ML algorithms, with the highest values in bold.

Frequency [Hz]	KNN [%]	SVM [%]	DT [%]	RF [%]	GB [%]
1	88.98	74.35	91.18	94.28	95.24
2	93.68	78.47	91.97	96.08	96.88
5	96.81	85.17	94.99	96.94	98.06
10	97.93	88.56	95.25	98.23	98.23
20	**98.36**	90.93	96.29	98.30	**98.55**
50	98.30	**91.55**	95.99	**98.55**	98.25
100	98.30	90.76	96.45	98.31	98.42
200	98.17	89.70	**96.55**	98.31	98.52

Table 9. Comparison of the AUROC across the ML algorithms, with the highest values in bold.

Frequency [Hz]	KNN [%]	SVM [%]	DT [%]	RF [%]	GB [%]
1	95.97	86.02	92.13	98.72	99.12
2	97.73	90.17	92.79	99.26	99.61
5	99.03	93.83	95.52	99.60	99.87
10	99.49	96.14	95.73	99.85	**99.93**
20	**99.50**	97.35	96.66	99.85	99.92
50	99.44	**97.66**	96.44	99.87	**99.93**
100	99.36	97.13	96.83	99.86	**99.93**
200	99.45	96.43	**96.93**	**99.90**	**99.93**

4.2. Sensors' Sampling Rate Effect

Regarding the sensors' sampling rate, the trend is that the higher the rate the better the results, which is intuitive since more data are considered when creating the feature vector. However, SVM has a different behavior than the other three, as shown in Figure 4. This shows the variation of the different metrics of each algorithm over the sensors' sampling rate. It peaks with a sensors' sampling rate of 20 Hz, indicating that the higher sampling rate does not necessarily improve performance. Especially since a high sampling rate comes with disadvantages such as more computational costs and higher battery consumption. Moreover, the results do not suggest that increasing the sampling rate any further would make a meaningful improvement. In our case, the performance no longer increases significantly after reaching 50 Hz. This sampling rate is in fact the typical one used in the reviewed literature, offering the best reported results (Table 1).

4.3. Multi-Class Approach Performance

The multi-class approach to identify different phases of falls achieved promising results with an accuracy close to 99% as shown by Figure 5 for two algorithms. The figure presents also the variability of the results over each fold of the cross-validation for each algorithm. The RF and GB algorithms consistently produced good results over the different metrics except for a single fold, which is seen as an outlier in Figure 5a–d. One explanation might be that it is related to data of a subject who performed the ADLs and falls differently to other subjects. The DT algorithm has a the biggest variability across the algorithms followed by KNN. The variability is low, close to 5% from which a high confidence on the algorithms can be inferred. This is the desired behavior for the type of application, where consistency in minimizing both SE and SP is important to facilitate adoption and usefulness of the FDS. Furthermore, the results of this experiment also confirm the expectation about ensemble learning algorithms performance, which had been observed in the results presented in Section 4.1.

Figure 6 presents a deeper insight of the classification results with the confusion matrices of each split of the k-fold cross-validation for the KNN algorithm. The accuracy of this algorithm is the median amongst all algorithms' accuracies, therefore it is useful to discuss in depth. We can see that the pre-fall and post-fall phases were consistently correctly classified. The main source of misclassifications comes from the other two classes, i.e., ADL and impact. This negative tendency is stronger in the SVM and DT algorithms but lessened in the RF and GB ones. These confusion matrices are interesting because the patterns of misclassifications are consistent to that expected in a binary detection (i.e., ADL vs. fall). Therefore, an approach could involve removing data associated to the correctly-identified phases of pre-fall and post-fall and treat the problem as a binary classification. However, having correctly isolated and identified these phases, these could be used as a supplementary input to confirm the prediction. Suppose for a given sample, a pre-fall and a post-fall are correctly identified, but the impact is predicted as an ADL. Then, by the mean of a threshold on a confidence interval, the misclassified impact could be overridden and corrected. Another solution could simply consider the fact of identifying a pre-fall and a post-fall phase to always raise an alarm for an impact, given the high confidence of the prediction of both phases.

This novel approach usefulness lies on its provision of an added guarantee that the fall is correctly detected, by offering a mechanism to "fix" a potential misclassification. For a given fall sample, the algorithm should identify once each part of a fall, otherwise, it is identified that one or several classifications are incorrect. Additionally, the ability to recognize the pre-fall stage has many useful applications for fall prevention systems, including airbags for example. This could reduce the likelihood of injuries caused by falls.

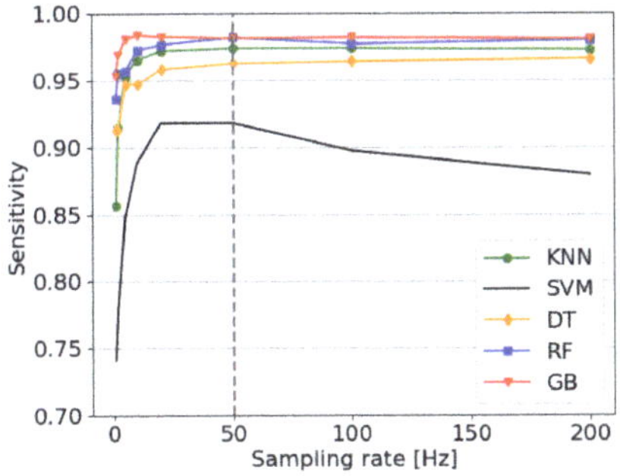

(**a**) Sensitivity variation.

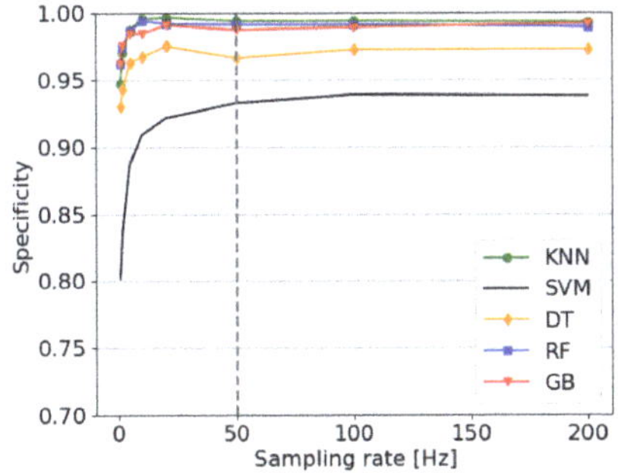

(**b**) Specificity variation.

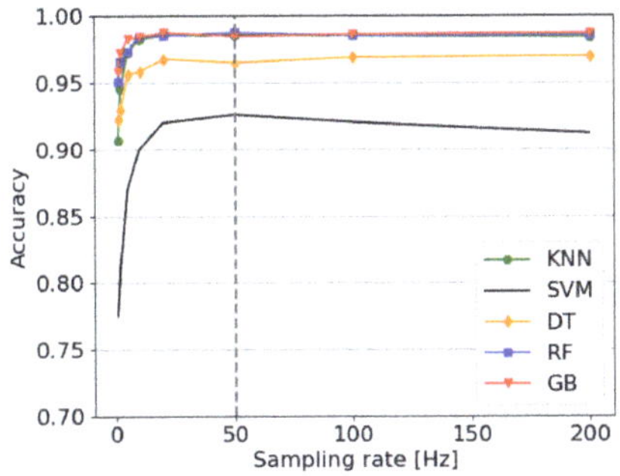

(**c**) Accuracy variation.

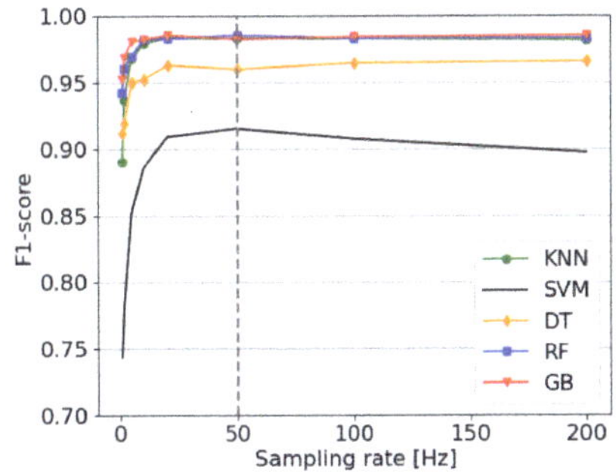

(**d**) F1-score variation.

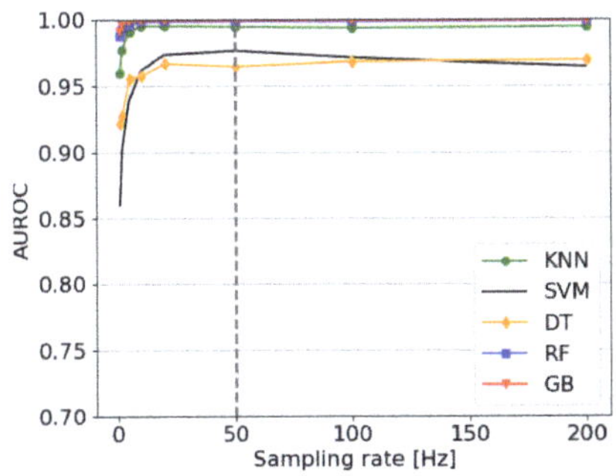

(**e**) AUROC variation.

Figure 4. Metrics variation over the sampling rates of five algorithms. The highest average metrics across all algorithms is obtained with a sampling of 50 Hz.

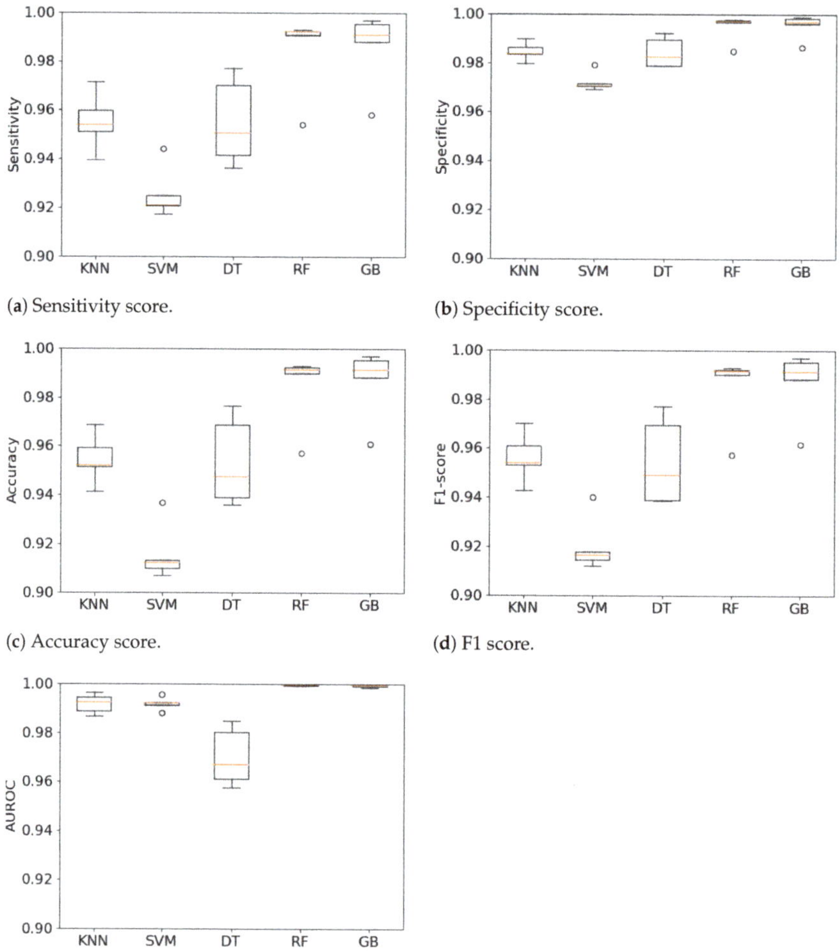

(**a**) Sensitivity score.

(**b**) Specificity score.

(**c**) Accuracy score.

(**d**) F1 score.

(**e**) AUROC score.

Figure 5. Comparison of various metrics including Sensitivity, Specificity, accuracy, F1 and AUROC of each k-fold split across the ML algorithms.

The obtained results are of high quality in terms of their accuracy, SE, SP, F-1 and AUROC. This may not be the case when applying the system on data collected on the wild, as we identified during the first experiment. As many other datasets in the FDS community, the *SisFall* dataset is highly discriminating between ADLs and falls. Because their samples lack realism, the studies under laboratory conditions will always outperform those in the real world. In particular, from inspecting *SisFall* data, subjects remained still after a fall, but it is unclear if an older person would act in this way during a real fall, particularly if there was no loss of consciousness.

In our experiment, the pre-fall part was very often correctly classified. However, under real conditions, misclassifications may have arisen (for example, as an ADL). This is due to the fact that, in reality, falls are *unexpected* events occurring perhaps in the middle of an ADL. Therefore, the pre-fall phase may be very short, following immediately from the ADL part of the sample. Whereas in the *SisFall* dataset (as shown in Figure 3) the pre-fall part is not an ADL, instead, the subject is "preparing" to fall (i.e., the fall is not unexpected). In addition, the setup of the experiment in the wild will not be the same as in the lab.

It would lack the annotation, and therefore the behavior of the algorithm may not be the same (in particular, with regards to dividing samples). With real-life non-annotated data, it is unknown whether the received data is a fall, and hence a sample associated to an ADL would also be divided into various parts. This would require further investigation.

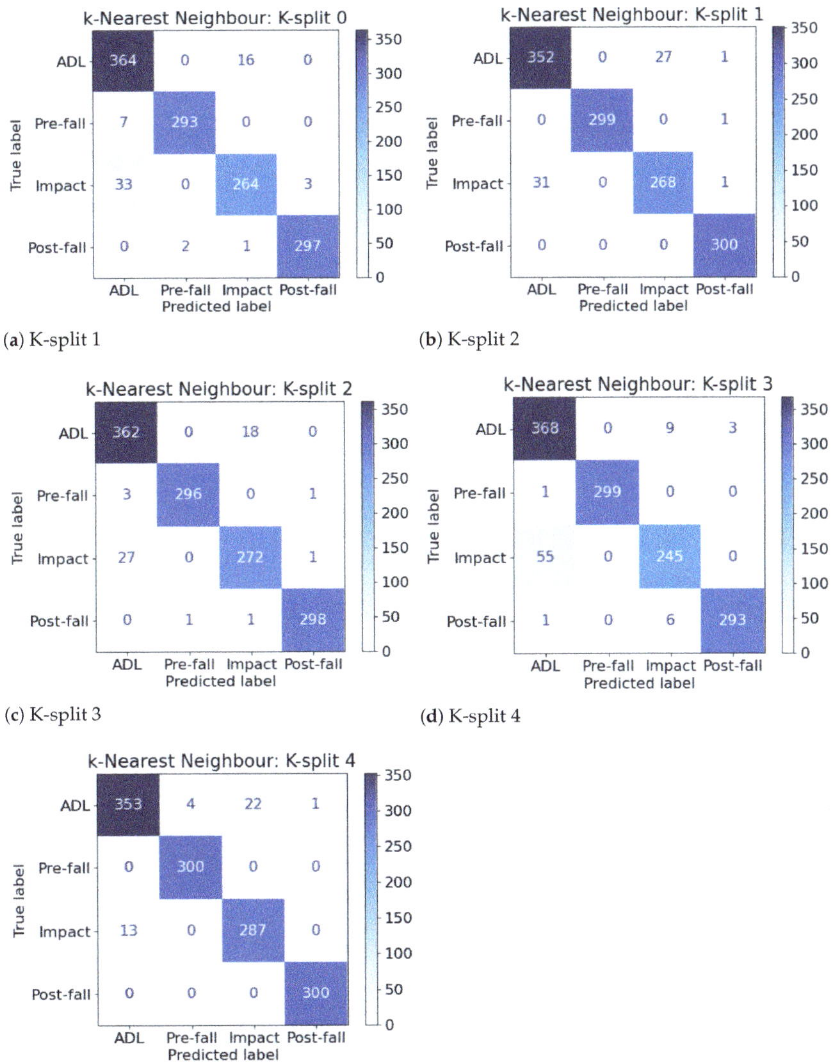

(**a**) K-split 1 (**b**) K-split 2

(**c**) K-split 3 (**d**) K-split 4

(**e**) K-split 5

Figure 6. Confusion matrices of the k-Nearest Neighbor Machine Learning algorithm whose accuracy is the median amongst all other algorithms' accuracies.

5. Conclusions and Future Work

In this paper, we present our development of a Fall Detection System (FDS) using wearable technologies, to investigate and answer the following three research questions:

RQ1 *What is the difference in performance across various types of Machine Learning (ML) algorithms in a FDS?*

Our FDS implemented several ML algorithms for comparison: k-Nearest Neighbors, Support Vector Machine, Decision Trees, Random Forest and Gradient Boosting. Our results are an improvement over those reported by Musci et al. [36] and Sucerquia et al. [31], with a final Sensitivity and Specificity over 98%. The system is reliable as we were able to test it on a large dataset containing several thousands of Activities of Daily Living (ADLs) and falls. We obtained these results using various ML algorithms which we were able to compare. We observed that ensemble learning algorithms perform better than lazy or eager learning ones. We also further investigated the effect of the sensors' sampling rate on the detection rate.

RQ2 *What is the effect of the sensors' sampling rate on the fall detection?*
We discovered a tendency that a high sampling rate usually produces better results than a lower one. However, it is not necessary to have an extremely high sampling rate (i.e., in the several hundreds). We recommend using a sampling rate of 50 Hz because it produces improved results with any algorithm while keeping a rather low computational cost.

RQ3 *What is the difference in performance across various types of ML algorithms by adopting a multi-class approach for identifying phases of a fall?*
We found that the multi-class approach to identify the phases of a fall showed promising results with an accuracy close to 99%. In addition, it includes key features which are the possibility for improved performance by adding subsequent logic to the ML algorithm to address possible misclassifications. Given this performance, we would advocate this multi-class approach as being useful in a different contexts such as fall prevention systems.

There is scope for future work. With the high computation resources available nowadays, it would be interesting to explore Deep Learning (DL) algorithms. In our case however, the size of the cleaned dataset is insufficient for this method to be appropriate given the requirements of DL. The much larger OPPORTUNITY dataset [45] for ADLs has been shown as appropriate for the use of the DL methods [46]. There is a study [36] using Recurrent Neural Networks but there are other algorithms available such as Convolutional Neural Networks with the advantage of automatic feature extraction from time series [46]. This reduces the number of steps to implement and removes the question of how many and which features are needed to be extracted. Additionally, it would be very interesting to reproduce the experiment on the sensors' sampling rate but with DL algorithms. The results may be different from traditional ML algorithms. The *SisFall* dataset allows plenty of experiments. However, the lack of falls data availability in realistic settings is a common challenge in FDS studies, which also affected our study. In particular currently available datasets with falls in realistic settings (such as in [26]) are far too small for ML approaches to be successful, most particularly, for the state-of-the-art DLs.

Further work would benefit from exploring the use of a multi-class approach for FDS using realistic datasets in order to compare against the performance in the lab and further address any misclassification issues arising in that context. The results presented in this work suggest this is worthwhile doing, and the use of such a system shows promise to make a difference in assisting people sustaining falls.

Author Contributions: Conception, design, experimentation, N.Z.; supervision, review, edition, A.W. and P.B. All authors have read and agreed to the published version of the manuscript.

Funding: This research was funded by HES-SO University of Applied Sciences and Arts Western Switzerland.

Institutional Review Board Statement: Not applicable.

Informed Consent Statement: Not applicable.

Data Availability Statement: Not applicable.

Acknowledgments: The authors wish to express their gratitude to Juan Ye from the School of Computer Science at the University of St Andrews; Adam Prugel-Bennett and Jonathon Hare from the

University of Southampton for their insightful comments on early stages of this work; the anonymous reviewers for their interesting and constructive comments.

Conflicts of Interest: The authors declare no conflict of interest.

Abbreviations

The following abbreviations are used in this manuscript:

ADL	Activity of Daily Living
ALS	Assisted-Living System
AUROC	Area Under the Receiver Operating Characteristics Curve
BDM	Bayesian Decision Making
DL	Deep Learning
DT	Decision Tree
FDS	Fall Detection System
FN	False Negative
FP	False Positive
GB	Gradient Boosting
KNN	K-Nearest Neighbor
ML	Machine Learning
RF	Random Forest
SE	Sensitivity
SP	Specificity
SVM	Support Vector Machine
TN	True Negative
TP	True Positive

References

1. Rubenstein, L.Z. Falls in older people: Epidemiology, risk factors and strategies for prevention. *Age Ageing* **2006**, *35*, ii37–ii41. [CrossRef]
2. World Health Organization. *WHO Global Report on Falls Prevention in Older Age*; OCLC: Ocn226291980; World Health Organization: Geneva, Switzerland, 2008.
3. Sadigh, S.; Reimers, A.; Andersson, R.; Laflamme, L. Falls and Fall-Related Injuries Among the Elderly: A Survey of Residential-Care Facilities in a Swedish Municipality. *J. Community Health* **2004**, *29*, 129–140. [CrossRef]
4. Wild, D.; Nayak, U.S.; Isaacs, B. How dangerous are falls in old people at home? *Br. Med. J. (Clin. Res. Ed.)* **1981**, *282*, 266–268. [CrossRef] [PubMed]
5. Rashidi, P.; Mihailidis, A. A Survey on Ambient-Assisted Living Tools for Older Adults. *IEEE J. Biomed. Health Inform.* **2013**, *17*, 579–590. [CrossRef] [PubMed]
6. Hawley-Hague, H.; Boulton, E.; Hall, A.; Pfeiffer, K.; Todd, C. Older adults' perceptions of technologies aimed at falls prevention, detection or monitoring: A systematic review. *Int. J. Med Inform.* **2014**, *83*, 416–426. [CrossRef] [PubMed]
7. Noury, N.; Fleury, A.; Rumeau, P.; Bourke, A.K.; Laighin, G.O.; Rialle, V.; Lundy, J.E. Fall detection—Principles and Methods. In Proceedings of the 2007 29th Annual International Conference of the IEEE Engineering in Medicine and Biology Society, Lyon, France, 22–26 August 2007; pp. 1663–1666. [CrossRef]
8. Zurbuchen, N.; Wilde, A.; Bruegger, P. A Comparison of Machine Learning Algorithms for Fall Detection using Wearable Sensors. In Proceedings of the The 2nd International Conference on Artifical Intelligence in Information and Communication, Fukuoka, Japan, 19–21 February 2020.
9. Mubashir, M.; Shao, L.; Seed, L. A survey on fall detection: Principles and approaches. *Neurocomputing* **2013**, *100*, 144–152. [CrossRef]
10. Yu, X. Approaches and principles of fall detection for elderly and patient. In Proceedings of the HealthCom 2008—10th International Conference on e-health Networking, Applications and Services, Singapore, 7–9 July 2008; pp. 42–47. [CrossRef]
11. Abbate, S.; Avvenuti, M.; Bonatesta, F.; Cola, G.; Corsini, P.; Vecchio, A. A smartphone-based fall detection system. *Pervasive Mob. Comput.* **2012**, *8*, 883–899. [CrossRef]
12. Bourke, A.K.; O'Brien, J.V.; Lyons, G.M. Evaluation of a threshold-based tri-axial accelerometer fall detection algorithm. *Gait Posture* **2007**, *26*, 194–199. [CrossRef]
13. Chan, A.M.; Selvaraj, N.; Ferdosi, N.; Narasimhan, R. Wireless patch sensor for remote monitoring of heart rate, respiration, activity, and falls. In Proceedings of the 2013 35th Annual International Conference of the IEEE Engineering in Medicine and Biology Society (EMBC), Osaka, Japan, 3–7 July 2013; pp. 6115–6118. [CrossRef]
14. Kangas, M.; Konttila, A.; Lindgren, P.; Winblad, I.; Jämsä, T. Comparison of low-complexity fall detection algorithms for body attached accelerometers. *Gait Posture* **2008**, *28*, 285–291. [CrossRef] [PubMed]

15. Yuwono, M.; Moulton, B.D.; Su, S.W.; Celler, B.G.; Nguyen, H.T. Unsupervised machine-learning method for improving the performance of ambulatory fall-detection systems. *BioMed. Eng. OnLine* **2012**, *11*, 9. [CrossRef]
16. Bourke, A.K.; Lyons, G.M. A threshold-based fall-detection algorithm using a bi-axial gyroscope sensor. *Med Eng. Phys.* **2008**, *30*, 84–90. [CrossRef]
17. Tang, M.; Ou, D. Fall Detection System for Monitoring an Elderly Person Based on Six-Axis Gyroscopes; In Proceedings of the 2018 3rd International Conference on Electrical, Automation and Mechanical Engineering (EAME 2018), Xi'an, China, 24–25 June 2018. [CrossRef]
18. Dinh, A.; Teng, D.; Chen, L.; Shi, Y.; McCrosky, C.; Basran, J.; Bello-Hass, V.D. Implementation of a Physical Activity Monitoring System for the Elderly People with Built-in Vital Sign and Fall Detection. In Proceedings of the 2009 Sixth International Conference on Information Technology: New Generations, Las Vegas, NV, USA, 27–29 April 2009; pp. 1226–1231. [CrossRef]
19. Wang, J.; Zhang, Z.; Li, B.; Lee, S.; Sherratt, R.S. An enhanced fall detection system for elderly person monitoring using consumer home networks. *IEEE Trans. Consum. Electron.* **2014**, *60*, 23–29. [CrossRef]
20. Fudickar, S.J.; Lindemann, A.; Schnor, B. Threshold-based Fall Detection on Smart Phones. In Proceedings of the HEALTHINF, Angers, France, 3–6 March 2014; pp. 303–309. [CrossRef]
21. Medrano, C.; Igual, R.; Plaza, I.; Castro, M. Detecting Falls as Novelties in Acceleration Patterns Acquired with Smartphones. *PLoS ONE* **2014**, *9*, e94811. [CrossRef]
22. Hwang, J.Y.; Kang, J.M.; Jang, Y.W.; Kim, H.C. Development of novel algorithm and real-time monitoring ambulatory system using Bluetooth module for fall detection in the elderly. In Proceedings of the The 26th Annual International Conference of the IEEE Engineering in Medicine and Biology Society, San Francisco, CA, USA, 1–5 September 2004; Volume 1, pp. 2204–2207. [CrossRef]
23. Choi, Y.; Ralhan, A.S.; Ko, S. A Study on Machine Learning Algorithms for Fall Detection and Movement Classification. In Proceedings of the 2011 International Conference on Information Science and Applications, Jeju Island, Korea, 26–29 April 2011; pp. 1–8. [CrossRef]
24. Gjoreski, H.; Lustrek, M.; Gams, M. Accelerometer Placement for Posture Recognition and Fall Detection. In Proceedings of the 2011 Seventh International Conference on Intelligent Environments, Nottingham, UK, 25–28 July 2011; pp. 47–54. [CrossRef]
25. Aziz, O.; Robinovitch, S.N. An Analysis of the Accuracy of Wearable Sensors for Classifying the Causes of Falls in Humans. *IEEE Trans. Neural Syst. Rehabil. Eng.* **2011**, *19*, 670–676. [CrossRef]
26. Bagalà, F.; Becker, C.; Cappello, A.; Chiari, L.; Aminian, K.; Hausdorff, J.M.; Zijlstra, W.; Klenk, J. Evaluation of Accelerometer-Based Fall Detection Algorithms on Real-World Falls. *PLoS ONE* **2012**, *7*, e37062. [CrossRef]
27. Özdemir, A.T.; Barshan, B. Detecting Falls with Wearable Sensors Using Machine Learning Techniques. *Sensors* **2014**, *14*, 10691–10708. [CrossRef]
28. Vilarinho, T.; Farshchian, B.; Bajer, D.G.; Dahl, O.H.; Egge, I.; Hegdal, S.S.; Lønes, A.; Slettevold, J.N.; Weggersen, S.M. A Combined Smartphone and Smartwatch Fall Detection System. In Proceedings of the 2015 IEEE International Conference on Computer and Information Technology; Ubiquitous Computing and Communications; Dependable, Autonomic and Secure Computing; Pervasive Intelligence and Computing, Liverpool, UK, 26–28 October 2015; pp. 1443–1448. [CrossRef]
29. Casilari, E.; Oviedo-Jiménez, M.A. Automatic Fall Detection System Based on the Combined Use of a Smartphone and a Smartwatch. *PLoS ONE* **2015**, *10*, e0140929. [CrossRef]
30. Gibson, R.M.; Amira, A.; Ramzan, N.; Casaseca-de-la Higuera, P.; Pervez, Z. Multiple comparator classifier framework for accelerometer-based fall detection and diagnostic. *Appl. Soft Comput.* **2016**, *39*, 94–103. [CrossRef]
31. Sucerquia, A.; López, J.D.; Vargas-Bonilla, J.F. SisFall: A Fall and Movement Dataset. *Sensors* **2017**, *17*, 198. [CrossRef]
32. Hsieh, C.Y.; Liu, K.C.; Huang, C.N.; Chu, W.C.; Chan, C.T. Novel Hierarchical Fall Detection Algorithm Using a Multiphase Fall Model. *Sensors* **2017**, *17*, 307. [CrossRef]
33. Krupitzer, C.; Sztyler, T.; Edinger, J.; Breitbach, M.; Stuckenschmidt, H.; Becker, C. Hips Do Lie! A Position-Aware Mobile Fall Detection System. In Proceedings of the 2018 IEEE International Conference on Pervasive Computing and Communications (PerCom), Athens, Greece, 19–23 March 2018; pp. 1–10. [CrossRef]
34. Krupitzer, C.; Sztyler, T.; Edinger, J.; Breitbach, M.; Stuckenschmidt, H.; Becker, C. Beyond position-awareness—Extending a self-adaptive fall detection system. *Pervasive Mob. Comput.* **2019**, *58*, 101026. [CrossRef]
35. Casilari, E.; Lora-Rivera, R.; García-Lagos, F. A Study on the Application of Convolutional Neural Networks to Fall Detection Evaluated with Multiple Public Datasets. *Sensors* **2020**, *20*, 1466. [CrossRef]
36. Musci, M.; De Martini, D.; Blago, N.; Facchinetti, T.; Piastra, M. Online Fall Detection using Recurrent Neural Networks. *arXiv* **2018**, arXiv:1804.04976.
37. SISTEMIC: SisFall Dataset. 2017. Available online: http://sistemic.udea.edu.co/en/investigacion/proyectos/english-falls/ (accessed on 16 December 2020).
38. Nyan, M.N.; Tay, F.E.H.; Murugasu, E. A wearable system for pre-impact fall detection. *J. Biomech.* **2008**, *41*, 3475–3481. [CrossRef]
39. Casilari, E.; Santoyo-Ramón, J.A.; Cano-García, J.M. Analysis of Public Datasets for Wearable Fall Detection Systems. *Sensors* **2017**, *17*, 1513. [CrossRef]
40. Casilari, E.; Santoyo-Ramón, J.A.; Cano-García, J.M. UMAFall: A Multisensor Dataset for the Research on Automatic Fall Detection. *Procedia Comput. Sci.* **2017**, *110*, 32–39. [CrossRef]

41. Micucci, D.; Mobilio, M.; Napoletano, P.; Micucci, D.; Mobilio, M.; Napoletano, P. UniMiB SHAR: A Dataset for Human Activity Recognition Using Acceleration Data from Smartphones. *Appl. Sci.* **2017**, *7*, 1101. [CrossRef]
42. Pedregosa, F.; Varoquaux, G.; Gramfort, A.; Michel, V.; Thirion, B.; Grisel, O.; Blondel, M.; Prettenhofer, P.; Weiss, R.; Dubourg, V.; et al. Scikit-learn: Machine Learning in Python. *J. Mach. Learn. Res.* **2011**, *12*, 2825–2830.
43. Witten, I.H.; Frank, E.; Hall, M.A.; Pal, C.J. *Data Mining: Practical Machine Learning Tools and Techniques*; Morgan Kaufmann: Burlington, MA, USA, 2017; ISBN:978-0-12-804291-5.
44. Fawcett, T. An introduction to ROC analysis. *Pattern Recognit. Lett.* **2006**, *27*, 861–874. [CrossRef]
45. Chavarriaga, R.; Sagha, H.; Calatroni, A.; Digumarti, S.T.; Tröster, G.; Millán, J.d.R.; Roggen, D. The Opportunity challenge: A benchmark database for on-body sensor-based activity recognition. *Pattern Recognit. Lett.* **2013**, *34*, 2033–2042. [CrossRef]
46. Ordóñez, F.J.; Roggen, D. Deep Convolutional and LSTM Recurrent Neural Networks for Multimodal Wearable Activity Recognition. *Sensors* **2016**, *16*, 115. [CrossRef]

sensors

MDPI

Article

IoTCrawler: Challenges and Solutions for Searching the Internet of Things

Thorben Iggena [1,*], Eushay Bin Ilyas [1], Marten Fischer [1], Ralf Tönjes [1], Tarek Elsaleh [2], Roonak Rezvani [2], Narges Pourshahrokhi [2], Stefan Bischof [3], Andreas Fernbach [3], Josiane Xavier Parreira [3], Patrik Schneider [3], Pavel Smirnov [4], Martin Strohbach [4], Hien Truong [5], Aurora González-Vidal [6], Antonio F. Skarmeta [6], Parwinder Singh [7], Michail J. Beliatis [7], Mirko Presser [7], Juan A. Martinez [8], Pedro Gonzalez-Gil [6], Marianne Krogbæk [9] and Sebastian Holmgård Christophersen [9]

1 Faculty of Engineering and Computer Science, University of Applied Sciences Osnabrück, 49076 Osnabrück, Germany; e.bin-ilyas@hs-osnabrueck.de (E.B.I.); m.fischer@hs-osnabrueck.de (M.F.); r.toenjes@hs-osnabrueck.de (R.T.)
2 Centre for Vision, Speech and Signal Processing, University of Surrey, Guildford GU2 7XH, UK; t.elsaleh@surrey.ac.uk (T.E.); r.rezvani@surrey.ac.uk (R.R.); n.pourshahrokhi@surrey.ac.uk (N.P.)
3 Siemens AG Austria, 1210 Vienna, Austria; bischof.stefan@siemens.com (S.B.); andreas.fernbach@siemens.com (A.F.); josiane.parreira@siemens.com (J.X.P.); patrick-schneider@siemens.com (P.S.)
4 AGT International, 64295 Darmstadt, Germany; PSmirnov@agtinternational.com (P.S.); MStrohbach@agtinternational.com (M.S.)
5 NEC Labs Europe, 69115 Heidelberg, Germany; hien.truong@neclab.eu
6 Information and Communication Engineering Department, University of Murcia, 30100 Murcia, Spain; aurora.gonzalez2@um.es (A.G.-V.); skarmeta@um.es (A.F.S.); pedrog@um.es (P.G.-G.)
7 Department of Business Development and Technology, Aarhus University, 7400 Herning, Denmark; parwinder@btech.au.dk (P.S.); mibel@btech.au.dk (M.J.B.); mirko.presser@btech.au.dk (M.P.)
8 Odin Solutions, R&D Department, 30820 Murcia, Spain; jamartinez@odins.es
9 City of Aarhus, 8000 Aarhus, Denmark; mkrog@aarhus.dk (M.K.); sech@aarhus.dk (S.H.C.)
* Correspondence: t.iggena@hs-osnabrueck.de

Citation: Iggena, T.; Bin Ilyas, E.; Fischer, M.; Tönjes, R.; Elsaleh, T.; Rezvani, R.; Pourshahrokhi, N.; Bischof, S.; Fernbach, A.; Xavier Parreira, J.; et al. IoTCrawler: Challenges and Solutions for Searching the Internet of Things. *Sensors* **2021**, *21*, 1559. https://doi.org/10.3390/s21051559

Academic Editor: Paolo Bellavista

Received: 27 January 2021
Accepted: 18 February 2021
Published: 24 February 2021

Abstract: Due to the rapid development of the Internet of Things (IoT) and consequently, the availability of more and more IoT data sources, mechanisms for searching and integrating IoT data sources become essential to leverage all relevant data for improving processes and services. This paper presents the IoT search framework IoTCrawler. The IoTCrawler framework is not only another IoT framework, it is a system of systems which connects existing solutions to offer interoperability and to overcome data fragmentation. In addition to its domain-independent design, IoTCrawler features a layered approach, offering solutions for crawling, indexing and searching IoT data sources, while ensuring privacy and security, adaptivity and reliability. The concept is proven by addressing a list of requirements defined for searching the IoT and an extensive evaluation. In addition, real world use cases showcase the applicability of the framework and provide examples of how it can be instantiated for new scenarios.

Keywords: Internet of Things; search; security; privacy; reliability; IoT search framework; IoT data sources

1. Introduction

During the last years, the Internet of Things (IoT) has grown massively and is still growing because of the availability of cheap sensor devices and more and more widespread IoT frameworks, increasing the number of devices and services. This leads to new possibilities for use cases and scenarios in the IoT (e.g., http://www.ict-citypulse.eu/scenarios/ accessed on 24 February 2021). These scenarios range from agriculture, Industry 4.0, to smart cities and many others. A quite common problem for all of these domains is the search and discovery of available IoT devices, which is the main purpose of the IoTCrawler framework.

To realize the envisaged IoT search framework, a two-layered approach is foreseen, containing the Discovery and Processing Layer and the Search and Orchestration Layer. The term Discovery refers to the process of connecting new data sources to the framework. This may require a step to extract additional information from other databases named Crawling. The Processing refers to actions to ease up and enhance the later search. Processing includes the Indexing, i.e., preparing ordered references to discovered data sources for faster access; the Semantic Enrichment (SE), i.e., the deduction of new data, either describing higher-level context or the data stream itself. The Search and Orchestration Layer becomes active when a search process is started. Search refers to the act of finding suitable data sources in the system by an application and includes a ranking mechanism to sort out the results to fit best the specific use case. Creating the ability for an application to receive live observations from a data stream is done during the orchestration step. When designing an IoT search framework, there are several issues to be considered: volume (the amount of data), heterogeneity (different kinds of data sources), dynamics (changes in data sources/environments) and security and privacy (e.g., IoT data sources measuring sensitive data). By analysing these issues, a number of general requirements for an IoT search platform can be derived:

- **R-1** **Scalability:** Coming from the issue of Volume, a requirement for scalability arises when designing products for the IoT. The huge amount of available, and often heterogeneous, data sources, which have to be considered for the process of search, leads to a challenge of scalability. All components and solutions in this environment have to be designed to work with large scale data. As a result, the machine initiated search shall be answered within a reasonable time.
- **R-2** **Semantics and Context for Machine Initiated Search:** Newly emerging search models require to tackle the search problems based on the human- and machine originated users' contexts and requirements such as location, time, activity, previous records and profile. The search results are targeted to be based on emerging IoT application models, where search can be initiated without human involvement. The generation of higher-level context, such as traffic conditions, e.g., from low-level observations, can enhance the search functionality for applications that require information on trends and profiles about sensory data. Generated data from IoT deployments are largely multivariate, and therefore require aggregation methods that can preserve and represent its key characteristics, while reducing the processing time and storage necessities.
- **R-3** **Discovery and Search:** To provide a well performing and responsive IoT search framework, the entire process needs to be considered as a two stages approach, namely Discovery and Search. In the first stage, knowledge about available IoT devices and the data streams they provide has to be crawled. The goal is to build up a data repository containing available information about the data streams. In the second stage, while processing a search request, the potential data streams, satisfying the search query, are then extracted from the repository. Before being returned to the requester, the list of candidates needs to be ranked, to allow the application to use the best fitting data streams.
- **R-4** **Security and Privacy by Design:** It is vital that Privacy and Security are addressed from the beginning in a design phase and through all the development of a project. It requires authentication, access control and privacy mechanisms in order to provide a controlled environment where providers can specify the access policy attached to their data, and even broadcast it in a privacy preserving manner, so that only legitimate consumers are able to access the information.

While traditional IoT middleware platforms allow users to search for particular IoT devices, they still require manual interaction to integrate data sources into a use case. As the number of IoT devices has increased profoundly in last couple of years, many middlewares have also surfaced to introduce more flexibility and functionality to IoT solution providers. Middlewares like Kaa (https://www.kaaproject.org/ accessed on 24 February 2021) and SiteWhere (https://sitewhere.io/ accessed on 24 February 2021) provide features like data storage, data analysis, device management along with the tools to analyse infrastructure and optimise computation or provide additional functionality like digital twins in Kaa. MainFlux (https://www.mainflux.com/ accessed on 24 February 2021) and OpenRemote (https://openremote.io/ accessed on 24 February 2021) employ protocol and device agnostic strategies to ease the connectivity of devices. Distributed Services Architecture (DSA) (http://iot-dsa.org/ accessed on 24 February 2021) provide solutions for the devices to communicate in a decentralised manner. With all these different middlewares, there is still a lack of searching mechanisms that facilitates both Machine-to-Machine (M2M) and Machine-to-Human (M2H) communication. The main goal of IoTCrawler is to provide tools that answer search queries according to user's preferences such as sensor types, location, data quality. For better M2M communication, automated context dependent access is provided based on a machine initiated semantic search. IoTCrawler also monitors these IoT devices and informs the users about changes in data quality and the availability of new relevant sensors to provide flexibility and additional information. Moreover, IoTCrawler envisions a platform which can provide any user an easy access to open data while also facilitating private users such as industries and businesses. For this, research has been conducted to implement strategies which ensure that private data stays protected and is only provided to the authenticated user.

This paper provides an overview of the IoT search framework IoTCrawler, which is able to crawl IoT data sources and provides an interface to allow for human- as well as machine-initiated search requests. The IoTCrawler framework consists of a series of loosely coupled components and is thoroughly designed to address the identified requirements. The components are designed to be used individually or as a whole framework to allow the search for IoT data sources in a fast, stable and secure way.

The remainder of this paper is organised as follows. Section 2 presents related work in regarding the solutions and components of the IoTCrawler framework. Section 3 describes the idea of IoTCrawler as a search framework for data sources in the IoT, while Sections 4 and 5 depict the two layered approach and present the enablers for the discovery and the enablers for the search layer in detail, including solutions to address the presented requirements. Section 6 provides an overall evaluation of several IoTCrawler framework instances running for certain use cases in real-world environments. Finally, Section 7 concludes the paper.

2. Related Work

The question of search and discovery in the internet is not new. The developed techniques range from the well-known and widely deployed Domain Name System (DNS), the Lightweight Directory Access Protocol (LDAP), to decentralised Distributed Hash Table (DHT). However, none of them address all the challenges in the IoT domain introduced in Section 1. This section highlights the current technical state of topics relevant for a search engine in the IoT.

2.1. Search over Discovered Metadata

A number of approaches for managing IoT metadata and performing a search over it can be found in literature. In the Dyser search engine [1], a query-based search mechanism is used for tracking the states of physical entities in real-time. Using a typical link-traversing approach, it performs crawling and maintains the actual state of dynamically changing metadata. Another service for semantic search and sensor discovery among the Web of Things (WoT) is DiscoWoT [2]. Using a RESTful approach, it enables the integration of WoT entities. The service is based on extensible discovery strategies. Along with that, it allows publishers to semantically annotate WoT sources. The Thingful engine (https://www.thingful.net/ accessed on 24 February 2021) uses a ranking algorithm over geographical indexed resources. A map-based Web UI is provided for verified sensors with locations. A contextual search allows to query sensors based on their type and location and nearby surroundings. A wrapping approach for integrating real-time data sources is applied in a platform called Linked Stream Middleware (LSM) [3], which uses Semantic Web technology for integrating real-time physical sensory data. For annotating and visualising data, the platform exposes a web UI with a SPARQL endpoint for querying. A predefined taxonomy includes location, physical context, accuracy and other metadata used for displaying types of sensing devices. SPARQL 1.1 with federation extension is used for federating queries from distributed endpoints [4]. WOTS2E [5]—a search engine for a Semantic WoT proposes a novel method for discovering WoT devices and services and semantically annotated data related to IoT/WoT. The engine relies on results of traditional search engines (e.g., Google), where it crawls Linked Data endpoints (SPARQL), which are semantically analysed. For the relevant endpoints, metadata will be extracted and stored in the service description repository, used later by IoT applications such as WoT index. In [6], authors analyse the state-of-art literature about IoT search engines and conclude that the most influencing (citing) contributions were done around 2010. This explains the fact that major references might look obsolete in 2021. Along with that, authors outlined two major functionalities performed by IoT search engines (content discovery and search over it), proposed a so-called meta-path methodology, identified 8 types of meta-paths and classified search mechanisms of existing IoT search engines. According to their classification (combinations of R, D, S, F), search mechanisms of IoTCrawler are able to consider the following assets: aspects of streams representatives (R) and stream observations (Dynamic Content, D), semantics of sensors and sensing devices (Representatives of IoT things possessing streams, again R). Due to the use of ontologies (IoTStream, Sosa) and extensible GraphQL-based querying mechanism [7], a submission of new information models is not a problem for the IoTCrawler metadata storage. For example, one of the crawling mechanisms [8] uses the DogOnt ontology [9] and enriches the Metadata Repository (MDR) and searches over it by the following assets: (a) types of sensors and sensing devices; (b) types of electrical appliances connected to energy-metering smart home sensors. Submission of new ontologies and extension of search mechanism with their facets share the same principles and would easily let IoTCrawler for cover functionality aspects (F) of IoT things. Considering that, we can conclude that the search mechanism of IoTCrawler covers the most of the proposed meta-path categories (except of microsensors level, S) and competes the search capabilities of engines belonging to them. Together with other capabilities (such as security and publish-subscribe, virtual sensors) IoTCrawler framework outperforms capabilities of pure search IoT search engines.

2.2. Semantics, Ontologies and Information Models for Interoperability

Over the past decade, a number of efforts have been made to define information models for IoT using ontologies and semantic annotations, although since these ontologies are developed by different entities, they are bound to be a diverge in semantics, since the IoT domain is quite broad in general. An important focus of IoTCrawler is the description of sensors and IoT data streams. Regarding sensors, one of the main initiatives made in this field is the W3C SSN ontology [10]. It defines an ontology for describing `Sensors` and `Observations`, but also expands to `Systems`, `Deployments` and `Processes`. SOSA [11] was created as an extension to SSN to simplify the ontology and to separate `Sensors` and `Observations` from other concepts that are deemed relevant for `Sensor` and `Observations` management. IoT-lite [12] was an effort to bind the core concepts of SSN with IoT concepts that were not covered by it, such as the concept of `Service`, but to support the scalability of annotations to IoT resources in a minimalist manner. The Stream Annotation Ontology (SAO) [13] is another effort which extends SSN to address sensor data streams. It employs a class taxonomy for stream analysis techniques, which is useful for high granularity. For this reason, the IoT-Stream [14] ontology was developed to serve the framework by carrying the principles that were adopted for IoT-lite to data streams, in the sense that stream annotations should be annotated as minimally as possible to support scale in the context of IoT data, but also to be flexible to increase the granularity of annotation as needed by the system.

2.3. Security and Privacy in IoT

Security and privacy cover different areas such as authentication, authorisation, integrity, as well as confidentiality to name a few. In the scope of IoT, Abomhara and Køien [15] identified three different core aspects: privacy for humans, confidentiality of business processes and third-party dependability. They also classified different attacks related to eavesdropping communications, which together with traffic analysis techniques allow attackers to identify information with special roles and activities in IoT devices and data. Nevertheless, they state that there are still open issues related to privacy in data collection, sharing and management, as described by Riahi et al. [16]. Another security aspect which has gained a lot of attention in both academia and industry, is the combination of authentication and identity management. This is widely acknowledged in the literature, such as the works of Mahalle et al. [17] or Bernal et al. [18], the latter associates the term privacy-preserving to identity management with the objective of representing not only users, but also devices or services. These aspects, together with the access control, have been also dealt in different EU research projects, such as Smartie, SocIoTal or CPaaS.io, where the integration of these technologies are also proved as an appropriate solution for different domains such as smart buildings or smart cities. These projects also propose the use of access control mechanisms based on eXtendible Access Control Markup Language (XACML) [19], even in a decentralised manner by using Attriute-Based Access Control (ABAC) [20], and to deal with privacy over the data by using encryption techniques based on attributes, such as Perez et al. [21] which composes an identity.

Hwang [22] also raises the well-known concern regarding the security threats related to IoT, for example the possibility to overwhelm a system by means of a few IoT attackers using Denial-of-Service (DoS)-based attacks [23]. The most remarkable point from this paper's perspective is that, as Hwang states, a demand exists for security solutions capable of supporting multi-profile platforms with different security levels. On the other hand, Hernandez-Ramos et al. [24] address the issue of security and privacy from the point of view of the smart city. In this work, the necessity of having a mechanism for empowering citizens to manage their security and privacy by tools such as access control management, as well as decentralised data sharing, are addressed. This idea is endorsed also in another research work [25] where they describe a future data-driven society requiring a harmonised vision of cybersecurity.

2.4. Reliability in IoT

In the past, reliability in IoT has been handled by diverse techniques and solutions, from quality analysis to algorithms for fault detection and recovery, or replacement of faulty data sources. The term Quality of Information (QoI) determines the "fitness for use" of an information that is being processed [26]. It has been originally described as a quality indicator in the context of database systems [27], but has also been used in several frameworks for information processing. The authors of [28] proposed a framework for data translation and identity resolution for heterogeneous data sources including QoI. In comparison to other frameworks, their framework relies on linked data sets instead of real-time data. Other frameworks using QoI are shown in [29] for dealing with security in the context of healthcare including QoI or [30], which deals with streaming data that are stored into a database. For later analysis, they also store calculated QoI bundled to the data. A subscription system for data streams, which are selected on their data quality, is proposed in [31]. Puiu et al. [32] focused on real-time information processing with integrated semantic annotation [33] and QoI calculation for fault-recovery and event processing. Whereas all of these solutions integrate QoI and some of them provide real-time capabilities and semantics, they are bound to specific domains and none of these solutions are flexible enough to work as a decoupled solution supporting different IoT sensors.

As a result of the recent popularity of IoT, different platforms are trying to integrate large numbers of IoT devices in their systems. For this reason, there is already some research done for fault detection in IoT systems. IoTRepair [34] is a fault diagnosis system for IoT systems. Its diagnosis is facilitated by developer configuration files along with user preferences and works by monitoring the states of each sensor and how they correlate with the states of their neighbours. Power and Kotonya [35] provide an architecture with micro-services for fault diagnosis, through event handling and online machine learning, as a two-step approach. To provide a reasonable sensor value in case of faults, different imputation techniques are defined in the literature. Izonin et al. [36] developed a missing data recovery method by using Adaboost regression on transformed sensor data through Itô decomposition and compared the results with other algorithms like Support Vector Regression (SVR), Stochastic Gradient Descent (SGD) regressor, etc. Liu et al. [37] defined a procedure to deal with large patches of faulty data in uni-variate time-series data. Al-Milli and Almobaideen [38] proposed a recurrent Jordan neural network with weight optimisation through genetic algorithms. Most of the techniques that are used for the detection and recovery of faults are computationally expensive techniques that would evidently become a burden on the processing units with the increase of devices in IoT systems. In contrast to the aforementioned approaches for a search engine for the IoT that can be used in cross-domain scenarios, an objective approach to calculate the quality of received information is presented in this work.

2.5. Indexing of Discovered Resources

The large volumes of heterogeneous and dynamic IoT data sources that are available nowadays should be indexed in a distributed and scalable way in order to provide fast retrieval to user queries [39]. Depending on the attributes to be indexed, different techniques are required. For location attributes, the work in [40] proposed a framework that supports spatial indexing of the geographic values of data collected from sensing devices based on geohash (Z-order curve). Barnaghi et al. [41] combines the use of geohashing and the semantic annotation of sensor data for creating a spatio-temporal indexing. Before applying the k-means clustering algorithm to distribute data in the repository and allow data query, dimensionality reduction is performed to the geohash vectors by means of Singular Value Decomposition (SVD). An index structure is proposed in [42]. The process starts by clustering the resources based on their spatial characteristics and creating a tree structure in each cluster, where each branch represents a type of resource (e.g., humidity or CO_2 sensors). The most notable works that are used for indexing time series are Symbolic Aggregate Approximation (SAX) and its variants (e.g., iSAX 2.0 [43] and adaptive iSAX [44]). A great deal of IoT data can be considered as a time-series, since by nature each observation will have a timestamp associated to it. These methods consider that the data follow a Gaussian distribution and use z normalisation processing, by which the magnitude of data vanishes. Since IoT data do not necessarily follow the Gaussian distribution and/or due to concept drift, the data distribution may change over time, SensorSAX [45] adapts the window size of the data according to its standard deviation in a online manner. Another work that is relevant in this sense is Blocks of Eigenvalues Algorithm for Time Series Segmentation (BEATS) [46], since it uses a non-normalized algorithm for constructing the segment representation of the time-series raw data. The mentioned methods, derived from SAX, are used to convert raw sensor data into symbolic representations and to infer higher level abstractions (for example, dark rooms or warm environments).

2.6. Ranking of Search Results

While the index cares for fast retrieval of search results, users and applications might still face the problem of sorting through a potentially large number of search results. Ranking mechanisms can help to sort and prioritise resources and services by selecting the most suitable one. In the Web domain, Google's PageRank [47] is probably one of the most notable ranking algorithms. PageRank explores the links among Web pages to assign scores to documents, which are used in combination with text similarity metrics in the context of Web document search. In the IoT domain, on the other hand, the definition of similarity varies and resources can relate to each other based on a number of different features such as their type or their location. Not only the number of features for IoT resources can vary, but also the notion of similarity itself. Therefore, IoT ranking requires a multi-objective decision-making process in which the criteria to be considered are heavily dependent on the application and the domain. There exists work that already explores the multi-criteria nature of IoT domains for assigning ranking scores [39]. Guinard et al. [48] propose a ranking method for IoT resources which takes into account the resources' type (e.g., temperature), their multi-dimensional attributes (e.g., location) and/or the Quality of Service (QoS) (e.g., latency), and applies different ranking strategies for multi-criteria evaluation with different criteria weights which are determined by the query (e.g., 40% for location, 40% for resource type and 20% for network latency). The work in [49] ranks sensor services based on two different QoS categories in Wireless Sensor Network (WSN), namely network-based (bandwidth, delay, latency, reliability and throughput) and sensor-based (accuracy, cost and trust). Other works incorporate user feedback/rating into their ranking mechanisms [50,51]. In IoTCrawler, we have devised a ranking method which can be tailored to the different applications.

3. Search Framework for IoT

In contrast to web search engines, a search engine for the IoT is used mainly by other machines or applications that need information to work properly. While a human user has the ability to assess the usability of a search result to his needs, a machine is not able to do so. It is expected that all search results returned satisfy the search query, as there is no objective way to decide between them. Therefore, an IoT search engine should rank the results beforehand, even without specifically stated requirements within the search query. For this, it should use all available information about the IoT device, such as long-term availability and reliability. Search results for a human can be presented in different ways. Not only text-based results, but also images, tables and videos are popular ways to transfer knowledge. A machine, in contrast, requires not only a fixed endpoint, but also predefined data formats. It needs to know beforehand how to interpret a received search result as well as the IoT data stream.

Like with any conventional search engine, looking for available resources at the time a search request was issued is not feasible. To provide search results in a timely manner, a data repository or database about the data sources needs to be built in advance. To further decrease the search time, the data within the database needs to be setup with appropriate indices. For example, as the location of a device is an important factor when searching the IoT device, providing indices related to the location can significantly improve the search. Before all of that, the search engine needs to be aware of IoT devices. This is probably the most challenging task since there exists a variety of different IoT devices and configuration possibilities. In addition, the IoT domain is more dynamic than the World Wide Web. While web servers usually remain online and stationary over a long period of time, the IoT devices may appear and disappear frequently. Thus, once an IoT device has been identified and integrated into the search engine's database, it needs to be monitored for availability and stream quality. At the same time, the environmental context of the IoT device can change, which needs to be captured to provide additional search criteria.

For the IoTCrawler framework we adopted the search concept for IoT into the following two steps: (a) by presenting the Crawling and Processing Layer and (b) by presenting an incoming search request into the Search and Orchestration Layer. The parts labelled with a number (1–5) belong to the former layer and the ones labelled with an alphabetical character (A–D) belong to the later layer.

The **Crawling and Processing Layer** is the "online" part of the framework. It is constantly running and responsible for integrating new data sources into the framework. In this first step (1), data sources of different kinds are found and integrated in the MDR level. The federated MDR is the anchor point of the IoTCrawler framework (cf. Section 4.2), and holds metadata information for all data streams available in the framework. In step (2), the IoTCrawler information model is applied. IoTCrawler features an extensive Information Model based on the Next Generation Service Interface for Linked Data (NGSI-LD) standard and centred around the concept of IoTStreams [14]. The model provides the basis for the information stored in the MDRs and the integration of heterogeneous data sources. Both steps enable other parts of the framework to handle heterogeneous data sources. After the integration of new data sources, the SE comes into play (3) to further add new information to the data sources. The SE component enriches known data sources with new information extracted from the received data. The SE includes a quality analysis component that adds QoI (cf. Section 4.4.1) as well as a Pattern Extractor (PE) (cf. Section 4.4.2), which analyses data and provides higher level information.

In parallel, the enriched data (streams) are monitored (4) to enable the Fault Detection (FD) and Fault Recovery (FR) solutions of the framework. The Monitoring component ensures a constant user experience by detecting faulty streams and providing data recovery mechanisms (cf. Section 4.3). In addition, it features a virtual sensor creator to replace faulty data streams by an ML-based virtual copy. In the last step (5), within this layer, the search indices are created, allowing data sources to be found in the search process in a fast manner. The Indexing component is directly supporting the search of data streams by building indexes for the stream types and their attributes, such as locations (cf. Section 4.5).

The **Search and Orchestration Layer** contains components for handling search and subscription requests coming from IoT applications or individual users (A). The Orchestrator (B) is the main entry point for any user or application that wants to search for IoT devices (cf. Section 5.4). It organises the search process and orchestrates the needed data streams. The Orchestrator utilises the Search Enabler component (C), to resolve context-aware GraphQL requests to NGSI-LD requests and thus providing an easy-to-use interface hiding the complex NGSI-LD query mechanisms. For subscription requests coming from IoT applications, the Orchestrator can process the information gathered from the Search Enabler and is able to provide an endpoint to receive notifications about the stream properties, e.g., detected faults. NGSI-LD requests are redirected to the Ranking (D) component, which uses the built indices, given (user) constraints, and enriched information to rank the found data sources before they are sent back to the user or application.

All steps, in both upper and lower layers, are constantly supported by IoTCrawler's Privacy and Security components (cf. Section 5.1) to continuously ensure restricted access to IoT data sources for legitimate users (indicated with a * in Figure 1).

IoTCrawler enables users and applications to search for data sources, while addressing the challenges mentioned before. Due to the loose coupling of components via publish and subscribe APIs and the design of the single components, the framework is designed to reach high scalability (**R-1**).

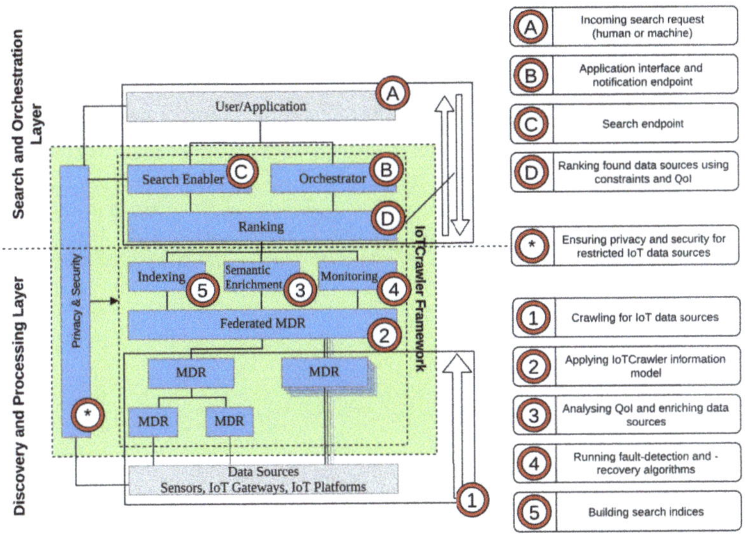

Figure 1. IoTCrawler addressing search in Internet of Things (IoT).

4. Enablers for Discovery and Processing Layer

This section addresses enablers for the discovery in the IoTCrawler framework, introduced in Section 3. A detailed description for each enabler is provided and complemented by an evaluation on the enabler's performance.

4.1. Information Model

The IoTCrawler information model is built upon standards. It follows the NGSI-LD standard and combines it with well-known ontologies, to reflect IoT use cases in the context of IoTCrawler and to address the requirement of semantics (R-2) to provide machine readable results. The choice of NGSI-LD is justified by several factors: being based on an standard makes it easier to inter-operate with, to integrate with other technologies and to maintain and evolve. Added to that, NGSI-LD supports semantics from the ground-up, which is one of the core strengths of IoTCrawler and an enabler for some of the functionalities that it provides. NGSI-LD provides not only the incorporation of semantic information to the data, but also a core information model and a common API to interact with that information (commonly called context). It was chosen as the main anchor point for the interactions between the components in IoTCrawler, greatly reducing and simplifying the number of different APIs to implement and keep track of, as well as data formats and models. This "common language" not only serves an internal purpose to simplify and optimise, but also makes IoTCrawler components easier to be integrated outside of IoTCrawler itself, and has already allowed to integrate existing components (like the MDR) seamlessly into IoTCrawler. The model has been designed to capture a domain that focuses primarily on sensors and stream observations. To achieve this and following best practises [52], concepts were reused from the SOSA ontology [11]. To enable search based on phenomena, the `ObservableProperty` is also reused. The `Platform` class is used to capture where the `Sensor` is hosted on. In addition to SOSA, the SSN ontology is used to capture what `Systems` sensors belong to and where they are deployed. Although SOSA captures concepts for sensors and observations, the concept of streams is missing, which is a fundamental aspect for IoTCrawler as it involves stream processing. For this, the IoT-Stream ontology provides the concept by defining an `IotStream` [14]. The `IotStream` class represents the data stream that is `generatedBy` the sensor as an entity. It also extends the SOSA ontology by defining a subclass of the `Observation` class, `StreamObservation`. This has been done to extend the temporal properties of an `Observation` to include windows as well as time points. For accessing the `Service` exposing the `IotStream`, the `Service` class from the IoT-lite ontology is used, and this enables direct invocation of the data source. The IoT-Stream ontology also provides concepts for `Analytics` and `Events`, which represent aspects of the semantic enrichment process. Moreover, with regard to the semantic enrichment process, IoT-Streams link to external concepts that capture QoI information about the streams. The QoI ontology provides this, which captures aspects of quality such as `Age`, `Artificiality`, `Completeness`, `Concordance`, `Frequency` and `Plausibility` [53]. An important aspect to any entity is location. Here, the NGSI-LD meta-model which defines a GeoProperty is used. The main classes and relationships of the IoTCrawler model are illustrated in Figure 2.

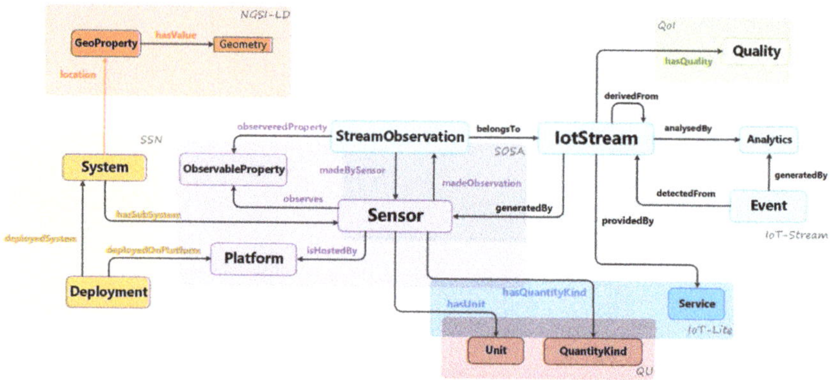

Figure 2. IoTCrawler information model.

4.2. Federation of Metadata Repositories

A key enabler in the IoTCrawler framework is the federation of multiple MDRs. The MDRs stores all available metadata information gathered by the discovery process. Considering the requirements from Section 1, the MDR has not only to support the IoT search as a whole, but also to address the requirements for scalability (R-1) and semantics to allow machines and applications to use available IoT data sources (R-2).

In addition to other technologies, e.g., triple stores or relational databases, IoTCrawler has chosen to use the NGSI-LD standard, which not only defines a data model for context information forming the basis for IoTCrawler's data model (see Section 4.1), but also defines an API, which will be used by consumers and providers alike, to access information. Among the API functionalities offered by the MDR are: the direct query and publish/subscribe mechanisms, which allow context consumers to receive notifications whenever new information is made available in the system. This publish/subscribe mechanism is extensively used in IoTCrawler for communication and synchronisation between different components, which will subscribe to context information relevant for their purpose, and will publish the processed information to make it available to other components.

NGSI-LD brokers can be interconnected in different ways to achieve scalability. The most best performing deployment configuration of NGSI-LD brokers, which is used in the IoTCrawler framework, is the federated one as shown in Figure 3. It consists of a federation of brokers, in which all information of the different federated brokers is accessible automatically through the federation broker. This last broker acts as the central point of IoTCrawler's architecture and is the key in making IoTCrawler horizontally scalable and well performing. This allows all other components in IoTCrawler to use the MDR in a scalable and standardised way and, being based on the NGSI-LD standard, not only makes the MDR inter-operable and compliant to standards, but also allows for the use of different already existing implementations.

For our current deployment, we have used Scorpio (https://github.com/ScorpioBroker/ScorpioBroker accessed on 24 February 2021) because it is the only implementation which considers a federated scenario. Nevertheless, in the frame of this paper we have focused our metrics on a single instance of this broker, obtaining both latency and scalability metrics. To do so, we have deployed a virtual machine with the following features: 8 CPU cores and 28 GB of RAM inside a Google cloud. Latency has been evaluated over the different operations provided by the MDR, specifically: entity management, publication/subscription and context provisioning.

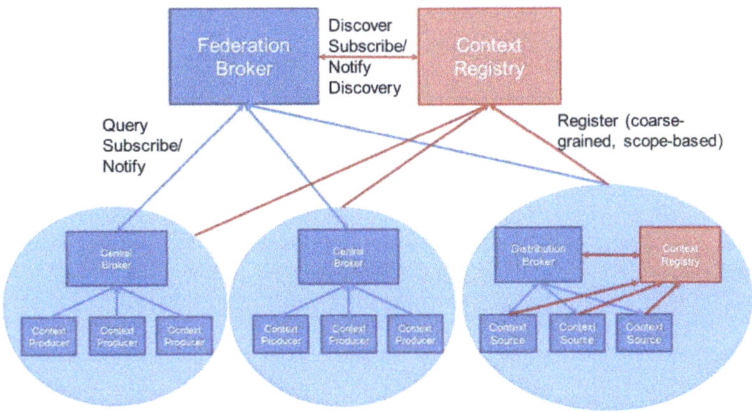

Figure 3. Federated broker architecture [54].

Measurements show that the most time consuming operation is the process of getting entities specified by their ID, which takes around 800 ms. This operation should not be so cumbersome and we think that the low performance associated with this task could be due to the maturity of this software. Apart from this operation, the rest of the operations take from 17 to 37 ms to perform, which is a more affordable processing time. Regarding subscription management, the creation of subscriptions is a heavier task taking up to 270 ms, whereas the other operations take only about 17 ms. Finally, context provisioning tasks, which comprise the registration of the information coming from context providers pointing at the end-point services provided by them, take more time compared to the previous tasks. Nevertheless, the registration and deletion of context providers are operations which are usually executed once per context provider. By contrast, the operations to obtain context providers take about 100 ms.

Finally, regarding the scalability metric, we have focused on the CPU and memory resources consumed by the instance of the NGSI-LD broker according to a specific range of simultaneous connections (2, 4, 8, 16, 32, 64, 128, 256, 512 and finally 1024). In addition, we have repeated this process four times. The results of these tests are presented in Figure 4, depicting that the CPU resources' consumption follows a logarithmic curve where the steepness of the slope is lowered from 8 simultaneous communications on. On the other hand, we can see that the increase in simultaneous communication does not impair the memory resources notably.

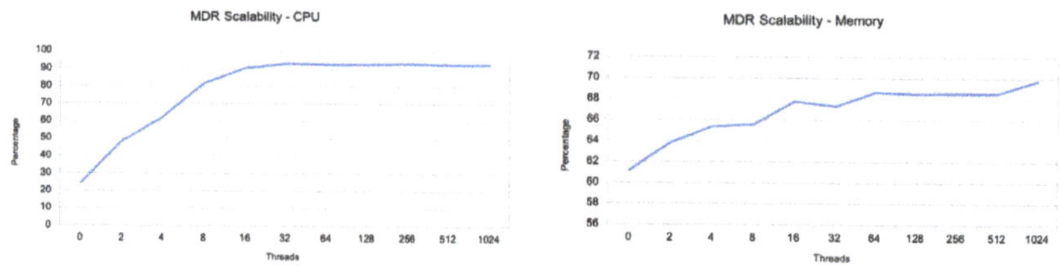

(a) Use of CPU Resources (b) Use of Memory Resources

Figure 4. Metadata Repository (MDR) scalability assessment.

4.3. Monitoring

IoTCrawler allows aforementioned participants to connect their sensors to the system, to make them available for a broader audience. As sensors are often deployed in environments where their operation cannot be controlled or even guaranteed, and given that many of them are battery powered and have a limited life span, it is to be expected that their reliability might fluctuate over time. It is therefore important to observe the performance of the sensors. For this, IoTCrawler has developed the Monitoring component, which is responsible for observing the incoming data streams of different sensors, detecting possible faults in the data, and, if possible, providing counter measures to mitigate them. The proposed monitoring concept with its different subcomponents provides an extensive set of features for addressing issues of dynamics in IoT environments.

4.3.1. Fault Detection and Fault Recovery

The Fault Detection (FD) component monitors the data streams that are available to the IoTCrawler framework and follows a two-layered approach. In the first layer, the component categorises faults as definite faults (due to hardware issues) or as anomalies, which could occur because of brief environmental factors, an unexpected behaviour detected through learned patterns. These anomalies can be categorised as faults, if they persist for a longer period of time. To cater to the needs of most of the sensor streams, the FD component uses different algorithms, e.g., the Prophet algorithm [55] for time series analysis and stochastic algorithms which determines the likelihood for a value to occur based on the previous observations of the sensor. The FD component subscribes to new data streams that become available through the MDR. Through the metadata, the FD determines which approach should be used. This is differentiated based on how much information is provided in metadata. The MDR is then notified in case of faults and trigger the recovery mechanism. To deal with faulty sensors, IoTCrawler has developed a two-stage counter measure. The Fault Recovery (FR) mechanism is a first response to handle missing sensor observations by imputing artificially generated sensor readings. The goal here is to have a quick solution to provide uninterrupted data streams for the applications using them. For long-lasting faults, the FD can issue the deployment of a virtual sensor to replace the broken one.

In the case of multiple sensors, we employ a knowledge-based Bayesian Maximum Estimation (BME) for imputing an identified faulty value [56]. BME is a mapping method for spatiotemporal estimation that allows various knowledge bases to be incorporated in a logical manner—definite rules for prior information, hard (high precision) and soft (low precision) data into modelling [57]. More details about this algorithm can be checked in [56].

To evaluate the working of FD and FR, an instance of FD is presented in the example below. Sensors deployed in three different parking areas in the city of Murcia are integrated into the IoTCrawler framework. These sensors record the information about the number of free parking spots in their respective parking lots with an update interval of 2 min. A model was trained on the data of several days from the parking areas to learn the normal behaviour. As an instance of the results, Figure 5 shows the original data for one day from each sensor, each along with one injected anomaly.

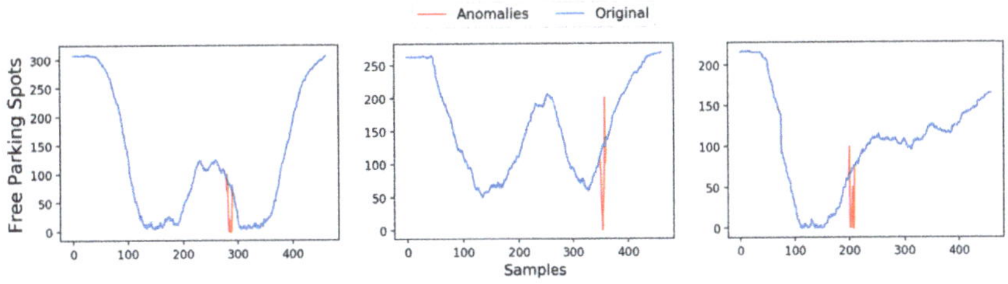

Figure 5. Data with injected anomalies at different instances.

The algorithm detected two anomalous patches in each instance. The first detected patch consists of the initial samples where the values do not change and the second anomalous patch is the actual anomalies. The first fault in each instance, caused by the sensor, is considered to be in stuck-at fault condition as this behaviour was not observed in the training set. The stuck-at condition is fulfilled when a sensor repeats an observation more times than was observed in the training set.

For the stuck-at fault, an estimated value cannot be interpolated by the neighbouring sensors, as all of the sensors show the same faulty behaviour at the same instance. The second anomalous patch in each sensor occurs when the data from another sensor is normal at those time instances. A recovery value is then generated using data from sensors with normal behaviour and BME as the interpolation technique (explained above). Results can be seen in Figure 6.

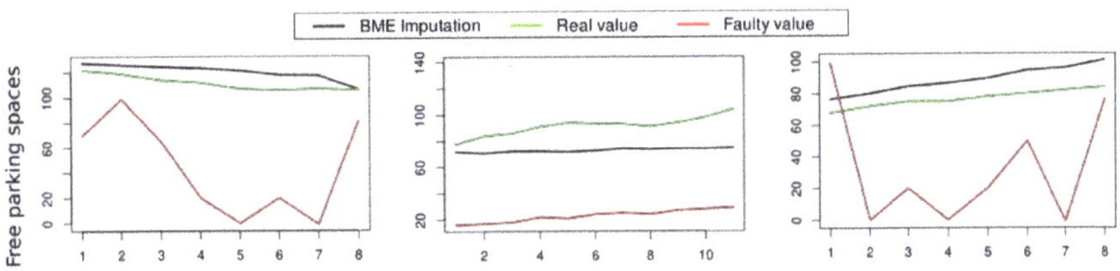

Figure 6. Comparison of actual and recovery values at the anomalous patch.

4.3.2. Virtual Sensor

To replace a faulty sensor in the longer term, IoTCrawler provides the virtual sensor component. A virtual sensor is capable of providing artificial sensor observations for a longer time as it is trained on larger data sets with different algorithms. As a result of the FR mechanisms as a first response, virtual sensors are allowed to train for a longer period of time, hence allowing the algorithms to learn more patterns which also make them capable of learning data drift. The concept of virtual sensors is that it takes historical data from a broken sensor and its correlating sensors and use the relationship to predict the values in place of the broken sensor. For instance, in the case of a broken temperature sensor, a virtual sensor can be trained to project the temperature at the failed sensor's location using nearby temperature sensors as predictors. To achieve this, the component searches for neighbouring sensors that can be used as predictors in the ML model. The correlation between the broken sensor and each candidate is calculated to train only with the most promising data sets. Via a grid search approach, the most promising ML model is selected.

To test the component, different scenarios are considered and the results are documented in [58]. The viability of virtual sensors has been shown in different environments by considering neighbouring sensors with the same and different sensor types than the faulty one. Models selected through grid searching along with models created through ensembling were used to make the predictions, both of which showed promising solutions. The results show that a fully autonomous deployment of virtual sensors is possible, although it should be mentioned that their effectiveness highly depends on the availability of correlating surrounding sensors.

4.4. Semantic Enrichment

The IoTCrawler framework is capable of adding new meta-information to known data and data sources. For this purpose, the Semantic Enrichment (SE) is being used. Currently, the component contains two parts, but can be extended further: the QoI Analyser and the Pattern Extractor, where the first one is responsible to add information about QoI to a data stream and the second one to extract patterns and therefore learn additional information from a stream.

4.4.1. QoI Analyser

The QoI Analyser is responsible to annotate data streams within the MDR with additional QoI metadata. By combining metadata and predefined QoI metrics, it is possible to rate incoming data from data streams and therefore to annotate these streams with QoI. This additional information about quality enables other components of the framework to provide (better) results, especially the Monitoring (cf. Section 4.3) and the Ranking (cf. Section 5.2) components.

An important step is the definition of QoI metrics that are available within the IoTCrawler framework. Currently, the QoI Analyser supports five QoI metrics that have been defined: Completeness, Age, Frequency, Plausibility, Concordance and Artificiality. For details and calculation of the QoI metrics, we refer to [59–61]. To integrate the results of the QoI calculation an ontology has been created and integrated into the information model as shown in Section 4.1.

A main anchor point for the integration is the publish/subscribe interface provided by the MDR. Figure 7 provides an overview of the interactions of the QoI Analyser and the IoTCrawler framework. When a data source is registered or updated at the MDR, the registration contains additional metadata, e.g., a detailed description of the data sources properties and its characteristics. This allows to adopt the QoI calculation to changes in the metadata or to connect to a new data endpoint description to access data. Finally, the QoI Analyser calculates the QoI for each known data source and adds the results to the metadata. This allows other IoTCrawler components as well as third-party users to access the additional information.

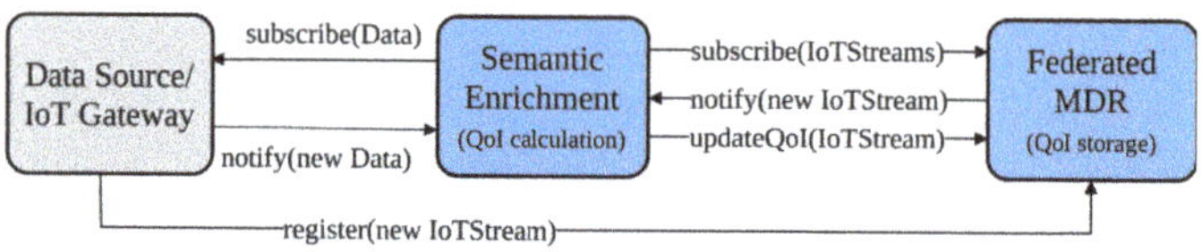

Figure 7. Semantic Enrichment (SE)–MDR communication.

For the following experiment, data from the city of Aarhus, Denmark are used. The data set named "CityProbe" is a real-time data source, which consists of 24 sensors that are mounted on light poles. The devices are solar powered and provide different sensor values, e.g., humidity, temperature, rain or CO. These data are analysed and it is shown how the QoI Analyser detects increasing or decreasing quality of the incoming data. For the experiment, the metadata annotation has been set to the following values: The range for the measured temperature has been set to −30 °C to 40 °C, which depicts a common temperature range for a northern country, whereas the humidity ranges from 0% to 100%. These ranges are used for the calculation of the Plausibility metric by checking if the observations remain in the defined ranges. Figure 8 shows an analysis of two sensor devices for temperature and humidity data for a time span of one week. The first graph depicts the measured values, whereas the second one shows the calculated Plausibility values. Figure 8a shows some suspicious temperature and/or humidity peaks in the minus area. From a human point of view, they can assumed to be wrong. In addition, the Plausibility metric decreases as it can be seen in Figure 8b. This example shows a use case of a decreasing QoI metric. A possible subscriber of the QoI, e.g., the Monitoring, can now react to the dropping information quality. In case of the Monitoring, it is now possible to initialise a more complex FD or FR algorithm or to create a new virtual sensor instance.

With the QoI Analyser, it is possible to identify data streams with decreasing quality. As an example, the Frequency metric is able to detect if a data stream does not provide data in the annotated time interval. Of course, it is not possible to directly detect the reason for a decreasing Frequency as IoTCrawler has no access to the sensor devices, but it provides the results of the QoI calculation to other components, which can then react to a changing QoI, e.g., by selecting an alternative data source. With that, the QoI Analyser enhances the Reliability in IoT environments. The QoI annotations also give objective criteria to choose between data streams, especially when the search is performed by a non-human system (Requirement R-2).

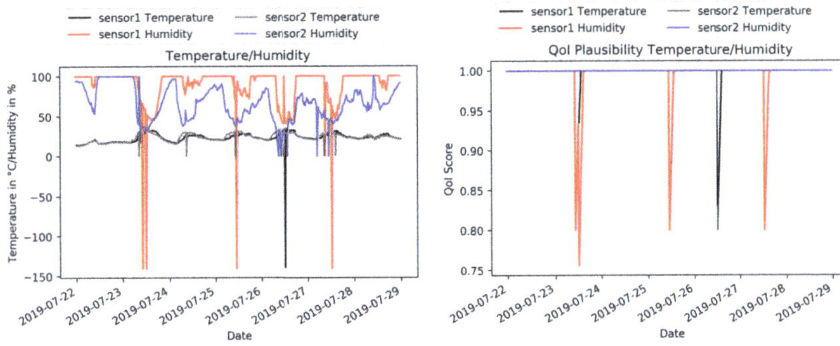

(**a**) Humidity and Temperature (**b**) Plausibility for Humidity and Temperature

Figure 8. Aarhus CityProbe sensor's data and Plausibility.

4.4.2. Pattern Extractor

To allow context-based search (Requirement R-2), the Pattern Extractor (PE) module enables the generation of higher-level context The context itself would be defined by the domain(s) of interest of the deployment, e.g., traffic congestion levels or personal health activity monitoring. The PE relies on a pre-training process in which it creates a set of clusters, each corresponding to a state or event. The PE analyses annotated IoT data streams that are pushed to the Metadata Repository to detect `Events`, by employing a data analysis technique. A subscription to the MDR is made for `StreamObservations` that have a certain property, and can also include spatial and temporal filters. `iot-stream:StreamObservations` of `iot-stream:IotStreams` that meet the requirements are then pushed as notifications the PE component. The PE temporarily stores a certain number of observations that correspond to the time window pre-defined by the deployer. The output of the analysis is a textual label that interprets the pattern of data. The label is then encapsulated in an `iot-stream:Event` instance, along with the start and end times of the window in question, and published to the MDR.

The algorithm for pattern extraction is based on aggregating observations from a time window for pattern representation. Observations are grouped in time windows of predetermined size. On each window, Lagrangian Pattern Representation (LPR) [62,63] is applied to determine the patterns. Patterns are then clustered and grouped using Gaussian Mixture Models (GMM). The number of clusters depends on the number of expected events for a specific scenario. A label representing the pattern is given to each cluster. Label nomenclature is defined by the topical domain ontology for the specific use case.

In the PE component, there are two models that represent patterns [63]. K-means clustering was used for the first approach of representing patterns and our model applied to some data sets from UCR Time-series Classification Archive [64], which is known as a benchmark data set for clustering and classification methods. The data sets Arrowhead, Lightning7, Coffee, Ford A and Proximal Phalanx Outline Age Group from the time-series archive were used. The Arrowhead data set contains shapes of projectile points in time series. Lightning7 has data of time-domain electromagnetic from lightnings. The Coffee data set contains data from measurements of infrared radiation interaction with coffee beans, which is used to verify the coffee species. Ford A has measurements of car engine noise and Proximal Phalanx Outline Age Group has observations from radiography images from hands and bones. Silhouette coefficient was used to evaluate the model. Silhouette is a measure of how separated the constructed clusters are from each other. To evaluate the clustering technique in the real-world scenario, we need to use a measurement to evaluate the separation of the clusters as we do not have the true classes. The results were compared by using K-means on raw data without Lagrangian representation. Table 1 proves that our method improves the clustering results of these data sets.

Table 1. Silhouette evaluation of Lagrangian representation using k-means.

Model/Data Set	Arrow Head	Lightning 7	Coffee	Ford A	Proximal
Raw Data k-means	0.47	0.12	0.33	0.05	0.46
Lagrangian k-means	0.67	0.57	0.69	0.56	0.62

The measurements for the above data sets were conducted using a machine with a 4.00 GHz 4-core CPU and 32 GB of RAM. In the case of the time series in the Ford A data set, the averaging processing time for applying LPR on it was between 400–500 milliseconds. Figure 9 shows the relative comparison of the clustering algorithm processing time applied to each data set.

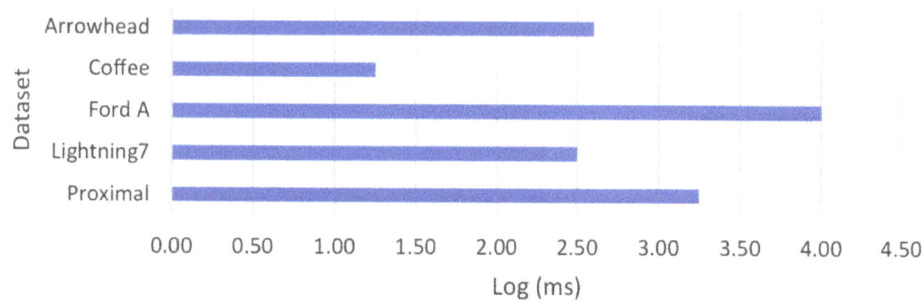

Figure 9. Processing time of the clustering algorithm for different data sets.

For the evaluation of Principal Component Analysis (PCA)-Lagrangian representation, the method was applied to both synthetic and real-world data. GMM was then used for clustering. We generated a synthetic data set using a multivariate Gaussian distribution and generated a time series including 2400 samples with four dimensions with three different Gaussian distributions which have the same covariance matrix and different mean vectors. Each distribution had 800 samples. In addition, another data set was generated by adding white noise with Signal-to-Noise Ratio (SNR) of 0.01. The results of the Silhouette coefficient are 0.87 for data w/o noise and 0.47 for data with noise.

For a real-world scenario, we used air quality data from Aarhus' open data. We used air quality data from a period of two months with a sampling frequency of every five minutes. The data have two dimensions; Nitrogen-dioxide (NO_2) and Particulate Matter (PM). There are three different clusters: low risk, medium risk and high risk. We evaluated the results using Silhouette coefficient and compared the results. The results are shown in Table 2.

Table 2. Results of Silhouette coefficient for the Aarhus data set.

Method	Silhouette Coefficient
PCA-Lagrangian + GMM	0.69
Raw data + GMM	0.46
Lagrangian scaling + GMM	0.45
PCA + GMM	0.39

The proposed algorithms for pattern extraction allow to extract high level events directly from the IoTCrawler framework (R-2). They also reduce the need for external applications to subscribe to raw data and decrease the amount of transferred data, improving scalability (R-1).

4.5. Indexing

Indexing provides a means for clients to search for IoT entities efficiently. It focuses on IoT streams and sensors, where queries can be based on sensor type and absolute or relative location. To initiate the process of indexing, a platform manager needs to register a MDR with the Indexing component. In turn, this will trigger the subscription to sensors and streams at the registered MDR. As the metadata descriptions are updated at the MDR, the Indexing component will be notified and then index the sensors and streams based on location. For scalability, the Indexing component can be configured so that the persistence it relies on (MongoDB) can be shared (see Figure 10).

Indexing exposes a query interface which complies with the NGSI-LD specification. Upon querying by a client, entities that relate specifically to sensors, IoT streams, location points or QoI will be responded to directly. Else the query is the forwarded to the MDR for complete query resolution. The approach enhances the query resolution performance significantly, as co-located entities and common types are indexed and grouped, allowing reduced latency in query processing.

The indexing technique applied is based on a geospatial approach defined by Janeiko et al. [65]. The index is a tripartite whereby two of the indices link `iot-stream:IoTStream` and `sosa:Sensor` entities to a geo-partition key. The other index contains the actual data and is also geo-partitioned. The partition key is determined by intersecting the location of the sensor represented as GeoJSON objects with predefined GeoJSON polygons representing geographical regions. The index contains the entities in the form of a graph, whereby linked entities are stored as a single entry. Here, the IoTStream entity is the root entity with all other indexed entities are nested within it, hence any query for any entity must be linked to an IoTStream entity. The structure defined allows to construct compound indices, which accelerates nested queries. By providing its indices for search, the Indexing component addresses scalability R-1.

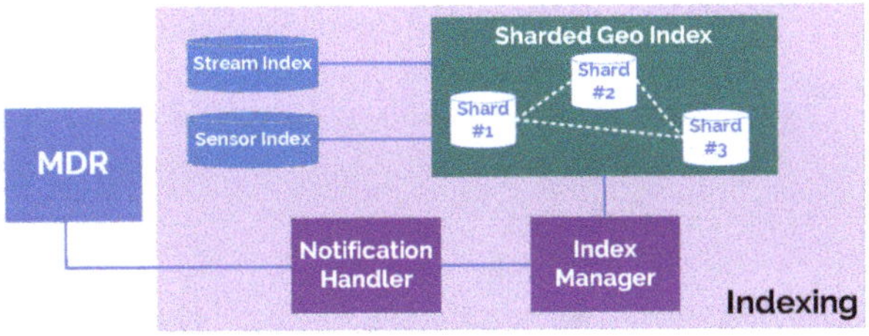

Figure 10. Indexing sensors and IoTStreams.

The Indexing component is responsible for creating and updating the metadata indices to allow fast search and retrieval of the metadata stored in the MDR, using geospatial indexing. The initial approach for geospatial indexing IoT Streams and Sensors was to use geohash, whereby the location is represented by a string of characters with a predefined length reflecting the granularity of the bounding box the entity will be associated with. A new approach has been taken to maintain the exact location of the entity by using a Quad Search Tree. The main KPI that is applicable is the latency and retrieval time. The Indexer partitions the notifications from MDR broker notifications for stream or sensor data location by country. Latency and retrieval time can be measured based on: a data set's size or number of entities, i.e., streams and sensors, a number of countries or a number of concurrent requests.

Therefore, the approach to evaluation will be applied to a data set with different sizes, multiple countries. Data sets were randomly generated which covered entities located within 6 countries. In terms of hardware, the experiment was conducted on a computer with an Intel CORE i7 CPU of 6 cores, 1.9 GHz and 32 GB RAM. Concurrency tests were performed using the Apache Bench tool. Two sets of tests were performed. Each set had a number of entities stored in the indexer. For each set, two sets of concurrency tests were performed: one with 100 requests (the graphs show the total time for all requests) with a concurrency of 10; and 10,000 requests with a concurrency of 1000. Regarding the query response time, two factors are measured, the total time for the response, the wait time and the time the indexer receives the requests and responds, irrespective of the connection time.

Between the 3 sets of tests, the wait- and total response times show a gradual increase with respect to the number of stored entities. What is also noted is that for the last set of concurrency requests, a significant change in delay is observed, especially in the case of requests with a concurrency of 1000. Figure 11a,b show the response times for requests with increasing concurrency. The plots have been smoothed out with a moving average of 10 and 20, respectively.

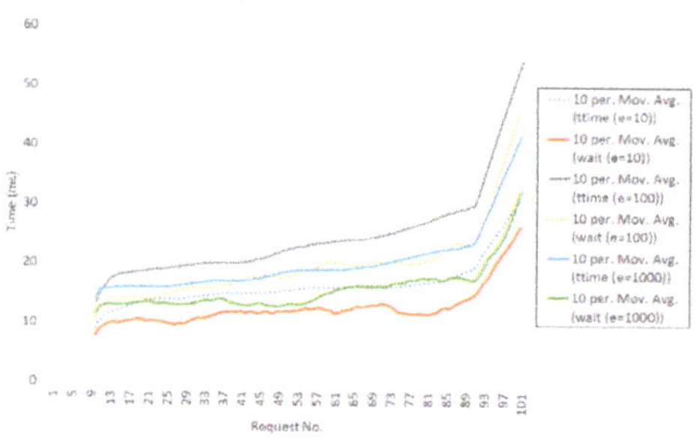

(**a**) 100 Requests and Concurrency of 10

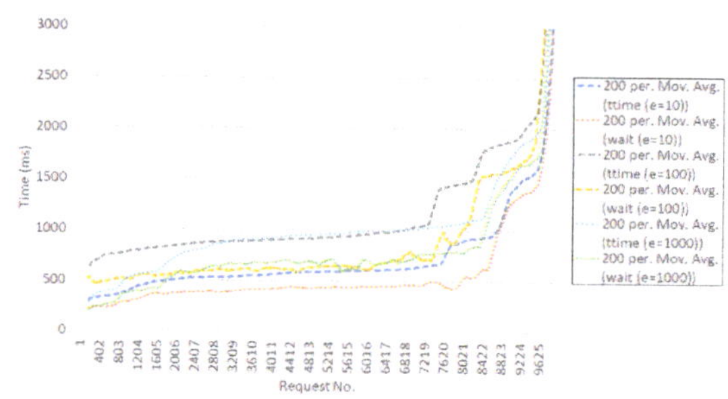

(**b**) 10000 Requests and Concurrency of 1000

Figure 11. Indexing response times.

5. Enablers for Search and Orchestration Layer

This section addresses the enablers for search in the IoTCrawler framework. Based on the description in Section 3, all enablers will be shown in detail, including experimental results and evaluations.

5.1. Privacy and Security

The IoTCrawler framework places security and privacy as a traversal pillar interacting with the different layers of its architecture (cf. Figure 1). This pillar comprises: Identity Management (idM), access control management, for both intra-domain and inter-domain and finally privacy from a data point of view. Starting with idM, this component is responsible for handling the different identities that are registered in the IoTCrawler framework. An identity, which can be a user, device or service comprises different attributes such as: name, email, role and organisation, to name a few. They are quite important for the definition of access control and privacy encryption policies as we will see below. Another important function carried out by the idM is that of authentication. Any entity registered in the system must perform the login operation due to the exposed API. In our case, we have selected the FIWARE KeyRock GE (https://fiware-idm.readthedocs.io/en/latest/ accessed on 24 February 2021), which exposes an OAuth2 API.

To deal with this heterogeneous landscape, we have designed a comprehensive approach where we are combining a distributed authorisation solution called Distributed Capability-Based Access Control (DCapBAC) with Distributed Ledger Technology (DLT), specifically Hyper Ledger Fabric (https://www.hyperledger.org/use/fabric accessed on 24 February 2021) and the use of smart contracts. DCapBAC decouples traditional authorisation solutions, such as XACML framework, in two different phases: authorisation request and access. For that, a new component, called Capability Manager (CM), is introduced. It is the end-point for the authorisation requests and it also issues an authorisation token called Capability Token (CT) after validating the authorisation request by communicating with the XACML framework. Regarding the access phase, the XACML Policy Enforcement Point (PEP) is moved as a Proxy located close to the server where resources are stored. In this case, CT acts as a proof of possession which allows the PEP Proxy to validate it easily without querying any other third party. This CT contains all details regarding the resources to be accessed, the access mode among others.

DLT provides numerous advantages in term of resilience, and traceability because of its consensus approach where all nodes of the network must agree on global policies. For this reason, in IoTCrawler, an additional step is taken as showcased in Figure 12, by introducing the Blockchain as an added element in the security process; by storing policies in the Blockchain, as well as CTs that can later be revoked; and thus need to be checked by the PEP Proxy in the Blockchain for validity.

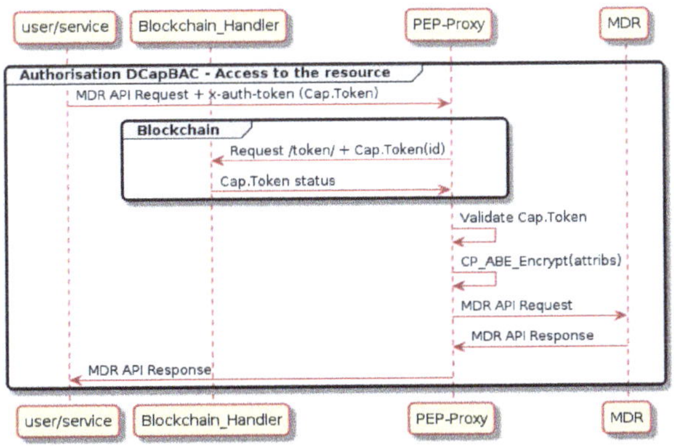

Figure 12. Policy Enforcement Point (PEP) Proxy interaction diagram.

Access control components are integrated into the Blockchain to enhance security and scalability. By leveraging Blockchain, several issues of current access control systems can be overcome.

- Untrustworthy entities: First, Policy Administration Point (PAP) might be subject to an attack and perform malicious actions such as updating a policy against the resource owner's will. Having a Blockchain helps avoid misbehaviour of PAP. The access control policy's integrity is checked by registering and checking its meta-data, such as the hash value managed by the Blockchain network. Second, policy evaluation done by Policy Decision Point (PDP), which could be manipulated by an untrusted PAP. The Blockchain ensures this misbehaviour to be detectable.

- Auditability: The verifiable property of Blockchain allows detecting if an access control service falsely denied access to a subject that the policy would grant or if the access control service granted a permission while the policy was not satisfied.
- Revocability: The attribute-based access control model that we have in this framework assumes, once a subject has granted an access permission, that the subject will receive an access token. It is challenging to revoke the token once it has been misused or stolen. Blockchain resolves this issue by executing a token smart contract to invalidate the vulnerable token.
- Fault tolerance: Access control components are distributed among peers over the Blockchain network. Such components are PAP, PDP and CM, among others. By having functions executed as smart contracts and invoked by a peer of the network, it avoids becoming a single point of failure as it would be the case with traditional PAP, PDP or CM.
- Integrity: New changes may cause disruption of such services and therefore they should be done cautiously. No single individual can introduce changes. This property is essential in the network where the participants often do not trust each other.

To address the scalability requirements R-1, we carefully design the security components so that only critical parts are executed on-chain and other parts can be done off-chain. Policy and capability managing operations are on-chain with policy enforcement and identity management can be done off-chain or access to another service. In addition, we carefully select the consensus algorithm, which is one of the core parts of the Blockchain, so that it provides efficient throughput and latency performance. As a result, security and privacy enablers provide by-design secure access to IoT data thanks to the DCapBAC access control model in privacy-preserving using attribute-based encryption. DCapBAC is coupled with Blockchain to provide distributed trust among untrusted domains by agreeing on common policies and ensuring policies' integrity. In addition, Blockchain offers transparency, auditability and fault tolerance to access control. Our chosen Blockchain deployment with sufficient consensus algorithm ensures low overhead, in another word, high scalability.

For the evaluation of these components we have measured the latency associated to each of the operations that these components perform to grant authentication and authorisation, as well as the performance metrics linked to the CPU and memory consumption of these operations by increasing the number of simultaneous requests up to 2048 connections. We ran the benchmark experiment on a server with Intel Xeon E-2146G CPU, 32GB RAM, in a local network environment.

5.1.1. Identity Management and Authentication Evaluation

Starting with Keyrock, we have evaluated the latency on two different sets of operations, the first one is related to the generation, information retrieval and deletion of the Identity Management Token, while the second set is focused on the user point of view, providing information about common operations related to user management. For the evaluation of this metric, we have launched 100 executions of these operations to provide the average latency value and 95% confidence intervals as presented in the following graphs and tables. As we can see in Figure 13a, the delay obtained for the difference is really low, reaching up to 30 ms for the generation of the idM token. This operation, compared to the others, is the heaviest one because it comprises the different mathematical operations required to generate the token. The most common authentication operations usually triggered via web interface or REST API are shown in Figure 13b. In light of these results, we can also state that user operations last about 30 ms, which is reasonable in terms of latency.

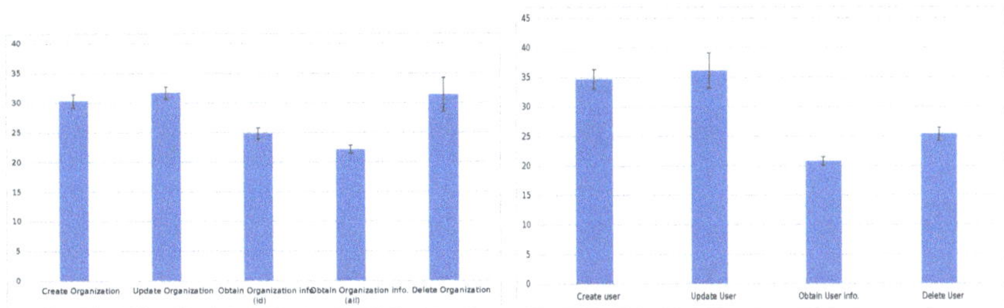

(**a**) Token Operations (**b**) User Operations

Figure 13. Delay of Identity Management Operations (units in milliseconds).

Regarding the scalability aspect, we have assessed the performance of the idM in terms of CPU and memory consumption resources. The objective was to provide a trend as the number of requests increases, so that we can provide an estimation for a higher number of simultaneous requests. Therefore, to achieve this goal we have launched different number of simultaneous connections: 2, 4, 8, 16, 32, 64, 128, 256, 512, 1024 and 2048. Additionally, we have repeated the experiments 4 times. More specifically, we have employed a query for authenticating a user. According to Figure 14a,b, we can state that the CPU resources managed by the idM remain stable as the simultaneous requests increase. Regarding the memory resources, we can see that up to 256 simultaneous connections, the increase is less than 1.5%. From that number on, the memory resources increase again about 1.5%. Therefore, we can state that it is able to handle a large number of communications.

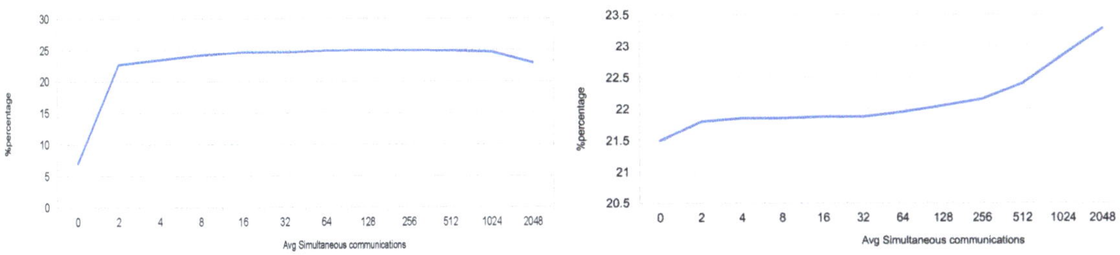

(**a**) CPU Consumption vs. Number of Simultaneous Communications (**b**) Memory Consumption vs. Number of Simultaneous Communications

Figure 14. Identity Management Resources consumption.

5.1.2. Authorisation Evaluation

Authorisation addresses two different scenarios, intra-domain and inter-domain scenarios. Regarding the former, DCapBAC has been implemented. From this point of view, we have assessed different metrics with the objective of measuring the time to grant access to a user to a specific resource, and also to measure the performance in terms of simultaneous connections.

Consider thePDP request, which is the XACML validation process that is performed by the PDP after receiving the authorisation request coming from the CM. This takes around 200 ms. The CT generation considers the previous task, and includes also the processing time required by the CM to issue a CT, which takes about 1.6 s. Finally, the overall authorisation process from the point of view of the clients from the moment they issue an authorisation request, to the moment they receive the authorisation answer together with the CT was measured with around 1.65 s. Since the token includes a validity period to the resources to be accessed, issuing a CT is not required for every access.

The DLT operates using Hyperledger Fabric framework with the Kafka consensus algorithm. The most essential and critical factor that affects the overall performance of a Blockchain network is the ordering service. Ordering service is a part of the consensus protocol. It generates a unique ordered sequence of transactions in a block and the block is delivered to nodes. We measured the Blockchain latency and throughput as the primary performance metrics for Blockchain, with various parameter settings of network size (number of ordering nodes) and block size (blocks committed to the Blockchain of each transaction). Ordering latency is the time a transaction needs to wait for the ordering service until its order in a block is assigned. Ordering throughput is the capacity the service can handle a certain number of transactions per second. Figure 15a,b show benchmark results of our Blockchain network. With small size network (7 ordering nodes), latency is very low (less than 0.5 s for 1000 concurrent clients). When the size of the network increases, the latency also increases (at size of 151 nodes, the latency at 1000 concurrent clients is 3 s). The same pattern for throughput performance. Throughput drops when network size grows (at size of 151 nodes, throughput at 1000 concurrent clients goes below 500 transaction/second). These benchmark results show the tradeoff between latency/throughput performance and network resilience against faulty requests. When the network size is larger it is more resilient to tolerate faulty nodes, however, it bears higher latency and lower throughput.

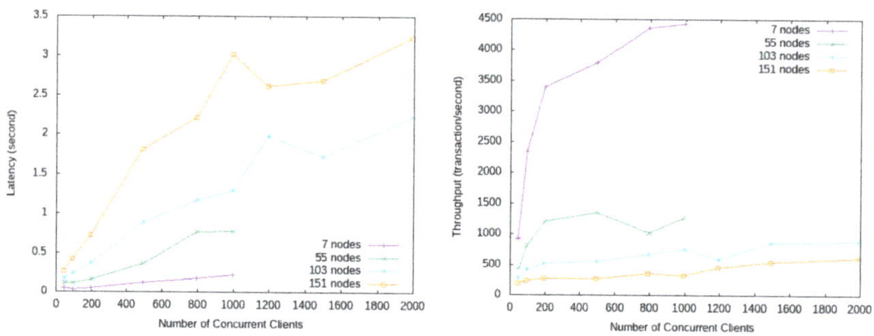

(**a**) Latency for Blockchain Transactions (**b**) Throughput for Blockchain Transactions

Figure 15. Latency and throughput performance for Blockchain transactions.

We have excluded the cost for the execution of transactions in the benchmark. Instead, we vary the block sizes which simulate various application scenarios in practice. Figure 15a,b show results at a typical block size of 100 bytes. For other block sizes, the patterns of latency and throughput hold the same. The overall performance of Blockchain network needs to consider also the transaction execution time, smart contract invocation time and transaction validation cost, which are application dependent.

5.2. Ranking

The Ranking component implements ranking mechanisms for IoT resources. Its purpose is to aid users and applications to not only find a set of resources relevant to their needs, but also to select the best or most appropriate one(s) from that set. There are multiple criteria for ranking IoT resources such as data type, proximity, latency and availability. Therefore, IoTCrawler's Ranking component supports application-dependent, multi-criteria ranking. Within the IoTCrawler framework, the Ranking component is available to the Search Enabler component to facilitate entity discovery. The Ranking component relies on an NGSI-LD compliant endpoint as a backend, which is often times the Indexing component but could also be any NGSI-LD broker. Upon receiving a query request and its ranking criteria, the Ranking component initially forwards the query to the underlying index or broker to get the set of `IoTStreams` entities matching the query. A ranking function then computes, for each result, a ranking score, according to the ranking criteria. The score is then attached to each `IoTStream` result as an additional property. The ranking criteria specifies the relevance of different properties to the application. The current ranking function computes a weighted average of the QoI values of a `IoTStream` entity, where the weight values are specified in the ranking criteria, but it can be easily adapted to other ranking criteria. In this way, the ranking is addressing the requirement for search R-3 by successfully ordering the search results.

The Ranking component offers an extended NGSI-LD interface, where ranking criteria can be specified in addition to the query. To avoid any influence of indexing strategies implemented by the Indexing component and be able to focus on the performance of the Ranking component itself, we have evaluated the Ranking component in a simplified architecture consisting only of the Ranking component and an NGSI-LD broker. Although the Ranking component supports horizontal scaling (adding more instances behind a load balancer for better scalability) due to its stateless implementation, in this evaluation we have only tested on a single instance. To assess the scalability of the component, multiple queries have been sent, both directly to the broker and to the ranking + broker combination. For the ranking + broker combination, we used a single ranking weight as the ranking criterium, that means that results were sorted based on the value of a single property. We have varied the number of concurrent query requests and measured the latency in retrieving the results. Each request returned 1000 entities, where the entities' size was approximately 7 kB.

The results shown in Figure 16 indicate that the Ranking component introduces a small latency in retrieving the results, but it can nevertheless scale with the volume of query requests.

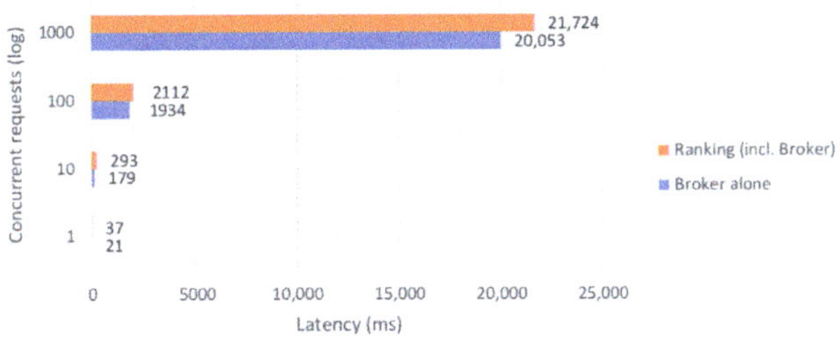

Figure 16. Ranking latency.

5.3. *Search Enabler*

The Search Enabler component is responsible for providing functionally rich query language and the search interface for seeking over metadata of discovered sensors and streams. Using GraphQL technology, the IoTCrawler search component offers end-to-end functionality for performing complex queries, allowing users to access data coming from distributed large-scale IoT deployments. Any complex GraphQL query is decomposed and resolved via a corresponding number of atomic NGSI-LD queries, as it is prescribed by the NGSI-LD standard. The schema-based approach of GraphQL allows to describe key entities (see IoTStream Ontology [14]) and the relationships between them. A compiled schema becomes a basis for query parser/validator engine and for a GUI, where users can design their queries. To comply with the linked data approach, all types and their properties in the schema are annotated with type URIs according to the IoTCrawler data model. Annotations describe hierarchical relations (equal to the `subclassOf`) between types, which are considered during query resolution process. This allows to be fully compliant with ontologies used for data modelling. For example, to describe a set of sensors hosted by a platform, a correct definition in terms of SOSA ontology would be: "system hosted by a platform", which means that sensors, actuators and others subtypes belong to the more generic type used in this statement. Use of types and subtypes and considering their relations during query resolution process is an exclusive feature of the Search Enabler component developed for the IoTCrawler platform. Another exclusive feature developed for IoTCrawler is the resolution of nested filters made on top of NGSI-LD. Nested filters are equivalent to join clauses in traditional query languages (e.g., SPARQL), where multiple entity types can be returned or used as filters in a query. The recursive query resolution processor carefully passes through all the types used as filters or output fields and initiates the corresponding number of NGSI-LD requests. GraphQL queries designed and tested via GraphiQL (GUI) might be integrated into IoT applications and executed programmatically. Results are returned in machine-interpretable JSON format. Alongside the GraphQL-based search, the IoTCrawler is equipped with a rule-/pattern-based generator and mapping mechanism for generating filter conditions [7]. As a result, a state-based context model empowers GraphQL queries with context-based reasoning. The described Search Enabler's search functionality is performed on top of the federated metadata infrastructure, which employs security and privacy-aware mechanisms.

The Search Enabler component offers a GraphQL interface, where search queries expressed in GraphQL are resolved via HTTP-requests over NGSI-LD interface. Since NGSI-LD allows to query only one type of entities per request, complex GraphQL queries (requesting more than one data type) require a corresponding number of NGSI-LD requests to be performed. The number and order of subsequent requests are prescribed by Search Enabler according to a structure of a GraphQL query. For example, a simple query of stream identifiers (`streams{ id }`) would be resolved by a single NGSI-LD request for entities with type `iot-stream:IotStream` (query #1). The extension of the query by the names of sensors (query #2) requires an additional resolution step: one NGSI-LD request for each sensor ID associated with the stream from the list of query #1. Further extension of query #2, e.g., by the names of properties observed by sensors, requires an additional resolution step: one NGSI-LD request for each property IDs associated with sensors. In case different sensors observe the same property, the Search Enabler avoids duplicating NGSI-LD requests.

For performance benchmarking, four different GraphQL queries have been selected. The difference between queries is in their complexity (requesting from 1 to 4 different entity types), which would require a different number of NGSI-LD requests to be performed. The expected number of NGSI-LD requests N depends on (1) a number of requested types T and if $T > 1$, then on (2) a number of unique entities R of subsequent types referenced in the results set. More formally, it is described as follows:

$$N = (T - 1) * R + 1 \quad (1)$$

The caching mechanism avoids duplicating requests, so a real number of them might be significantly lower than was expected. During the experiments, we have measured the average GrahpQL query execution times and summarised the execution time of the corresponding NGSI-LD queries. Dependency on a number of results is demonstrated via limiting them within the range 1–500 with step size 100. Each experiment was repeated 10 times and the average times were calculated. In Figure 17, an average query execution time depending on number of results is demonstrated. Figure 18 represents a GraphQL query execution time against the summarised execution time of the corresponding NGSI-LD requests. From Figure 18d, it can be seen that GraphQL query execution goes faster than execution of the corresponding NGSI-LD requests. This can be explained by a particular query's structure, where two types (observable properties and platforms) can be resolved in parallel. In the case of no parallel type resolutions (Figure 18a–c), the overhead of GraphQL engine is not higher than 0.2 s (1% of the overall query execution time). For complex queries with parallel type resolution, the overhead is mitigated at all. Experiments have been done using the NGSI-LD broker (Scorpio) running on Intel NUC i5-5250U with 8 GB of RAM. The Search Enabler and GraphQL client were running on a laptop Intel Core i7-5600U with 16GB of RAM, both were connected to a 1 GB/s local network.

The Search Enabler solves the machine-initiated search challenge (R-2) by providing programmatic interfaces (APIs), to which remote IoT applications can send search requests and get results back in an automated way.

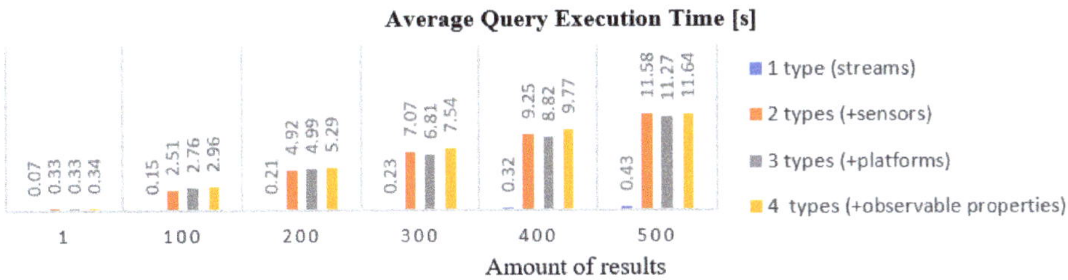

Figure 17. GraphQL query execution times.

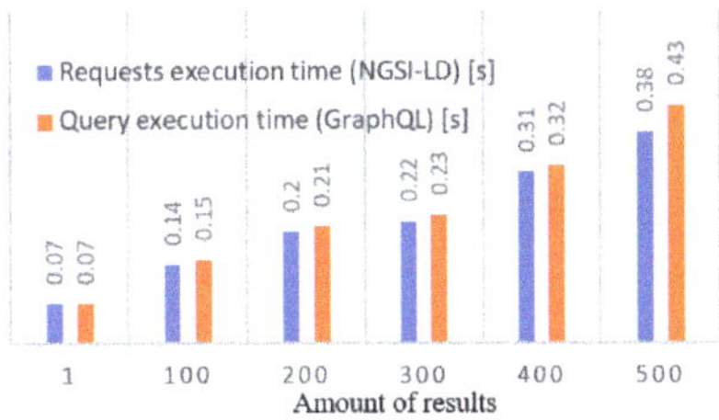

(**a**) 1 Type (Streams)

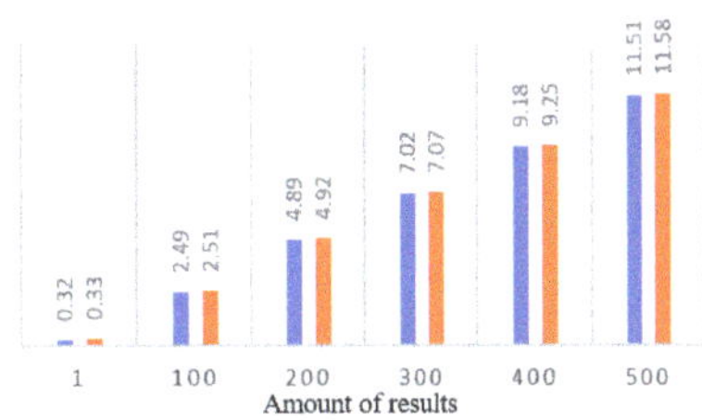

(**b**) 2 Types (+Sensors)

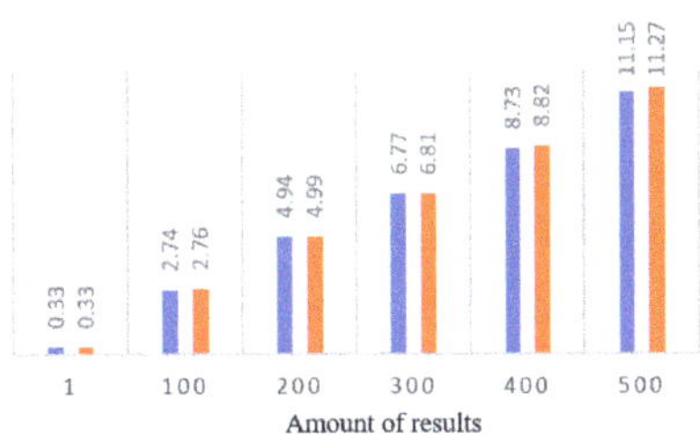

(**c**) 3 Types (+Observable Properties)

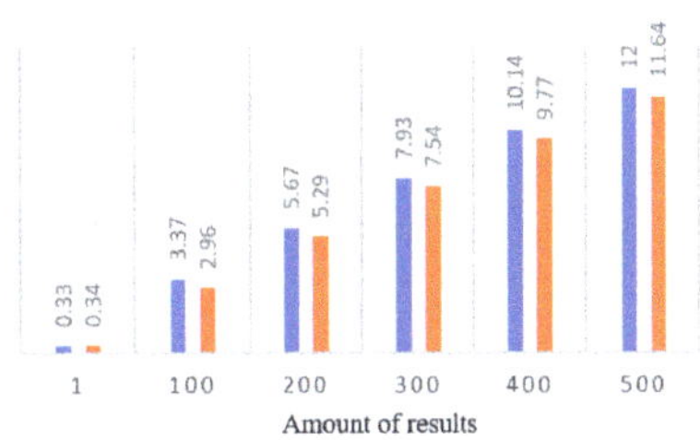

(**d**) 4 Types (+Platforms)

Figure 18. Requests execution time (Next Generation Service Interface for Linked Data (NGSI-LD)) vs. query execution time (GraphQL).

5.4. Orchestrator

The Orchestrator component is targeted to be the mediating component for IoT applications, which are expected to be running outside the IoTCrawler platform, interacting with it via interfaces of the Orchestrator. The Orchestrator is an endpoint, which forwards all metadata requests to a Ranking component and subscription requests are forwarded directly to the MDR. At the same time, the Orchestrator provides its endpoint for receiving notifications coming from the MDR. Without it, applications would have to expose their own REST endpoint, which is often not possible (e.g., for apps running on mobile devices or in private networks). The Orchestrator mitigates that by providing its own endpoint (not exposed to the public) and redirecting all incoming notifications to a dedicated queue in a publicly available publish-subscribe service (Advanced Messaging and Queuing Protocol (AMQP)). It is enough for an IoT application to subscribe to a queue in the messaging service to get notified immediately. The described publish-subscribe mechanism also allows the setup to notify the IoT applications about stream failures detected by the monitoring component.

The Orchestrator implements the NGSI-LD interface and redirects incoming NGSI-LD requests to two components: MDR and Ranking. Entity subscription requests are analysed, modified (if required) and forwarded to MDR. The metadata/discovery requests are forwarded directly to the Ranking component, which allows it to rank the results of metadata requests according to a specified ranking criteria. As a result, the Orchestrator hides two IoTCrawler components under a single NGSI-LD interface—one of the interfaces used by IoTCrawler applications.

The evaluation of the Orchestrator component consists of measuring a dependency of performance characteristics (throughput and latency) on the number of parallel connections—IoT applications, running remotely. In this experiment, the Orchestrator component is working on top of Djane Broker—a lightweight NGSI-LD broker, which is less functional than Scorpio. The benchmarking process has been conducted using a single Intel Xeon machine (4 cores, 16 GB Ram). Each value was obtained by averaging the values of 10 repetitive experiments. Results can be seen in Figure 19. The number of parallel clients varied within the range 64–1024, where each client performed intensive and non-intensive workloads. For the non-intensive workload (1 request by each of 64–1024 parallel clients), the maximal average throughput is around 400 requests per second when the latency is less than 0.2 s. For intensive workloads (100 consecutive requests by each of 64–1024 clients), the maximal average throughput increases up to 1200 requests per second with the average latency increased to 1 s.

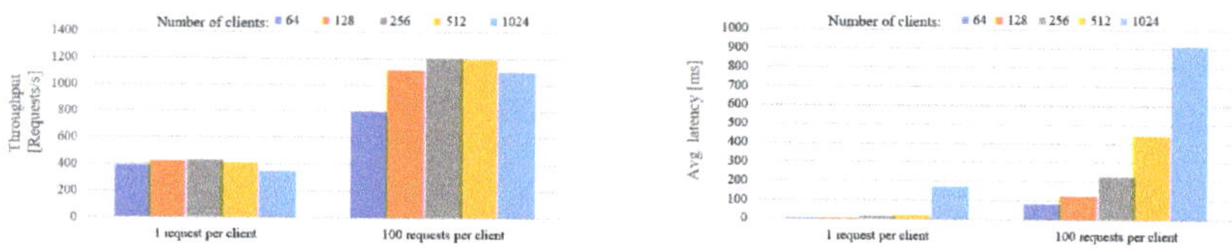

(**a**) Throughput (requests per second) (**b**) Average Latency (ms)

Figure 19. Orchestrator performance.

6. Application Domain Instantiation

This section presents two application examples of how IoTCrawler is being instantiated in real-world scenarios. Other scenarios for different domains are under development and will be part of a future publication.

6.1. Smart Home—Semantic Integration Focus

The target of the smart home use case was to understand the challenges which smart-home owners are facing when deploying and using their smart home devices. We have implemented an energy insight dashboard and tested it in a longitudinal study with end users in an early stage of the project. The energy insight dashboard was built with the objective to provide smart home users insights about their energy consumption and thereby to reduce their energy costs and carbon footprint. This was achieved by collecting energy measurements from smart plugs and other smart energy meters. The web-based application includes various aggregated and real-time views of the energy data as well as information about the usage frequencies of appliances attached to the smart plugs.

Evaluation: As part of IoTCrawler, we extended the dashboard to a public test bed running 24 h a week for almost a year. More than 60 homes and 3400 devices were connected during that period. Power users have more than two hundred devices connected to a smart home. Thus, we realised that managing these devices, which include knowing their locations, and for smart plugs, what kind of appliances are connected to which, created a considerable challenge for smart home owners. More importantly, the heterogeneity of devices with respect to their communication technologies, APIs and the gateways to which they are connected, makes it hard to develop smart home applications that run seamlessly with different vendors. As a response to tackling this challenge, we integrated an early version of an IoTCrawler feature for semantic annotation in which we used machine learning to detect device types, their locations and connected appliances in real-time [8]. We conducted a survey to validate the benefits of IoTCrawler features. Most of the respondents indicated that comparing and analysing energy usage is a benefit of the Energy Insights Dashboard (77%). On the second rank, respondents indicated that the automatic device detection feature is a benefit of the Energy Insights Dashboard (41%).

Further conversations with smart home owners and application developers have shown that IoTCrawler has the potential to be an effective IoT platform. For example, smart home users will be able to keep their data on their own hardware (located in private networks) and federate it into the IoTCrawler for processing by third-party analytical services. A Blockchain-based security mechanism (part of IoTCrawler) enables data owners to grant access to certain analytical services the similar way as a smart phone user grants access to certain mobile apps. Analytical service developers are considered responsible for managing their processing infrastructure and federating the processing results back to IoTCrawler. The core of IoTCrawler consists of the NGSI-LD standard together with a number of semantic ontologies, which makes data and metadata models more structured and understandable by independent service developers, which opens a potential for service compositions. As a result, raw data owners (smart-home users) will be authorised to access the intermediate (if needed) and final processing results calculated out of their data.

Encouraged by these findings, we further developed crawling and semantic annotation mechanisms to reduce time and effort when integrating smart home and other IoT and stream data into IoTCrawler. As IoTCrawler provides a common, semantic abstraction for finding and accessing the respective data streams, it becomes much easier to develop smart home applications. Consequently, we developed the "What's happening at home" prototype that is fully implemented on top of the IoTCrawler infrastructure and interacts with the Orchestrator, Search Enabler, Ranking and Security components. The application detects users' activities based on the energy consumption of appliances attached to smart plugs. Activities are modelled in terms of Home Activity ontology (http://sensormeasurement.appspot.com/ont/home/homeActivity accessed on 24 February 2021), which is partly described in one of the GraphQL schemas (https://github.com/IoTCrawler/Search-Enabler/blob/master/src/resources/schemas/homeActivity.graphqls accessed on 24 February 2021) used by the Search Enabler. The schema allows applications to filter households by type or location of detected activities (considering privacy policies). The developed application prototype demonstrates the separation between functionality and benefits from the granularity of the IoTCrawler data model by dealing with sensors and their streams.

6.2. Smart Parking—Security and Privacy Focus

Finding a free parking spot can be very cumbersome in populated cities with the collateral effects of having more vehicles circulating in the city, such as the increase in noise and pollution. In IoTCrawler, we provide a solution to alleviate this problem by offering a parking recommendation service, which allows the user to define the destination, time of arrival and the affordable walking distance. This solution takes advantage of IoTCrawler by gaining a way of representing the information homogeneously, allowing the new information to be introduced without any modification to our solution. More specifically, this solution uses Indexing and Ranking components to retrieve an ordered list of parking sites and parking meter information. Additionally, we allowed the data providers to specify different access policies, as an exercise for proving the security capabilities of our IoTCrawler platform, which the latter will affect the consumers in terms of the visibility of the information depending on the consumer's attributes.

Evaluation: The SmartParking Most Valuable Product (MVP) is being tested in the City of Murcia, in the south-east of Spain. Previous to this solution, the City of Murcia had devoted efforts in research and development based on IoTCrawler, in order to incorporate and integrate promising solutions that would undertake the different challenges with respect to working with data from competing parking providers and regulated parking zones. The previous system was inspired by the participation in the CPAAS.IO project by the University of Murcia, where a solution for parking was devised, using technology derived from the FIWARE ecosystem: FogFlow. The parking solution based on FogFlow, utilised small "edge" devices that were to be installed in different parking locations, charged with the task of gathering data and performing local computations (such as aggregation or availability evaluation). This way, the system leveraged edge computing to enable quick and efficient data transfer, while relying on cloud resources for the heavy-lifting and edge workload-management centralisation. This solution already involved the use of NGSI interfaces for data access, which later on eased the transition to the next iteration, based on IoTCrawler. Some of the difficulties faced by the FogFlow approach were caused by some locations that already had online systems in place. They had special interfaces and connectors, which had to be developed in order to adapt the information and make it available to the rest of the system. In some cases, security and privacy were an issue, as providers wanted to be in control of what and was shared when with the system, and furthermore, how that information was to be accessed later by different parking solutions.

Those gaps have been successfully addressed by the IoTCrawler architecture, which provides a better and broader fit to the parking scenario, by introducing security through fine-detail policies that allow us to define how and whom is allowed to access or produce data. It also considers different ways of which data are to be incorporated into the system, be it directly from NGSI-LD enabled devices connecting to the parking system, through adaptation of other devices or even integrating entire existing systems through connectors and gateways. SmartParking leverages this security, providing a way to discriminate which end-users can access certain information. This way, a user could have permission to access specific parking alternatives. Although in our current implementation this functionality is only utilised by two fictitious users "Juan", who has access to private parking, and "Pedro", who has access to both parking and regulated parking zones. This functionality will allow us to introduce special user roles, such as medical professionals, who would have additional access to parking information for special private parking lots close to their hospital, or city officials that would have access to parking in official buildings, students having access to parking information in the city campus, etc. Furthermore, the security components of the platform would easily allow to define other flows of information coming from the end-users themselves, beyond the classical star ratings. This could mean the ability of claiming parking spaces, updating parking availability in zones with no (or poor) sensory information and it even opens up for future social/collaborative parking solutions, in which end-users can temporarily offer others their domestic parking lot while at work.

SmartParking, through IoTCrawler, copes with the diversity of data existing in the system, by using semantic technologies, such as those found in the semantic web. The extended usage in IoTCrawler of the NGSI-LD standard both for APIs and data modelling, allows the precise representation of information coming from different parking providers and allows for successful searches over highly diverse data. In a similar way to the previous FogFlow solution, which had a local scalability strategy based on the usage of edge devices as part of a distributed system, the IoTCrawler solution allows for the distribution of information through distributed MDRs, but it also provides a federation strategy that allows for broader and more diverse architectures, in which existing parking platforms can be integrated into IoTCrawler's framework, enabling the federation with other parking systems. This federation capability, paired with the Indexing and Ranking components of IoTCrawler, as well as security components, allows for scaling beyond the local city to upper tiers, such as regional or national levels.

Finally, IoTCrawler integrates monitoring, fault-detection and fault-recovery mechanisms, providing useful data regarding the availability and reliability of the parking information contained in the system that can be directly used as part of the parking recommendation system with no further development needed. In short, the IoTCrawler approach for the SmartParking solution in Murcia, by far outperforms (feature-wise) the previous solution based in FogFlow, by accounting for the security aspect of data access, the diversity of data and the integration of existing solutions while allowing for greater scalability and flexibility to adapt and adopt new strategies and ideas, making it, in a way, future-proof.

7. Conclusions and Future Work

This paper presents the IoT search framework IoTCrawler, which allows for the search of data sources in the IoT. It features a domain-independent and layered design and provides solutions for crawling, indexing and searching of IoT data sources. Key enablers supporting the search process ensure privacy and security, scalability and reliability.

We started out the paper by presenting, several issues regarding an IoT search framework listed and analysed to build the basis for our requirements. These requirements have been successfully addressed by the IoTCrawler framework and its components. The loosely coupled components allow for different instantiations of the framework without blocking the search process. The scalability of the discovery and search enablers has also been evaluated to fulfil requirement R-1. With the adaptation of well-known ontologies and standards, an information model has been created to ensure a reliable basis for semantic annotation and context provision. This and the integration of standardised query interfaces enables the framework to be used for machine-initiated search queries R-2.

Requirement R-3 is addressed by designing the framework in a layered approach, which allows the discovery layer to work independently from the search layer. This enables crawling and discovery of new data sources, constantly semantically enriching and monitoring the data sources as well as building indices to speed up incoming search requests. In addition, it makes it possible to include existing solutions, it offers interoperability and overcomes data fragmentation and heterogeneity. As data sources in the IoT are often of private or restricted nature, security and privacy have to be considered R-4. Through the integration of an extensive security and privacy component, from design time on into the architecture of the framework, this requirement is successfully addressed.

To showcase the capabilities and applicability of IoTCrawler, two real-world instantations in different domains have been realised, featuring the search process in a smart home environment and the search in a Smart City use case. In future work, it is planned to enrol the IoTCrawler framework to further use cases covering other domains. This will bring "real" results and present how the framework could increase the benefits gained by the IoT.

Author Contributions: Conceptualization, T.I. and M.F.; methodology, T.I., T.E., J.X.P., H.T., J.A.M., P.G.-G. and P.S. (Pavel Smirnov); software, T.I., E.B.I., M.F., P.G.-G., A.G.-V., T.E., P.S. (Pavel Smirnov), J.A.M., S.B., A.F., N.P. and R.R.; validation, M.J.B., P.S. (Parwinder Singh), A.F., R.R. and N.P.; formal analysis, T.I., E.B.I., A.G.-V., T.E., R.R. and N.P.; investigation, E.B.I., R.R., A.F., H.T., A.G.-V., J.A.M. and P.S. (Pavel Smirnov); resources and data curation, M.K. and S.H.C.; writing—original draft preparation, T.I., E.B.I., M.F., T.E., J.X.P., P.S. (Patrik Schneider), H.T., A.G.-V., P.S. (Parwinder Singh), M.J.B., J.A.M., P.G.-G. and P.S. (Pavel Smirnov); writing—review and editing, R.T. and M.S.; visualization, T.I., E.B.I., P.G.-G. A.G.-V., J.A.M., H.T. P.S. (Pavel Smirnov), J.X.P., T.E.; supervision, M.S. and M.P.; project administration, A.F.S.; funding acquisition, A.F.S. and R.T. All authors have read and agreed to the published version of the manuscript.

Funding: This work has been funded by the EU Horizon 2020 Research and Innovation program through the IoTCrawler project under grant agreement number 779852.

Institutional Review Board Statement: Not applicable.

Informed Consent Statement: Not applicable.

Data Availability Statement: No new data were created in this study. Data sharing is not applicable to this article. Where available, source of data has been referenced in text.

Conflicts of Interest: The authors declare no conflict of interest.

References

1. Ostermaier, B.; Römer, K.; Mattern, F.; Fahrmair, M.; Kellerer, W. A real-time search engine for the web of things. In Proceedings of the 2010 Internet of Things (IOT), Tokyo, Japan, 29 November–1 December 2010; pp. 1–8.
2. Mayer, S.; Guinard, D. An extensible discovery service for smart things. In Proceedings of the Second International Workshop on Web of Things, San Francisco, CA, USA, 16 June 2011; pp. 1–6.

3. Le-Phuoc, D.; Quoc, H.N.M.; Parreira, J.X.; Hauswirth, M. The linked sensor middleware–connecting the real world and the semantic web. *Proc. Semant. Web Chall.* **2011**, *152*, 22–23.
4. Le-Phuoc, D.; Dao-Tran, M.; Parreira, J.X.; Hauswirth, M. A native and adaptive approach for unified processing of linked streams and linked data. In *International Semantic Web Conference*; Springer: Berlin/Heidelberg, Germany, 2011; pp. 370–388.
5. Kamilaris, A.; Yumusak, S.; Ali, M.I. WOTS2E: A search engine for a Semantic Web of Things. In Proceedings of the 2016 IEEE 3rd World Forum on Internet of Things (WF-IoT), Reston, VA, USA, 12–14 December 2016; pp. 436–441.
6. Tran, N.K.; Sheng, Q.Z.; Babar, M.A.; Yao, L.; Zhang, W.E.; Dustdar, S. Internet of Things search engine. *Commun. ACM* **2019**, *62*, 66–73. [CrossRef]
7. Smirnov, P.; Strohbach, M.; Schneider, P.; Gonzalez Gil, P.; Skarmeta, A.F.; Elsaleh, T.; Gonzalez, A.; Rezvani, R.; Truong, H. D5.2 Enablers for Machine Initiated Semantic IoT Search. *IoTCrawler* **2020**. Available online: https://iotcrawler.eu/index.php/project/d5-2-enablers-for-machine-initiated-semantic-iot-search/ (accessed on 24 February 2021).
8. Strohbach, M.; Saavedra, L.A.; Smirnov, P.; Legostaieva, S. Smart Home Crawler: Towards a framework for semi-automatic IoT sensor integration. In Proceedings of the 2019 Global IoT Summit (GIoTS), Aarhus, Denmark, 17–21 June 2019; pp. 1–6.
9. Bonino, D.; Corno, F. Dogont-ontology modeling for intelligent domotic environments. In *International Semantic Web Conference*; Springer: Berlin/Heidelberg, Germany, 2008; pp. 790–803.
10. Compton, M.; Barnaghi, P.; Bermudez, L.; García-Castro, R.; Corcho, Ó.; Cox, S.; Graybeal, J.; Hauswirth, M.; Henson, C.; Herzog, A.; et al. The SSN ontology of the W3C semantic sensor network incubator group. *J. Web Semant.* **2012**, *17*, 25–32. [CrossRef]
11. Janowicz, K.; Haller, A.; Cox, S.J.; Le Phuoc, D.; Lefrançois, M. SOSA: A lightweight ontology for sensors, observations, samples, and actuators. *J. Web Semant.* **2019**, *56*, 1–10. [CrossRef]
12. Bermúdez-Edo, M.; Elsaleh, T.; Barnaghi, P.; Taylor, K. IoT-Lite: A lightweight semantic model for the internet of things and its use with dynamic semantics. *Pers. Ubiquitous Comput.* **2016**, *21*, 475–487. [CrossRef]
13. Kolozali, S.; Bermudez-Edo, M.; Puschmann, D.; Ganz, F.; Barnaghi, P. A knowledge-based approach for real-time iot data stream annotation and processing. In Proceedings of the 2014 IEEE International Conference on Internet of Things (iThings), and IEEE Green Computing and Communications (GreenCom) and IEEE Cyber, Physical and Social Computing (CPSCom), Taipei, Taiwan, 1–3 September 2014; pp. 215–222.
14. Elsaleh, T.; Enshaeifar, S.; Rezvani, R.; Acton, S.T.; Janeiko, V.; Bermudez-Edo, M. IoT-Stream: A Lightweight Ontology for Internet of Things Data Streams and Its Use with Data Analytics and Event Detection Services. *Sensors* **2020**, *20*, 953. [CrossRef] [PubMed]
15. Abomhara, M.; Køien, G.M. Security and privacy in the Internet of Things: Current status and open issues. In Proceedings of the 2014 International Conference on Privacy and Security in Mobile Systems (PRISMS), Aalborg, Denmark, 11–14 May 2014; pp. 1–8.
16. Riahi, A.; Challal, Y.; Natalizio, E.; Chtourou, Z.; Bouabdallah, A. A systemic approach for IoT security. In Proceedings of the 2013 IEEE International Conference on Distributed Computing in Sensor Systems, Cambridge, MA, USA, 20–23 May 2013; pp. 351–355.
17. Mahalle, P.; Babar, S.; Prasad, N.R.; Prasad, R. Identity management framework towards internet of things (IoT): Roadmap and key challenges. In *International Conference on Network Security and Applications*; Springer: Berlin/Heidelberg, Germany, 2010; pp. 430–439.
18. Bernal Bernabe, J.; Hernandez-Ramos, J.L.; Skarmeta Gomez, A.F. Holistic privacy-preserving identity management system for the internet of things. *Mob. Inf. Syst.* **2017**, *2017*. [CrossRef]
19. Mazzoleni, P.; Crispo, B.; Sivasubramanian, S.; Bertino, E. XACML policy integration algorithms. *ACM Trans. Inf. Syst. Secur. (TISSEC)* **2008**, *11*, 1–29. [CrossRef]
20. Hernández-Ramos, J.L.; Jara, A.J.; Marin, L.; Skarmeta, A.F. Distributed capability-based access control for the internet of things. *J. Internet Serv. Inf. Secur. (JISIS)* **2013**, *3*, 1–16.
21. Pérez, S.; Rotondi, D.; Pedone, D.; Straniero, L.; Núñez, M.J.; Gigante, F. Towards the CP-ABE application for privacy-preserving secure data sharing in IoT contexts. In *International Conference on Innovative Mobile and Internet Services in Ubiquitous Computing*; Springer: Berlin/Heidelberg, Germany, 2017; pp. 917–926.
22. Hwang, Y.H. Iot security & privacy: Threats and challenges. In *Proceedings of the 1st ACM Workshop on IoT Privacy, Trust, and Security*; Association for Computing Machinery: New York, NY, USA, 2015; doi: 10.1145/2732209.2732216. [CrossRef]
23. Garcia, N.; Alcaniz, T.; González-Vidal, A.; Bernabe, J.B.; Rivera, D.; Skarmeta, A. Distributed real-time SlowDoS attacks detection over encrypted traffic using Artificial Intelligence. *J. Netw. Comput. Appl.* **2021**, *173*, 102871. [CrossRef]
24. Hernandez-Ramos, J.L.; Martinez, J.A.; Savarino, V.; Angelini, M.; Napolitano, V.; Skarmeta, A.; Baldini, G. Security and Privacy in Internet of Things-Enabled Smart Cities: Challenges and Future Directions. *IEEE Secur. Priv.* **2020**. [CrossRef]
25. Hernandez-Ramos, J.L.; Geneiatakis, D.; Kounelis, I.; Steri, G.; Fovino, I.N. Toward a Data-Driven Society: A Technological Perspective on the Development of Cybersecurity and Data-Protection Policies. *IEEE Secur. Priv.* **2019**, *18*, 28–38. [CrossRef]
26. Juran, J.; Godfrey, A.B. *Quality Handbook*, 5th ed.; McGraw-Hill: Irwin, NY, USA, 1999; Volume 173.
27. Wang, R.Y.; Strong, D.M.; Guarascio, L.M. Beyond accuracy: What data quality means to data consumers. *J. Manag. Inf. Syst.* **1996**, *12*, 5–33. [CrossRef]
28. Mendes, P.N.; Mühleisen, H.; Bizer, C. Sieve: Linked data quality assessment and fusion. In *Proceedings of the 2012 Joint EDBT/ICDT Workshops*; Association for Computing Machinery: New York, NY, USA, 2012; pp. 116–123.

29. Sicari, S.; Rizzardi, A.; Grieco, L.; Piro, G.; Coen-Porisini, A. A policy enforcement framework for Internet of Things applications in the smart health. *Smart Health* **2017**, *3*, 39–74. [CrossRef]
30. Klein, A.; Do, H.H.; Hackenbroich, G.; Karnstedt, M.; Lehner, W. Representing data quality for streaming and static data. In Proceedings of the 2007 IEEE 23rd International Conference on Data Engineering Workshop, Istanbul, Turkey, 17–20 April 2007; pp. 3–10.
31. Kothari, A.; Boddula, V.; Ramaswamy, L.; Abolhassani, N. DQS-Cloud: A Data Quality-Aware autonomic cloud for sensor services. In Proceedings of the 10th IEEE International Conference on Collaborative Computing: Networking, Applications and Worksharing, Miami, FL, USA, 22–25 October 2014; pp. 295–303.
32. Puiu, D.; Barnaghi, P.; Toenjes, R.; Kümper, D.; Ali, M.I.; Mileo, A.; Parreira, J.X.; Fischer, M.; Kolozali, S.; Farajidavar, N.; et al. Citypulse: Large scale data analytics framework for smart cities. *IEEE Access* **2016**, *4*, 1086–1108. [CrossRef]
33. Iggena, T.; Fischer, M.; Kuemper, D. Quality Ontology. Available online: http://purl.oclc.org/NET/UASO/qoi (accessed on 8 January 2021).
34. Norris, M.; Celik, B.; Venkatesh, P.; Zhao, S.; McDaniel, P.; Sivasubramaniam, A.; Tan, G. IoTRepair: Systematically Addressing Device Faults in Commodity IoT. In Proceedings of the 2020 IEEE/ACM Fifth International Conference on Internet-of-Things Design and Implementation (IoTDI), Sydney, NSW, Australia, 21–24 April 2020; pp. 142–148.
35. Power, A.; Kotonya, G. A Microservices Architecture for Reactive and Proactive Fault Tolerance in IoT Systems. In Proceedings of the 2018 IEEE 19th International Symposium on "A World of Wireless, Mobile and Multimedia Networks" (WoWMoM), Chania, Greece, 12–15 June 2018; pp. 588–599.
36. Izonin, I.; Kryvinska, N.; Tkachenko, R.; Zub, K. An approach towards missing data recovery within IoT smart system. *Procedia Comput. Sci.* **2019**, *155*, 11–18. [CrossRef]
37. Liu, Y.; Dillon, T.; Yu, W.; Rahayu, W.; Mostafa, F. Missing Value Imputation for Industrial IoT Sensor Data With Large Gaps. *IEEE Internet Things J.* **2020**, *7*, 6855–6867. [CrossRef]
38. Al-Milli, N.; Almobaideen, W. Hybrid Neural Network to Impute Missing Data for IoT Applications. In Proceedings of the 2019 IEEE Jordan International Joint Conference on Electrical Engineering and Information Technology (JEEIT), Amman, Jordan, 9–11 April 2019; pp. 121–125.
39. Fathy, Y.; Barnaghi, P.; Tafazolli, R. Large-scale indexing, discovery, and ranking for the Internet of Things (IoT). *ACM Comput. Surv. (CSUR)* **2018**, *51*, 1–53. [CrossRef]
40. Zhou, Y.; De, S.; Wang, W.; Moessner, K. Enabling query of frequently updated data from mobile sensing sources. In Proceedings of the 2014 IEEE 17th International Conference on Computational Science and Engineering, Chengdu, China, 19–21 December 2014; pp. 946–952.
41. Barnaghi, P.; Wang, W.; Dong, L.; Wang, C. A Linked-Data Model for Semantic Sensor Streams. In Proceedings of the IEEE International Conference on Green Computing and Communications and IEEE Internet of Things and IEEE Cyber, Physical and Social Computing, Beijing, China, 20–23 August 2013; pp. 468–475. [CrossRef]
42. Fathy, Y.; Barnaghi, P.; Enshaeifar, S.; Tafazolli, R. A distributed in-network indexing mechanism for the Internet of Things. In Proceedings of the IEEE 3rd World Forum on Internet of Things (WF-IoT), Reston, VA, USA, 12–14 December 2016; pp. 585–590. [CrossRef]
43. Camerra, A.; Palpanas, T.; Shieh, J.; Keogh, E. iSAX 2.0: Indexing and mining one billion time series. In Proceedings of the 2010 IEEE International Conference on Data Mining, Sydney, NSW, Australia, 13–17 December 2010; pp. 58–67.
44. Zoumpatianos, K.; Idreos, S.; Palpanas, T. Indexing for Interactive Exploration of Big Data Series. In *Proceedings of the 2014 ACM SIGMOD International Conference on Management of Data*; Association for Computing Machinery: New York, NY, USA, 2014; pp. 1555–1566. [CrossRef]
45. Ganz, F.; Barnaghi, P.; Carrez, F. Information Abstraction for Heterogeneous Real World Internet Data. *IEEE Sens. J.* **2013**, *13*, 3793–3805. [CrossRef]
46. Gonzalez-Vidal, A.; Barnaghi, P.; Skarmeta, A.F. Beats: Blocks of eigenvalues algorithm for time series segmentation. *IEEE Trans. Knowl. Data Eng.* **2018**, *30*, 2051–2064. [CrossRef]
47. Brin, S.; Page, L. Reprint of: The Anatomy of a Large-Scale Hypertextual Web Search Engine. *Comput. Netw.* **2012**, *56*, 3825–3833. [CrossRef]
48. Guinard, D.; Trifa, V.; Karnouskos, S.; Spiess, P.; Savio, D. Interacting with the SOA-Based Internet of Things: Discovery, Query, Selection, and On-Demand Provisioning of Web Services. *IEEE Trans. Serv. Comput.* **2010**, *3*, 223–235. [CrossRef]
49. Yuen, K.K.F.; Wang, W. Towards a ranking approach for sensor services using primitive cognitive network process. In Proceedings of the 4th Annual IEEE International Conference on Cyber Technology in Automation, Control and Intelligent, Hong Kong, China, 4–7 June 2014; pp. 344–348. [CrossRef]
50. Niu, W.; Lei, J.; Tong, E.; Li, G.; Chang, L.; Shi, Z.; Ci, S. Context-Aware Service Ranking in Wireless Sensor Networks. *J. Netw. Syst. Manag.* **2014**, *22*, 50–74. [CrossRef]
51. Xu, Z.; Martin, P.; Powley, W.; Zulkernine, F. Reputation-Enhanced QoS-based Web Services Discovery. In Proceedings of the IEEE International Conference on Web Services (ICWS 2007), Salt Lake City, UT, USA, 9–13 July 2007; pp. 249–256. [CrossRef]
52. Noy, N.F.; McGuiness, D.L. *Ontology Development 101: A Guide to Creating Your First Ontology*; Stanford Knowledge Systems Laboratory Technical Report KSL-01-05 and Stanford Medical Informatics Technical Report SMI-2001-0880; Standford University: Standford, CA, USA, 2001.

53. Iggena, T.; Kuemper, D. Quality Ontology for IoT Data Sources. Available online: https://w3id.org/iot/qoi (accessed on 8 January 2021).
54. Bees, D.; Frost, L.; Bauer, M.; Fisher, M.; Li, W. *NGSI-LD API: For Context Information Management*; ETSI White Paper No. 31; ETSI: Valbonne, France, 2019.
55. Taylor, S.; Letham, B. *Prophet: Forecasting at Scale*; Facebook Research: Menlo Park, CA, USA, 2018.
56. González-Vidal, A.; Rathore, P.; Rao, A.S.; Mendoza-Bernal, J.; Palaniswami, M.; Skarmeta-Gómez, A.F. Missing Data Imputation with Bayesian Maximum Entropy for Internet of Things Applications. *IEEE Internet Things J.* **2020**. [CrossRef]
57. Christakos, G.; Li, X. Bayesian maximum entropy analysis and mapping: A farewell to kriging estimators? *Math. Geol.* **1998**, *30*, 435–462. [CrossRef]
58. Ilyas, E.B.; Fischer, M.; Iggena, T.; Tönjes, R. Virtual Sensor Creation to Replace Faulty Sensors Using Automated Machine Learning Techniques. In Proceedings of the 2020 Global Internet of Things Summit (GIoTS), Dublin, Ireland, 3 June 2020; pp. 1–6.
59. Kuemper, D.; Iggena, T.; Toenjes, R.; Pulvermueller, E. Valid.IoT: A framework for sensor data quality analysis and interpolation. In Proceedings of the 9th ACM Multimedia Systems Conference, Amsterdam, The Netherlands, 12–15 June 2018; Association for Computing Machinery: New York, NY, USA, 2018; pp. 294–303.
60. Iggena, T.; Bin Ilyas, E.; Tönjes, R. Quality of Information for IoT-Frameworks. In Proceedings of the 2020 IEEE International Smart Cities Conference (ISC2), Piscataway, NJ, USA, 28 September–1 October 2020; pp. 1–8.
61. González-Vidal, A.; Alcañiz, T.; Iggena, T.; Ilyas, E.B.; Skarmeta, A.F. Domain Agnostic Quality of Information Metrics in IoT-Based Smart Environments. In *Intelligent Environments 2020: Workshop Proceedings of the 16th International Conference on Intelligent Environments, Madrid, Spain*; IOS Press: Amsterdam, The Netherlands, 2020; Volume 28, p. 343.
62. Rezvani, R.; Enshaeifar, S.; Barnaghi, P. Lagrangian-based Pattern Extraction for Edge Computing in the Internet of Things. In Proceedings of the 5th IEEE International Conference on Edge Computing and Scalable Cloud, Paris, France, 21–23 June 2019.
63. Rezvani, R.; Barnaghi, P.; Enshaeifar, S. A New Pattern Representation Method for Time-series Data. *IEEE Trans. Knowl. Data Eng.* **2019**. [CrossRef]
64. Chen, Y.; Keogh, E.; Hu, B.; Begum, N.; Bagnall, A.; Mueen, A.; Batista, G. The UCR Time Series Classification Archive. 2015. Available online: www.cs.ucr.edu/~eamonn/time_series_data/ (accessed on 2 January 2021).
65. Janeiko, V.; Rezvani, R.; Pourshahrokhi, N.; Enshaeifar, S.; Krogbæk, M.; Christophersen, S.; Elsaleh, T.; Barnaghi, P. Enabling Context-Aware Search using Extracted Insights from IoT Data Streams. In *2020 Global Internet of Things Summit (GIoTS), Dublin, Ireland*; IEEE: New York, NY, USA, 2020; pp. 1–6.

sensors

MDPI

Article

Energy Management Expert Assistant, a New Concept

Matias Linan-Reyes *, Joaquin Garrido-Zafra, Aurora Gil-de-Castro and Antonio Moreno-Munoz

Departamento de Ingeniería Electrónica y de Computadores, Escuela Politécnica Superior de Córdoba, Campus de Rabanales, Universidad de Córdoba, Edificio Leonardo Da Vinci, E-14071 Cordoba, Spain; p22gazaj@uco.es (J.G.-Z.); agil@uco.es (A.G.-d.-C.); amoreno@uco.es (A.M.-M.)
* Correspondence: matias@uco.es

Abstract: In recent years, interest in home energy management systems (HEMS) has grown significantly, as well as the development of Voice Assistants that substantially increase home comfort. This paper presents a novel merging of HEMS with the Assistant paradigm. The combination of both concepts has allowed the creation of a high-performance and easy-to-manage expert system (ES). It has been developed in a framework that includes, on the one hand, the efficient energy management functionality boosted with an Internet of Things (IoT) platform, where artificial intelligence (AI) and big data treatment are blended, and on the other hand, an assistant that interacts both with the user and with the HEMS itself. The creation of this ES has made it possible to optimize consumption levels, improve security, efficiency, comfort, and user experience, as well as home security (presence simulation or security against intruders), automate processes, optimize resources, and provide relevant information to the user facilitating decision making, all based on a multi-objective optimization (MOP) problem model. This paper presents both the scheme and the results obtained, the synergies generated, and the conclusions that can be drawn after 24 months of operation.

Keywords: home energy management systems (HEMS); Internet of Things (IoT); artificial intelligence (AI); Voice Assistant; machine learning (ML); big data

Citation: Linan-Reyes, M.; Garrido-Zafra, J.; Gil-de-Castro, A.; Moreno-Munoz, A. Energy Management Expert Assistant, a New Concept. *Sensors* **2021**, *21*, 5915. https://doi.org/10.3390/s21175915

Academic Editor: Geoff Merrett

Received: 27 July 2021
Accepted: 27 August 2021
Published: 2 September 2021

1. Introduction

The constant advancement of ICT opens up great opportunities to improve systems' functionality, performance, and efficiency. Technologies such as IoT, Big Data, AI, WIFI 6, or 5G could come together to enhance the capabilities of equally emerging systems, oriented for use in the home, such as HEMS and Voice Assistants.

The residential sector is a key element in the context of both energy savings and people's well-being. The restructuring of the energy sector through Smart Grids and Microgrids [1], as well as the arrival of the 5G network [2–4] and WIFI 6, will allow an exponential development of connected devices and appliances, opening the doors of the network to the IoT both in the industrial sector and in the home [5].

Bringing all these elements together in the home environment can be challenging, but they form the fundamental structure of the Home Energy Management Expert Assistant (HERMES) system. However, it can offer us a revolution that can positively impact climate change, energy efficiency, or quality of life. Each of these elements separately already offers solutions to specific problems. The literature shows considerable evidence of this use and its benefits, as shown below, focusing on the most relevant, emphasizing the latest efforts and advances in applying the methodologies.

Home Energy Management Systems (HEMS)

HEMS are hardware and software systems that enable advanced control of energy-using systems and devices in the home, continuously analyzing data to provide real-time information on the energy performance of the home [6], creating data streams (both external and internal such as weather, electricity price, or sensors) and making decisions for energy efficiency improvement [7], peak demand management and demand response, so their specifications include the necessary integration (monitoring and control) and communication with all smart home devices, sensors, relays, and appliances regardless of their communication protocols [8]. A proper implementation would allow a reduction of about 35% of the total electricity bill, prioritizing load consumption based on the cost of energy [9]. Other studies, such as the ACEEE study not focused exclusively on energy cost, set this saving at a maximum of 17% [7].

Nevertheless, in practice, these systems have some limitations, mainly including interoperability between devices, lack of training of the users themselves, doubts about security or limitations of commitment to the customer, as well as the lack of studies showing the real possibilities of savings [5,7,10,11], as well as lacking true intelligence and the ability to manage demand peaks and demand response. To develop an efficient HEMS, it is necessary to know the characteristics and requirements of each of the technologies that will allow a complete communication and configuration of all the devices [12–15]. One solution to interoperability would be implementing a widely consolidated Building Management System (BMS) [16]. However, they are very closed systems, mostly proprietary solutions that can only be upgraded by the system manufacturer with relatively high costs, ranging from $25 to $70 per square meter of housing [17]. On the market, we can find a wide offer of both Open-Source and Proprietary HEMS [8,18]: Open-Source Home Assistant [19] and OpenHAB (based on Eclipse SmartHome™) [20,21] that have a large number of protocols and configurable devices, BEMOSS [22] built-in Python on VOLTTRON [23], ioBroker [24], Open Energy Management (OGEMA) [25,26], and Open remote [27], among others.

In this paper, we will consider the following architecture of an advanced HEMS system (Figure 1):

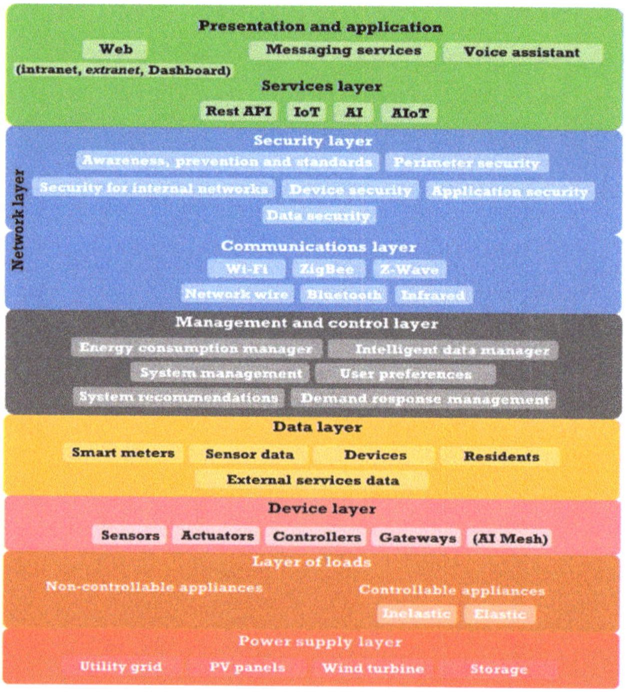

Figure 1. The generic architecture of a HEMS system.

We continue in the next section by detailing the main elements that will make up our HERMES system.

2. Materials and Methods

In addition to the energy management system described in the introduction, the HERMES system is developed integrating the following elements:

2.1. IoT: Smart Devices

Depending on the context, there have been many definitions of IoT [28–31]. A good approximation to its current definition could be: "System of devices, machines, or everyday objects provided with unique identifiers (UID) with the ability to transfer data through a network without the need for interaction between people or between people and computers".

In recent years the number of Internet-connected devices has exploded, to such an extent that forecasts have become outdated [32]. There is currently no consensus on their number, offering figures that range from CISCO's 50 billion [33] to Intel's 200 billion [34]. In parallel to this growth, their price has been reduced by more than 90%, coining the concept of "democracy of devices" [35].

Not only devices or things but also household appliances are beginning to join the world of elements connected to the network, and its growth will be almost total in the coming years with the arrival of 5G and the IPv6 protocol without address exhaustion problems (2^{128} addresses).

Therefore, IoT is a significant challenge for HEMS as it has to interact with a large number of "smart devices" with a wide variety of protocols, in addition to the large amount of data they will generate. In this regard, some emerging technologies can help address this challenge: big data, cloud computing, and AI, as also postulated in [17,36].

2.2. Big Data

Traditionally big data refers to the concept of an amount of data that exceeds the capacity of conventional software to be captured, managed, and processed in a reasonable amount of time. Nowadays, the concept extends to analyzing user behavior, extracting value from stored data, and formulating predictions through the patterns observed. A first approximation to this definition was given in 2012 by [37].

Big data uses the following characteristics described by the three V's: volume, variety, and velocity, and several other characteristics including veracity, value, and the identification of nonlinear systems (from large data sets) to reveal relationships or to make predictions of outcomes and behaviors [38–40].

From the large amount of data generated in a smart home, whether internal (through the IoT network integrated into the home) or external (such as weather or electricity prices), it would be interesting to improve energy efficiency, to analyze these data and extract all the relevant information they can provide to the system, so the use of big data can be an essential tool, providing great value for the optimization of home resources and user comfort. However, the storage, processing, and analysis of this large volume of continuously generated data, while maintaining their security and privacy, is a significant challenge for HEMS.

To this end, HERMES uses various strategies to process volumes of data, store the information periodically and in real-time, and process it to obtain an analysis and projection of the data to trigger specific automated actions without user intervention. It also offers information that is provided to the user through the Expert Assistant to guide decision making or as information on predictions or patterns detected through machine learning (ML).

2.3. Cloud Computing

Cloud computing is the resources and services of the computer system accessed through a network without direct active management by the user. Initially, the services were focused on data storage and computing power, but, today, the user's services and systems cover virtually any need.

Cloud computing offers advantages and disadvantages that must be assessed by the user when implementing or not implementing these systems. From the point of view of HEMS, we can highlight the following advantages:

- Reduction of costs and implementation times;
- Reduction of scalability problems in cases where the system must grow;
- The user can focus on the system's functionality and not on the technical aspects of the infrastructure;
- Access to services from anywhere;
- System portability and protection against data loss. If the local system suffers damage or failure, the data or services in the cloud remain secure and loss-free;
- Transparent updates for the user, as long as the vendor maintains this commitment and the local system is not affected by version incompatibilities;
- Software installation is avoided or reduced;
- Local system requirements in terms of computational capacity are reduced. By deriving computing services and processes to the cloud, a lighter hardware system is required. This, in turn, leads to a benefit due to reducing local consumption by requiring equipment with lower performance;
- Security is often a critical factor in these services as providers can equip their systems with the latest technologies in the face of the limitations faced by a single user, both in terms of technological capacity and knowledge.

On the contrary, some drawbacks can be very critical to the viability of the system:

- Absolute dependence on the commitment or continuity of the service provider: discontinuity or modification of services may critically affect the HEMS system;
- Fixed fee for the use of the services;
- Lifetime dependence on external suppliers;
- Small systems are more vulnerable than more extensive infrastructures, especially concerning:
    - Downtime;
    - Technical interruptions from suppliers, which are unavoidable and can occur at critical moments;
    - More limited bargaining power, leading to limited customization;
- Dependence on external network access versus a HEMS system based on a local network isolated from the Internet;
- Aspects such as security, privacy, or confidentiality may be exposed or compromised.

Although a priori cloud computing is possibly an inevitable tool if we want to develop a truly competitive HEMS, we must be very aware of some of the limitations and implications that its use may entail, so we must adopt hybrid strategies between functional HEMS through local networks isolated from the Internet and HEMS based on cloud computing. Therefore, we are committed to systems that take full advantage of the functionality of the isolated network and to ensuring that the contracted services, which are based on or use the Internet, do not pose a risk or functional disruption to the system.

In this regard, the design of the HERMES system presents a dual functionality with communicating vessels between the own network (partially isolated) and the contracted cloud computing services. In addition, to protect the system's security, various levels of protection have been planned according to its exposure to the Internet. HERMES can maintain operational functionality if, for security reasons, it is decided or required to isolate the system from the Internet.

2.4. *Artificial Intelligence (AI), Expert System (ES), and Machine Learning (ML)*

AI is the ability of a man-made system to interpret and analyze data, learn from that data, and use that new knowledge to perform actions or tasks. This definition is an evolution of the one given by Andreas Kaplan and Michael Haenlein [41]. Another agent-based approach defines it as: "Computational intelligence is the study of the design of intelligent agents. An intelligent agent is a system that acts intelligently: What it does is appropriate for its circumstances and its goal, it is flexible to changing environments and changing goals, it learns from experience, and it makes appropriate choices given perceptual limitations and finite computation" [42]. The definition is not trivial and has evolved over the years to encompass very diverse disciplines with applications in virtually all scientific fields [41–46], such as expert systems that emulate the behavior or responses that a human expert in an area of knowledge would give.

There is no doubt that IA is a fundamental tool in the present and future development of HEMS. IA encompasses a multitude of technologies, some of which are shown in Figure 2:

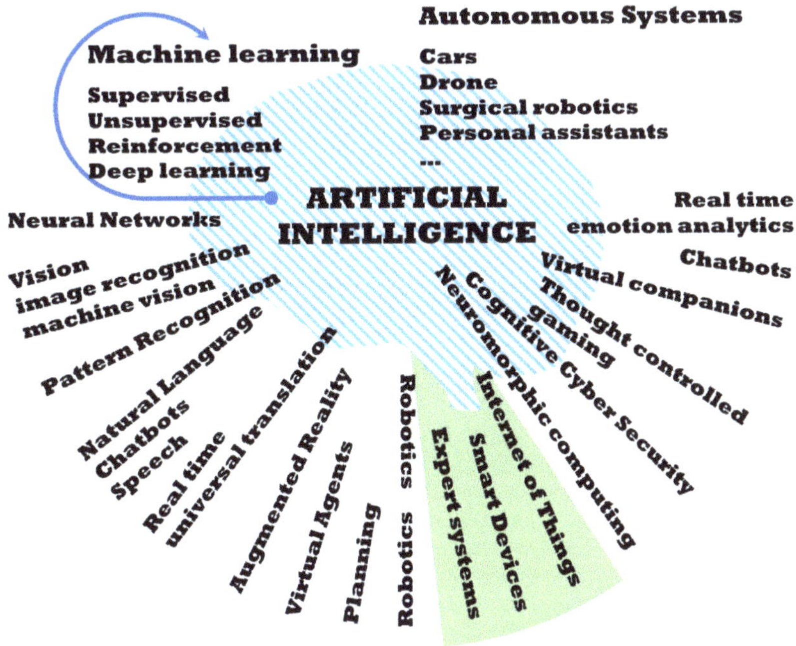

Figure 2. AI Technologies.

The HERMES system has been developed integrating ML, natural language processing, expert systems, and speech, without ruling out other technologies in the future such as image recognition (vision) for more advanced analysis of presence [47–49] with a higher level of personalization of interactions.

2.5. Virtual Assistant

A Virtual Assistant [50] or Voice Assistant is a software agent that can interpret human speech and certain commands and respond with synthesized voice, tasks, or services. Other definitions can be found in M. B. Hoy [51]. The development of natural dialogues between humans and machines is one of the goals of AI [50,52]. Voice assistants are here to stay [53], not only because of their benefits for people with specific needs or older adults [54–57] but also because they have been shown to bring benefits such as social cohesion [58] or improve comfort and allow the user to interact in a very natural way with machines, since speech is the main mode of communication for humans [59]. In our case, the last of the pillars that make up the HERMES system is precisely the assistant but endowed with greater intelligence, as we will indicate below.

From a HEMS perspective, voice assistants have several notable handicaps. First, their intelligence is limited in terms of energy efficiency, as verbal commands and functionality are focused on activating or deactivating devices. However, our HERMES system integrates a bidirectional communication channel to the virtual assistant (Figure 3) both with the system as a whole and with the residents, connecting the intelligence of the system with the user, becoming an "intelligent assistant" beyond the function and intelligence of these systems, complementing the functionality of the HEMS. This dual bidirectional channel represents a qualitative leap in the functionality and interaction of the HEMS system with the users.

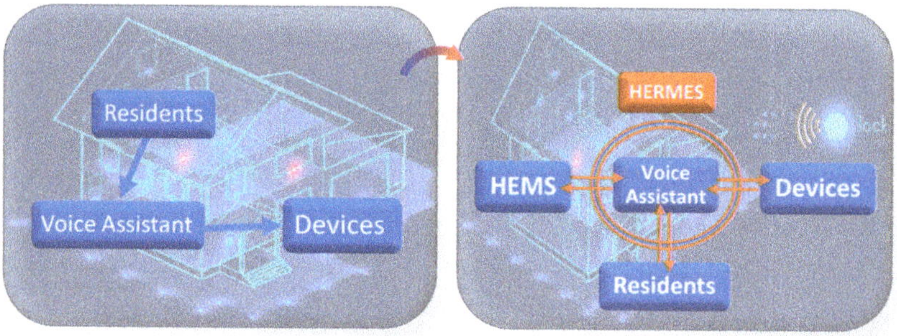

Figure 3. HERMES: Bi-directional dual-channel formed between HEMS-Voice Assistant-Residents. Background: © 123fr.com.

Secondly, another important handicap is the vulnerability presented by these devices; for example, any user can issue verbal commands: "Open the door" or "buy this and send it to such address") [60–62]. In this case, the integration of the voice assistant in the HERMES system is done keeping in mind that this type of vulnerability cannot be fraudulently passed on, the system itself is the one that filters them. In this regard, HERMES detects the presence at the home of all the usual residents and identifies them so that certain commands can only be executed if at least one of them is at home or if the presence simulator has been activated, which can be activated remotely by the residents for a limited time. Other avenues that could be explored to avoid vulnerabilities could be the identification of users by smart cameras or by their voice profile [63,64]. In this way, all commands, or those that we consider critical to the system, can be filtered to prevent a local or network intruder from exploiting them.

2.6. Results from Knowledge

The benefit of applying advanced and complex systems must be realized from the knowledge acquired from the collection of data and the application to balanced models, not forced, that allow the creation of precise and effective forms counting at all times on the users. Otherwise, the system will lack practical application, falling into the dynamics of a good theoretical study without a practical route. Therefore, the system has been developed in different phases, data collection being the first of them, from which practical solutions have been channeled. For this reason, the system has been developed in different phases, the first of which is data collection. Based on the data, practical solutions have been proposed, focusing on economics but adapting to the users, which allows long-term habits to be established, quality of life to be maintained, and ensures that the system is applied as it is beneficial.

From this dynamic of results from knowledge, the integration of all these systems in one (HEMS + IoT + Big Data + Cloud Computing + AI + Voice Assistant = HERMES) has led us to an ES with a multitude of possibilities in the field of energy efficiency and well-being of residents. This work shows the development of a new system with the following objective: "developing a comprehensive model for smart home consumption management assisted by an ES (HERMES)".

The development of this objective was based on the following pillars: results from knowledge, energy savings, usability, user assistance, comfort, privacy, and security. The following lines of analysis and development were proposed:

- Obtaining data to develop the best possible system for the objectives pursued;
- Cloud integration provides the system with scalable computational capacity, access and management of information flows, extension to big data analysis, AI, and ML;
- Usability: "Home" system of interaction with the user that allows triggering complex services from a simple and accessible functionality:
    - Actions: Programming household appliances and devices. Presence detection and habit analysis;
    - Warnings: Advice and recommendations for savings based on detected habits or pre-established patterns;
    - Alerts and maintenance of equipment and appliances;
    - Integration with voice technology.
- Measures to ensure user privacy and system security;
- Interoperability of electronic systems that allow the implementation of the integral model;
- System specifications to enhance efficient energy consumption management under IoT architecture;
- Estimated energy savings and user benefits. Differences between HERMES, HEMS systems, and a non-connected home.

3. Materials and Methods

It is not easy to make accurate predictions of electricity demand, microgeneration, or appliance usage in domestic environments. Factors such as the type of billing (five main energy billing approaches can be found in the literature [65–67]), weather conditions, or assumed habits and routines of users involve in themselves elements of uncertainty that are difficult to predict, so that deviations on forecasts of electricity consumption, microgeneration, or the operational needs of household appliances, can compromise the planning of HEMS. These uncertainties may result in situations where contracted power limits are required to be exceeded with consequent limitations or penalties, or the comfort level of residents may be affected. Therefore, in decision making, the value of past and present data must be prioritized over future data, with dynamic (stochastic) programming approaches [65,68,69].

If we add to this uncertainty the diversity of load types and their different scheduling possibilities, HEMS design strategies can be approached from multiple perspectives [65,68–71]. Before focusing in more depth on our development, we will review some of the discussed aspects to settle and show the fundamentals of the HERMES system presented in this paper.

3.1. Classification of Load Types

There is no consensus on the classification of load types, so we propose a new model that will be useful for our work and is based on several classifications that focus on the characteristics of the loads [65,67], but to which we add the user's decision capability through the wizard or by programming so that some devices can change category based on the user's decision.

Classification of devices or systems according to their load scheduling (Figure 4):

1. Non-controllable loads. Their operation cannot be programmed, changed, or reprogrammed by a HEMS. They usually provide added value, and users control some of them. Televisions, stereos, computers, or appliances such as refrigerators or lighting without control systems fall into this category. Appliance standby would be included in this class;
2. Controllable loads. The HEMS system has some control over them or through the user in a given time horizon. A traditional HEMS system could not control the loads

through the user; in this regard, control is one of the contributions of the HERMES system. In turn, within this category, we can divide the loads into elastic or inelastic;

a. Inelastic. Once initiated or required to operate, it must complete a full cycle;
   i. Uninterruptible loads. Once started, they must run a complete cycle continuously; only the corresponding start time can be programmed. In this category, we can find dishwashers, washing machines, or dryers, among other appliances;
   ii. Interruptible loads. Once started, they can be interrupted but must be reconnected to complete the full cycle. These are usually constant-drain devices. Examples include plug-in hybrid electric vehicles and other rechargeable devices or external batteries, and the electric boiler;
b. Elastic. Loads with the capacity to be able to adjust power consumption in the middle of an operation;
   i. Variable loads with alteration of comfort. Energy consumption can be adjusted in the middle of an operation but leads to loss of comfort and may require subsequent compensation. These are usually systems whose operation is maintained according to a reference defined by the residents, so their temporary variation by the HEMS may affect comfort. Ventilation, heating, or cooling are examples of this category;
   ii. Variable loads without alteration of comfort. Energy consumption can be adjusted in the middle of an operation without significant loss of comfort or subsequent compensation. For example, dimming of artificial lighting by compensating with daylight.

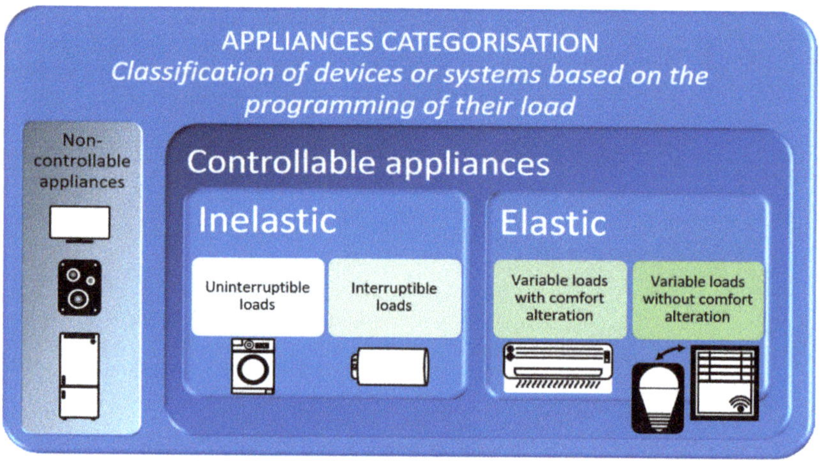

Figure 4. Classification of equipment or systems based on the programming of their load.

In the HERMES system, the above examples of appliances could change category (temporarily or permanently) depending on the user's decision-making. An extreme example could be the refrigerator defined a priori as an uncontrollable load. The user can instruct the HEMS assistant to turn it off for a short period that does not jeopardize food preservation, making it controllable, inelastic, and interruptible. This example can be used to avoid a peak demand as long as there is no other controllable load to bridge the peak demand.

Based on the above structure, Table 1 is a classification of the main household appliances.

This classification is flexible and dynamic since the system can adjust specific parameters according to the characteristics of the residents or according to different scenarios. The system has general and appliance-specific parameters that it can readjust (see Table 2) to adapt to a dynamic classification of appliances. In certain cases, this adjustment is shared by the system and the users, as could be the case for the air conditioning temperature. This behavior thus allows the system to adapt to the characteristics of different user groups and different scenarios (seasons of the year, vacation absences). Users can adjust these parameters within a range and even set the air-conditioning switch-on temperature by voice. The system acts accordingly to maintain comfort but reduce consumption, for example, after a period of operation, raising or lowering the cooling/heating temperature.

3.2. The Preamble of the HERMES System

HERMES system scheduling is performed to manage a present and future time horizon based on past and present data. Both a continuous representation of time and a discretization into the minute, hourly, daily, weekly, and monthly intervals are used. For example, once a month, a heating cycle above 60 °C is completed in the electric boiler to eliminate possible Legionella outbreaks. This scheduling pursues the reduction of the consumption of household appliances and the shifting of loads (shifting to optimize expenditure and their optimal time of operation) to reduce electricity billing [72–74] and maintain or increase the comfort of residents [73,75]. Regarding billing optimization, the appliance scheduling technique based on mathematical optimization is suitable for small-sized problems such as individual dwellings instead of other less demanding techniques for larger problems, as we will discuss later. By contrast, in terms of comfort, the evaluation of resident comfort is a very complex task from a scheduling point of view due to how personal the perception and subjectivity of comfort can be, as well as the inconveniences of having to schedule appliances outside the preferred time window, maintain a certain order (washing machine before dryer) or accept unwanted elastic load modulations. As a step before implementing the HERMES system (as of 27 October 2019), daily usage profiles were recorded over an extended period (from 16 February 2019 to 26 October 2019) to characterize and minimize potential drawbacks that could affect comfort.

Table 1. Classification of the main household appliances according to their load.

Uncontrollable Loads	**Controllable Loads**			
	Inelastic		**Elastic**	
	Uninterruptible Loads	**Interruptible Loads**	**Variable Loads with Alteration of Comfort**	**Variable Loads without Altering Comfort**
Television	Washing Machine	Electric Vehicle	Air Conditioning	Natural + artificial light
Sound equipment	Dishwasher	Phone Charger	Heating System	Automatic opening of windows
Computer	Dryer Machine	Battery/Energy Storage	Fan	
Fridge-freezer	Oven	Water Heater	Stove	
Lighting		Water Pump (Well, Pool)		
Standby		Vacuum Cleaner (robot)		
Microwave				
Vacuum Cleaner				
Iron				
Cooker pot				
Cooker Hood				
Hair dryer				
Toaster, Blender, Kettle				

In addition, given that household demand cannot be predicted with complete accuracy, we rely on a consumption profile characterized by minimizing the elements that introduce a certain degree of uncertainty, reinforced by two-way communication with residents to whom, on the one hand, electricity prices are provided a day in advance, as well as other statistics, and on the other, the system analyzes the use of household appliances and recommends their use based on history, coordination, and the use of appliances. This minimizes the problems of stochastic optimization, and although not all the elements that are a source of uncertainty and their derived problems (consumption peaks with penalties or loss of comfort) are avoided, they are reduced, and with experience, the residents themselves and the system learn and converge towards an increasingly optimal situation in terms of billing and comfort. However, deviations of one from the other are allowed, although both are the ultimate goal, so that the system is constantly evolving around the optimal balance at all times, maintaining an MOP [76,77] that is very competitive with other techniques [65] of setting a single objective and the rest as constraints. MOP has allowed us to satisfy both consumer and system objectives [78].

Table 2. Programmable parameters associated with loads of each appliance. System managed control: "•" or "■". Resident-managed control: "○" or "□".

Appliance	Parameters	Description
Variable loads without altering comfort		
Automatic opening of roller shutters—Automatic opening of roller shutters	•○ % of shutter opening. ○ Activate or deactivate the automatic shutter opening control.	•○ Normally residents will control the % opening of the blinds, but if automatic opening control is active, the system will open the blinds based on outside natural light and whether or not residents are present in the room.
Variable loads with alteration of comfort		
■□ Climate control-Air Conditioning + Heating System	■□ On and off. □ Initial temperature □ Time in minutes until the system automatically applies a second regulation. □ Adjustment of the degree variation (+1, −1, +2, −2) for the second regulation. ■ Order of execution of the second regulation. □ Annulment of the Order of execution of the second regulation. ■ Automatic shutdown in the absence of residents for more than a specified time.	Residents set the temperature. The system resets the temperature (▲▼) after a few minutes to reduce consumption without affecting comfort. ■ Automatic shutdown in case there is no resident in the house (thanks to the GPS tracking controlled by the System) or the presence detector in the room does not detect movement for more than 1 h.
•○ Fan-Fan	•○ On and off. •○ Power.	•○ Usually, users will control its on, off, and power, but the system can turn it off or lower its power if the contracted consumption limit in the electricity tariff is exceeded.
■□ Stove-Stove	■□ On and off. □ Power	■□ Normally users will control its on, off, and power, but the system can turn it off if the contracted consumption limit in the electricity tariff is exceeded.
Interruptible loads		
•○ Electric boiler	•○ On and off. •○ 6 + 1 operating time slots. ○ Temperature targets for each activation band. •○ Adjusting the water heating curve. • Observation of permanent consumption. • Long-term disconnection	•○ Absolute management of thermos operation by both residents and the system. • Switched off during prolonged periods of absence of residents (vacations). Switched back on several days before return.
■□ Vacuum cleaner (robot)-Vacuum Cleaner (robot)	□ Switching on and off ■ Recharge control	■□ Normally users will control its power on and off, but a power-on time and recharge time can be programmed.
•○ Electric Vehicle •○ Battery-Energy Storage •○ Water Pump (Well, Pool)	•○ Recharge time (on/off).	•○ Recharging would take place at the cheapest hours. Note: Not applied or scheduled to the study dwelling in the article.

Table 2. *Cont.*

Appliance	Parameters	Description
Uninterruptible loads		
■□ Washing Machine	■□ Power-on time. ■□ Permanent consumption observation.	■□ Manual or system-programmed ignition at the cheapest time between 7:00 and 11:00 AM.
•○ Dishwasher-Dishwasher.	•○ Power-on time. •○ Permanent consumption observation.	•○ Manual or system-programmed ignition at the cheapest time for the next 12 h (24 h).
■□ Dryer Machine	■□ Power-on time.	■□ Manual or system-programmed ignition. Note: Not applied or programmed to the study dwelling in the article.
■□ Oven	■□ Power-on time.	•○ Manual or system-programmed ignition Note: Not applied or programmed to the article study dwelling.
Uncontrollable loads		
•○ Television •○ Sound equipment •○ Computer. •○ Refrigerator/Fridge-freezer. •○ Light Spots/lighting. •○ Microwave •○ Vacuum Cleaner •○ Iron •○ Cooker pot •○ Cooker Hood •○ Hair dryer •○ Toaster •○ Kettle •○ Blender	○ On and off Observation of general consumption. Notification by the Assistant. • Rate information to residents on an hourly, strip, and daily basis.	○ Manual switching on and off by residents. • Warning of excessive consumption (via Assistant and Telegram) in the absence of residents or exceeding the contracted power limit. • Notification (via Wizard, Telegram, and control panel) of the electricity tariff.

Using this bidirectional technique, which not only brings benefits but has also allowed us to limit uncertainties, it has been possible to implement stochastic dynamic programming with at most two levels of estimation: the target variable plus an additional level with stochastic variables, which greatly increases the accuracy of the predictions as will be seen in the results section. The following references show up to six different strategies for stochastic optimization: stochastic optimization, robust optimization, chance-constrained optimization, stochastic dynamic programming, stochastic fuzzy optimization, and stochastic model, which generates synthetic consumption profiles [65,68,79–85]. For example, in [79] a stochastic energy consumption scheduling algorithm based on time-varying prices known in advance (similar to the one used in the HERMES system) is described as achieving a 24% to 41% reduction in simulations in billing costs. However, in HERMES, we have opted for a mixed model (with some elements with deterministic programming and others with stochastic programming), which has allowed us to obtain very similar reductions but with real data, not simulated, of up to 42% in absolute values (see Table 8) and 24% with counterbalanced data (see Table 6). Other techniques achieve reductions from 8% to 35% of the electricity bill [9], the optimization-based residential energy management (OREM) technique being the most efficient [86] based on dividing the days into time slots, very similar to the time of use (ToU) scheme and the one proposed in this article, scheduling the operating time of the appliances in the minimum tariff time slot, but in our case minimizing the delays of the OREM technique by shifting the loads in a very efficient way combining the strategy with other techniques.

The various techniques employed in HEMS scheduling to find the optimal operating time of household appliances can be grouped into five categories [65,87]: mathematical optimization; heuristic and metaheuristic methods; model-based predictive control; ML; and game theory approaches. Each of these techniques has strengths for certain types of loads versus weaknesses for all other loads, and in almost all cases, the benefits provided drop drastically when uncertainties manifest themselves in a practical or worst-case form. The main source of uncertainty comes from the residents themselves, who are often influenced by external factors that are difficult to predict (changes in routines, illness, cancellation of a meeting) or varying perceptions and subjectivity. Based on this, HERMES decided to use a mixed model of techniques that would allow the residents to make their own decisions or let the system decide independently under MOP, using techniques such as mathematical optimization, heuristics, or ML.

The subsection "Equations" shows mathematical elements used and developed under a tree structure for decision making and resident assistance. Further on, the mutual learning process between the system and the residents will become evident by adapting the system to the residents' habits and the residents' system, which makes it possible to achieve the percentages of reductions indicated above. This feedback has allowed very significant improvements in the first year, which were further improved in the second year and again in the third year. This continuous improvement highlights the bidirectional interaction of the system with the residents, which would be difficult to achieve by applying a single technique and without the expert assistant to interact with and guide the residents.

As indicated, to highlight the potential of the wizard, HEMS has been developed on a mixed model of techniques [87] to improve energy use through load scheduling, in which uncertainties have been minimized so that these models must be able to admit the interaction of several agents that would become the elements of uncertainty as well as load scheduling. The objective of all HEMS is to optimize consumption, so they require scheduling over a future time horizon, for which household demands and electricity generation cannot be accurately predicted, requiring adequate consumption profiles, representative, and incorporating a certain degree of uncertainty management. For all these reasons, their efforts are focused precisely on optimizing consumption profile predictions. In our case, to highlight the assistant's potential, we have reduced the uncertainties to a scenario in which the development of HEMS is already considered sufficiently mature, with the assistant being a differentiating element and allowing us to show its potential. In this regard, to optimize consumption profile predictions, we will adopt a dark box model (modeling and forecasting frameworks based on data analysis schemes) as opposed to white-box models (classical and transparent modeling tools based on solving physical equations) and gray box models as a combination of white box and dark box [87,88]. We have limited the uncertainties to the demand area without incorporating electricity microgeneration and setting variable but known day-ahead prices. To obtain the prediction of household consumption, the element used for data analysis was based on ML techniques. Other data analysis techniques could have been applied (see [87]). A very accurate and robust model has been obtained using stochastic data of only two levels: The target or output variable plus an additional level on certain input variables of the ML itself. Since the HERMES system combines different techniques, e.g., deterministic programming for MOP or stochastic programming for consumption estimation, we have tried to simplify it, while trying not to harm the pursued objectives, as we will see in the next section.

3.3. Deployment of the HERMES System and Involved Instruments

The basis for developing the HERMES system to optimize savings and comfort is collecting past and present data and forecasting certain elements to create a robust and elastic energy use model. The only essential future data are the hourly kW price (€/kWh) 24 h in advance; thus, the chosen tariffs allow us to predict their value quickly; however, if they were not known, they could be obtained from prediction models with very accurate approximations. The weather forecast and the presence of the residents in the home are also necessary but not essential future data.

Given the complexity of our system, and the need to obtain data, its implementation has been gradual, following "natural growth" towards the proposed objectives. Figure 5 shows the principal elements and services of the HERMES system.

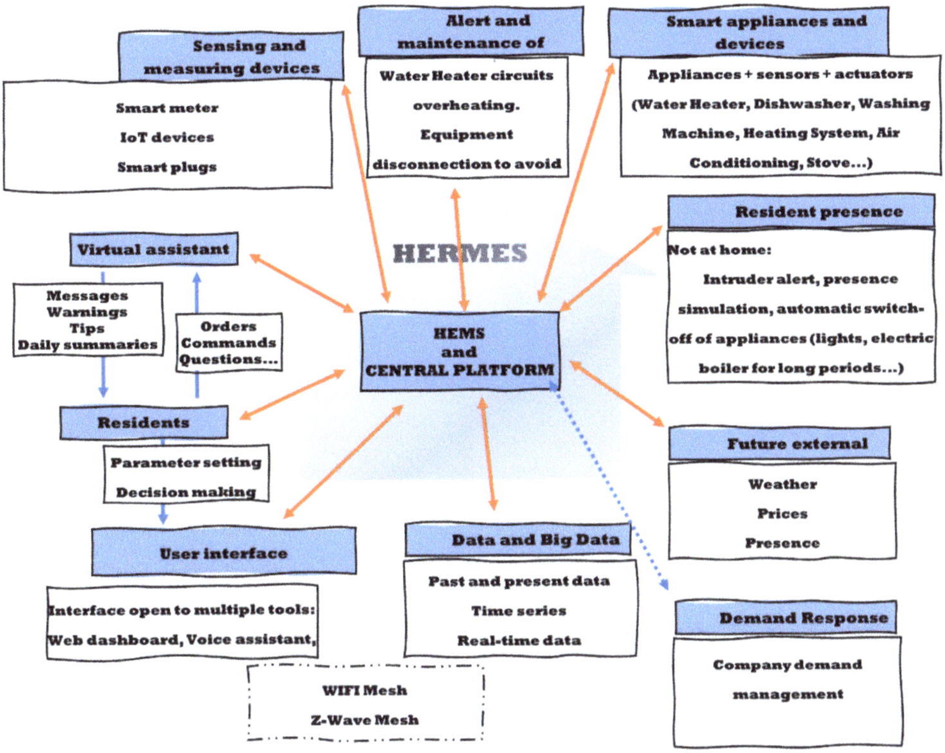

Figure 5. The general framework of HERMES system elements and services.

Each of the elements shown in Figure 5 has been developed considering analysis, characterization, development of operating models, improvements achieved, deployment of the models and implementation, review of results, and return to the previous phase as necessary.

3.4. *Programming and Multi-Objective Optimization (MOP) of the HERMES System*

As discussed above, the HERMES system is based on MOP scheduling in order to (1) reduce electricity bills by reducing appliance consumption and shifting loads and (2) 'maintain or increase residents' comfort. The system sends the residents the next day's hourly rate by instant messaging; they can also consult it at any time through the HERMES user panel or consult it through the wizard. In Figures 6 and 7 we can see different data of the 2.0DHA electricity tariff. Although there are two well-defined time slots, there are significant daily variations in prices for each hour, so the system uses the daily prices in its programming, using any tariff as long as the prices are known or estimated with daily anticipation. For each day, the system selects the optimal time zone.

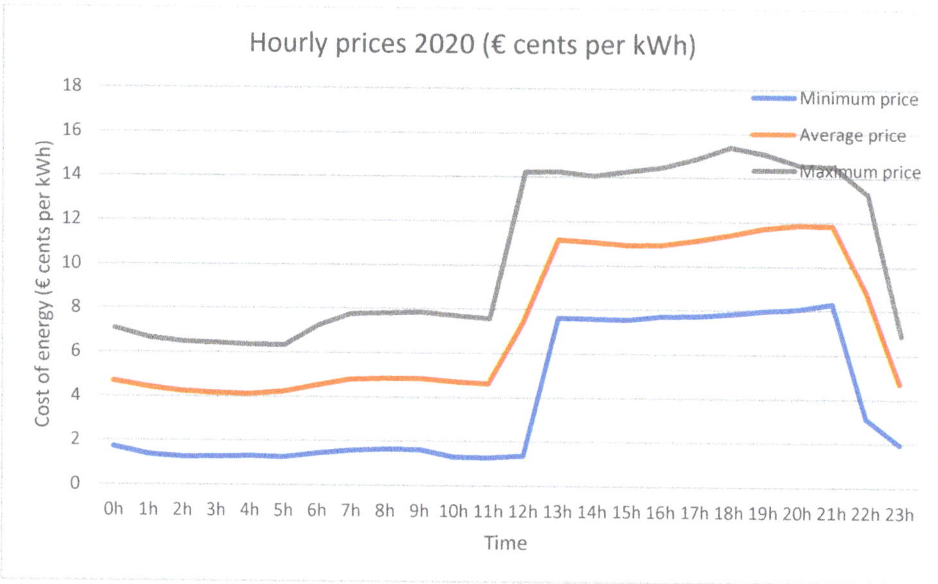

Figure 6. Average hourly prices with the range of fluctuations during 2020.

Figure 7. Example graph sent daily to the user with the next day's prices (example of 11 December 2020).

3.4.1. Equations

HERMES optimal scheduling is modelled as an MOP problem [89]). In this model, the first objective (f_1) is related to the minimization of the monthly bill, so the scheduling can shift loads either on time scales of minutes, hours, or even between days, with a monthly horizon for the optimization, instead of a daily horizon as most HEMS schedules usually present. Therefore, the shifting of loads is allowed even between days, as long as it does not impair the comfort of the residents, so that the first objective can be formulated as:

$$f_1 = \sum_{d=1}^{PT} \sum_{h=1}^{HD} \left[p_{d,h}^{Tariff} E_{d,h} \right] \tag{1}$$

where:

- $E_{d,h}$ = Energy consumed by the household in kWh during the hour of the day h of the day d of the tariff period PT;
- $p_{d,h}^{Tariff}$ = Price in €/kWh of the hourly cost of energy term in each hour of the day h of the day d of the tariff period PT for the contracted tariff ($Tariff$);

Equation (1) excludes fixed costs and taxes not associated with consumption. The choice of the tariff is important because it determines both the variable costs and part of the fixed costs, so, initially, a study was made to determine which tariff was the most suitable for the habits of the residents and the potential of the HERMES system. This choice led to a first saving in the monthly bill without affecting comfort, as reflected in the results (see the first part of Table 6).

To minimize the value of the function f_1 several resources and constraints must be considered:

$$P_a = P_a^{UNC} + P_a^{Con,Ine} + P_a^{Con,Ela} \leq \begin{cases} P_{aMax}^{Tariff} \\ P_{aMax,\ penalty}^{Tariff} \end{cases} \tag{2}$$

Equation (2) establishes that the active power at any instant of time P_a expressed in kW cannot exceed the maximum contracted power P_{aMax}^{Tariff} nor the one higher than this one of penalty $P_{aMax,\ penalty}^{Tariff}$ (in the case under study, this limit is set at 105% [90] of the maximum contracted power P_{aMax}^{Tariff}). The P_a is the sum of all household loads, consisting of uncontrollable loads P_a^{UNC} and controllable inelastic $P_a^{Con,Ine}$ and elastic $P_a^{Con,Ela}$ loads. This constraint affects the HEMS scheduling, which is oriented to avoid reaching the maximum allowed and the penalty level. However, due to the freedom of the residents, in case of reaching the first level (5.5 kW), the assistant warns the residents, and in case of exceeding the penalty level (5.775 kW) the system can act by disconnecting elastic loads, and the warning of the assistant to the residents is of greater emphasis.

One relevant aspect is the ability of the system to schedule shiftable loads within the entire known price period PT_{Known} guaranteeing the comfort of the residents and, extending the scheduling horizon beyond the 24 h with which HEMS normally work:

$$PT_{Known} \geq 24 \text{ hours} \tag{3}$$

This condition allows the system to increase consumption on days whose prices are lower than adjacent days, i.e., the system reschedules the loads when it obtains the prices for each hour of the following day, also taking into account the prices for the rest of the current day. After obtaining the prices, the system performs sorting by prioritizing the cheapest hours:

$$Array\left(p_{d,d+1,h}^{Tariff}\right) : \left[\ Min\left(p_{d,d+1,h}^{Tariff}\right) \ \ldots \ Max\left(p_{d,d+1,h}^{Tariff}\right) \ \right] = \left[\ p_1 \ \cdots \ p_m \ \right] \tag{4}$$

where $p_1 \leq \cdots \leq p_m$; and assigning specific controllable loads (such as the electric boiler $P_a^{E.Boiler}$, dishwasher $P_a^{Dishwasher}$ or batteries $P_a^{Baterias}$) to the cheapest hours according to the energy required by each load (for example $E_{E.Boiler} > E_{d,h} > E_{Batteries}$) while maintaining the constraint (2) and comfort:

$$\begin{array}{l} p_1 \overset{d,h}{\rightarrow} \alpha_{11} P_a^{E.Boiler} + \alpha_{12} P_a^{Dishwasher} + \ldots + \alpha_{1n} P_a^{Batteries} \leq P_{aMax}^{Tariff} \\ p_2 \overset{d,h}{\rightarrow} \alpha_{21} P_a^{E.Boiler} + \alpha_{22} P_a^{Dishwasher} + \ldots + \alpha_{2n} P_a^{Batteries} \leq P_{aMax}^{Tariff} \\ p_m \overset{d,h}{\rightarrow} \alpha_{m1} P_a^{E.Boiler} + \alpha_{m2} P_a^{Dishwasher} + \ldots + \alpha_{mn} P_a^{Batteries} \leq P_{aMax}^{Tariff} \end{array} \quad (5)$$

where α_{ij} form a matrix $m \times n$ of binary coefficients associated with each controllable load as a function of the energy required by each load (α_{11} is associated with a load whose energy is greater than the equivalent for α_{12} and so on until the α_{1n}; the next row corresponding to p_2 represents loads whose work duration extends beyond the hour h associated with the minimum price p_1. The constraint is given by P_{aMax}^{Tariff} forcing the HEMS scheduling to set to zero those loads whose energy is lower, i.e., to those coefficients α_{ij} of higher columns, so that if the power P_{aMax}^{Tariff} is exceeded, they would not be activated until the next cheapest hour or once the appliances with higher loads have finished their operation (or the sum of the loads already allows incorporating a new lower load). An example matrix for three controllable appliances might look like the following:

$$\alpha = \begin{bmatrix} 1 & 1 & 0 \\ 1 & 0 & 1 \\ 0 & 0 & 1 \\ 0 & 0 & 0 \\ \vdots & \vdots & \vdots \\ 0 & 0 & 0 \end{bmatrix} \quad (6)$$

The operation of the loads is not limited to whole hours, and they can be longer or shorter periods, so the coefficient α_{21} set to 1 in this example does not imply that the associated appliance is the second full hour working; it only indicates that it requires more than one hour to complete its cycle. In some cases, the order of the hours is not relevant, whereas in others it is, so the system also takes into account this limitation for each appliance. The matrix also shows that for the first cheapest hour, the system can only activate two controllable appliances to ensure that condition (2) is not violated or that the third appliance requires two consecutive hours to run its program, and that any other sum of consecutive hours would offer a higher sum price. $p_x + p_y > p_2 + p_3$. Similarly, if during that period residents activate any other loads (uncontrollable, controllable inelastic, or elastic) that compromise condition (2), the system will warn the residents and ultimately displace the elastic loads that can be displaced at that time. In the following results section, Figures 16–18 show how loads of the appliances are concentrated in the least cost hours. Three zones are distinguished corresponding to (1) the zone where the system works without interference from residents, usually night hours; (2) another optimal zone where both the system and the residents activate loads; (3) and a third one associated with residents' comfort where the system tries not to schedule loads because they correspond to the highest prices and informs the residents through the wizard.

One last remarkable resource to achieve condition (1) has been the development of a cumulative hourly consumption forecast for the next day $f_{1,d,h}^{Forecast}$, understood as the expected consumption based on past consumption under similar conditions. It is based on the use of ML linear regression, offering a forecast based on historical consumption data of residents over a long period comprising a total of 9946 h (over 20,000 h for a second version). It uses data such as weather (both historical and forecast data), as well as the percentage of presence of residents in the house, the day of the week and month, and the price of the kWh. The project is developed in the subsection Consumption estimation (Machine

Learning), more information is provided in the Data Availability Statement. Based on this forecast, the following can be established:

$$\begin{gathered}\left(f_{1,d,h} \leq 0.75 f_{1,d,h}^{Forecast}\right);\ f_{1,d,h} \leq 0.95 f_{1,d,h}^{Forecast} \\ 0.95\ f_{1,d,h}^{Prevision} \leq f_{1,d,h} \leq 1.05\ f_{1,d,h}^{Prevision} \\ f_{1,d,h} \geq 1.05\ f_{1,d,h}^{Forecast};\ \left(f_{1,d,h} \geq 1.25\ f_{1,d,h}^{Forecast}\right)\end{gathered} \quad (7)$$

For each hour of the day, three levels are established that compare the actual accumulated consumption of the day $f_{1,d,h}$ and the accumulated forecast for that same hour $f_{1,d,h}^{Forecast}$ so that the user can consult through the wizard if their consumption is lower, higher, or close to the forecast. Lower and upper limits are also established in which the system informs the residents through the Wizard without waiting for the consultation; this would be in cases where the deviation is significant, set by default at a deviation of 25% of the expected amount. The residents can modify the margins established in Equation (7). In this way, a reinforcement message is established when consumption is lower than expected and a "warning" in cases where consumption is higher than expected.

The information provided by the ML could also be beneficial in cases where alternative or complementary energy sources or systems to the public power grid are used. We are referring to microgrids in which energy management would be based on a different dynamic and in which generation and consumption forecasting through the ML would become much more important for the objective stated in (1).

3.4.2. Strategy for Comfort and f_1 Optimization

If the first objective is related to the minimization of the monthly bill f_1, the second objective of the MOP programming of the HERMES system is associated with the residents' comfort, trying to maintain a balance between both objectives because, on many occasions, they are opposed to each other. Since the HEMS does not have direct access to the uncontrollable loads, both the optimization of function f_1 and comfort is usually focused on the controllable loads. However, in our case, thanks to the work of the Assistant, residents are more aware of the costs associated with the loads, so there is bi-directional feedback, and the system gains some influence over the uncontrollable loads, allowing optimization of both objectives, f_1 and comfort more efficiently.

Comfort has a significant amount of resident subjectivity, and its programming can compromise the hardware resources of the system, so to avoid or alleviate these drawbacks, we chose to change the comfort penalty (discomfort) function typically used in HEMS [65,73,91,92] to a parameter approach adjustable by both residents and the system so that residents could vary these parameters to fit their conception of comfort and the system would balance them to optimize the f_1 function within ranges that do not compromise comfort. However, this concept would only be practical if the parameters were associated with each appliance; it would not make sense if they were global, as we would return to the concept of a global comfort penalty function. At the same time, it would not be necessary to define individual comfort functions for each appliance because the residents are part of their programming through the parameter settings, so the programming must be very well calibrated, which requires more extended testing periods in the implementation of the system and a certain flexibility. This strategy avoids the two problems associated with comfort: subjectivity, as users can adjust the parameters within a range, and programming is simplified as it is customized for each appliance; however, it requires a longer testing period.

The following two examples (responsible for a large part of the electricity bill [9]) show the potential of this approach to comfort: The electric boiler and the air conditioning system:

- The two main problems of the electric water heater are that it runs out of hot water or that it consumes at very high or non-optimal cost hours. In a traditional HEMS, this situation should penalize the overall comfort function, although it might not anticipate the problem or optimize consumption to the maximum. In our case, three groups of parameters have been created to optimize consumption and comfort, solving both problems. The first group selects the time slots in which the thermos flask is allowed to be turned on. The second group sets the temperature targets for each activation band. Finally, the third group adjusts the water heating curve. This third group is continuously adjusted thanks to the temperature sensor inside the tank and determines exactly how long it takes for the boiler to heat the water to the desired values. In this way, if very low temperatures are reached after use, the system responds by increasing the heating time and raising the maximum temperature of each range. The system adjusts these parameters automatically, ensuring the hot water supply and shifting the load to the optimal time slots (see Figure 8). However, residents can readjust the parameters to suit their comfort (maximum heating temperature and the number of heating hours). This set of parameters covers the complete characterization of the water heater, making it possible to cater for particular scenarios such as, for example, completely switching off the electric water heater during prolonged absences by disabling all operating slots, or from time to time run a heating cycle to 60–65 °C to eliminate possible Legionella outbreaks.

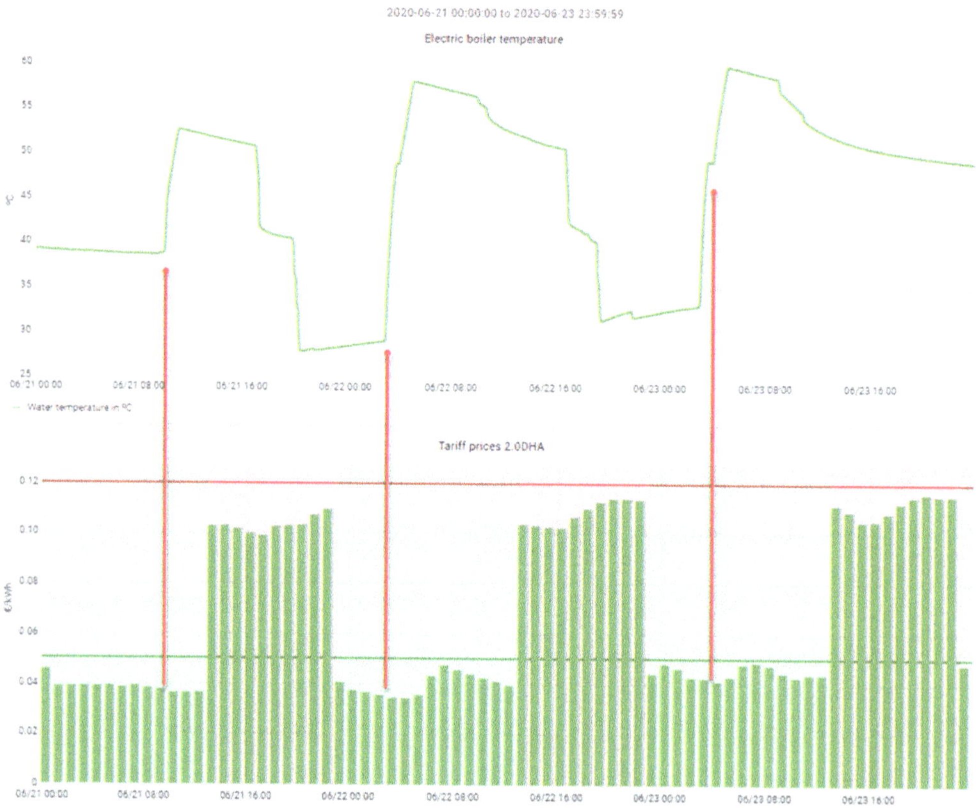

Figure 8. Heating of the water in the electric boiler during the most economical hours. The heating cycle of the boiler is displayed, and the heating hours are marked. Data from 21 June 2020 00:00:00 to 23 June 2020 23:59:59.

- For air conditioning, the HERMES system controls several parameters and employs the following strategy to optimize consumption and maintain comfort: after a certain time after switching on the climate, the system automatically lowers or raises the temperature to reduce consumption while maintaining comfort. The parameters used in this strategy are again three: initial temperature when the heating or cooling is turned on, the time in minutes until the system automatically applies the second temperature regulation (to reduce consumption and which could depend secondarily on other parameters such as outdoor temperature, indoor temperature or whether or not the residents come from outside), and finally, the third parameter would be the difference in degrees of the new temperature. The adjustment of the parameters is again dynamic depending on what the system requires and the residents' preferences.

In the programming of this strategy for the optimization of the comfort and f_1 the following parameters associated with loads of each appliance have been used (Table 2):

In the previous paragraphs, the strategies and scheduling of the two main objectives of the HERMES system have been detailed. However, in this multi-objective structure, others could have been added, such as the reduction of CO_2 emissions in tune with the reduction of the bill, demand response, and others. However, in terms of CO_2 emissions, any system that minimizes consumption already contributes to reducing emissions, and if it also concentrates consumption in off-peak hours where renewable and less polluting energies tend to prevail, the reduction is even more significant. HERMES provides both benefits.

Therefore, this section can be concluded by indicating that the residents set the comfort levels they desire, and the system optimizes the objectives of minimizing the monthly bill and maximizing comfort, keeping the balance between the two.

4. Results

This section shows, evaluates, and interprets the results of the HERMES system. It is developed under the proposed MO model integrating the Assistant.

4.1. HERMES System Deployment and Infrastructure

The system has been deployed in a single-family house with four residents with an average annual pre-installation consumption of 6346 kWh and powered exclusively by the electricity grid. The deployment of the HERMES system was carried out in several phases taking into account the characteristics of the house, which has a kitchen, living room, three bedrooms, attic, and three bathrooms.

In the first phase, the passive elements of the house that could affect comfort and thermal insulation were analyzed, something basic but usually a factor that is not taken into account in most installations with HEMS [5]. Some thermal leaks were detected that could be easily solved, such as installing a weather strip (see Figure 9) on the access door to the house, improving the thermal insulation from the outside.

Figure 9. Weatherstripping to improve the thermal insulation of the house.

Several groups of sensors, actuators, and various smart home hubs were deployed to achieve efficient control of consumption and comfort in the home. While one group of sensors collected environmental data such as temperature, humidity, or lighting, the second group of sensors collected data on the presence or door opening, and a third group collected data on loads such as power or energy. As for actuators, there were smart switches and sockets for load control. In addition, some appliances already incorporated IoT management. Finally, the various Smart home hubs allowed communication with all sensors and actuators covering various protocols: WIFI, Z-Wave, Zigbee, BLE, and Infrared.

Following the architectural model given in Figure 1, Figure 3, and Figure 5, a comprehensive infrastructure of devices and systems was deployed for the physical implementation of the HERMES system, as shown in Table 3 and Figure 10:

Table 3. IoT sensors and actuators in the main loads of the home.

Appliance	Sensors/Actuators	Description
Variable loads without altering comfort		
Automatic opening of roller shutters-Automatic opening of roller shutters	WiFi shutter switch	Allows raising and lowering of blinds with percentage function by programming
Variable loads with alteration of comfort		
Air Conditioning-Air Conditioning + Heating System	Universal Remote Control with WiFi and IR. Programmable. Temperature sensor in the room Presence sensor in the room System consumption sensor	Thanks to the controller, the System controls all the functions of the Air Conditioning and Heating System. The temperature, presence, and consumption sensor allows the system to perform a secondary adjustment.
Fan-Fan	IoT built-in from the factory	It is linked to the Assistant and the system for voice control and automation.
Stove-Stove	WiFi Smart Plug	On/off control.
Interruptible loads		
Electric water heater	WiFi Smart Plug with consumption and power measurement. Wifi temperature sensor inside the water tank.	The system controls the on, off, and actual temperature of the water in the tank.
Vacuum Cleaner (robot)-Vacuum Cleaner (robot)	Factory-integrated IoT. WiFi Smart Plug.	The system (or residents) can activate and deactivate it. The optimal recharging time is programmed via the smart plug.
Uninterruptible loads		
Washing Machine	WiFi Smart Plug with consumption and power measurement.	After being manually programmed, the System (or the residents) decides the switch-on time.
Dishwasher-Dishwasher	WiFi Smart Plug with consumption and power measurement. Door opening sensor.	After being manually programmed, the System (or the residents) decides the switch-on time.
Uncontrollable loads		
Television Sound equipment Computer Refrigerator/Fridge-freezer Light Spots/lighting Microwave Vacuum Cleaner Iron Cooker pot Cooker Hood Hair dryer Toaster Kettle Blender	-	Manual turn-on and turn-off by residents. -Warning of excessive consumption (via Assistant and Telegram) in case of absence of residents or exceeding the contracted power limit. -Notification (via Wizard, Telegram, and control panel) of the electricity tariff.

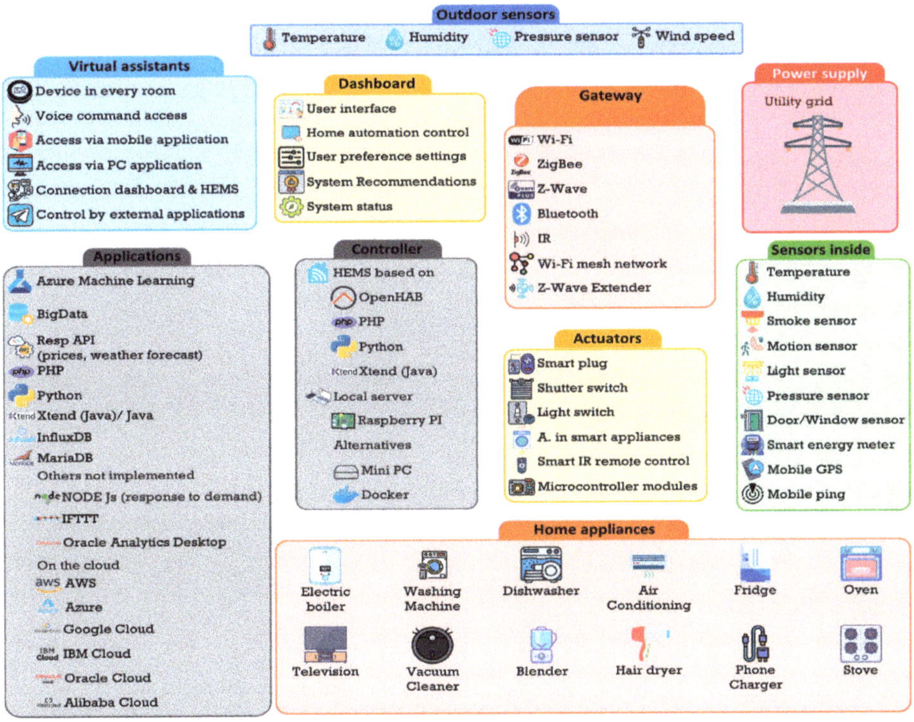

Figure 10. HERMES system infrastructure.

In addition to the sensors and actuators indicated in the table above, the system has sensors for presence, temperature, humidity, twilight, outdoor weather station, door opening in certain rooms and windows, and general consumption meters (energy and power) in the house, as well as consumption meters in certain appliances and meters in two additional areas of the house (lighting + plugs and kitchen).

The deployed infrastructure enables interoperability between devices, event synchronization, real-time (and historical) data logging, analysis and visualization, and present and long-horizon decision making by both the system and the residents, maintaining or improving comfort and cost reduction.

4.2. Voice Assistant and Control Panel

The Voice Assistant provides relevant information to residents to safeguard the balance between both objectives (f_1 and comfort) and accepts voice commands to inform or act on specific subsystems. It is the central core of communication with the residents, although they also have a control panel that offers both information (current and historical data) and the possibility of configuring most of the system parameters. The main interactions of the Wizard (Table 4), an extract of the Control Panel with options for setting some parameters (Figure 11**)**, and several data access interfaces (Figures 11–13) are detailed below.

4.3. Phases of Incorporation of HERMES System Functionalities and Change in Residents' Habits

Finally, we present a series of data to conclude with the achievements in terms of consumption reduction (cost evolution graphs, load shifting, prices, consumptions, invoices) and comfort improvement (process automation, commands, automated actions).

As we will see later in the subsection "Net load shifting", residents have a wide margin of improvement for consumption reduction based on shifting controllable (and some uncontrollable) loads to hours with lower prices. The system will try to approach the state of minimum consumption while maintaining comfort. Residents are provided with more information to make decisions they might not have considered before, allowing them to achieve an optimal balance between comfort and electricity bills by adjusting the parameters to their preferences at any time. The information provided by the system through the information panels, or the Voice Assistant keeps users constantly informed of the influence of their consumption habits on their electricity bills.

Table 4. Main voice interactions with the Assistant.

Command	Type (Residents Request Information/System Informs about Triggering Events)	Description
"turn on/off/regulate device"	Residents/System	Residents control more than 80 functions (turn on, turn off, raise the temperature by one degree) of the different devices connected in the home. In some cases, the system detects that a device has been switched on so that under certain conditions, it acts automatically to reduce consumption while maintaining comfort (e.g., it raises the cooling temperature by one degree after a few minutes of operation).
"price", "power", "consumption", "daily consumption", "cheapest washing machine/hour"...	Residents	Residents can ask at any time for data related to consumption and expenditure: Price or active power being consumed at that moment to know the impact of connected appliances, the next cheapest hours, the accumulated consumption per hour, daily or monthly.
"room temperature", "outside temperature", "thermos temperature", "probability of rain"....		Residents can know the data from the sensors connected in the house through the voice assistant or the probability of rain to make decisions based on these conditions and the electricity tariff to reduce consumption and maintain comfort.
"Departure or arrival home" (GPS + ping Wifi + door sensor).	System	The system detects if a Resident arrives or leaves the house by issuing a welcome message or checking if there are devices or unwanted presences.
"Power warnings"	System	The System monitors the active power level, informing Residents if the contracted power limit is reached or exceeds 105%, which would incur penalties.
"Price and consumption/expense notices"	System	The System reports at each start of a time slot with a different energy price, except during night hours (peak, flat or off-peak). In the event of higher or lower than expected consumption, the reports and responses to automatic warnings and queries made by the Residents to the Assistant are modified.
"Notices on ways to save"	Residents/System	A compendium of tips with saving techniques. The advice offered is random unless an inappropriate use of an appliance is detected (e.g., forced turning on of the electric boiler or continued use of the washing machine during peak rate hours). The system has a calendar, so some responses and warnings change depending on whether it is a national holiday or a weekend, or if adverse weather conditions are expected, or a very high consumption prediction estimated by ML.
"Text-to-speech and social networks"	Residents/System	Residents can send any command to the Assistant through their mobile application by voice commands or by text through social networks that the Assistant receives and executes. The System uses social networks to send text and text-to-speech messages to Residents with the help of the Assistant.

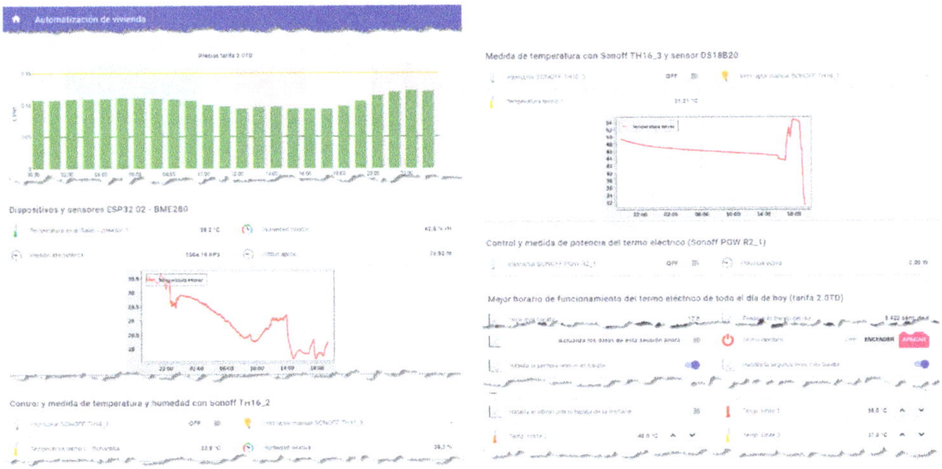

Figure 11. Control Panel excerpt (OpenHAB).

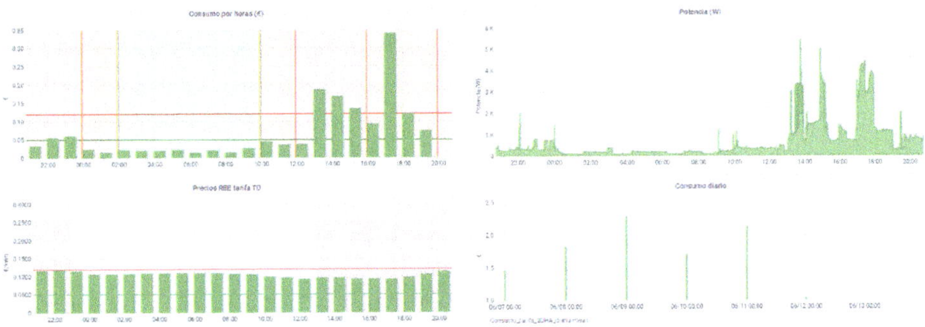

Figure 12. Data dashboard extract (Grafana + InfluxDB + MariaDB).

Figure 13. Example of text messages and images sent by the system to residents (Telegram).

The following figures (Figures 14–18) show how the daily distribution of loads has changed in line with prices and the impact these changes have had on bills. Both the system and the residents have been adapting to each other to achieve the above-mentioned optimal balance. The data have been divided into four periods (see Table 5): (0) P0 or previous. (1) P1 or first period where the system was still being implemented, and the optimal tariff was determined according to the residents' habits and HERMES' potential. In this period, the system did not yet allow load shifting, but it did offer information on their consumption. It was determined that it was necessary to move from the 2.0A tariff without time discrimination to the 2.0DHA tariff, distinguishing two price bands. (2) P2 or the second period starts with the new tariff, consumption management and allows the displacement of some loads. (3) P3 or third period where the system is implemented with the total operational capacity to displace all dispatchable loads and is ready to readjust the cooling/heating temperature to optimize consumption and comfort managed by the Wizard.

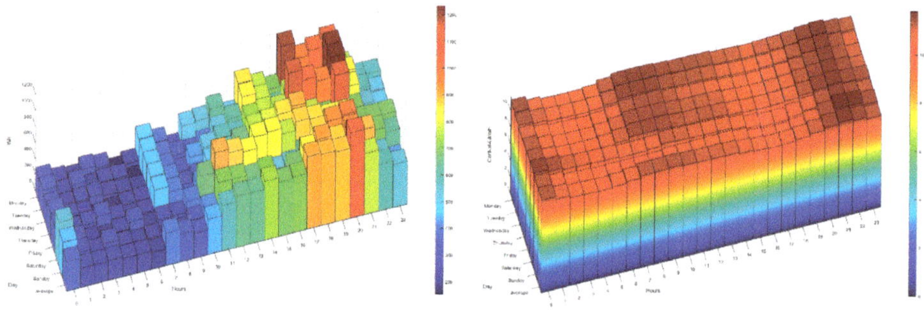

(**a**) (**b**)

Figure 14. P0 and P1. (**a**) Average consumption (Wh) and (**b**) average prices (€ cents per kWh) during P0 and P1 (tariff 2.0A).

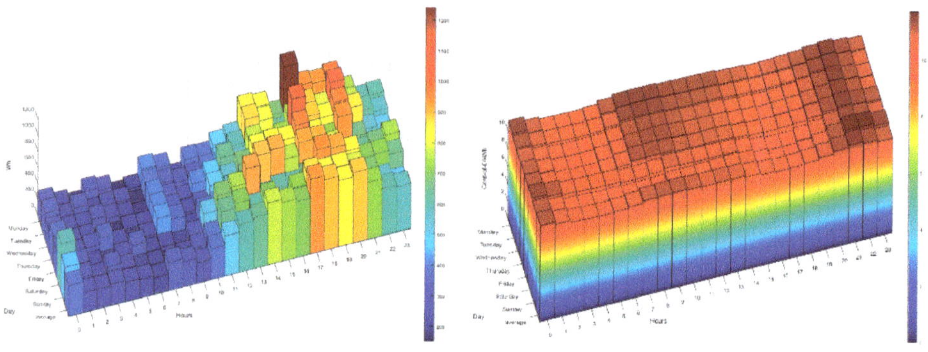

(**a**) (**b**)

Figure 15. P1. (**a**) Average consumption (Wh) and (**b**) average prices (€ cents per kWh) during the first period (tariff 2.0A).

During P1 (first period), the HERMES system infrastructure was developed and started to work effectively from P2 (second period), with full development in P3 (third period). During P1, the residents already have information on their consumption, but the system cannot shift loads. The maximum consumption coincides with the most expensive hours. The pattern of P2 and P3 is very different from that of P1 (see Figures 14–18), mainly due to the shifting of loads to the cheapest price hours, optimizing the monthly electricity bills as shown below (see Figure 24). Consumption has shifted from being centered from 17 h to 20 h, coinciding with the most expensive hours, to being divided into two and three bands of specially reduced prices, centered from 02 h to 04 h, from 10 h to 12 h, and 23 h, coinciding with the average of the lowest prices. Above all, this adjustment stands out in the third period, where the load shifting is optimized to reduce the bill while maintaining comfort, being very significant to see how the consumption needs are reduced in the most expensive hours (from 19 h to 21 h) and concentrated in the cheapest ones while maintaining a certain balance due to the maintenance of the residents' comfort.

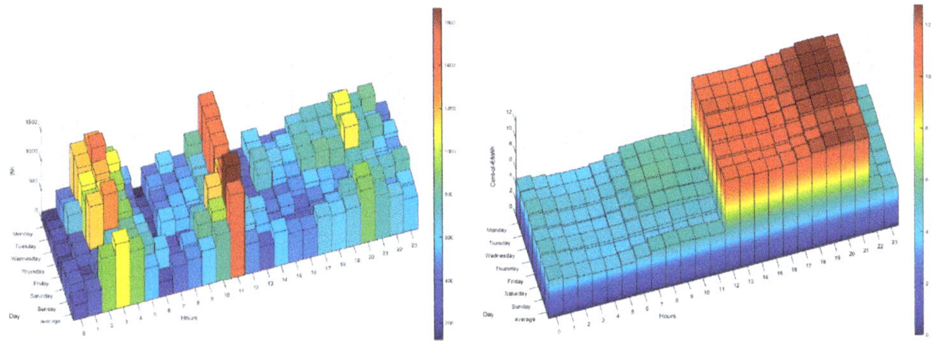

(**a**) (**b**)

Figure 16. P2. (**a**) Average consumption (Wh) and (**b**) average prices (€ cents per kWh) during the second period (tariff 2.0DHA).

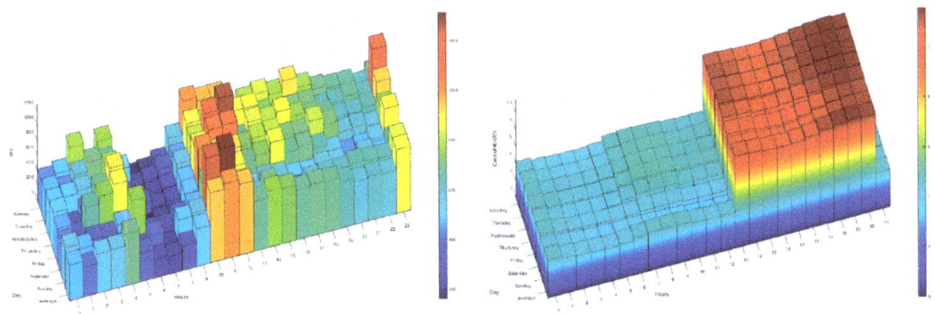

(**a**) (**b**)

Figure 17. P3. (**a**) Average consumption (Wh) and (**b**) average prices (€ cents per kWh) during the third period from 29 March 2020 to 24 October 2020 (tariff 2.0DHA).

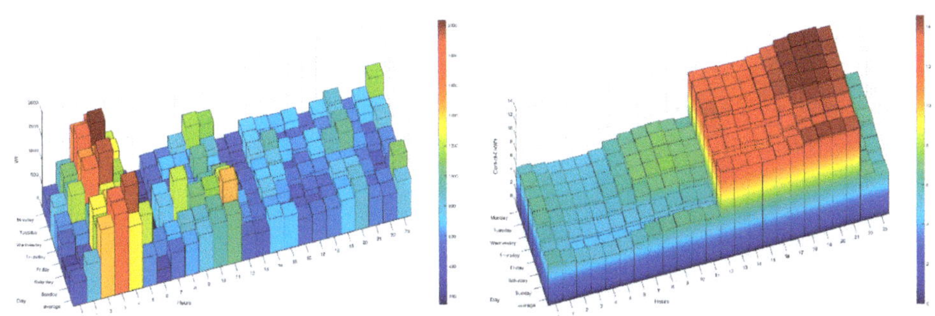

(**a**) (**b**)

Figure 18. P3. (**a**) Average consumption (Wh) and (**b**) average prices (€ cents per kWh) during the third period from 25 October 2020 to 06 February 2021 (tariff 2.0DHA).

Table 5. Phases of incorporation of HERMES system functionalities.

Periods		Validity (Day-Month-Year)	Incorporation of HERMES System Functionalities
P0	Previous	<31-03-2019	None
P1	First	31-03-2019 to 26-10-2019	Consumption information wizard
P2	Second	04-11-2019 to 28-03-2020	Consumption management and load shifting (electric boiler and washing machine). Change of optimal electricity tariff for the HERMES system.
P3	Third	29-03-2020 to 06-02-2021	Load shifting (dishwasher) and cooling temperature control

4.4. Net Load Displacement

It would be necessary to compare the real load distribution with respect to a scenario with no load shifting and no change inhabits to quantify the savings provided by the Hermes system. From the recorded data, two scenarios can be distinguished, one formed by periods P0 and P1 in which there were no load shifts or changes in habits, and another scenario formed by periods P2 and P3 in which HERMES has carried out load shifts, and there is some adaptation of the residents' habits to the time slots with lower prices.

From the first scenario (no-load shifting and no change in habits), an "average load distribution" has been obtained for each hour of the day so that the load shifting for any given day can be calculated by obtaining the difference of loads to the average distribution. A distinction is made between shifts that produce savings (above average loads at economic hours or below average at expensive hours) and those that do not produce savings (below average loads at economic hours or above average at expensive hours). The difference between the displacements that produce savings minus those that do not produce savings gives us the measure of the net displacement of loads, this being positive when savings are produced and negative when cost overruns are produced, and the greater the displacement, the greater the savings, balanced by the price per kWh and total consumption, so although it offers a measure of displacement, it does not offer a direct measure of the savings that will be calculated as will be explained later. Figure 19 shows the "average load distribution" for the scenario without load shifting, and the load distribution for the day 23 October 2020 has been added as an example to obtain the net displacement for that day:

Figure 19. Average load distribution for each hour of the day and the scenario without load shifting or habit adaptation (P0P1 series). The load distribution for day 23 October 2020 has been added as an example to obtain the net load shifting.

The calculation of the "net load shifting" (shifts that produce savings: Add; shifts that produce cost overruns: Subtract) for that day, following the procedure indicated in the previous paragraph, the net balance is positive and has a value of 10.69 kWh:

- Savings-producing displacements: Above-average loads at economical hours or below-average loads at expensive hours;
- Commuting that does not produce savings: Below-average loads at inexpensive hours or above-average loads at expensive hours;
- Economic hours for the day 23 October 2020: 0 h–12 h and 23 h;
- Expensive hours for day 23 October 2020: 13 h–22 h.

Suppose we extend this calculation to all the days of the different billing periods (periods indicated in the first column of Table 6). In that case, we obtain the following graph with the net load shifts per billing month, obtaining an average daily net shift of 5.61 kWh for the billing range 23–37, which is equivalent to 35.8% of the average daily consumption established at 15.68 kWh (481.88 kWh for each billing month). In Figure 20, two zones can be distinguished, one with negative shifts where there were no savings and covers from invoice 15 to 21, and the other from 23 to 37 where all net shifts are positive, which indicates the correct operation of the HERMES system. Even invoice 22 already has a positive shift, although it was not enough to obtain significant savings; that month was the one in which HERMES started operating. It is also shown how during the summer months of July and August (invoices 31 and 32), the system is less efficient since the most intense use of refrigeration coincides with the most expensive hours and represents a significant part of the total consumption.

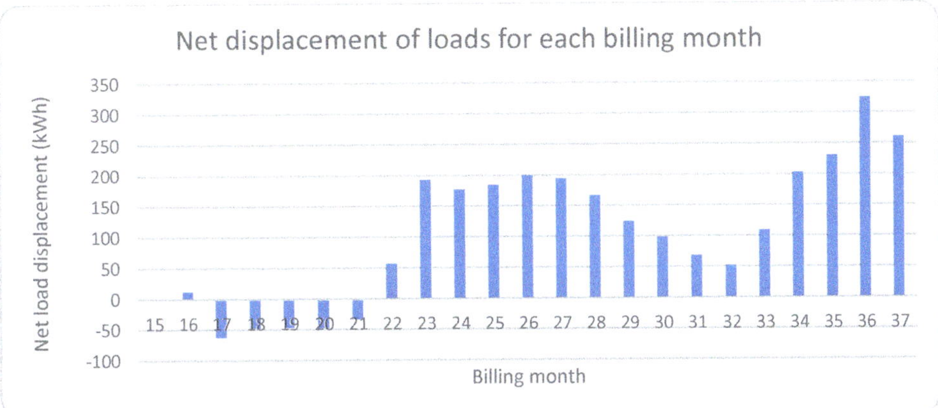

Figure 20. Net load shifting (kWh) per billing month. A positive net balance is obtained from billing 22 due to the performance of the HERMES system.

4.5. Calculation of Balanced Savings Obtained by HERMES

Once the net load shifting has been obtained, the savings calculation will partly follow the data obtained previously, but taking into account the total consumption of each day and the prices for each hour of that day. We started from the scenario with no load shifting or change of habits, in which electricity tariffs did not influence residents' habits since the behavior pattern was based on comfort. Based on this pattern, the actual daily consumption, and the two most favorable tariffs (2.0A and 2.0DHA), an expense model is obtained for P0 and P1, as shown in Table 6. The first model, P0P1 2.0A, during billing 15 to 22 only has a mean deviation of ±0.32 € to the actual monthly billed behavior, validating its use as an estimate for subsequent billings. If we modify the model for the 2.0DHA tariff, we obtain the third column of Table 6 (P0 and P1 2.0DHA) that offers lower costs simulating a scenario in which residents prioritize comfort but would have contracted the 2.0DHA tariff. Next, we will compare both models for actual consumption to determine the savings generated by the HERMES system after its implementation.

Table 6. Cost of energy consumed (€) by billing months for the scenario without load shifting or behavioral adaptation (P0P1 series) and actual billed cost. The data are divided into modeled and real data.

Billing Periods (Day-Month-Year)	Modeling		Real	
	P0 and P1 2.0A	P0 and P1 2.0DHA	Invoiced [1] (Energy)	Contracted Rate
Invoice 15: 09-03-2019 to 07-04-2019	44.22	40.52	44.26	2.0A
Invoice 16: 07-04-2019 to 07-05-2019	41.18	37.70	40.38	2.0A
Invoice 17: 08-05-2019 to 07-06-2019	41.08	37.57	40.78	2.0A
Invoice 18: 08-06-2019 to 06-07-2019	37.03	33.85	36.91	2.0A
Invoice 19: 07-07-2019 to 05-08-2019	35.71	32.75	35.05	2.0A
Invoice 20: 06-08-2019 to 06-09-2019	60.18	54.78	60.19	2.0A
Invoice 21: 07-09-2019 to 06-10-2019	37.91	34.53	37.67	2.0A
Invoice 22: 07-10-2019 to 03-11-2019	31.15	28.44	30.78	2.0A
Subtotal before Hermes (€)	**328.46**	**300.14**	**326.02**	
HERMES system implementation				
Invoice 23: 04-11-2019 to 09-12-2019	50.77	46.41	38.54	2.0DHA
Invoice 24: 10-12-2019 to 09-01-2020	49.25	44.50	36.99	2.0DHA
Invoice 25: 10-01-2020 to 07-02-2020	55.93	50.82	43.51	2.0DHA
Invoice 26: 08-02-2020 to 07-03-2020	39.82	35.86	28.50	2.0DHA
Invoice 27: 08-03-2020 to 10-04-2020	42.40	37.92	30.76	2.0DHA
Invoice 28: 11-04-2020 to 09-05-2020	27.74	24.36	18.38	2.0DHA
Invoice 29: 10-05-2020 to 06-06-2020	30.37	26.99	22.54	2.0DHA
Invoice 30: 07-06-2020 to 06-07-2020	35.95	32.25	30.56	2.0DHA
Invoice 31: 07-07-2020 to 08-08-2020	61.84	55.78	50.49	2.0DHA
Invoice 32: 09-08-2020 to 06-09-2020	45.00	40.73	39.57	2.0DHA
Invoice 33: 07-09-2020 to 06-10-2020	42.19	38.33	32.53	2.0DHA
Invoice 34: 07-10-2020 to 08-11-2020	40.75	36.91	29.62	2.0DHA
Invoice 35: 09-11-2020 to 07-12-2020	50.53	46.15	30.89	2.0DHA
Invoice 36: 08-12-2020 to 11-01-2021	89.31	82.04	74.02	2.0DHA
Invoice 37: 12-01-2021 to 06-02-2021	59.28	54.56	40.56	2.0DHA
Data from 04-11-2019 to 06-02-2021:				
Total (€)	721.13	653.61	547.46	
Average energy billed per month (€)	48.08	43.57	36.50	
Average monthly savings (%)	24.08%	16.24%		
Average monthly savings (€)	11.58	7.08		
Monthly savings with taxes (€)	14.73	9.00		
Average daily savings (€)	0.3859	0.2359		
Daily with taxes (€)	0.4909	0.3000		

[1] Actual data provided by the electric company.

From billing period 23 to 37, the average monthly savings in energy billed would be from 7.08 € to 11.58 €, i.e., a reduction of between 16.24% to 24.08% in energy billed compared to models without load shifting (see Table 6). If we consider taxes (excise tax of 5.11269632% for VAT of 21%: *1.0511269632*1.21), the average saving in each invoice would be from 9.00 € to 14.73 € (from 0.3 to 0.5 € per day) since the implementation of the HERMES system.

If we represent these data graphically, we obtain Figure 21:

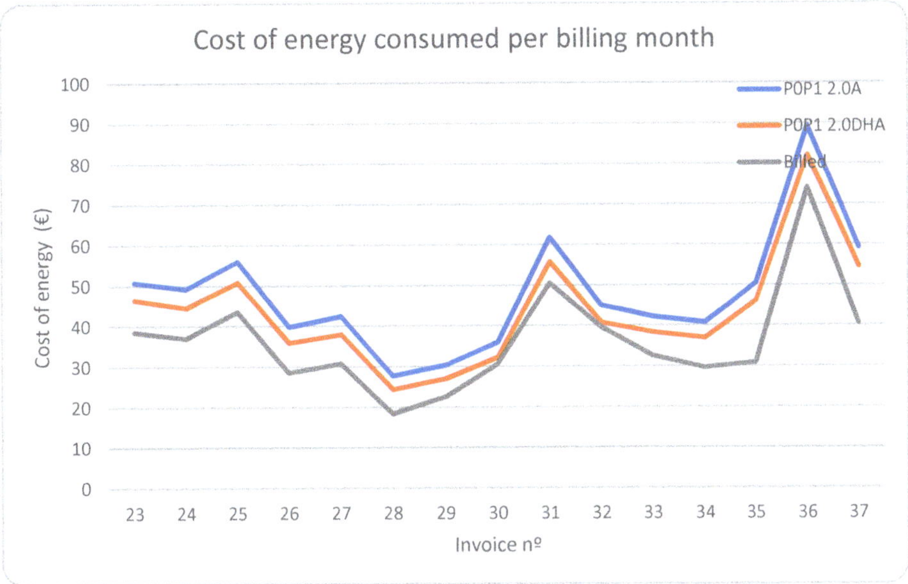

Figure 21. Cost of energy consumed (€) by billing months for the scenario without load shifting or habit adaptation (P0P1 series) and the actual cost billed since implementing the HERMES system.

Finally, in Figure 22, we compare the cost of energy consumed daily for the two regulated price tariffs, tariff 2.0A and 2.0DHA, from the first period to the third period.

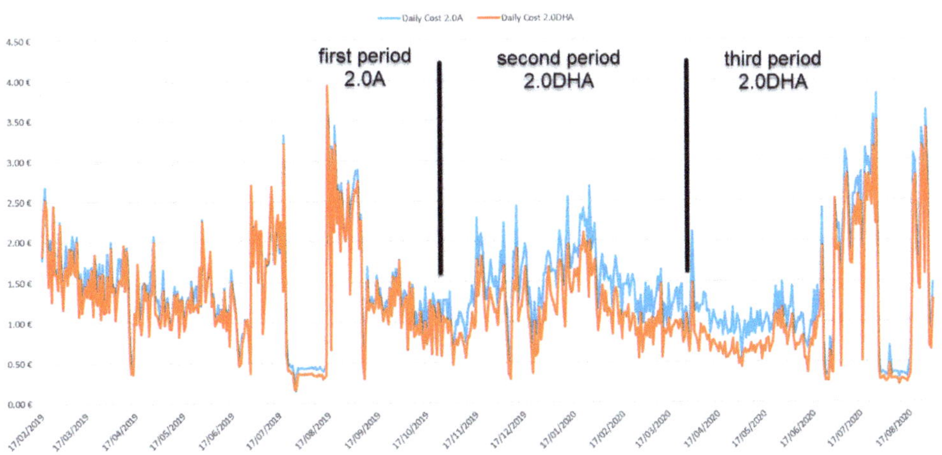

Figure 22. Comparison of the cost of energy consumed daily between the 2.0A (blue) and 2.0DHA (orange) tariffs from 17 February 2019 to 31 August 2020.

We can see how graphically the cost in both tariffs is very similar during the first period. From the second period onwards, the 2.0DHA tariff was contracted, which implied a change in certain habits of the residents to adapt to the new tariff. In addition, from this second period onwards, the system already managed consumption and load shifting, which made it possible to optimize the time slots with lower prices, achieving a very significant reduction in the daily cost compared to the 2.0A tariff.

4.6. Billing Expenses in Absolute Values without Balancing

Independently of the studies and models discussed above, we can conclude the savings analysis by detailing the bills issued by the electricity company, although in this case, the results are not balanced against price variations (tariff 2.0DHA: 2018: 0.1025 €/kWh; 2019: 0.0898 €/kWh; 2020: 0.0739 €/kWh), different annual temperature cycles (average Tmean: 2018: 18.2 °C; 2019: 18.8 °C; 2020: 19.2 °C) or different annual consumptions (total per year: 2018: 6346 kWh; 2019: 5211 kWh; 2020: 5644 kWh). However, it is of interest to show them given that the variations in conditions between 2019 and 2020 have not been very significant and yet show a remarkable reduction in bills even though the reduction in consumption has not been so significant (see Tables 7 and 8, Figures 23 and 24), mainly due to comfort requirements (higher consumption). Despite these demands, all months from the first period (31 March 2019 to 26 October 2019) present lower bills than the previous period (from 1 January 2018 to 31 March 2019), with a reduction of 18.3% where residents were unaware of their consumption details; as the first period progresses, the reduction in the bill is increasingly significant. This reduction is very striking with the entry of the second period (from 4 November 2019 to 28 March 2020). During this phase, the sum of the bills amounts to 357.1 € compared to 639.8 € during the same period a year earlier; the saving is 282.7 €, reducing 44.2% in the electricity bill. Finally, the bills from the third period (from 29 March 2020 to 31 August 2020) add up to 327.5 € compared to 427.7 € in the first period (reduction of 23.4%) or 515.3 € in the previous period (reduction of 36.5%) during those same months (from April to July), which shows the efficiency of the system able to continue optimizing periods when the system was partially implemented and already showing good performances as it was the first period. Table 7 (energy consumed) and Table 8 (monthly billing) also show the annual variations, including all periods.

Table 7. Monthly energy billed: previous period P0; first period P1 (yellow); second period P2 (green); third period P3 (blue).

Month	Billed Monthly Energy (kWh)		
	2018	2019	2020
January	760	784	538
February	635	468	422
March	518	383	482
April	449	356	367
May	353	369	365
June	398	337	419
July	358	309	625
August	690	583	473
September	420	362	406
October	399	296	418
November	625	457	369
December	741	507	760
Total per year (kWh)	6346	5211	5644
Variation compared to 2018	0	−1135	−702
Variation compared to 2018 (%)	0%	−17.88%	−11.06%
Annual invoice (€)	1387.16	1082.32	801.19
Variation € compared to 2018 (%)	0%	−21.98%	−42.24%

Table 8. Monthly bill: previous period P0; first period P1 (yellow); second period P2 (green); third period P3 (blue).

Month	Monthly Amount (€)		
	2018	**2019**	**2020**
January	150.02	160.92	79.09
February	128.21	99.51	59.99
March	99.83	85.92	66.96
April	95.66	83.04	47.13
May	87.46	83.55	51.59
June	91.94	76.57	63.43
July	89.00	75.24	91.23
August	151.23	109.25	74.08
September	**101.68**	78.57	65.93
October	98.72	78.70	64.70
November	137.92	78.58	63.04
December	155.49	72.47	74.02 [1]
Total per year (€)	1387.16	1082.32	801.19
Variation compared to 2018	0	−304.84	−585.97
Variation compared to 2018 (%)	0%	−21.98%	−42.24%
Annual energy billed (kWh)	6346	5211	5644
Variation kWh/year compared to 2018 (%)	0%	−17.88%	−11.06%

[1] HERMES System upgrades from 20 December 2020 to 11 January 2021 for system maintenance (change from Raspberry Pi3B+ to 4B, upgrade to Raspbian Buster, Java 11, OpenHAB 3, fixed and removed security bugs, update of certain parts of the programming due to version changes and new syntax...).

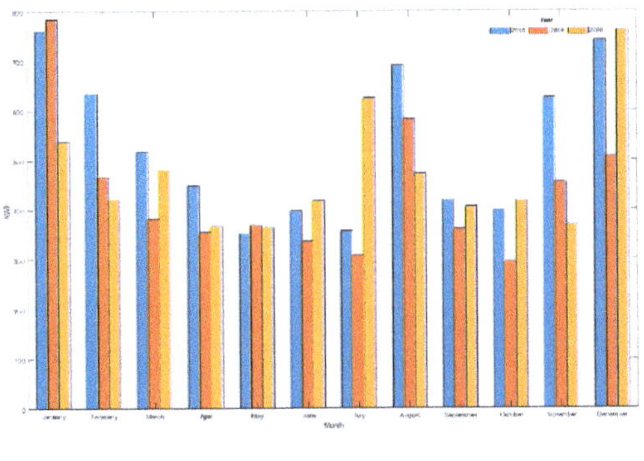

(a)

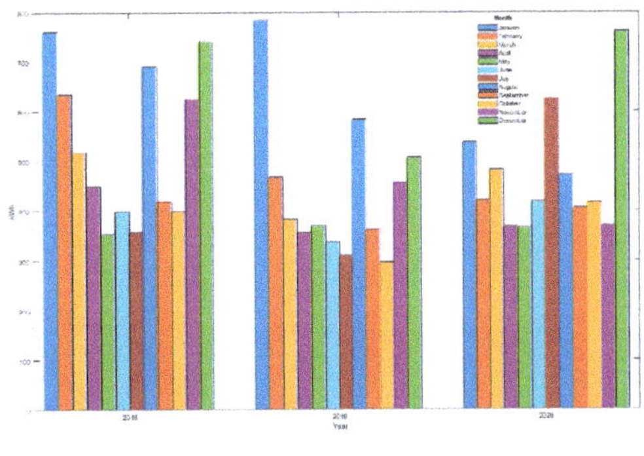

(b)

Figure 23. (**a**) For each month, the comparison of electricity consumed (kWh) is sorted by year. (**b**) For each year, a comparison of electrical energy consumed (kWh) is sorted by month.

The following figure shows a comparison of the data in Table 7:

The following figure shows a comparison of the data in Table 8:

Since the implementation of the system (a process developed during the first period), there has been practically no reduction in energy consumption in the home (see Figure 23), so comfort has not been sacrificed. However, the electricity bill has been reduced (see Figure 24); that is, the comfort of the residents has been maintained (and even improved) (thanks to the Wizard), and the loads have been shifted to reduce the monthly bill significantly.

The Assistant has significantly improved the residents' sense of comfort by allowing them to voice-control most of the charges. Thanks to this positive impact, the additional (and primary) function of the Assistant of being able to transfer information to the residents (and to the system) to reduce the amount of the bills has been easily assimilated by the users, so the impact has been very positive and relevant, favoring the feeling of comfort and the reduction of the electric bill of up to 42%.

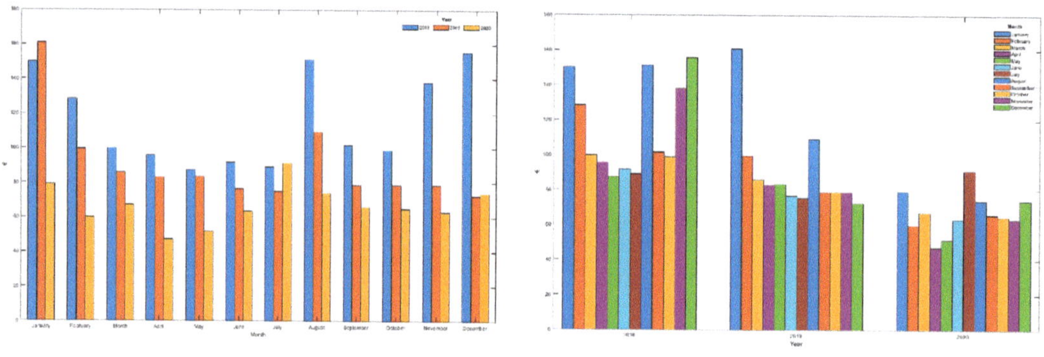

(**a**) (**b**)

Figure 24. (**a**) For each month, a comparison of the amount of the electricity bill (€) is sorted by year. (**b**) For each year, a comparison of the amount of the electricity bill (€) is sorted by month.

4.7. Consumption Estimation (Machine Learning)

Finally, we show a comparison between what was consumed and the consumption estimate in Figure 25, which allows residents to detect habits that may increase spending when actual consumption consistently exceeds the estimate or beneficial habits when actual consumption is lower than the prediction.

For ML development, several regression algorithms were used to train the model. Given the characteristics of the data and the desired outcome, the algorithms offering the most accurate predictions were boosted decision tree regression (BDTR) and decision forest regression (FDR), as opposed to linear regression or neural network regression [93–95]. In our case, after multiple pieces of training with different data structures, the BDTR algorithm has provided excellent accuracy (coefficient of determination: 0.9842; relative absolute error: 0.1085; mean absolute error: 452.399) at the cost of moderate training times. The BDTR algorithm is very sensitive to overfitting, so care must be taken in setting up the algorithm.

The consumption forecast obtained by ML is very accurate because it handles a large number of variables, so if the same conditions are repeated, the consumption should be similar. Although there may be discrepancies, the long-term trend should show a high correlation between the forecast and the actual consumption, which made the BDTR algorithm an optimal candidate because it is based on the creation of a set of regression trees through boosting, which means that each tree depends on previous trees. The algorithm learns by adjusting the residual value of the trees preceding it, so boosting tends to improve accuracy by creating series of trees incrementally and selects the optimal tree by an arbitrary differentiable loss function.

The study data in this paper cover periods extending before, during, and after the confinement period due to COVID'19. Residents remained during the first confinement period (15 March 2020 to 20 June 2020) in the home and through mid-August 2020, with consumption increasing significantly during July due to high-temperature weather. In general, it was expected that consumers would be much higher than normal during the confinement period because the residents remain in the home all the time, which should translate into higher consumption. Figure 23 shows that the consumption from March to July 2020 is higher than the previous two years; however, the bills during that period (see Figure 24) were lower than the previous years (except July 2020). This highlights two relevant aspects, on the one hand, consumption should have increased significantly, but the system as a whole has been able to control these unfavorable conditions, and on the other hand, bills should have been much higher than in the same period of previous years, but again, the system has been able to manage the loads by reducing the energy impact to bills with lower amounts than in the previous two years. The system's efficiency is very relevant, as, without it, we could have expected these bills to have increased very significantly.

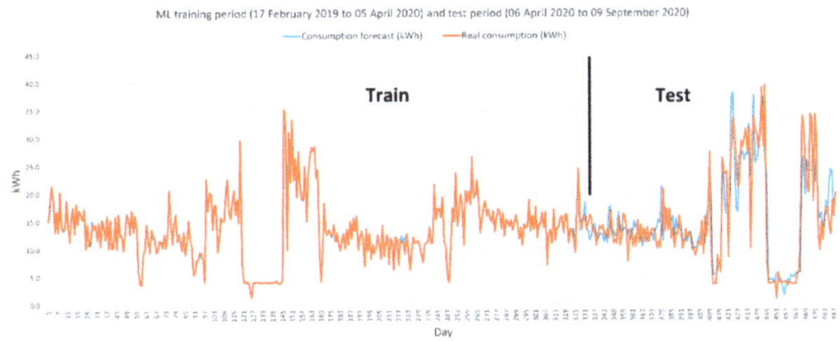

(**a**)

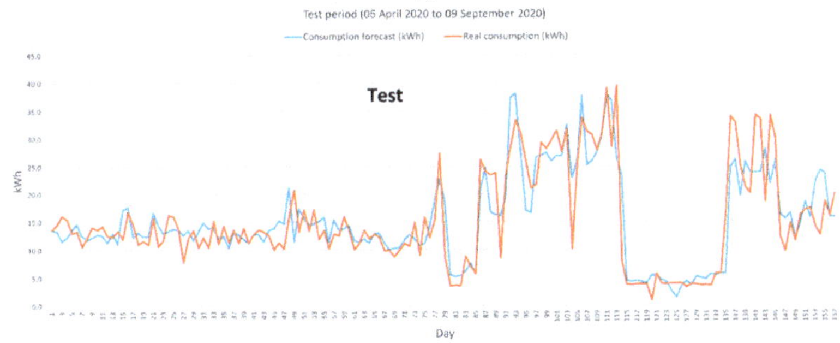

(**b**)

Figure 25. Comparison between estimated consumption and actual consumption in Wh. (**a**) ML Table (17 February 2019 to 5 April 2020) and test period (6 April 2020 to 9 September 2020). (**b**) Test period (6 April 2020 to 9 September 2020).

5. Discussion

Intelligent energy management is a recurring and widely discussed topic in the scientific community. The continuous incorporation of new hardware and software elements is achieving increasingly complex and efficient goals. In this paper, we have presented a novel approach at the crossroads between energy management systems and Voice Assistants. The research is focused on residential environment but could be extended to energy communities, commercial buildings, or microgrids benefiting both customers (energy savings and comfort) and utilities (support of demand side management role in enhancing the flexibility of local energy systems). It combines energy management system, Voice Assistant, IoT, AI, and big data in a single ecosystem to create a novel Energy Management Expert Assistant that learns and adapts to users while improving system efficiency without sacrificing comfort. The system has been developed and implemented in a real pilot, allowing it to evaluate and optimize the decisions taken and improve during its implementation. This practical implementation has required a development that has been spread over two years. It integrates numerous IoT sensors and actuators, thus a large amount of data have been collected and stored in time series and relational databases. The implementation has been developed in three phases (P0–P1, P2, and P3) to optimize the development of the system. In the first period (P1), the habits of the residents were monitored, which made it possible to create a base model to optimize the decisions made by the system within acceptable comfort ranges for the users. The incorporation of the Virtual Assistant has maximized the results obtained. In this phase, we also optimized the best location and type of sensors and actuators to improve comfort and incentivize the participants. Two more phases were developed, being the third one where the system is already in full performance, and the best results are obtained. In this paper, we have presented the data up to this third period.

The work provides new developments in several lines of interest with real experimental results (not simulated) for which a measured deployment of sensors, actuators, as well as the development of IoT applications, recording of large amounts of data, visualization and processing of the data generated, modelling, ML, IoT intelligent environments, ES, and obtaining patterns has been required. It has been developed to obtain energy savings, cost reduction, comfort improvement, and social projection.

In addition to the above benefits, if this energy management system were widely adopted, it could provide interesting value-added elements for both users and utilities. Some of these elements could be: (1) adaptation of residents to routines suggested by the Wizard that allow to modify consumption habits and reduce the amount of bills, (2) load-shifting to the valley times, therefore (3) reducing consumption at peak times, (4) allowing the reduction of total peak demand for distribution grid congestion alleviation, (5) a more flexible response to demand from two levels of action: a first level that would be managed by our system (local) without significantly affecting the comfort of users, and a second level in which it is the aggregator or the utility (external system) which, through a demand response policy, act on the consumption of household appliances, potentially affecting the comfort of users, and (6) social work by reducing consumption and therefore emissions of greenhouse gases or assistance to specific groups with special needs served by the Assistant: elimination of barriers in the home, a sense of companionship, natural connection with the outdoors, information and advice on consumption, etc.

Finally, it should be noted that the ML data and its model have been published (see Data Availability Statement), and given the information it can generate, we believe it will be a fundamental tool for optimizing energy consumption and comfort as a future continuation of this work. Future research directions would focus on adding new elements of power generation, storage, demand response, power quality, greater flexibility of the system to shorten adaptation times for users and vice versa, and consumption prediction to optimize the use of these energy sources, minimize expenditure and maximize comfort. The Wizard will continue to be a fundamental element after the good results achieved during its use in the present work.

Author Contributions: Conceptualization, M.L.-R.; data curation, M.L.-R.; formal analysis, M.L.-R.; funding acquisition, A.M.-M.; investigation, M.L.-R.; methodology, M.L.-R.; project administration, A.M.-M.; resources, J.G.-Z. and A.G.-d.-C.; software, M.L.-R.; supervision, A.M.-M.; validation, J.G.-Z. and A.G.-d.-C.; visualization, J.G.-Z. and A.G.-d.-C.; writing—original draft, M.L.-R.; writing—review and editing, M.L.-R. All authors have read and agreed to the published version of the manuscript.

Funding: This research was funded by the Project IMPROVEMENT (grant SOE3/P3E0901) co-financed by the Interreg SUDOE Programme and the European Regional Development Fund (ERDF), and partially funded by the Spanish Ministry of Economy and Competitiveness under Project TEC2016-77632-C3-2-R.

Institutional Review Board Statement: Not applicable.

Informed Consent Statement: Not applicable.

Data Availability Statement: The Machine Learning and associated data presented in this study are available in Experiments from Azure AI Gallery (https://gallery.azure.ai/experiments accessed on 23 July 2021) under the title "Predicting the energy consumption of a house (BDTR-E1d)". Public link: https://gallery.cortanaintelligence.com/Experiment/Predicting-the-energy-consumption-of-a-house-BDTR-E1d accessed on 23 July 2021.

Conflicts of Interest: The authors declare no conflict of interest.

References

1. Dobakhshari, A.S.; Azizi, S.; Ranjbar, A. Control of microgrids: Aspects and prospects. In Proceedings of the 2011 International Conference on Networking, Sensing and Control, Delft, The Netherlands, 11–13 April 2011; pp. 38–43. [CrossRef]
2. Borkar, S.; Pande, H. Application of 5G Next Generation Network to Internet of Things. In Proceedings of the 2016 International Conference on Internet of Things and Applications (IOTA), Pune, India, 22–24 January 2016; pp. 443–447. [CrossRef]
3. Alsulami, M.M.; Akkari, N. The Role of 5G Wireless Networks in the Internet-of-Things (IoT). In Proceedings of the 2018 1st International Conference on Computer Applications & Information Security (ICCAIS), Riyadh, Saudi Arabia, 4–6 April 2018; p. 8. [CrossRef]
4. Hannaidh, B.O.; Fitzgerald, P.; Berney, H.; Lakshmanan, R.; Coburn, N.; Geary, S.; Mulvey, B. Devices and Sensors Applicable to 5G System Implementations. In Proceedings of the 2018 IEEE MTT-S International Microwave Workshop Series on 5G Hardware and System Technologies (IMWS-5G), Dublin, Ireland, 30–31 August 2018; pp. 1–3. [CrossRef]
5. Rinaldi, K.S.; Bunnen, E. Redefining Home Performance in the 21st Century. Available online: https://web.archive.org/web/20200131171346/http://www.homeperformance.org/sites/default/files/HPC_Smart-Home-Report_201810.pdf (accessed on 10 March 2019).
6. Opportunities for Home Energy Management Systems (HEMS) in Advancing Residential Energy Efficiency Programs. Available online: https://web.archive.org/web/20200927004525/https://neep.org/opportunities-home-energy-management-systems-hems-advancing-residential-energy-efficiency-programs (accessed on 10 March 2019).
7. King, J. Energy Impacts of Smart Home Technologies. Available online: https://www.aceee.org/research-report/a1801 (accessed on 10 March 2019).
8. Zandi, H.; Kuruganti, T.; Vineyard, E.A.; Fugate, D.L. Home Energy Management Systems: An Overview. In Proceedings of the 9th International Conference on Energy Efficiency in Domestic Appliances and Lighting (EEDAL '17), Irvine, CA, USA, 13–15 September 2017; pp. 606–614.
9. Mahapatra, B.; Nayyar, A. Home energy management system (HEMS): Concept, architecture, infrastructure, challenges and energy management schemes. *Energy Syst.* **2019**. [CrossRef]
10. Gun, Y.J.; Vasquez, J.C.; Guerrero, J.M.; Samovich, N.; Vanya, S.; Oravec, V.; Garcia-Castro, R.; Serena, F.; Poveda-Villalon, M.; Radojicic, C.; et al. An Open Virtual Neighbourhood Network to Connect IoT Infrastructures and Smart Objects—VICINITY. In Proceedings of the 2017 Global Internet of Things Summit (GIoTS), Geneva, Switzerland, 6–9 June 2017; pp. 134–139. [CrossRef]
11. Ford, R.; Karlin, B.; Sanguinetti, A.; Nersesyan, A.; Pritoni, M. Assessing Players, Products, and Perceptions of Home Energy Management. Available online: https://www.researchgate.net/publication/320864681 (accessed on 11 July 2020).
12. Gungor, V.C. Smart Grid Technologies: Communication Technologies and Standards. *IEEE Trans. Ind. Inform.* **2011**, *7*, 529–539. [CrossRef]
13. Robotina Platform. Available online: https://web.archive.org/web/20190809154308/https://robotinarox.io/robotina-platform/ (accessed on 10 March 2019).
14. Shakeri, M. Implementation of a novel home energy management system (HEMS) architecture with solar photovoltaic system as supplementary source. *Renew. Energy* **2018**, *125*, 108–120. [CrossRef]
15. Shareef, H. Review on Home Energy Management System Considering Demand Responses, Smart Technologies, and Intelligent Controllers. *IEEE Access* **2018**, *6*, 24498–24509. [CrossRef]
16. Kejriwal, S.; Mahajan, S. Smart Buildings: How IoT Technology Aims to Add Value for Real Estate Companies. Available online: http://web.archive.org/web/20200714113546/https://www2.deloitte.com/content/dam/Deloitte/nl/Documents/real-estate/deloitte-nl-fsi-real-estate-smart-buildings-how-iot-technology-aims-to-add-value-for-real-estate-companies.pdf (accessed on 11 July 2020).
17. Guang-Yu, J. IoT, Big Data, Cloud Computing, Artificial Intelligence/Machine Learning, BIM and Digital Twins in Building Automation and Management Applications. Available online: https://web.archive.org/web/20210726100646/https://www.sleb.sg/UserFiles/Resource/Technology%20Review%20Report/IoT,%20Big%20Data,%20Cloud%20Computing,%20Artificial%20Intelligence%20and%20Machine%20Learning,%20BIM%20and%20Digital%20Twins%20in%20Building%20Automation%20and%20Management%20Applications.pdf (accessed on 11 July 2020).
18. Jerabandi, M.; Kodabagi, M.M. A Review on Home Automation System. In Proceedings of the 2017 International Conference On Smart Technologies for Smart Nation (SmartTechCon), Bengaluru, India, 17–19 August 2017; pp. 1411–1415. [CrossRef]
19. Home Assistant. Available online: https://www.home-assistant.io/ (accessed on 12 March 2019).
20. Open Hab. Available online: https://www.openhab.org/ (accessed on 12 March 2019).
21. Eclipse SmartHome—A Flexible Framework for the Smart Home. Available online: https://www.eclipse.org/smarthome/ (accessed on 23 February 2019).
22. Building Energy Management Open Source Software (BEMOSS™). Available online: http://www.bemoss.org/ (accessed on 10 March 2019).
23. VOLTTRON™. Available online: https://volttron.org/ (accessed on 10 March 2019).
24. ioBroker. Available online: http://www.iobroker.net/ (accessed on 10 March 2019).
25. OGEMA (Open Gateway Energy Management). Available online: https://www.ogema.org/ (accessed on 10 March 2019).

26. Zillgith, M.; Nestle, D.; Wagner, M. Security Architecture of the OGEMA 2.0 Home Energy Management System. In Proceedings of the International ETG-Congress 2013; Symposium 1: Security in Critical Infrastructures Today, Berlin, Germany, 5–6 November 2013; pp. 1–6.
27. OpenRemote. Available online: http://www.openremote.com/ (accessed on 10 March 2019).
28. Ray, P.P. A survey on Internet of Things architectures. *J. King Saud Univ.—Comput. Inf. Sci.* **2018**, *30*, 291–319. [CrossRef]
29. Li, Y.; Huang, Y.; Zhang, M.; Rajabion, L. Service selection mechanisms in the Internet of Things (IoT): A systematic and comprehensive study. *Clust. Comput.* **2019**, *23*, 1163–1183. [CrossRef]
30. Krotov, V. The Internet of Things and new business opportunities. *Bus. Horiz.* **2017**, *60*, 831–841. [CrossRef]
31. Saarikko, T.; Westergren, U.H.; Blomquist, T. The Internet of Things: Are you ready for what's coming? *Bus. Horiz.* **2017**, *60*, 667–676. [CrossRef]
32. Nordrum, A. Popular Internet of Things Forecast of 50 Billion Devices by 2020 Is Outdated. Available online: https://spectrum.ieee.org/tech-talk/telecom/internet/popular-internet-of-things-forecast-of-50-billion-devices-by-2020-is-outdated (accessed on 12 July 2020).
33. Earle, N. 50 Billion Things, Coming to a Cloud Near You. Available online: https://web.archive.org/web/20210726070602/https://blogs.cisco.com/news/50-billion-things-coming-to-a-cloud-near-you?dtid=osscdc000283 (accessed on 12 July 2020).
34. A Guide to the Internet of Things. How Billions of Online Objects Are Making the Web Wiser. Available online: https://web.archive.org/web/20200729014237/https://www.intel.com/content/www/us/en/internet-of-things/infographics/guide-to-iot.html (accessed on 29 July 2020).
35. Pureswaran, V.; Brody, P. Device Democracy: Saving the Future of the Internet of Things. Available online: https://web.archive.org/web/20190727040755/https://www.ibm.com/downloads/cas/Y5ONA8EV (accessed on 23 February 2019).
36. Zafar, U.; Bayhan, S.; Sanfilippo, A. Home Energy Management System Concepts, Configurations, and Technologies for the Smart Grid. *IEEE Access* **2020**, *8*, 119271–119286. [CrossRef]
37. Snijders, C.; Matzat, U.; Reips, U.D. "Big Data": Big Gaps of Knowledge in the Field of Internet Science. *Int. J. Internet Sci.* **2012**, *7*, 1–5.
38. Billings, S.A. *Nonlinear System Identification: NARMAX Methods in the Time, Frequency, and Spatio-Temporal Domains*; John Wiley & Sons: Hoboken, NJ, USA, 2013; pp. 1–555. [CrossRef]
39. Mayer-Schönberger, V.; Cukier, K. *Big Data: A Revolution That Will Transform How We Live, Work, and Think*; Houghton Mifflin Harcourt: Boston, MA, USA, 2013; p. 242.
40. Lee, I. Big data: Dimensions, evolution, impacts, and challenges. *Bus. Horiz.* **2017**, *60*, 293–303. [CrossRef]
41. Kaplan, A.; Haenlein, M. Siri, Siri, in my hand: Who's the fairest in the land? On the interpretations, illustrations, and implications of artificial intelligence. *Bus. Horiz.* **2019**, *62*, 15–25. [CrossRef]
42. Poole, D.; Mackworth, A.; Goebel, R. *Computational Intelligence: A Logical Approach*; Oxford University Press: Oxford, NY, USA, 1998.
43. McCarthy, J. What is Artificial Intelligence? Available online: https://web.archive.org/web/20200730225611/http://www-formal.stanford.edu/jmc/whatisai/node1.html (accessed on 12 July 2020).
44. Nilsson, N.J.; Nilsson, N.J. *Artificial Intelligence: A New Synthesis*; Morgan Kaufmann: Burlington, MA, USA, 1998.
45. Russell, S.; Norving, P. *Artificial Intelligence: A Modern Approach*; Addison Wesley: Boston, MA, USA, 2018; p. 1132.
46. Contreras, I.; Vehi, J. Artificial Intelligence for Diabetes Management and Decision Support: Literature Review. *J. Med. Internet Res.* **2018**, *20*, e10775. [CrossRef]
47. Liu, Q.; Yang, X.; Deng, L. An IBeacon-Based Location System for Smart Home Control. *Sensors* **2018**, *18*, 1897. [CrossRef]
48. Yang, D.; Xu, B.; Rao, K.; Sheng, W. Passive Infrared (PIR)-Based Indoor Position Tracking for Smart Homes Using Accessibility Maps and A-Star Algorithm. *Sensors* **2018**, *18*, 332. [CrossRef] [PubMed]
49. Singh, S.; Aksanli, B. Non-Intrusive Presence Detection and Position Tracking for Multiple People Using Low-Resolution Thermal Sensors. *J. Sens. Actuator Netw.* **2019**, *8*, 40. [CrossRef]
50. Kepuska, V.; Bohouta, G. Next-Generation of Virtual Personal Assistants (Microsoft Cortana, Apple Siri, Amazon Alexa and Google Home). In Proceedings of the 2018 IEEE 8th Annual Computing and Communication Workshop and Conference (CCWC), Las Vegas, NV, USA, 8–10 January 2018; pp. 99–103.
51. Hoy, M.B. Alexa, Siri, Cortana, and more: An introduction to voice assistants. *Med Ref. Serv. Q.* **2018**, *37*, 81–88. [CrossRef]
52. McTear, M.F. Spoken dialogue technology: Enabling the conversational user interface. *ACM Comput. Surv.* **2002**, *34*, 90–169. [CrossRef]
53. Voice assistants steal the show. *New Sci.* **2018**, *237*, 7. [CrossRef]
54. Reis, A.; Paulino, D.; Paredes, H.; Barroso, I.; Monteiro, M.J.; Rodrigues, V.; Barroso, J. Using intelligent personal assistants to assist the elderlies An evaluation of Amazon Alexa, Google Assistant, Microsoft Cortana, and Apple Siri. In Proceedings of the 2018 2nd International Conference on Technology and Innovation in Sports, Health and Wellbeing (Tishw), Thessaloniki, Greece, 20–22 June 2018; pp. 1–5. [CrossRef]
55. Dojchinovski, D.; Ilievski, A.; Gusev, M. Interactive home healthcare system with integrated voice assistant. In Proceedings of the 2019 42nd International Convention on Information and Communication Technology, Electronics and Microelectronics (Mipro), Opatija, Croatia, 20–24 May 2019; pp. 284–288. [CrossRef]

56. Friedman, N.; Cuadra, A.; Patel, R.; Azenkot, S.; Stein, J.; Ju, W.; Assoc Comp, M. Voice Assistant Strategies and Opportunities for People with Tetraplegia. In Proceedings of the 21st International ACM SIGACCESS Conference on Computers and Accessibility (ASSETS '19), Pittsburgh, PA, USA, 28–30 October 2019; pp. 575–577. [CrossRef]
57. O'Brien, K.; Liggett, A.; Ramirez-Zohfeld, V.; Sunkara, P.; Lindquist, L.A. Voice-Controlled Intelligent Personal Assistants to Support Aging in Place. *J. Am. Geriatr. Soc.* **2020**, *68*, 176–179. [CrossRef]
58. Lee, K.; Lee, K.Y.; Sheehan, L. Hey Alexa! A Magic Spell of Social Glue?: Sharing a Smart Voice Assistant Speaker and Its Impact on Users' Perception of Group Harmony. *Inf. Syst. Front.* **2020**, *22*, 563–583. [CrossRef]
59. Tiwari, V.; Hashmi, M.F.; Keskar, A.; Shivaprakash, N.C. Virtual home assistant for voice based controlling and scheduling with short speech speaker identification. *Multimed. Tools Appl.* **2020**, *79*, 5243–5268. [CrossRef]
60. Alepis, E.; Patsakis, C. Monkey Says, Monkey Does: Security and Privacy on Voice Assistants. *IEEE Access* **2017**, *5*, 17841–17851. [CrossRef]
61. Lei, X.Y.; Tu, G.H.; Liu, A.X.; Li, C.Y.; Xie, T. The Insecurity of Home Digital Voice Assistants—Vulnerabilities, Attacks and Countermeasures. In Proceedings of the 6th IEEE Conference on Communications and Network Security (CNS), Beijing, China, 30 May–1 June 2018.
62. Lei, X.; Tu, G.H.; Liu, A.X.; Ali, K.; Li, C.Y.; Xie, T. The Insecurity of Home Digital Voice Assistants—Amazon Alexa as a Case Study. *arXiv* **2017**, arXiv:1712.03327.
63. Link Your Voice to Your Devices with Voice Match. Available online: https://web.archive.org/web/20200901043712/https://support.google.com/assistant/answer/9071681?co=GENIE.Platform%3DAndroid&hl=en (accessed on 23 July 2020).
64. What Is the Voice Profile Setting Personalize Skills? Available online: https://web.archive.org/web/20210726090521/https://www.amazon.com/gp/help/customer/display.html/ref=hp_left_v4_sib?ie=UTF8&nodeId=GF4LMSZG6SUFHQUZ (accessed on 11 July 2020).
65. Leitao, J.; Gil, P.; Ribeiro, B.; Cardoso, A. A Survey on Home Energy Management. *IEEE Access* **2020**, *8*, 5699–5722. [CrossRef]
66. Rasheed, M.B.; Javaid, N.; Ahmad, A.; Khan, Z.A.; Qasim, U.; Alrajeh, N. An Efficient Power Scheduling Scheme for Residential Load Management in Smart Homes. *Appl. Sci.* **2015**, *5*, 1134–1163. [CrossRef]
67. Bayram, I.S.; Ustun, T.S. A survey on behind the meter energy management systems in smart grid. *Renew. Sustain. Energy Rev.* **2017**, *72*, 1208–1232. [CrossRef]
68. Beaudin, M.; Zareipour, H. Home energy management systems: A review of modelling and complexity. *Renew. Sustain. Energy Rev.* **2015**, *45*, 318–335. [CrossRef]
69. Palacios-Garcia, E.J.; Chen, A.; Santiago, I.; Bellido-Outeiriño, F.J.; Flores-Arias, J.M.; Moreno-Munoz, A. Stochastic model for lighting's electricity consumption in the residential sector. Impact of energy saving actions. *Energy Build.* **2015**, *89*, 245–259. [CrossRef]
70. EL Jaouhari, S.; Jose Palacios-Garcia, E.; Anvari-Moghaddam, A.; Bouabdallah, A. Integrated Management of Energy, Wellbeing and Health in the Next Generation of Smart Homes. *Sensors* **2019**, *19*, 481. [CrossRef]
71. Cao, Z.; O'Rourke, F.; Lyons, W.; Han, X. Home Energy Management System Incorporating Heat Pump Using Real Measured Data. *Sensors* **2019**, *19*, 2937. [CrossRef]
72. Chavali, P.; Yang, P.; Nehorai, A. A distributed algorithm of appliance scheduling for home energy management system. *IEEE Trans. Smart Grid* **2014**, *5*, 282–290. [CrossRef]
73. Anvari-Moghaddam, A.; Monsef, H.; Rahimi-Kian, A. Optimal Smart Home Energy Management Considering Energy Saving and a Comfortable Lifestyle. *IEEE Trans. Smart Grid* **2015**, *6*, 324–332. [CrossRef]
74. Golmohamadi, H.; Keypour, R.; Bak-Jensen, B.; Pillai, J.R. Optimization of household energy consumption towards day-ahead retail electricity price in home energy management systems. *Sustain. Cities Soc.* **2019**, *47*, 101468. [CrossRef]
75. Silva, B.N.; Khan, M.; Han, K. Load Balancing Integrated Least Slack Time-Based Appliance Scheduling for Smart Home Energy Management. *Sensors* **2018**, *18*, 685. [CrossRef]
76. Li, L.D.; Li, X.D.; Yu, X.H. A Multi-Objective Constraint-Handling Method with PSO Algorithm for Constrained Engineering Optimization Problems. In Proceedings of the 2008 IEEE Congress on Evolutionary Computation, Hong Kong, China, 1–6 June 2008; pp. 1528–1535. [CrossRef]
77. Wang, Y.; Cai, Z.; Guo, G.; Zhou, Y. Multiobjective optimization and hybrid evolutionary algorithm to solve constrained optimization problems. *IEEE Trans. Syst. Man Cybern. Part B (Cybern.)* **2007**, *37*, 560–575. [CrossRef] [PubMed]
78. Joo, I.Y.; Choi, D.H. Optimal Household Appliance Scheduling Considering Consumer's Electricity Bill Target. *IEEE Trans. Consum. Electron.* **2017**, *63*, 19–27. [CrossRef]
79. Chen, X.; Wei, T.; Hu, S. Uncertainty-Aware Household Appliance Scheduling Considering Dynamic Electricity Pricing in Smart Home. *IEEE Trans. Smart Grid* **2013**, *4*, 932–941. [CrossRef]
80. Nan, S.B.; Zhou, M.; Li, G.Y.; Xia, Y. Optimal Scheduling Approach on Smart Residential Community Considering Residential Load Uncertainties. *J. Electr. Eng. Technol.* **2019**, *14*, 613–625. [CrossRef]
81. Lokeshgupta, B.; Sivasubramani, S. Multi-objective home energy management with battery energy storage systems. *Sustain. Cities Soc.* **2019**, *47*, 101458. [CrossRef]
82. Chen, Z.; Wu, L.; Fu, Y. Real-Time Price-Based Demand Response Management for Residential Appliances via Stochastic Optimization and Robust Optimization. *IEEE Trans. Smart Grid* **2012**, *3*, 1822–1831. [CrossRef]

83. Zakariazadeh, A.; Jadid, S.; Siano, P. Smart microgrid energy and reserve scheduling with demand response using stochastic optimization. *Int. J. Electr. Power Energy Syst.* **2014**, *63*, 523–533. [CrossRef]
84. Powell, W.B. A unified framework for stochastic optimization. *Eur. J. Oper. Res.* **2019**, *275*, 795–821. [CrossRef]
85. Wang, C.S.; Zhou, Y.; Wu, J.Z.; Wang, J.D.; Zhang, Y.Q.; Wang, D. Robust-Index Method for Household Load Scheduling Considering Uncertainties of Customer Behavior. *IEEE Trans. Smart Grid* **2015**, *6*, 1806–1818. [CrossRef]
86. Erol-Kantarci, M.; Mouftah, H.T. Wireless sensor networks for cost-efficient residential energy management in the smart grid. *IEEE Trans. Smart Grid* **2011**, *2*, 314–325. [CrossRef]
87. Bourdeau, M.; Qiang Zhai, X.; Nefzaoui, E.; Guo, X.; Chatellier, P. Modeling and forecasting building energy consumption: A review of data-driven techniques. *Sustain. Cities Soc.* **2019**, *48*, 101533. [CrossRef]
88. Tardioli, G.; Kerrigan, R.; Oates, M.; James, O.D.; Finn, D. Data driven approaches for prediction of building energy consumption at urban level. *Energy Procedia* **2015**, *78*, 3378–3383. [CrossRef]
89. Sardahi, Y.H. *Multi-Objective Optimal Design of Control Systems*; UC Merced: Merced, CA, USA, 2016.
90. Article 8.1.b of Royal Decree 216/2014 of the Spanish Legislation. Available online: https://www.boe.es/eli/es/rd/2014/03/28/216 (accessed on 10 March 2019).
91. Althaher, S.; Mancarella, P.; Mutale, J. Automated demand response from home energy management system under dynamic pricing and power and comfort constraints. *IEEE Trans. Smart Grid* **2015**, *6*, 1874–1883. [CrossRef]
92. Ahmad, S.; Naeem, M.; Ahmad, A. Low complexity approach for energy management in residential buildings. *Int. Trans. Electr. Energy Syst.* **2019**, *29*, e2680. [CrossRef]
93. Likebupt. Algorithm & Module Reference—Azure Machine Learning. Available online: https://web.archive.org/web/20210701001309/https://docs.microsoft.com/en-us/azure/machine-learning/algorithm-module-reference/module-reference (accessed on 11 December 2020).
94. Criminisi, A.; Shotton, J.; Konukoglu, E. Decision Forests: A Unified Framework for Classification, Regression, Density Estimation, Manifold Learning and Semi-Supervised Learning. *Found. Trends Comput. Graph. Vis.* **2012**, *7*, 81–227. [CrossRef]
95. Oprea, S.-V.; Pîrjan, A.; Căruțașu, G.; Petroșanu, D.-M.; Bâra, A.; Stănică, J.-L.; Coculescu, C. Developing a Mixed Neural Network Approach to Forecast the Residential Electricity Consumption Based on Sensor Recorded Data. *Sensors* **2018**, *18*, 1443. [CrossRef] [PubMed]

sensors

MDPI

Article

Perturbed-Location Mechanism for Increased User-Location Privacy in Proximity Detection and Digital Contact-Tracing Applications

Elena Simona Lohan [1], Viktoriia Shubina [1,2,*] and Dragoș Niculescu [2]

1 Electrical Engineering Unit, Tampere University, 33720 Tampere, Finland; elena-simona.lohan@tuni.fi
2 Computer Science and Engineering Department, University Politehnica of Bucharest, 060042 Bucharest, Romania; dragos.niculescu@upb.ro
* Correspondence: viktoriia.shubina@tuni.fi

Abstract: Future social networks will rely heavily on sensing data collected from users' mobile and wearable devices. A crucial component of such sensing will be the full or partial access to user's location data, in order to enable various location-based and proximity-detection-based services. A timely example of such applications is the digital contact tracing in the context of infectious-disease control and management. Other proximity-detection-based applications include social networking, finding nearby friends, optimized shopping, or finding fast a point-of-interest in a commuting hall. Location information can enable a myriad of new services, among which we have proximity-detection services. Addressing efficiently the location privacy threats remains a major challenge in proximity-detection architectures. In this paper, we propose a location-perturbation mechanism in multi-floor buildings which highly protects the user location, while preserving very good proximity-detection capabilities. The proposed mechanism relies on the assumption that the users have full control of their location information and are able to get some floor-map information when entering a building of interest from a remote service provider. In addition, we assume that the devices own the functionality to adjust to the desired level of accuracy at which the users disclose their location to the service provider. Detailed simulation-based results are provided, based on multi-floor building scenarios with hotspot regions, and the tradeoff between privacy and utility is thoroughly investigated.

Keywords: location privacy; perturbation mechanism; proximity detection; digital contact tracing; multi-floor areas

Citation: Lohan, E.S.; Shubina, V.; Niculescu, D. Perturbed-Location Mechanism for Increased User-Location Privacy in Proximity Detection and Digital Contact-Tracing Applications. *Sensors* **2022**, *22*, 687. https://doi.org/10.3390/s22020687

Academic Editors: Suparna De and Klaus Moessner

Received: 11 December 2021
Accepted: 14 January 2022
Published: 17 January 2022

1. Introduction and Problem Statement

People are increasingly interconnected through their wireless devices, such as smartphones, smartwatches, and other wearable devices. Most of such devices are already capable of localization and sensing, either through Global Navigation Satellite Systems (GNSS) chipsets in outdoor scenarios or through IEEE802.11* (e.g., WiFi), Ultra-Wide Band (UWB), or Bluetooth Low Energy (BLE) chipsets in indoor scenarios. Many future wireless standards will also make localization and sensing as a part of the system design, such as emerging Sixth generation of cellular communications (6G) cellular communications [1], IEEE802.11bf WiFi upcoming standard [2], and UWB chipsets incorporated in modern smartphones [3].

Proximity-detection services based on wireless signals, and in particular based on BLE, have gained a significant interest in the past two years as they are enabling digital contract-tracing techniques [4] shown to be relevant in the context of COVID-19 disease management [5,6]. Magnetic-field proximity detection solutions have also been recently proposed in the context of digital contact tracing, for example, in [7].

Digital contact tracing is an approach that has been built according to the privacy-by-design concept to augment the manual ways of tracing the COVID-19-disease spread. By design, mobile and wireless gadgets equipped with BLE chipsets can transmit and receive anonymized signals with timestamps from nearby devices. This concept has become handy for digital contact-tracing purposes in the past year, since the BLE is a short-range technology that is particularly suitable for estimating close-range distances (e.g., less than 2 m) of the mobile phone users who crossed their paths. The BLE data with temporary identifiers, Received Signal Strength (RSS) values, and the timestamps of the encountered phones are therefore converted into the distance and time spent in proximity. Furthermore, there is a taxonomy [6,8] of centralized and decentralized decision-making approaches to handle data processing and inform the users about the risk of being exposed to the virus.

In the centralized approach [6,9], the logs from the mobile phone (or wearable bracelet) are encrypted and transferred to the cloud with a certain periodicity (e.g., once a day). Therefore in cases where the users opt-in to the protocol, the centralized server estimates the risk of being exposed and conveys this risk to the users. The majority of centralized approaches follow the data minimization principle and request to upload only relevant data, such as the temporary or ephemeral identities of the users who stayed within certain proximity for the time exceeding the set threshold. As an outcome, all computations for the risk scoring are made on the server-side, and the users only receive the notifications.

A different approach, known as decentralized or federated, delegates the risk scoring to own mobile devices or user edge devices, considering the logs are stored locally. Google and Apple adopted the consequent framework in their jointly designed Exposure Notifications protocol described in [10]. Here, only infected users, once confirmed being tested positive, upload the data to the cloud, whereas the rest of the users' devices download the data from the server and perform the risk estimates locally on their devices. The latter approach assumes that all data shared with the centralized server is subject to the user's consent.

As subjectively deemed in [6], based on end-user surveys, the users are more likely to perceive the decentralized decision-making approach as a better fit to preserve their location privacy due to the fact that the data is stored locally (typically for up to 21 days, unlike the server-side storage which can be much longer). However, there is no significant threat to the users' sensitive information in the centralized approach where the logs are encrypted and securely saved on a trusted server. The above-mentioned digital contact-tracing example demonstrates that the location privacy concerns take place in the context of sensitive information, such as one's whereabouts and identities of encountered contacts.

Location Privacy-Preserving Mechanisms (LPPM) intend to preserve the individual location privacy in scenarios where services request access to the users' spatial location [11]. Location-Based Services (LBS) that collect sensitive information of the users' locations, as described in the classification framework in [12], can benefit from implementing LPPM.

Other examples of proximity-based services are 'find-a-friend' applications [13] or other social-networking applications [14].

In all these proximity-based services, the utility of the services comes from a good detection probability (i.e., the probability to correctly detect two users in the vicinity of each other when they are neighbours, also known as sensitivity measure) as well as a low false-alarm probability (i.e., the probability of incorrectly detecting two users in the vicinity of each other when in fact they are far away). This utility is inherently in a tradeoff with the amount of location privacy that a user can have when disclosing his location.

In order to protect users' location privacy, many approaches have been proposed so far in the literature. For example, a comprehensive survey of location-privacy mechanisms has been recently provided in [15]. The authors in [15] divided the location-privacy mechanisms into three classes: the Geo-indistinguishability (GeoInd) class, the Local Differential Privacy (LDP) class, and private spatial-decomposition class. They also pointed out that the LDP mechanism is not directly applicable to location data, while the private spatial decomposition requires the presence of a trusted server.

Once LPPM have been implemented, it is necessary to evaluate their behavior and compare it with the initial state of the system. GeoInd refers to a privacy notion that preserves the user's precise location while revealing approximate geospatial area [16].

Furthermore, when a user disclose its location with a certain perturbation mechanism, this perturbation mechanism can yield GeoInd [17] if the traces of the user are disclosed with a certain radius and certain statistical distributions, such as when Laplacian or Gaussian random perturbations are applied to modify the true user location. The reported location will not reveal information to an adversary for distinguishing the ground truth location among neighboring devices [18].

The authors in [17], presented GeoInd as a possible notion to quantify privacy. They introduced the radius r, which corresponds to the level of privacy and showed that such radius is proportional to the location radius, i.e., the Euclidean distance between the true and perturbed locations. Consequently, the radius is increasing by adding controlled randomized (e.g., Laplacian) noise. The authors have encountered problems of discretization and truncation. In our paper we directly use the Euclidian distance between the true and perturbed locations as a measure of user location privacy and we study its tradeoff with the service utility.

Another location privacy-preserving approach in the literature, which is an adherent of Differential Privacy (DP), is the concept of the Private Spatial Decomposition presented in [19]. Private Spatial Decomposition refers to a gradient privacy-budget allocation scheme. The approach assumes a two-dimensional space and different privacy levels, and it is proved to achieve ϵ-differential privacy.

An additional aspect related to the location privacy is the choice of the privacy metric, which is still not unified in the current literature. Such a privacy metric serves to quantify the efficiency of a localization algorithm by exploring the privacy versus accuracy [20] or the privacy versus utility [21] tradeoffs. As above-mentioned, in this paper we measure the location privacy via the Root Mean Square Error (RMSE) between the perturbed location and the true user location.

The authors in [22] proposed a location-aware perturbation scheme for mobile environments, where the goal was to decrease the adversary's knowledge with added Laplacian noise. Using the Hilbert curve, each second location is projected on a map, thus reducing the overhead caused by the precision of the location estimates. To evaluate the performance and accuracy of the proposed algorithm, the authors in [22] used nearness, resemblance, and displacement metrics. As a common rule, lower levels of ϵ correspond to a higher privacy budget and effectively lower accuracy. For example, in [22], when the ϵ value reached 1.0, the number of points located within 1000 m of the actual positions were a high as 99.04 percent.

Albeit obfuscation mechanisms are growing in their popularity, they introduce errors to the localization system by altering the ground truth locations of the devices. Obfuscation mechanisms result in losing some of the performance, or in other words, the utility of the system. In [18], the authors designed a location obfuscation mechanism, where the GeoInd was satisfied. This work in [18] focused on achieving GeoInd for any pair of neighboring pairs of locations and they showed good results for privacy and utility in 2D spaces. Our work focuses on 3D spaced with multi-floor buildings.

To the best of our knowledge, studies investigating the optimal tradeoff between obfuscating or perturbing the user location (i.e., decreasing the granularity of the reported location) versus utility for proximity-detection applications are still not well explored in the current literature, especially when such a proximity-detection application is a digital contact-tracing solution. Moreover, multidimensional approaches, such as 3D scenarios, provide more freedom for the user to protect their location from an adversary and have not been studied a lot so far.

This paper proposes a new perturbation metric suitable for proximity-detection-based services and applications relying strictly on the relative distance between two users, but not needing absolute location information, offers a theoretical analysis of its properties, and demonstrates via extensive simulation-based results a very good tradeoff between privacy preservation and service utility. The proposed metric is based on a combination of mapping based on the argmax operator and Gaussian or Laplacian perturbations. For comparative purposes, the argmax-based metric is also compared with another metric, based on an argmin operator and Gaussian or Laplacian perturbations, and we show that it has a much better utility-privacy tradeoff than the argmin-based metric. It is to be noticed that the proposed argmax-based metric is only useful in the context of proximity-based services, when only the relative distance between users is needed, but not their absolute location. By contrast, the argmin-based metric would preserve its utility also for other location-based services (in addition to the proximity-based ones), at the expense of lower privacy protection compared to the argmax-based metric.

The remainder of the paper is organized as follows: Section 2 overviews various mechanisms for preserving location privacy in the literature and offers a classification of these mechanisms. Section 3 introduces the two proposed perturbation mechanisms, one based on argmax operator, suitable only for proximity-based services and another one based on argmin operator, suitable for all kinds of location-based services, but with lower privacy preservation levels than the one based on argmax operator. Section 4 offers a mathematical analysis of the proposed argmax operator and proves that it is able to offer GeoInd between users. Section 5 presents detailed simulation results in a 4-floor building with users located both within certain hotspot areas and outside hotspot areas. The presented results are easily scalable to any number of floors. Various configurations, in terms of building size, hotspot density, etc., are analyzed, and detailed results are presented in terms of user privacy and service utility. Finally, Section 6 summarizes the main findings and presents the conclusions.

2. Classification of Location-Privacy Mechanisms

A classification of location-privacy mechanisms from current literature is provided in Figure 1. The location privacy can be ensured by the server side, by the user side or can be applicable at both sides. A more elaborate explanation of each technique can be found in Table 1 and it is based also on the literature review provided in Section 1.

User-side location privacy mechanisms can be found for example in [23]. Privacy-preserving mappings solutions are born from optimal mappings to preserve privacy against statistical inference [24,25]. Noise perturbation mechanisms based on various noise types, such as Laplace and Gaussian noises are discussed for example in [26,27]. Dummy-location generation has been applied, for example, in [28].

Server-side location privacy mechanisms relying on spatial cloaking and k-anonymity mechanisms are described, for example, in [29–32]. Unlike in our paper, the assumptions in [32] are that the users communicate their location to the server with high accuracy; in our paper we assume that the users have full control to their location and choose to disclose it to the server with moderate-to-low accuracy, according to the chosen perturbation mechanisms, as explained later, in Section 3.

Private spatial decomposition solutions are discussed for example in [19]. Mix-zones solutions are addressed for example in [33,34]. Secure transformations are conceptually close to the privacy-preserving mappings done at the user/client side and they are addressed for example in [35]. Server-side solutions involve the trust in the service provider and they are susceptible to attacks of the server databases.

A privacy-preserving method that can be applied both at server and user sides is the encryption of location data, via various encryption mechanisms [36–38]. Even if encryption/decryption costs are quite affordable by nowadays mobile devices and smartphones, the encryption/decryption studies for location privacy available in the current literature point out that a main drawback of this approach is the relatively high delay [37] introduced in the data encryption/decryption processes, delay which may be not tolerable for many proximity-based services.

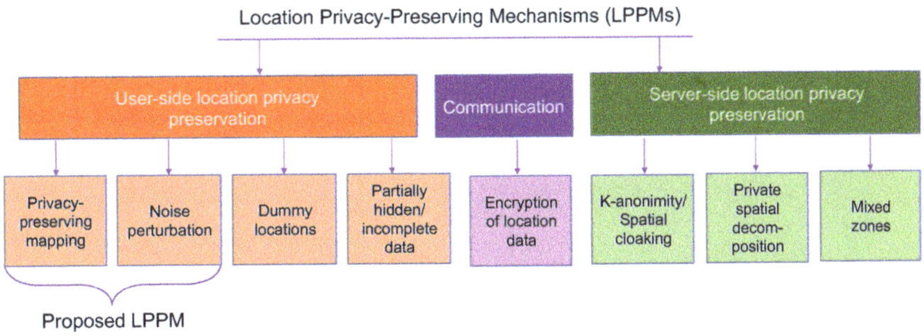

Figure 1. Three-fold classification of location-privacy mechanisms: starting from the edge device, a.k.a. user side (including two parts of the proposed privacy-preserving technique), communication part used for transferring data packets, and server-side perspective including the cases where the users' data is aggregated on the server.

Our proposed solutions, described in the next section, is a combination of a privacy-preserving mapping (two mappings provided) and a noisy perturbation (two noise distributions studied).

Table 1. Overview of LPPM in the literature.

Location-Preservation Area	Mechanism	Main Features	Refs.
User-side	Privacy-preserving mapping	Multiple initialization and data collection steps are required to build the initial map for further feature extraction and matching.	[24,25]
User-side	Noise Perturbation	The concept of adding noise from a sample distribution and modifying the reported locations of the users. This approach is easy to break in cases where the adversary has prior knowledge about the noise model in use.	[26,27,39]
User-side	Dummy locations	The mechanism is susceptible to inference attacks, easy to break with an application of heterogeneous location correlations.	[28,40,41]
User-side	Partially hidden (incomplete) data	This method assumes ditching or deliberately hiding non-essential pieces of data, which could reveal sensitive information of the users' whereabouts. This method is easy to break with an application of heterogeneous correlations.	[39]
Communication	Encryption	For security reasons, all data should be encrypted, consequently, this might cause insignificant delays in transferring the packets within a communication scheme [42].	[36–38]
Server-side	k-anonymity/ Spatial cloaking	Minimizes risks of re-identification of anonymized data; however, this approach is susceptible to privacy breaches, such as de-anonymization, in cases where the adversary has prior knowledge about individuals. To tackle the issue, such approaches as *t-closeness* and *l-diversity* were developed to augment the *k-anonymity* privacy protection [43,44].	[29,45]
Server-side	Private spatial decomposition	Via applications of the hierarchical decomposition, the location data is stored in clusters, being decomposed into small pieces.	[19,46]
Server-side	Mixed zones	This method aggregates the user data with common attributes and generalizes the location to set areas, having bigger radii than the ground truth location. Therefore, it is not providing a solid basis for preserving privacy as some data are still revealed.	[33,34]

3. Proposed Perturbed Location Mechanism

3.1. Scenario Definition, Hypotheses, and Preliminary Notations

We adopt a scenario when user devices are equipped with some form of an indoor localization engine, e.g., a combination of cellular-based positioning, WiFi/BLE-positioning, and other smartphone sensors-based positioning (barometers, gyroscopes, accelerometers), etc., which is already the state-of-the-art of indoor positioning. We also assume that each user u can have full control of his/her location data, modeled here via a 3D-location vector $\mathbf{x}_u \in \mathbf{B}$. It is also assumed that the used can choose the perturbation level with which he/she disclose own location data to a service provider. Thus, the user devices are able to apply a local perturbation mechanism $M(\mathbf{x}_u)$, before broadcasting the user location data to a service provider. Such service provider can be, for example, a centralized digital contact-tracing server which computes, based on the available perturbed locations $M(\mathbf{x}_u)$ the relative distances between any two users in the building and compares them to a safety threshold γ (e.g., $\gamma = 2$ m). The server stores such information in a database, together with timestamps and hashed users identities and when a user v informs the server that he or she has been detected with COVID-19, the server is able to find the information about all other users u that were in the vicinity of user v in a certain time window. For simplicity, we drop the time index in our model and look at snapshot decisions. Thus, if $||M(\mathbf{x}_u) - M(\mathbf{x}_v)|| \leq \gamma$, user u is informed by the contact-tracing server that he or she has been a 'close contact'. Above, $|| \cdot ||$ is the square root of the Euclidean norm (or the distance between two vectors).

Another example of a service provider relying on such proximity detection is a provider of a 'find a friend' service. Again, users can install an application which transmits to the service provider the hashed identities of themselves and their friends, and the server is keeping track of the $||M(\mathbf{x}_u) - M(\mathbf{x}_v)||$ distances, based on the perturbed location information transmitted by each user. If $||M(\mathbf{x}_u) - M(\mathbf{x}_v)|| \leq \gamma$, then the users u and v are informed that their friend is nearby, at a distance γ. Again, the threshold parameter γ can be user defined or server defined; most likely, for 'find-a-friend' application, γ can be higher (e.g., 5–10 m) than for a digital contact-tracing application.

Let us denote the perturbed 3D-location values via $\mathbf{y}_u$, with $\mathbf{y}_u = M(\mathbf{u}_u) \in \mathbf{B}$, with $\mathbf{B} \in \mathcal{R}^3$ being the building space, defined via a cube space with edges $[x_{min}\ x_{max}] \times [y_{min}\ y_{max}] \times [z_{min}\ z_{max}]$, where $x_{min}, x_{max}, y_{min}, y_{max}, z_{min}, z_{max}$ are the building edges (minimum and maximum, respectively) in the 3D space. It is assumed that the centralized digital contact-tracing server (which can be trusted or untrusted) has access to the building floor plans. It is also assumed that the server is dividing the whole building space into grid points $\mathbf{b} = [b_x, b_y, b_z] \in \mathbf{B}^3$, for example as shown in Figure 2 and that the set of grid points $\{\mathbf{b}|\mathbf{b} \in \mathbf{B}\}$ is transmitted to all users in the building, e.g., via cellular or WiFi connectivity. The grid step Δ_s is a parameter of the centralized server providing proximity-detection services or user digital contact tracing. With a Δ_s step it means that b_x for example can only take values in the interval $[x_{min} : \Delta_s : x_{max}]$.

3.2. Perturbation Metrics

Two perturbation metrics are proposed and investigated, as defined in Equations (1) and (2).

$$M_{argmin}(\mathbf{u}_u) = argmin_{\mathbf{b}\in\mathbf{B}}||\mathbf{b} - \mathbf{x}_u|| + \xi \quad (1)$$

where $|| \cdot ||$ is the distance between $\mathbf{b}$ and $\mathbf{x}_u$ vectors and ξ is a multivariate (3D) noise vector of zero mean (to be explained later in this section). Also,

$$M_{argmax}(\mathbf{u}_u) = argmax_{\mathbf{b}\in\mathbf{B}}||\mathbf{b} - \mathbf{x}_u|| + \xi \quad (2)$$

While the argmin operator is rather intuitive, stating that the user location is only slightly perturbed by mapping it to the nearest grid point and then adding a random noise to it, the argmax operator may seem less intuitive at a first glance. Indeed, with argmax operator, all users located, for example, at the extreme north-west of the building, will be mapped, after argmax operator, as being close to the extreme south-east of the building. As we are only focusing here on the proximity-detection type of application relying on the relative distance between users, such as digital contact tracing or find a friend, this mapping does not decrease the service utility, as nearby users (which were, for example, at the extreme north-west of the building) will still appear as nearby users after the mapping to the other side of the building.

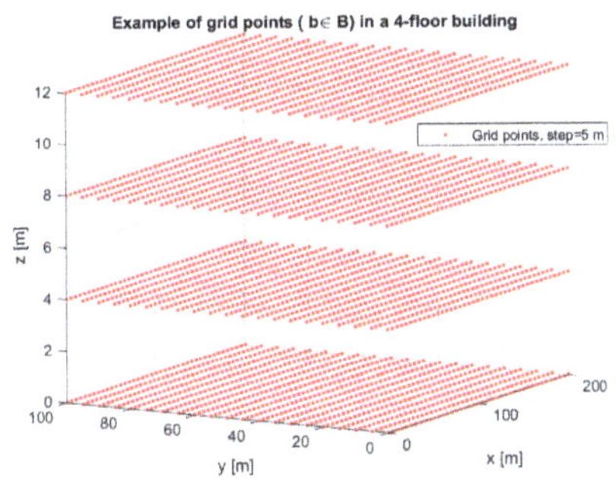

Figure 2. Example of mapping the whole building space **B** into grid points **b**, $\Delta_s = 5$ m for a 100×200 m^2 building with 4 floors and 4 m floor height .

In order for $M_{argmin}(\mathbf{u}_u)$ and $M_{argmax}(\mathbf{u}_u)$ metrics to remain inside the building space **B** and to offer plausible perturbed locations, an additional correction is done after the mappings in Equations (1) and (2), in such a way that the points that would fall outside the building edges, are re-mapped to the nearest point inside the building. In addition, if the perturbed z coordinate does not match any of the floor heights in the building, then the perturbed z-coordinate is mapped to the nearest floor level. Examples will be provided in Section 5.

The *argmin* metric in Equation (1) is mapping the true position to the nearest grid point in the building and it then applies a noise factor to it, while the *argmax* metric in Equation (2) is mapping the true position to the furthest grid point in the building and it then applies a noise factor to it. Clearly, on one hand, Equation (1) mapping preserves a minimum distance between the perturbed location and the true location, enabling various location-based services that require absolute user-location knowledge, but it acts quite poorly in terms of privacy preservation, as an attacker could still identify the approximate location of an user with an accuracy depending on the inverse of the standard deviation $1/\epsilon$ of the added multivariate noise ξ. On the other hand, the second proposed metric from Equation (2) is able to protect the user location privacy to a great extent (as the privacy increases when the distance between the perturbed location and original location increases), with an increased privacy level for larger/wider buildings, and, as we will show in Section 5, without destroying the usefulness of the services, meaning that an accurate contact tracing can be also achieved under a heavy protection of user's location privacy.

Regarding the added noise vector ξ, two multivariate noise distributions are considered, namely a Gaussian distribution of equal standard deviation in x, y, z dimensions of $1/\epsilon$, see Equation (3), and a Laplacian distribution of equal scale factor in x, y, z dimensions of $1/\epsilon$, see Equation (4). The zero-mean multivariate (3D) Gaussian noise is:

$$f_{Gauss}(\xi) = \frac{1}{(2\pi)^{1.5}|\Sigma|^{0.5}} exp(-0.5\xi^T\Sigma^{-1}\xi) \tag{3}$$

with $\Sigma = diag([\frac{1}{\epsilon}\ \frac{1}{\epsilon}\ \frac{1}{\epsilon}]) = \frac{1}{\epsilon}\mathbf{I}_3$ being a diagonal covariance matrix and $\mathbf{I}_3$ a unit matrix of dimension 3×3, and $|\Sigma| = \epsilon^{-3}$ being the determinant of Σ.

The zero-mean multivariate (3D) Laplacian noise is:

$$f_{Laplace}(\xi) = \frac{2}{(2\pi)^{1.5}|\Sigma|^{0.5}}(0.5\xi^T\Sigma^{-1}\xi)^{-0.5}K_v(\sqrt{2\xi^T\Sigma^{-1}\xi}) \quad (4)$$

where K_v is the modified Bessel function of second kind.

3.3. Private Proximity-Detection Architecture with the Proposed Mechanism

The wireless communication process between user/edge devices and the proximity-detection service is depicted in Figure 3. Users are assumed to be spread across a multi-floor space of commercial or commuting interest (e.g., shopping mall, commuting hall/airport/train station, etc.). Users' devices are supposed to be equipped with a localization engine, such as GNSS, WiFi, BLE or a combination of several localization methods. A proximity service provider is operating in the building of interest, with access to the building floor plans and able to send the floor-map coordinates **b** to all users interested in the proximity-based service or application. The coordinates can be provided as Earth Centered Earth Fixed (ECEF) coordinates, as (latitude, longitude, and altitude)-coordinates, or as local coordinates (x, y, z) and the mapping between any of these coordinate systems is assumed known both at the user side and at the server side. The user devices performs the location perturbation locally and sends the perturbed location to the server; the server processes in an aggregate form all the data based on the perturbed locations of the users inside the building and offers the proximity-based service to the users.

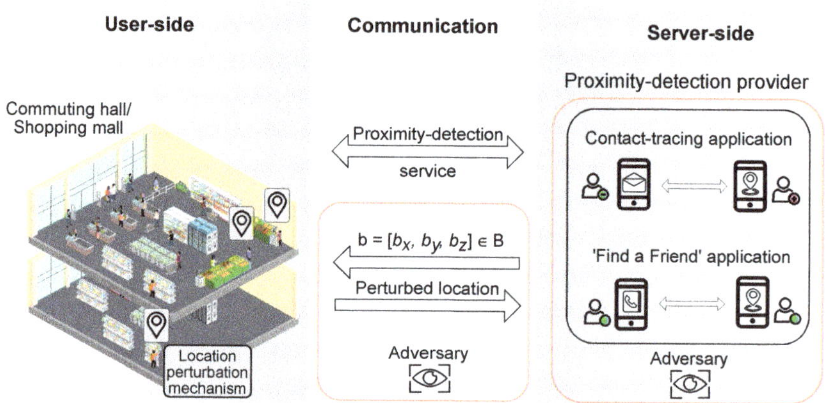

Figure 3. An illustration of the considered scenario: a building (e.g., a shopping mall) with users willing to use the digital contact-tracing and/or 'find-a-friend' applications. The 'Adversary' entity refers to any third party which aims to access the information about devices' whereabouts.

4. Theoretical Analysis of the Proposed Argmax Perturbed Location Mechanism

For simplicity, in this section we focus on the argmax metric from Equation (2) and we denote via $M(\cdot) = M_{argmax}(\cdot)$, with the observation that similar derivations can be obtained in a straightforward manner for argmin metric. Let denote by p_u the probability that an adversary finds out $\mathbf{x}_u$ by listening to $\mathbf{y}_u = M(\mathbf{x}_u)$. Then

$$\begin{aligned} p_u = proba(M(\mathbf{x}_u) = \mathbf{x}_u) &= proba(argmax_{\mathbf{b}\in\mathbf{B}}||\mathbf{b} - \mathbf{x}_u|| + \xi = \mathbf{x}_u) \\ &= proba(\xi = \mathbf{x}_u - argmax_{\mathbf{b}\in\mathbf{B}}||\mathbf{b} - \mathbf{x}_u||) \end{aligned} \quad (5)$$

If we denote via $\mathbf{a}_u \triangleq argmax_{\mathbf{b}\in\mathbf{B}}||\mathbf{b}-\mathbf{x}_u||$, under Gaussian-noise assumption, the above formula is determined by the Gaussian noise probability distribution function (PDF) from Equation (3) and it becomes equal to

$$p_u = \frac{\epsilon^3}{(2\pi)^{1.5}} exp(-0.5\epsilon||\mathbf{x}_u - \mathbf{a}_u||^2) \tag{6}$$

Similarly, if p_v is the probability that an adversary intercepts the perturbed location of user v, namely $M_{argmax}(\mathbf{x}_v)$ and maps it to the location of user u, after straightforward derivations (as above) and following the Gaussian noise assumption, we get

$$p_v = \frac{\epsilon^3}{(2\pi)^{1.5}} exp(-0.5\epsilon||\mathbf{x}_u - \mathbf{a}_v||^2) \tag{7}$$

with $\mathbf{a}_v \triangleq argmax_{\mathbf{b}\in\mathbf{B}}||\mathbf{b}-\mathbf{x}_v||$.

By dividing Equation (6) to Equation (7) and using Cauchy-Schwarz inequality, one gets

$$\begin{aligned} \frac{p_u}{p_v} &= exp\left(0.5\epsilon(||\mathbf{x}_u - \mathbf{a}_u||^2 - ||\mathbf{x}_u - \mathbf{a}_u||^2)\right) \\ &\leq exp\left(0.5\epsilon||\mathbf{a}_u - \mathbf{a}_v||^2\right) \\ &\leq exp\left(0.5\epsilon||\mathbf{x}_u - \mathbf{x}_v||^2\right) \end{aligned} \tag{8}$$

Thus, the proposed mechanism $M(\cdot)$ offers GeoInd type of user location privacy.

5. Simulation-Based Results

5.1. Simulation Scenarios and Performance Metrics

A 4-floor scenario with N_u users spread within the building, with most of them within couple of pre-defined hotspot areas was considered. Table 2 shows the main parameters used in the simulation model (additional parameters were investigated in some scenarios and they are specified in the figures' captions when different from those in Table 2). The users are assumed to transmit their perturbed location $M(\mathbf{x}_u)$ to a server provider offering a proximity-based service with a proximity threshold γ (i.e, the service is offered if the users are determined to be at a distance less than γ, based on their perturbed location transmitted to the server).

At each Monte Carlo run, another realization of users' random positions within the building is implemented. Two examples of the users distribution in the building during two Monte Carlo runs is shown in Figure 4.

Examples of perturbed locations during one Monte Carlo run with *argmin* metric (left plot) and *argmax* metric (right plot) are shown in Figure 5, for $\epsilon = 0.1$ and Laplacian noise.

A zoomed version of perturbed locations for one floor and with only 4 users is illustrated in Figure 6, this time showing both the scenario with no hotspots (left plot) and with hotspots (right plot). The squares show the perturbed location via *argmin* metric and the circles show the perturbed location via *argmax* metric.

The utility functions are defined as the probability of correctly detecting two users to be in close proximity to each other P_d, as well as the complement of the false alarm probability P_{fa}, meaning the probability to detect that two users are in close proximity to each other, when in fact they are not. Mathematically, P_d and P_{fa} are defined via

$$P_d = \frac{| \{(u,v) \in N_u \times N_u, u \neq v \mid \|M(\mathbf{x}_u) - M(\mathbf{x}_v)\| \leq \gamma \, and \, \|\mathbf{x}_u - \mathbf{x}_v\| \leq \gamma\} |}{| \{(u,v) \in N_u \times N_u, u \neq v \mid \|\mathbf{x}_u - \mathbf{x}_v\| \leq \gamma\} |} \quad (9)$$

and, respectively,

$$P_{fa} = \frac{| \{(u,v) \in N_u \times N_u, u \neq v \mid \|M(\mathbf{x}_u) - M(\mathbf{x}_v)\| \leq \gamma \, and \, \|\mathbf{x}_u - \mathbf{x}_v\| \geq \gamma\} |}{| \{(u,v) \in N_u \times N_u, u \neq v \mid \|\mathbf{x}_u - \mathbf{x}_v\| \geq \gamma\} |} \quad (10)$$

where $| \cdot |$ is the cardinal operator, N_u is the number of users inside the building, and P_d and P_{fa} correspond to detection probability (here also the sensitivity) and false positive rate in confusion-matrix terminology, respectively. Clearly, the proximity-based service utility increases when P_d increases and when P_{fa} decreases.

Table 2. Main simulation parameters (unless otherwise specified in plots' titles).

Parameter	Value [Unit]
Number of floors N_f	4 [-]
Building grid Δ_s	1 [m]
Building size	100×200 [m^2] horizontally 12 m vertically (4 m floor heights)
Number of users N_u	Variable, 100 or 1000 [-]
Privacy budget ϵ	Variable, between 10^3 and 10^2 [1/m]
Proximity threshold γ	Variable, 2 or 10 [m]
Number of hotspots per floor	Variable, between 2 and 4 [-]
Hotspot radius	Variable, between 4 and 10 [m]
Percentage of users within hotspot areas	80 [%]
Number of Monte Carlo runs	1000 [-]

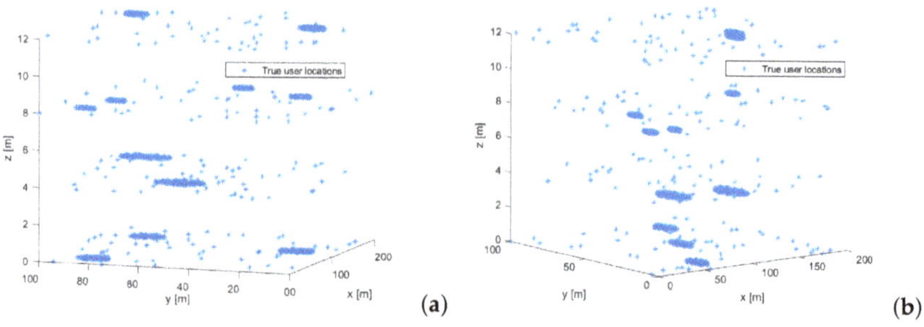

Figure 4. Two examples of users distribution within a 4-floor building during two Monte Carlo runs. (**a**) Monte Carlo run 1; (**b**) Monte Carlo run 2. In these runs, we allocated 80% of users are in hotspot areas and 20% of users are outside hotspot areas, uniformly distributed within the building.

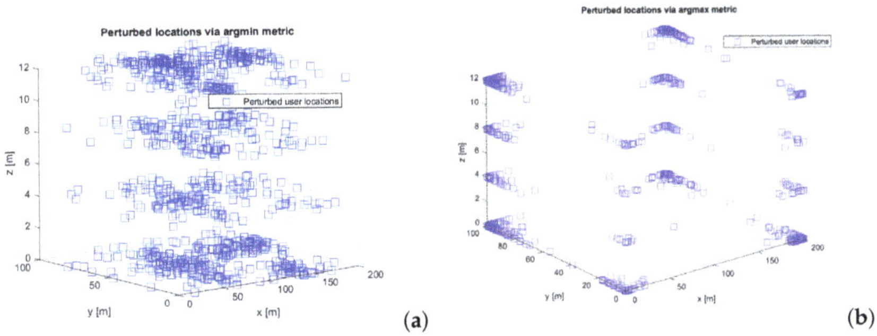

Figure 5. Examples of perturbed locations based on (**a**) $M_{argmin}(\cdot)$ and (**b**) $M_{argmax}(\cdot)$ metrics. $\epsilon = 0.1$ m, Laplace perturbation.

The ensured privacy level is proportional to the distance between the perturbed location and the true location, or the RMSE between $M(\mathbf{x}_u)$ and $\mathbf{x}_u$, namely

$$RMSE = \sqrt{\frac{1}{N_u}\sum_{u=1}^{N_u} ||M(\mathbf{x}_u) - \mathbf{x}_u||^2} \tag{11}$$

Clearly, the ensured privacy level is better when RMSE from Equation (11) is higher.

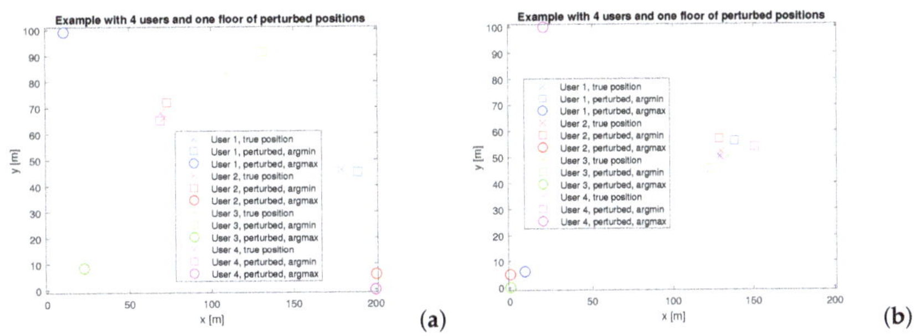

Figure 6. Two examples of perturbed location via argmin + Laplacian noise and via argmax + Laplacian noise. (**a**) users uniformly distributed over one floor; (**b**) users uniformly distributed within a circular hotspot of radius 5 m.

5.2. Comparison with State-of-the-Art Perturbation Mechanisms

Several obfuscation models have been proposed so far in the literature to protect the location information, as described in Section 2. Three of the most common ones, selected here as benchmarks are the uniform obfuscation [31], the Laplacian perturbation [47], and the Gaussian perturbation [48]. The uniform perturbation model from [31] was given for 2D case and it was based on the idea that a random vector shift is applied to the user location with a certain radius. The model from [31] extended to 3D scenarios can be written as

$$M_{uniform}(\mathbf{u}_u) = \mathbf{x}_u + \xi_u \tag{12}$$

where ξ_u is a 3D vector with elements $[\xi_{u,x}, \xi_{u,y}, \xi_{u,z}]$ given by

$$\xi_{u,x} = \mu cos(\theta) \tag{13}$$

$$\xi_{u,y} = \mu sin(\theta) \tag{14}$$

$$\xi_{u,z} = \mu tan(\alpha) \tag{15}$$

and μ, θ, and α are the random radius, azimuth, and elevation angles, respectively, drawn from the following three uniform distributions: μ $U(0, 1/\epsilon)$, θ $U(0, 2\pi)$, and α $U(0, 2\pi)$, where $U(a, b)$ stands for a uniform distribution in the interval $[a, b]$.

The Laplacian [47] and Gaussian [48] perturbations can be modeled as

$$M_{Laplace,Gaussian}(\mathbf{u}_u) = \mathbf{x}_u + \xi \tag{16}$$

where ξ is a Laplacian or a Gaussian noise, as given in Equations (4) and (3), respectively. The comparison with the three state-of-the-art algorithms described above, namely uniform obfuscation [31], Laplacian perturbation [47], and Gaussian perturbation [48] is shown in Figure 7.

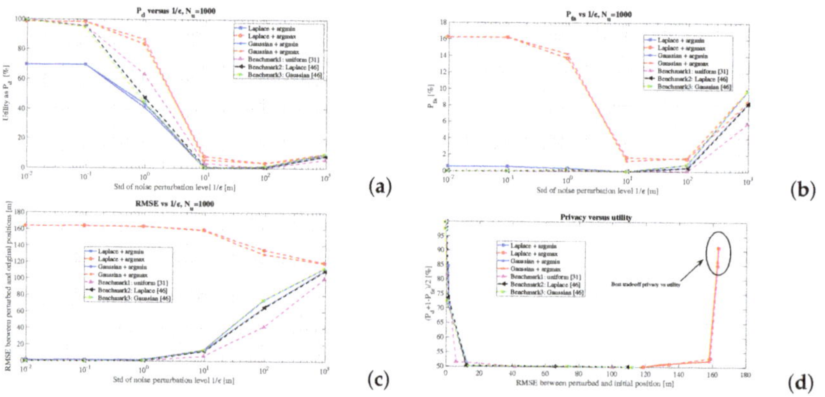

Figure 7. Comparison with state-of-the-art algorithms: (**a**) P_d versus the noise perturbation level; (**b**) P_{fa} versus the noise perturbation level; (**c**) RMSE between the perturbed location and original location versus the noise perturbation level; (**d**) utility versus privacy.

As seen in Figure 7, the argmax-based metric offers the best detection probability (upper left plot) and the best privacy level (lower left plot), but slightly worse false alarm probabilities (upper right plot) than the other four investigated algorithms, namely argmin-based and three bench,ark ones. The most important plot is however the one depicted in the lower right part of Figure 7, where the utility-privacy tradeoff is illustrated. For a fairer comparison, the utility here comprises the average between the P_d and $1 - P_{fa}$; the closest to 100% this value is, the higher utility we have; ideally, a best service would have $P_d = 1$ and $P_{fa} = 0$. The privacy level is given by RMSE; the higher the RMSE between the perturbed and true location is, the higher the privacy. Clearly, the argmax-based perturbation is a clear winner among all considered algorithms, as it can reach simultaneously high levels of privacy and high levels of utility of a proximity service relying in inter-users distance. It is to be emphasized that such utility pertains only to such proximity-based services relying on inter-user distances; other location-based services needing absolute location information would have a different utility, where our argmax-based algorithm would most likely perform poorer than the other approaches. In terms of argmin-based approach versus the three considered benchmark, there is very little difference in the utility-privacy tradeoff. For this reason and in order to keep clarity in the subsequent plots, we will focus from now on only on the comparisons between argmin- and argmax-based perturbations and on the deeper analysis of the argmax-based operator.

5.3. Privacy Level as a Function of ϵ Parameter

The RMSE between the transmitted perturbed location and the original location, as defined in Equation (11), is shown in Figure 8. A higher RMSE value means a higher user privacy level. There is no significant difference between the noise type ξ used in the perturbation mechanism, with the Laplacian noise giving slightly better results than the

Gaussian one in terms of privacy for the *argmax* metric, and the Gaussian noise giving slightly better results in terms of privacy for the *argmin* metric.

A very interesting finding is that by using an *argmax* metric, not only one achieves significantly higher privacy level than by using *argmin* metric (i.e., higher RMSE values), but also the noise level $1/\epsilon$ acts in an opposite manner on the *argmax* metric than on the *argmin* metric, meaning that a higher ϵ ensures more obfuscation in the argmin-based approach, but less obfuscation in the argmax-based approach. This points out that high levels of ϵ (or, equivalently low levels of the noise standard deviation) are giving better results in terms of privacy with the *argmax* metric than lower levels of ϵ. This is observed due to the fact that the users' location is already mapped far away from its initial location through the *argmax* operator, and it is enough to add only a small additional random perturbation in order to make difficult the 'guessing' of true user location $\mathbf{x}_u$ based on the disclosed perturbed location $M(\mathbf{x}_u)$ in case an attacker or eavesdropper gets access to the perturbed location.

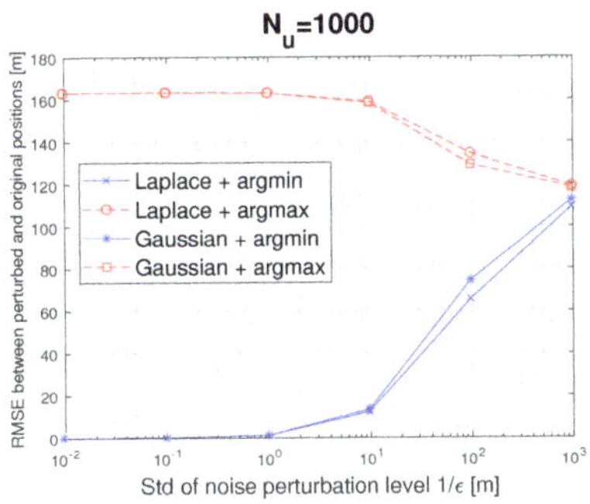

Figure 8. RMSE between the perturbed location and original location versus the noise perturbation level for two noise types (Laplacian and Gaussian) and two mapping metrics (argmin and argmax).

5.4. Utility Level as a Function of ϵ Parameter

Figure 9 shows the utility (i.e., the detection probability) as well as the false alarm probabilities in the presence of various perturbations (*argmin* versus *argmax* and Gaussian versus Laplacian noises).

Clearly, the *argmax* metric has higher utility at the expense of a moderately higher false alarm than the *argmin* metric. The differences between Gaussian and Laplacian noises are minor and therefore Gaussian perturbation is recommended to be used for simplicity. The best detection probabilities for a proximity-based application are achieved with ϵ values above 1 (or equivalently, standard deviation of the noise below 1 m). We can see from the left plot in Figure 9 that detection probabilities close to 100% are achievable with the proposed *argmax* metric, with moderate false alarms of about 16%. As the user privacy is highly preserved with an *argmax* metric and high enough ϵ values (see also Figure 8), the price to pay in terms of false alarm probabilities of up to 16% may seem reasonable for users desiring high location privacy. Indeed, the cost of a false alarm may be quite low to the user (e.g., user is incorrectly informed that a friend is nearby or user is incorrectly informed that he or she might have been close contact of a person confirmed with COVID-19 and thus he/she would take unnecessary, but also not-hurtful additional protection measures). However, the utility of a correct proximity detection in a proximity-based service is high and, as shown in the left plot of Figure 9, it is preserved with the M_{argmax} metric and an ϵ value above 1.

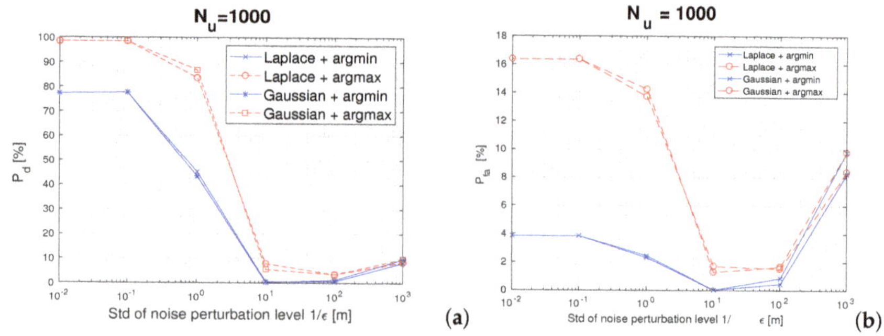

Figure 9. (**a**) Detection and (**b**) false-alarm probabilities versus the noise perturbation level for two noise types (Laplacian and Gaussian) and two mapping metrics (argmin and argmax). The proximity threshold γ was set to 2 m (e.g., for a digital contract-tracing application). A 4-floor building with 1000 users and 80% of them placed in hotspot areas.

5.5. Privacy-versus-Utility Tradeoffs

An illustration of the privacy-versus-utility tradeoff is shown in Figure 10, where the utility is defined as the correct detection probability P_d (see Equation (9)).

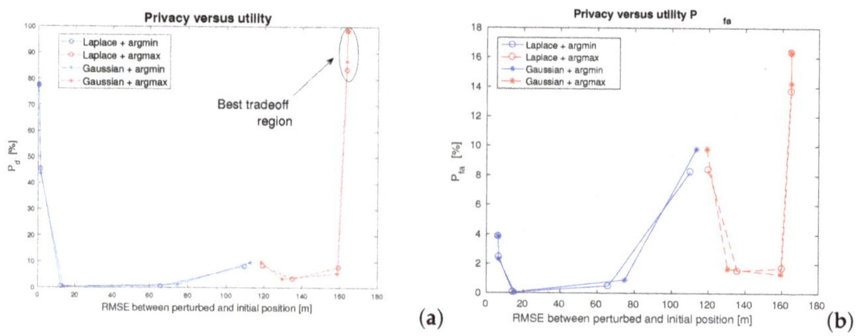

Figure 10. Privacy versus utility tradeoff. Proximity threshold $\gamma = 2$. The plots illustrate the behavior of the argmin vs argmax metrics against RMSE. (**a**) P_d as utility; (**b**) P_{fa} as utility

Figure 11 shows also the impact of the proximity threshold γ on the utility (detection probability) and false alarm probability. Two proximity thresholds were considered: $\gamma = 2$ m, useful for example for a digital contact-tracing service provider and $\gamma = 10$ m, useful for example for a 'find a friend' application in a shopping center. The proximity threshold choice does not change the main conclusions that *argmax* metric with an ϵ below 1 (i.e., a noise standard deviation above 1 m) offers the best tradeoff between utility and privacy. This threshold provides decent detection probabilities (higher than 90%) and moderately low false alarm probabilities (below 16%). The best tradeoff utility region is also illustrated in Figure 12, this time only for the *argmax* metric and two proximity thresholds.

Figure 13 shows that also the hotspot distribution of users has little bearing on the privacy-utility tradeoff, with best tradeoffs obtained again for *argmax* metric and a low ϵ value, mapping to high perturbed levels due to *argmax* operator. As in the $M_{argmax}(\cdot)$ metric, the user perturbed location is mapped to points far away from true user location, it is intuitive that higher RMSE values between the perturbed and true locations are obtained in the case with less users within the building hotspots, as seen in Figure 13 by comparing the 20% and 80% hotspot distributions.

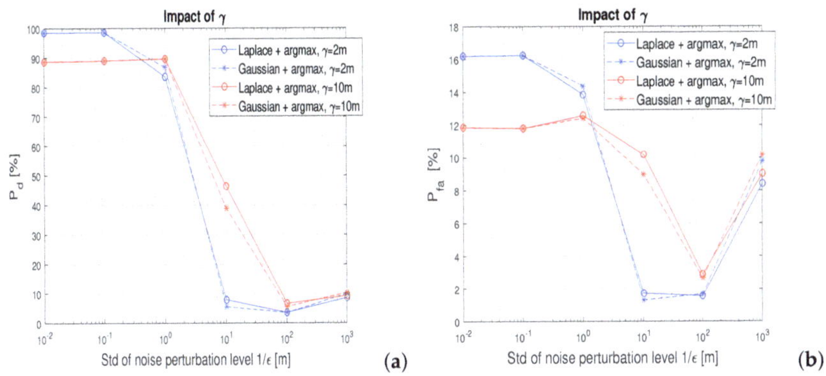

Figure 11. Impact of the proximity threshold on (**a**) detection P_d and (**b**) false-alarm rates P_{fa}.

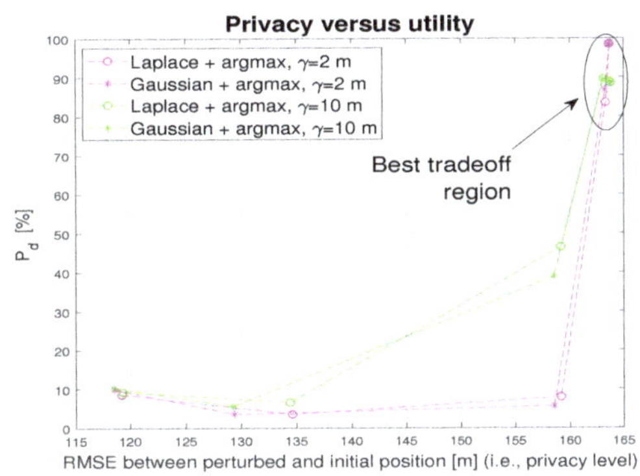

Figure 12. Privacy versus utility tradeoff. Argmax metric. Proximity thresholds $\gamma = 2$ m and $\gamma = 10$ m.

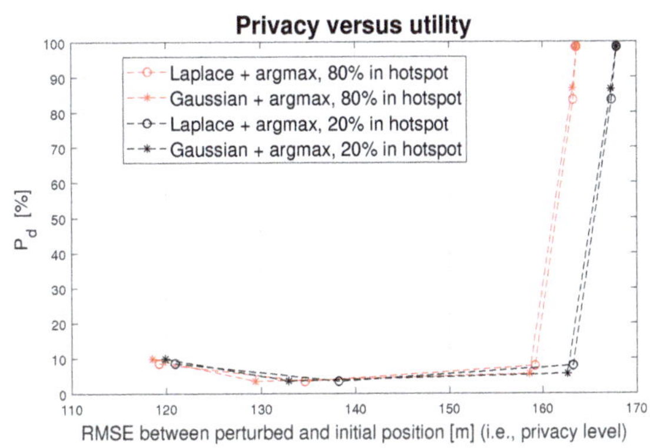

Figure 13. Privacy versus utility tradeoff in the presence of different hotspot distribution of users (80% of users within hostpots versus only 20% of users within the building hotspots). Argmax metric and $\gamma = 2$ m.

The impact of the grid step on the utility and the privacy level is shown in Figure 14. As mentioned above, the grid step influences the matrix $\mathbf{b} \in \mathbf{B}$ transmitted to the users within a building. For clarity purpose and because the noise type (Laplace versus Gaussian) has low impact, only the Gaussian noise perturbations are shown. Clearly, the impact of the step size is minimal on both the service utility (computed as the correct detection probability of close-by users within a threshold γ) and on the user privacy (computed as the RMSE between the disclosed perturbed location and the true user location). This fact eases the amount of data needed to be transferred from the service provider to the user, as the size of the building grid matrix **b** is decreasing when the grid step Δ_s is increasing. Nevertheless, the choice of the grid step Δ_s should take into account the building size (e.g., steps lower than 10% of maximum building length in a certain direction are recommended).

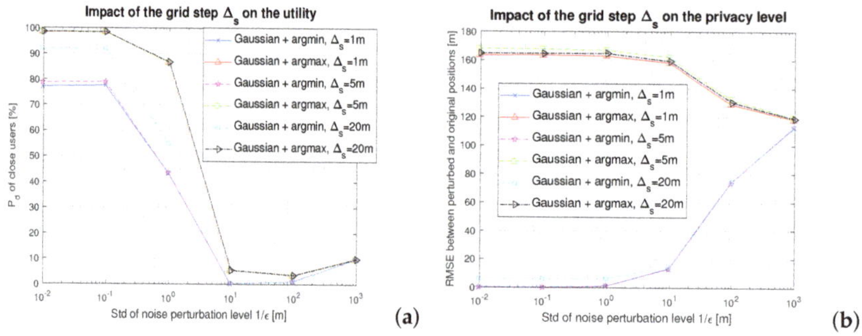

Figure 14. The impact of the grid step on the (**a**) utility and (**b**) privacy. A proximity service with $\gamma = 2$ m.

In Figure 15, the different building sizes are compared for a fixed number of users N_u. Here, the added noise in the perturbation yields similar results independent of its type. However, P_d levels are high up, as close to 100% for the largest building size, namely 20 × 20 m. Whereas the smallest building considered in the simulation, with the dimensions of 100 × 200 m, shows moderate P_d and P_{fa} levels, accordingly. One could translate the situation with a fixed number of users and varying building sizes into the density of the users, where a little space is offered to each user per se.

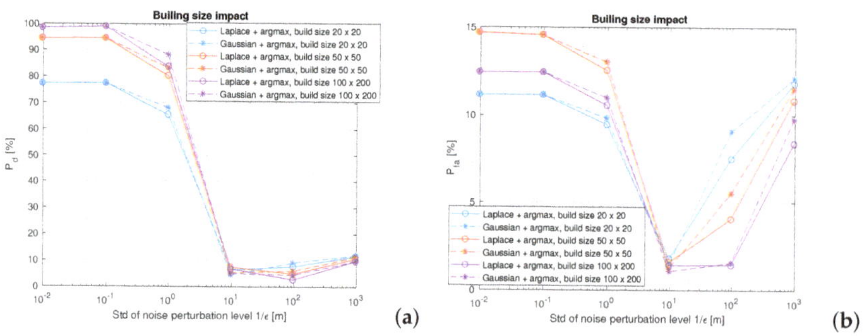

Figure 15. The impact of the building size on the application's utility. A proximity service with $\gamma = 2$ m, fixed $N_u = 1000$. (**a**) P_d and (**b**) P_{fa}.

Last but not least, Figure 16 shows that the number of users in the building has no impact on the utility-privacy tradeoff and the *argmax* metric with any of the two noise types (Gaussian or Laplacian) is able to attain very good tradeoff levels.

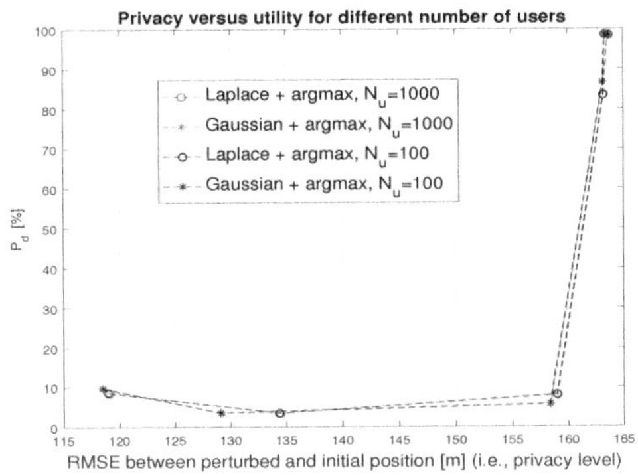

Figure 16. Privacy versus utility tradeoff in the presence of different number of users. Argmax metric and $\gamma = 2$ m.

6. Conclusions

This paper has proposed a local perturbation mechanism for preserving user-location privacy, while maintaining a high utility of proximity-detection-based services such as digital contact tracing or find-a-friend application. We would like to emphasize that the proposed argmax-based mechanism is useful only for applications relying strictly on the relative distance between any two users, such as digital contact tracing. However, the system loses its utility in the context of location-based services requiring absolute user location, such as finding the nearest shop or searching for a specific route in a mall.

The proposed mechanism is able to offer GeoInd and a very good privacy-utility tradeoff. It relies on the assumption that users have full control of the disclosure level of their location accuracy. Moreover, it is assumed that the service provider has access to the floor plans of the buildings of interests (e.g., a commuting hall, a shopping mall, etc.) and is transmitting the discretized grid map (in terms of x, y, z coordinates) of the building. to all users in the building.

We have provided detailed simulation-based results in a multi-floor building scenario, under different assumptions of user location distributions, grid map step size, hotspot distributions, and number of users in the building. We have also compared the proposed *argmax*-based metric with an *argmin*-based metric and other state-of-the-art metrics which would be useful in location-based services requiring absolute location information, not only relative location information as needed in proximity-based services. We have shown that argmax-based approach with a perturbation level $1/\epsilon$ between 1 and 10 cm offers the best tradeoff utility-privacy for proximity-based services, while argmin-based metric is more suitable for services requiring absolute location information. We have also shown that the number and distribution of users in a building, the random distribution type (Gaussian or Laplacian), as well as the building grid steps have little impact on the results. We were able to reach, via the argmax-based mechanisms, very good privacy levels (RMSE in the orders of the building sizes) with detection probabilities of the order of 90% and false alarm probabilities below 15%. The simulations have also shown that the service utility, measured as detection probability, which is slightly better for large buildings and low γ threshold than for small buildings and high γ threshold. At the same time, the false alarm probabilities are slightly better for small buildings and high γ threshold than for large buildings and low γ threshold. The γ threshold is highly dependent of the target proximity-based service (e.g., we considered $\gamma = 2$ m for digital contact-tracing applications and $\gamma = 10$ m for 'find-a-friend' type of applications).

Open challenges are related to mechanisms for ensuring full user control on local devices about his/her/their location information, the impact of the imperfect knowledge of the user location information (or true position), as well as the impact of imperfect floor-map knowledge (e.g., incorrect floor heights) from the proximity service provider's point of view.

Author Contributions: Conceptualization, E.S.L., V.S., and D.N.; methodology, E.S.L.; software, E.S.L. and V.S.; validation, E.S.L. and V.S.; formal analysis, E.S.L. and V.S.; writing—original draft preparation, E.S.L. and V.S.; writing—review and editing, E.S.L., V.S., and D.N.; visualization, V.S. and E.S.L.; supervision, E.S.L. and D.N.; project administration, E.S.L.; funding acquisition, E.S.L. All authors have read and agreed to the published version of the manuscript.

Funding: This work was supported by the funding from European Union's Horizon 2020 Research and Innovation programme under the Marie Skłodowska Curie grant agreement No. 813278 (A-WEAR: A network for dynamic wearable applications with privacy constraints, www.a-wear.eu). The work has also been supported by the Academy of Finland, project ULTRA (#328226), and by a grant from the Romanian National Authority for Scientific Research and Innovation, UEFISCDI project PN-III-P2-2.1-PED-2019-5413.

Institutional Review Board Statement: Not applicable.

Informed Consent Statement: Not applicable.

Conflicts of Interest: The authors declare no conflict of interest.

List of Acronyms

6G	Sixth generation of cellular communications
BLE	Bluetooth Low Energy
COVID-19	Coronavirus disease 2019
DP	Differential Privacy
ECEF	Earth Centered Earth Fixed
GNSS	Global Navigation Satellite Systems
GeoInd	Geo-indistinguishability
IEEE	Institute of Electrical and Electronics Engineers
LDP	Local Differential Privacy
LBS	Location-Based Services
LPPM	Location Privacy-Preserving Mechanisms
PDF	probability distribution function
RSS	Received Signal Strength
RMSE	Root Mean Square Error
UWB	Ultra Wide-Band

References

1. De Lima, C.; Belot, D.; Berkvens, R.; Bourdoux, A.; Dardari, D.; Guillaud, M.; Isomursu, M.; Lohan, E.S.; Miao, Y.; Barreto, A.N.; et al. Convergent Communication, Sensing and Localization in 6G Systems: An Overview of Technologies, Opportunities and Challenges. *IEEE Access* **2021**, *9*, 26902–26925. [CrossRef]
2. P802.11bf—Standard for Information Technology—Telecommunications and Information Exchange Between Systems Local and Metropolitan Area Networks—Specific Requirements—Part 11: Wireless LAN Medium Access Control (MAC) and Physical Layer (PHY) Specifications Amendment: Enhancements for Wireless Local Area Network (WLAN) Sensing. Available online: https://standards.ieee.org/project/802_11bf.html?utm_source=beyondstandards&utm_medium=post&utm_campaign=working-group-2020&utm_content=802 (accessed on 15 November 2021).
3. Brovko, T.; Chugunov, A.; Malyshev, A. Positioning Algorithm for Smartphone Based Staff Tracking. In Proceedings of the 2021 International Russian Automation Conference (RusAutoCon), Sochi, Russia, 5–11 September 2021; pp. 464–468. [CrossRef]
4. Flueratoru, L.; Shubina, V.; Niculescu, D.; Lohan, E.S. On the High Fluctuations of Received Signal Strength Measurements with BLE Signals for Contact Tracing and Proximity Detection. *IEEE Sens. J.* **2021**. [CrossRef]
5. Shubina, V.; Ometov, A.; Basiri, A.; Lohan, E.S. Effectiveness modelling of digital contact-tracing solutions for tackling the COVID-19 pandemic. *J. Navig.* **2021**, *74*, 853–886. [CrossRef]
6. Shubina, V.; Holcer, S.; Gould, M.; Lohan, E.S. Survey of Decentralized Solutions with Mobile Devices for User Location Tracking, Proximity Detection, and Contact Tracing in the COVID-19 Era. *Data* **2020**, *5*, 87. [CrossRef]
7. Bian, S.; Zhou, B.; Lukowicz, P. Social Distance Monitor with a Wearable Magnetic Field Proximity Sensor. *Sensors* **2020**, *20*, 5101. [CrossRef]

8. Vaudenay, S. Centralized or Decentralized? The Contact Tracing Dilemma. 2020. Available online: https://eprint.iacr.org/2020/531.pdf (accessed on 24 November 2021).
9. Castelluccia, C.; Bielova, N.; Boutet, A.; Cunche, M.; Lauradoux, C.; Le Métayer, D.; Roca, V. ROBERT: ROBust and privacy-presERving Proximity Tracing. Available online: https://hal.inria.fr/hal-02611265/document (accessed on 24 November 2020).
10. Leith, D.J.; Farrell, S. Contact tracing app privacy: What data is shared by europe's gaen contact tracing apps. In Proceedings of the IEEE INFOCOM 2021-IEEE Conference on Computer Communications, Vancouver, BC, Canada, 10–13 May 2021; pp. 1–10.
11. Jiang, H.; Li, J.; Zhao, P.; Zeng, F.; Xiao, Z.; Iyengar, A. Location privacy-preserving mechanisms in location-based services: A comprehensive survey. *ACM Comput. Surv. (CSUR)* **2021**, *54*, 1–36. [CrossRef]
12. Basiri, A.; Lohan, E.S.; Moore, T.; Winstanley, A.; Peltola, P.; Hill, C.; Amirian, P.; e Silva, P.F. Indoor location based services challenges, requirements and usability of current solutions. *Comput. Sci. Rev.* **2017**, *24*, 1–12. [CrossRef]
13. von Arb, M.; Bader, M.; Kuhn, M.; Wattenhofer, R. VENETA: Serverless Friend-of-Friend Detection in Mobile Social Networking. In Proceedings of the 2008 IEEE International Conference on Wireless and Mobile Computing, Networking and Communications, Avignon, France, 12–14 October 2008; pp. 184–189. [CrossRef]
14. Ye, A.; Chen, Q.; Xu, L.; Wu, W. The flexible and privacy-preserving proximity detection in mobile social network. *Future Gener. Comput. Syst.* **2018**, *79*, 271–283. [CrossRef]
15. Kim, J.W.; Edemacu, K.; Kim, J.S.; Chung, Y.D.; Jang, B. A Survey of differential privacy-based techniques and their applicability to location-Based services. *Comput. Secur.* **2021**, *111*, 102464. [CrossRef]
16. Chatzikokolakis, K.; Palamidessi, C.; Stronati, M. Geo-indistinguishability: A principled approach to location privacy. In Proceedings of the International Conference on Distributed Computing and Internet Technology, Bhubaneswar, India, 5–8 February 2015; pp. 49–72.
17. Andrés, M.E.; Bordenabe, N.E.; Chatzikokolakis, K.; Palamidessi, C. Geo-indistinguishability: Differential privacy for location-based systems. In Proceedings of the 2013 ACM SIGSAC Conference on Computer & Communications Security, Berlin, Germany, 4–8 November 2013; pp. 901–914.
18. Qiu, C.; Squicciarini, A.C.; Pang, C.; Wang, N.; Wu, B. Location privacy protection in vehicle-based spatial crowdsourcing via geo-indistinguishability. *IEEE Trans. Mobile Comput.* **2020**. [CrossRef]
19. Yan, Y.; Gao, X.; Mahmood, A.; Feng, T.; Xie, P. Differential private spatial decomposition and location publishing based on unbalanced quadtree partition algorithm. *IEEE Access* **2020**, *8*, 104775–104787. [CrossRef]
20. Shubina, V.; Ometov, A.; Andreev, S.; Niculescu, D.; Lohan, E.S. Privacy versus Location Accuracy in Opportunistic Wearable Networks. In Proceedings of the2020 International Conference on Localization and GNSS (ICL-GNSS), Tampere, Finland, 2–4 June 2020; pp. 1–6.
21. Chatzikokolakis, K.; Elsalamouny, E.; Palamidessi, C. Efficient utility improvement for location privacy. *Proc. Priv. Enhancing Technol.* **2017**, *2017*, 308–328. [CrossRef]
22. Zhang, X.; Huang, H.; Huang, S.; Chen, Q.; Ju, T.; Du, X. A context-aware location differential perturbation scheme for privacy-aware users in mobile environment. *Wirel. Commun. Mob. Comput.* **2018**, *2018*, 9173519. [CrossRef]
23. Agir, B.; Papaioannou, T.G.; Narendula, R.; Aberer, K.; Hubaux, J.P. User-side adaptive protection of location privacy in participatory sensing. *GeoInformatica* **2013**, *18*, 165–191. [CrossRef]
24. du Pin Calmon, F.; Fawaz, N. Privacy against statistical inference. In Proceedings of the 2012 50th Annual Allerton Conference on Communication, Control, and Computing (Allerton), Monticello, IL, USA, 1–5 October 2012; pp. 1401–1408. [CrossRef]
25. Salamatian, S.; Zhang, A.; Calmon, F.d.P.; Bhamidipati, S.; Fawaz, N.; Kveton, B.; Oliveira, P.; Taft, N. How to hide the elephant- or the donkey- in the room: Practical privacy against statistical inference for large data. In Proceedings of the 2013 IEEE Global Conference on Signal and Information Processing, Austin, TX, USA, 3–5 December 2013; pp. 269–272. [CrossRef]
26. Degue, K.H.; Ny, J.L. On Differentially Private Gaussian Hypothesis Testing. In Proceedings of the2018 56th Annual Allerton Conference on Communication, Control, and Computing (Allerton), Monticello, IL, USA, 2–5 October 2018; pp. 842–847. [CrossRef]
27. Hua, J.; Tong, W.; Xu, F.; Zhong, S. A Geo-Indistinguishable Location Perturbation Mechanism for Location-Based Services Supporting Frequent Queries. *IEEE Trans. Inf. Forensics Secur.* **2018**, *13*, 1155–1168. [CrossRef]
28. Zhao, M.; Zhu, X.; Niu, J.; Ma, J. A Semantic-Based Dummy Generation Strategy for Location Privacy. In Proceedings of the2019 International Conference on Networking and Network Applications (NaNA), Daegu, Korea, 10–13 October 2019; pp. 21–26. [CrossRef]
29. Shekhar, S.; Xiong, H., Location Perturbation. In *Encyclopedia of GIS*; Springer: Boston, MA, USA, 2008; pp. 630–630. [CrossRef]
30. Gruteser, M.; Grunwald, D. Anonymous Usage of Location-Based Services Through Spatial and Temporal Cloaking. Proceedings of the 1st International Conference on Mobile Systems, Applications and Services—MobiSys'03, San Francisco, CA, USA, 5–8 May 2003; 2003. [CrossRef]
31. Dini, G.; Perazzo, P. Uniform Obfuscation for Location Privacy. In *Data and Applications Security and Privacy XXVI*; Springer: Berlin/Heidelberg, Germany, 2012; pp. 90–105. [CrossRef]
32. Krumm, J. A survey of computational location privacy. *Pers. Ubiquitous Comput.* **2008**, *13*, 391–399. [CrossRef]
33. Xu, Z.; Zhang, H.; Yu, X. Multiple Mix-Zones Deployment for Continuous Location Privacy Protection. In Proceedings of the 2016 IEEE Trustcom/BigDataSE/ISPA, Tianjin, China, 23–26 August 2016; pp. 760–766. [CrossRef]

34. Khodaei, M.; Papadimitratos, P. Cooperative Location Privacy in Vehicular Networks: Why Simple Mix Zones are Not Enough. *IEEE Internet Things J.* **2021**, *8*, 7985–8004. [CrossRef]
35. Li, Y.; Li, S. A Real-Time Location Privacy Protection Method Based on Space Transformation. In Proceedings of the 2018 14th International Conference on Computational Intelligence and Security (CIS), Hangzhou, China, 16–19 November 2018; pp. 291–295. [CrossRef]
36. Pu, Y.; Luo, J.; Wang, Y.; Hu, C.; Huo, Y.; Zhang, J. Privacy Preserving Scheme for Location Based Services Using Cryptographic Approach. In Proceedings of the2018 IEEE Symposium on Privacy-Aware Computing (PAC), Washington, DC, USA, 26–28 September 2018; pp. 125–126. [CrossRef]
37. Jarvinen, K.; Leppakoski, H.; Lohan, E.S.; Richter, P.; Schneider, T.; Tkachenko, O.; Yang, Z. PILOT: Practical Privacy-Preserving Indoor Localization Using OuTsourcing. In Proceedings of the2019 IEEE European Symposium on Security and Privacy (EuroS P), Stockholm, Sweden, 17–19 June 2019; pp. 448–463. [CrossRef]
38. Gupta, S.; Arora, G. Use of Homomorphic Encryption with GPS in Location Privacy. In Proceedings of the 2019 4th International Conference on Information Systems and Computer Networks (ISCON), Mathura, India, 21–22 November 2019; pp. 42–45. [CrossRef]
39. Li, X.; Ren, Y.; Yang, L.T.; Zhang, N.; Luo, B.; Weng, J.; Liu, X. Perturbation-Hidden: Enhancement of Vehicular Privacy for Location-Based Services in Internet of Vehicles. *IEEE Trans. Netw. Sci. Eng.* **2020**, *8*, 2073–2086. [CrossRef]
40. Lu, H.; Jensen, C.S.; Yiu, M.L. Pad: Privacy-area aware, dummy-based location privacy in mobile services. In MobiDE'08 Proceedings of the Seventh ACM International Workshop on Data Engineering for Wireless and Mobile Access, Vancouver, BC, Canada, 13 June 2008; pp. 16–23.
41. Bindschaedler, V.; Shokri, R. Synthesizing plausible privacy-preserving location traces. In Proceedings of the 2016 IEEE Symposium on Security and Privacy (SP), San Jose, CA, USA, 22–26 May 2016; pp. 546–563.
42. AbdelWahab, O.F.; Hussein, A.I.; Hamed, H.F.; Kelash, H.M.; Khalaf, A.A. Efficient Combination of RSA Cryptography, Lossy, and Lossless Compression Steganography Techniques to Hide Data. *Procedia Comput. Sci.* **2021**, *182*, 5–12. [CrossRef]
43. Machanavajjhala, A.; Kifer, D.; Gehrke, J.; Venkitasubramaniam, M. l-diversity: Privacy beyond k-anonymity. *ACM Trans. Knowl. Discov. Data (TKDD)* **2007**, *1*, 3-es. [CrossRef]
44. Li, N.; Li, T.; Venkatasubramanian, S. t-closeness: Privacy beyond k-anonymity and l-diversity. In Proceedings of the 2007 IEEE 23rd International Conference on Data Engineering, Istanbul, Turkey, 15–20 April 2007; pp. 106–115.
45. Gedik, B.; and Liu, L. Protecting location privacy with personalized k-anonymity: Architecture and algorithms. *IEEE Trans. Mob. Comput.* **2007**, *7*, 1–18. [CrossRef]
46. Cormode, G.; Procopiuc, C.; Srivastava, D.; Shen, E.; Yu, T. Differentially private spatial decompositions. In Proceedings of the 2012 IEEE 28th International Conference on Data Engineering, Arlington, VA, USA, 1–5 April 2012; pp. 20–31.
47. Xu, Y.; Yang, G.; Bai, S. Laplace Input and Output Perturbation for Differentially Private Principal Components Analysis. *Secur. Commun. Networks* **2019**, *2019*, 9169802. [CrossRef]
48. Balle, B.; Wang, Y.X. Improving the Gaussian Mechanism for Differential Privacy: Analytical Calibration and Optimal Denoising. In Proceedings of the 35th International Conference on Machine Learning, Stockholmsmässan, Stockholm, Sweden, 10–15 July 2018; Volume 80, pp. 394–403.

sensors

MDPI

Article

Providing Fault Detection from Sensor Data in Complex Machines That Build the Smart City

Alberto Gascón [1], Roberto Casas [1,*], David Buldain [1] and Álvaro Marco [1,2]

[1] Aragon Institute of Engineering Research, University of Zaragoza, 50018 Zaragoza, Spain; algaroche@unizar.es (A.G.); buldain@unizar.es (D.B.); amarco@unizar.es (Á.M.)
[2] GeoSpatium Lab S.L., Carlos Marx 6, 50015 Zaragoza, Spain
* Correspondence: rcasas@unizar.es; Tel.: +34-976-762-856

Abstract: Household appliances, climate control machines, vehicles, elevators, cash counting machines, etc., are complex machines with key contributions to the smart city. Those devices have limited memory and processing power, but they are not just actuators; they embed tens of sensors and actuators managed by several microcontrollers and microprocessors communicated by control buses. On the other hand, predictive maintenance and the capability of identifying failures to avoid greater damage of machines is becoming a topic of great relevance in Industry 4.0, and the large amount of data to be processed is a concern. This article proposes a layered methodology to enable complex machines with automatic fault detection or predictive maintenance. It presents a layered structure to perform the collection, filtering and extraction of indicators, along with their processing. The aim is to reduce the amount of data to work with, and to optimize them by generating indicators that concentrate the information provided by data. To test its applicability, a prototype of a cash counting machine has been used. With this prototype, different failure cases have been simulated by introducing defective elements. After the extraction of the indicators, using the Kullback–Liebler divergence, it has been possible to visualize the differences between the data associated with normal and failure operation. Subsequently, using a neural network, good results have been obtained, being able to correctly classify the failure in 90% of the cases. The result of this application demonstrates the proper functioning of the proposed approach in complex machines.

Keywords: fault detection; sensor data; industry 4.0; data reduction; feature analysis; feature selection; indicators; artificial neural network

Citation: Gascón, A.; Casas, R.; Buldain, D.; Marco, Á. Providing Fault Detection from Sensor Data in Complex Machines That Build the Smart City. *Sensors* **2022**, *22*, 586. https://doi.org/10.3390/s22020586

Academic Editors: Suparna De and Klaus Moessner

Received: 15 December 2021
Accepted: 11 January 2022
Published: 13 January 2022

1. Introduction

Predictive maintenance is a recent technique, the result of the evolution of maintenance techniques over the years. Initially, the most commonly used maintenance systems were corrective. These systems carry out the relevant actions once the failure has occurred. With this approach, it may happen that the repair has to be postponed instead of being repaired on the spot due to a lack of readiness. Preventive maintenance, where maintenance activities are scheduled at periodic intervals to prevent component degradation, was introduced in the 1950s. However, as in the previous case, costs remain very high.

Given the growing demand for more reliable, safe and efficient industrial systems, the need to optimize these maintenance processes becomes evident. In the 1980s, some factories began to apply predictive maintenance techniques. They used sensors that continuously monitored the machines and sent alerts when predefined limits were exceeded. This significantly reduced scheduled maintenance activities and their associated costs. Currently, the use of large databases combined with machine learning techniques makes it possible to predict what is going to happen, when it is going to happen, and to alert the person in charge (Industrial Internet of Things, IIoT) [1,2]. In this way, a "just-in-time" maintenance that allows maximizing economic and productive performance is achieved. For this reason, as shown in Figure 1, predictive maintenance applications are having a great boom in the market, expecting, according to a PWC survey, a 3.6% reduction of the annual costs during 2020.

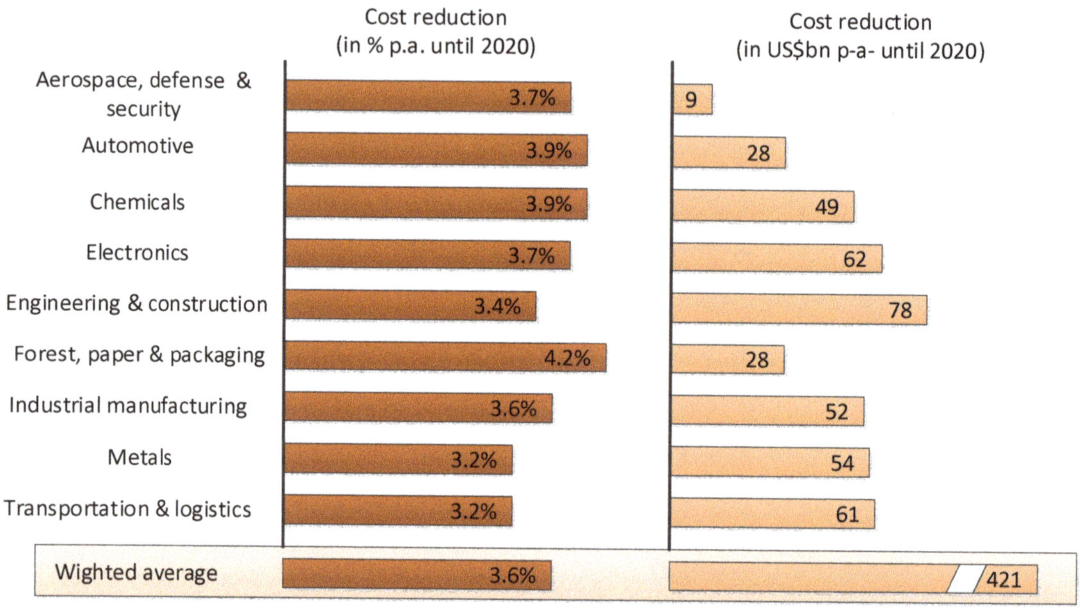

Figure 1. Annual cost reduction due to the incorporation of predictive maintenance techniques [1].

Predictive maintenance requires a great deal of dedication prior to installation. The problems and their causes must be identified in order to subsequently define and develop the monitoring system. This preliminary process could be structured in the following phases:

1. Detection of machines that suffer critical breakdowns for the production process.
2. Location of the machine element that produces the faults.
3. Identification of the causes that provoke the breakdowns (physical reasons why it breaks).
4. Definition of the variables to be monitored.
5. Selection of the sensors.
6. Data acquisition.
7. Data curation and extraction of indicators (features).
8. Data processing so that the system learns to detect failures.

In industrial environments, the most relevant machines that can be found most frequently are electric motors [3,4]. Within them, the elements that concentrate the highest number of failures are the rotating elements and the transmission mechanisms due to their fatigue wear. Figure 2 shows the result of an ABB study on the critical elements and the most common causes of failure in induction electric motors.

Once the critical machines, the elements that fail most frequently and the possible causes have been identified, the subsequent phases are carried out.

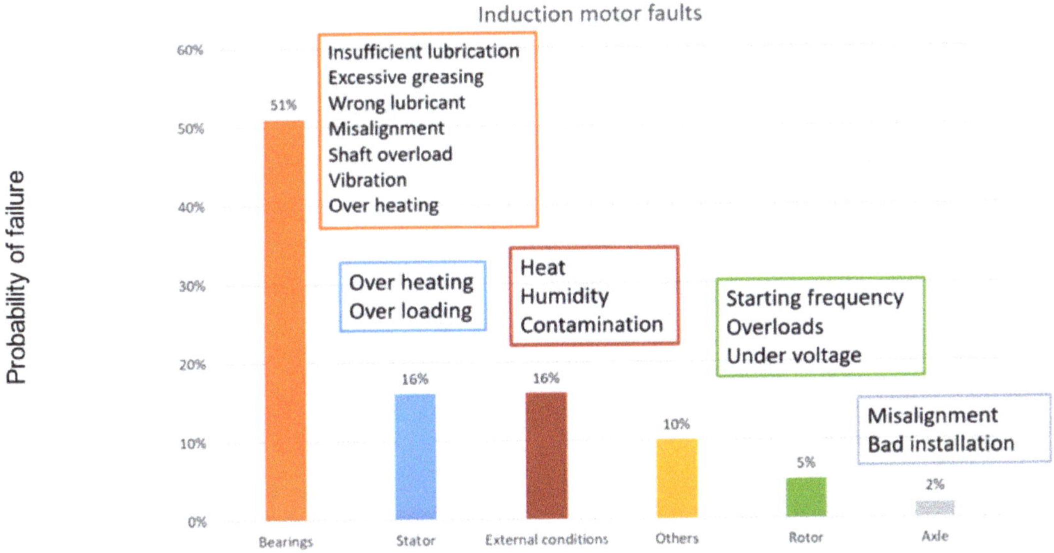

Figure 2. Representation of the critical elements of induction motors and their main causes of failure [3].

In contrast to the simple sensors and actuators that make up a large part of the common Internet of Things (IoT) scenarios in the smart city [5], there is an increasing number of applications that are made up of what we may call complex machines.

As illustrated in Figure 3, we define complex machine as a device that:

- Has a 24/7 operation operated by users without detailed knowledge of the operation of all the constituent parts of the machine.
- Integrates tens of sensors and actuators managed by several microcontrollers and microprocessors communicated by control buses.
- Requires energy from the mains to work, sometimes has a battery, but as a short-time backup.
- Has IP (Internet Protocols) connectivity.

Some examples of complex machines in the smart city are: household appliances, climate control machines, vehicles, elevators, cash counting machines, etc.

Technically both production lines and complex machines are made up by a network of controllers that integrate sensors and actuators. There are many works proposing predictive maintenance strategies in production lines [6] or industrial equipment [7,8]. These approaches gather all the data together in edge/fog devices [9] or in the cloud [10] and centrally analyzes them. In complex machines, this is not possible due to memory and computation restrictions of controllers and also to industrial bus bandwidth limitations.

Currently, there are different predictive maintenance strategies depending on whether they focus on physical aspects (physical model-based), on aspects of knowledge of the machine itself (knowledge-based), or if they are based on the use of large quantities data, pattern recognition, statistics, etc. (data-driven). This article proposes a fault detection methodology applicable to complex machines, trying to apply hybrid methodologies that combine the advantages of each strategy, adjusting them to the needs of complex machines. For this reason, special attention needs to be paid to preprocessing, seeking to minimize the number of data to be sent, so that malfunctions can be detected with the least amount of data possible. In this way, this strategy can be applied to machines with limited memory capacities, data transmission, etc.

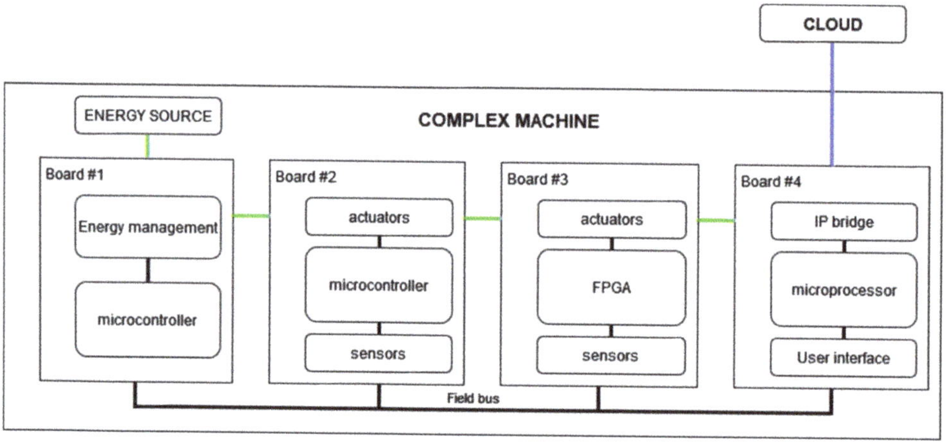

Figure 3. Block diagram of an example of a complex machine, green lines indicate energy flow, and blue lines indicate the data transmission to the cloud.

This paper is organized as follows. Section 2 presents the data processing methodology and its three different levels indicating the process applied in each case: sensor level—variable targeting, board level—embedded data curation and feature extraction, and machine level—feature integration and pattern finding. Then, Section 3 illustrates the testbench used to verify the system proposed and analyzes the results obtained on each layer. Finally, Section 4 provides conclusions.

2. Materials and Methods

2.1. Data Processing Methodology

A methodology for data processing consisting of three parts or levels will be proposed. The first level, called the sensor level, focuses on taking measurements using various types of sensors. The second level or board level, starts with the data obtained in the sensor level to carry out a processing that allows reducing the amount of data to be transmitted and extracting as much information as possible from them. The last level or machine level seeks to perform an analysis of the data of the board level in order to extract some results. In Figure 4, you can see the scheme of the proposed methodology.

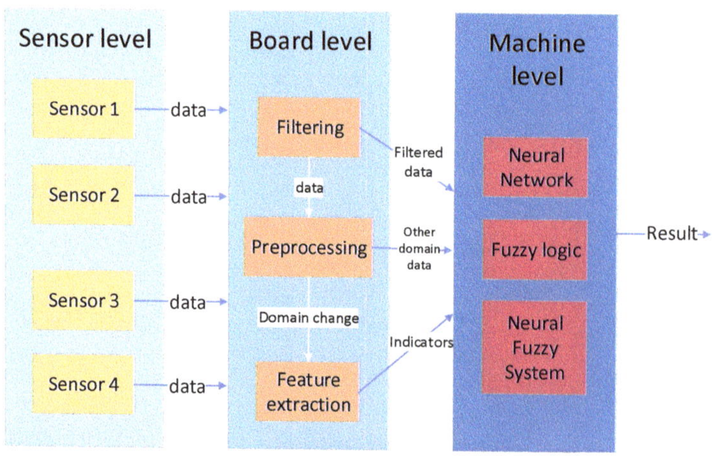

Figure 4. Scheme of the proposed methodology for data analysis.

2.1.1. Sensor Level—Variable Targeting

The variables to be monitored in this type of application can be grouped into the following groups: mechanical, electrical, audio, temperature and pressure [3].

The analysis of **mechanical variables** and specifically the analysis of vibrations are the most common. Depending on the frequency range of the vibrations to be measured, position sensors (0–10 kHz), speed sensors (10 Hz–1 kHz) or accelerometers (8 Hz–15 kHz) can be used [2]. However, the use of accelerometers is the most common, as has been seen in the vast majority of the articles consulted [11–19]. This type of analysis presents good results, since the most common faults always generate additional vibrations to those of the engine in normal operation. Thus, through its analysis, inappropriate behavior and even the type of failure can be identified [19].

Another common approach is the analysis of **electrical variables** [4,11,13–15,20–22]. In them, the values of the stator's motor currents and voltages are mainly monitored. The use of these variables is based on the fact that the consumptions of a damaged machine present variations compared to those of a "healthy" machine. In addition, the use of these measures has advantages, such as the possibility of measuring without having to access the interior of the motor, reducing the risk of damaging fragile parts and facilitating the installation of the sensors. On the other hand, it requires a great knowledge of the normal behavior of the machine and the different harmonics it presents due to construction characteristics or load variations. In addition, this knowledge of healthy functioning must be updated over time. Thus, applying techniques such as motor current signature analysis (MCSA), it is possible to detect anything from electrical failures, such as short circuits in the stator, to mechanical failures, such as eccentricities or rotor bar breaks [4].

In [2,23], **audio measurements** are used to detect bearing failures through the use of microphones. These types of measures are not so well-established, although they are gaining presence, as shown in [2]. The main cause is the contamination to which the audio signals are exposed in an industrial environment, requiring the use of techniques for their elimination. However, the possibility of obtaining them using microphones pointing towards the machine from the outside at between 2 and 10 cm is a clear advantage compared to vibration measurements [2].

These are the most commonly used types of measurements. However, others appear as complementary measures, such as **temperature measurements** [13–15,17,21] or **pressure measurements** [21], which are being used in very specific cases. A temperature increase makes possible to detect electrical and mechanical failures, since in the event of excessive friction or high electrical currents the elements tend to overheat. In addition, these types of measurements do not require complex processing, and faults can be detected by simply observing their values. Pressure measurements, for example, can be of great importance in the analysis of the motor of an air compressor (Air Booster Compressor).

2.1.2. Board Level—Embedded Data Curation and Feature Extraction

Each board has a smart controller that might have wide variety of computational and memory resources; from 8-bit microcontroller to an FPGA (Field Programmable Gate Array). To extract the most relevant characteristics and reduce the volume of data used, it is necessary to perform raw data filtering or pre-processing. In the case of predictive maintenance systems, preprocessing methods can be separated into three major groups according to whether they are, in the time domain [13,14,20–22], in the frequency domain [4,11,15–17,19,23] or in the time-frequency domain [12,18].

In the **temporal domain**, an attempt is made to reduce the number of data by filtering outliers and erroneous data [14,21]. Normalization [13,21,22] becomes relevant due to the use of various variables (mechanical, electrical, temperatures, pressures, etc.) that can take values on different scales. Once the data have been adjusted, they can be used as they are [21,22], or other indicators can be extracted from the parameters. In this second case, statistical indicators (such as maximum, minimum, mean, median, standard deviation, variance, gradients, kurtosis, skewness or crest factor [14,24]) or of another type (such as the principal components [13]) can be used.

On the other hand, in the **frequency domain**, the vast majority of cases use the fast Fourier transform (FFT) as an analysis method [4,15–17,19,23]. It provides useful information through the detection of the signal's frequency peaks and the detection of harmonics. This analysis is mainly applied when making vibration or sound wave measurements. The virtue of the FFT is that it allows decomposing a signal into individual periodic signals and establishing the relative intensity of each component, as can be seen in Figure 5. In this way, it is very easy to identify the faults, corresponding to peaks at unusual frequencies. In addition, it is a technique included in many electronic devices.

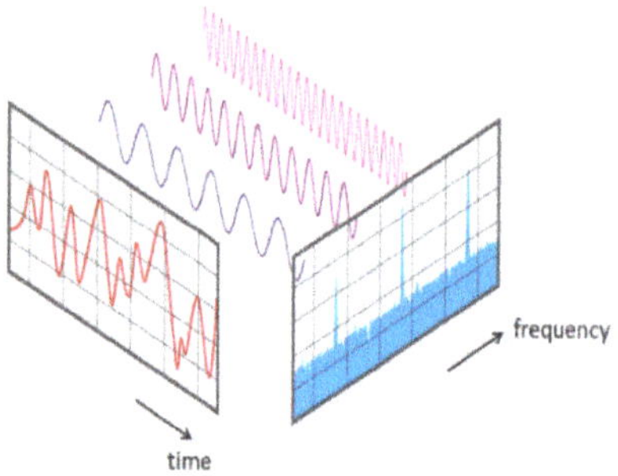

$$X_k = \sum_{n=0}^{N-1} x_n \cdot e^{-i2\pi kn/N} \qquad k = 0, 1, \ldots, N-1$$

Number of operations DFT: N^2

Number of operations FFT: $N \cdot \log N$

Figure 5. Graphical representation of the Fourier transform, equation and number of operations [25].

The FFT requires that the sampled signal contains a complete representation of the signal to be processed in the time domain or a periodic repetition. In cases where a complete cycle of the signal to be modulated is not captured, techniques such as the Hanning window are applied on the signal to mitigate possible reconstruction errors [16] (Figure 6).

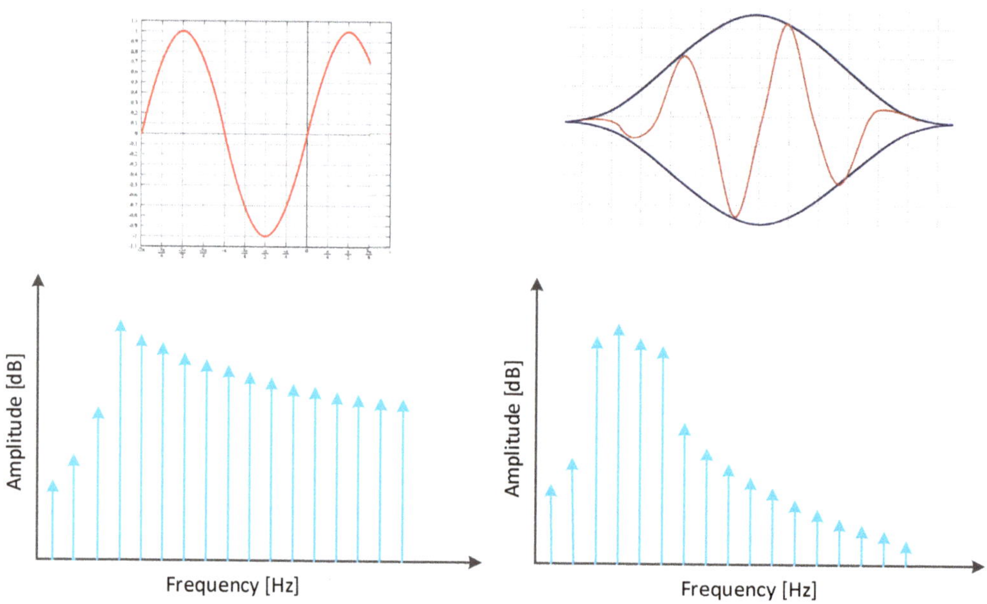

Figure 6. Representation of the Hanning window effect.

Finally, techniques in the **time-frequency domain** [12,18] provide a more realistic description of the state of the machine. The main advantage they provide is that they are capable of managing both stationary and non-stationary signals (limitation presented by the FFT). The most popular for vibration analysis in rotating machines is the wavelet transform (WT) [12].

This technique starts from an orthonormal wavelet located in time that multiplies the signal. It can be applied at different times and with different scales to analyze the high and low frequency components of the signal at different points. The signal is decomposed into the approximation and detail coefficients, allowing the identification of the different frequency contributions over time. An example can be seen in Figures 7 and 8.

$$X_\omega(a,b) = \frac{1}{|a|^{-\frac{1}{2}}}\int_{-\infty}^{\infty} \overline{\psi\left(\frac{t-b}{a}\right)}\, x(t)dt$$

$x(t)$: input signal

$\psi(t)$: orthonormal wavelet

a: binary dilation

b: binary position

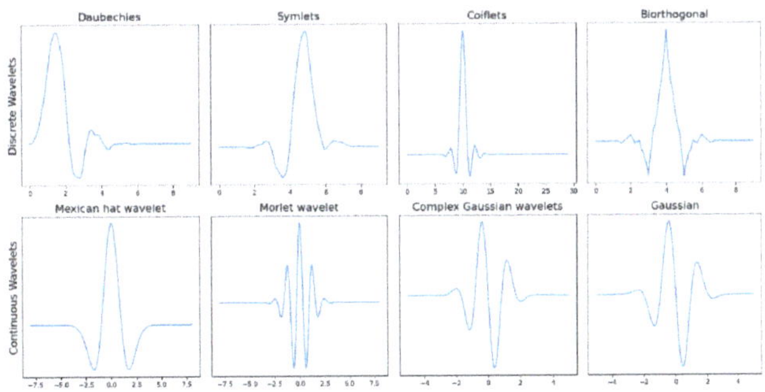

Figure 7. Wavelet transform equations and representation of different wavelets.

There are also other techniques such as the Short-Time Fourier Transform, STFT [18], which consists of dividing the signal into small time windows on which the Fourier transform is applied. In this way, it is possible to know the part of the signal in which each frequency appears, but it has a lower resolution than the WT.

In the case of audio signals, an additional preprocessing would be necessary to carry out the separation of the audio signal from the ambient noise. Some of the techniques used are BSS (Blind Source Separation) or TDSEP (Temporal Decorrelation source SEParation), which allow isolating a mixture of sounds from a specific process in real time [26].

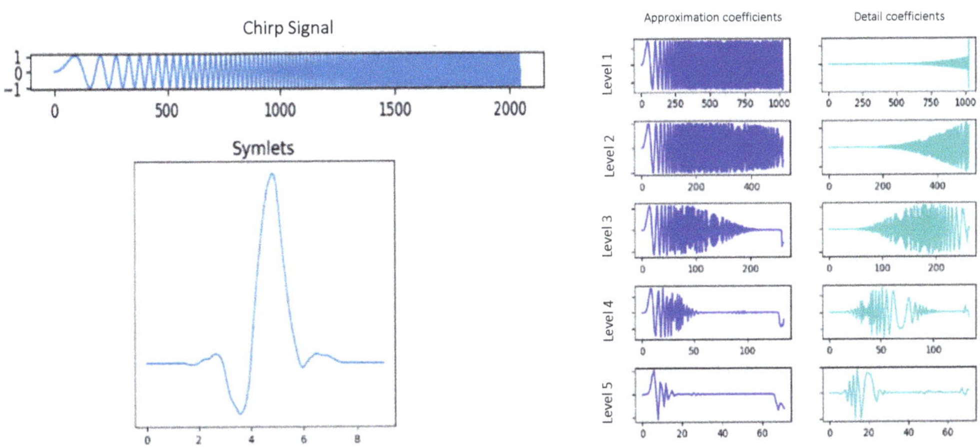

Figure 8. Coefficients of approximation and detail obtained by means of the WT of the signal displayed with a Symlets wavelet.

2.1.3. Machine Level—Feature Integration and Pattern Finding

Once all the desired indicators have been extracted, in order to make sense of these data, it is necessary to analyze them together in what will be called the machine level. This processing would consist of the procedure for identifying possible failures. Techniques for the detection and identification of failure mechanisms are based on pattern recognition. These can be applied following complex strategies, such as the use of neural networks and machine learning [11,13,14,18,20–23], or through simpler methods, such as the use of fuzzy logic [15,16] or visual analysis by specialist personnel [17].

Artificial neural networks (ANN) are very convenient for this type of task, as they are able to work with a large amount of data and manage non-linearity situations with a short response time [20]. However, actual failure data are scarce, and forced failure data acquisition can be expensive. Even so, supervised learning methods are commonly used [13,18,20–23], although a predictive maintenance implementation could be initiated with unsupervised or semi-supervised learning (with labeled and unlabeled data) [14].

Finally, **simpler techniques** such as fuzzy logic are also applied. In [15,16], a classification of the data is carried out based on the ranges in which they are found, using fuzzy classifiers (good, normal, bad...). This allows a greater interpretability, something that can be tricky with neural networks. However, this apparent simplicity presents a key point that can become a bottleneck, the definition of the ranges, which requires a great knowledge of the situation to be treated. Furthermore, since it has no learning capability, it is often used in combination with neural networks, generating the so-called neural fuzzy systems (NFS) [27].

Once the results of the machine level processing have been obtained, the data can be sent to the cloud in order to perform a normality model that considers a large amount of data from different machines. Thus, while in the machines, the neural networks have patterns at the local level, in the cloud, the patterns are at a global level, which allows a greater abstraction.

A complete example of a complex machine with the different layers can be seen represented in the diagram in Figure 9: in yellow, the lowest layer would be that of the sensor level; in orange, encompassing the previous one would be the board level; and finally, in red, encompassing the previous two, is the machine level.

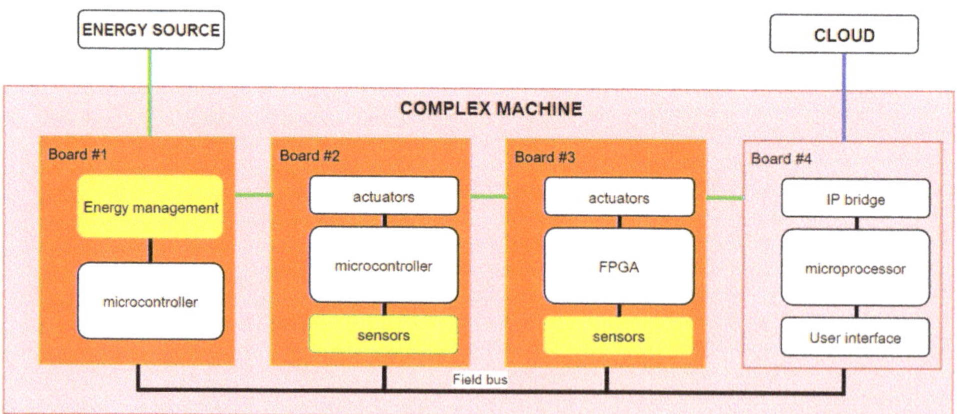

Figure 9. Diagram of an example of a complex machine with layer differentiation (Yellow, sensor level; orange, board level; red, machine level).

In this example, it can be seen how the first three boards (energy management, microcontrollers and FPGAs with actuators) communicate with a fourth, which is differentiated by the ability to communicate with the outside. Said communication can be both, with the user (Human Machine Interface) and with another computational element, proposed in the example through an IP bridge. This board also has processing capacity, along with large RAM and FLASH memories, so that it can be in charge of analyzing the results obtained. In this way, this fourth board is in charge of receiving possible orders from the user, performing an analysis of the data received from other boards and sending the data to a downstream processing unit (in the cloud in the proposed example).

3. Results and Discussion

3.1. Testbench Definition

The machine on which this methodology will be applied is a machine designed to count banknotes that will be used mainly in bank branches to be able to count cash and make deposits safely (Figure 10).

Figure 10. Image of the prototype used to test the methodology.

The complex machine is made up of a main board and three secondary boards: energy board, engines board and energy board (Figure 11), and the variables that are going to be monitored are shown in Table 1:

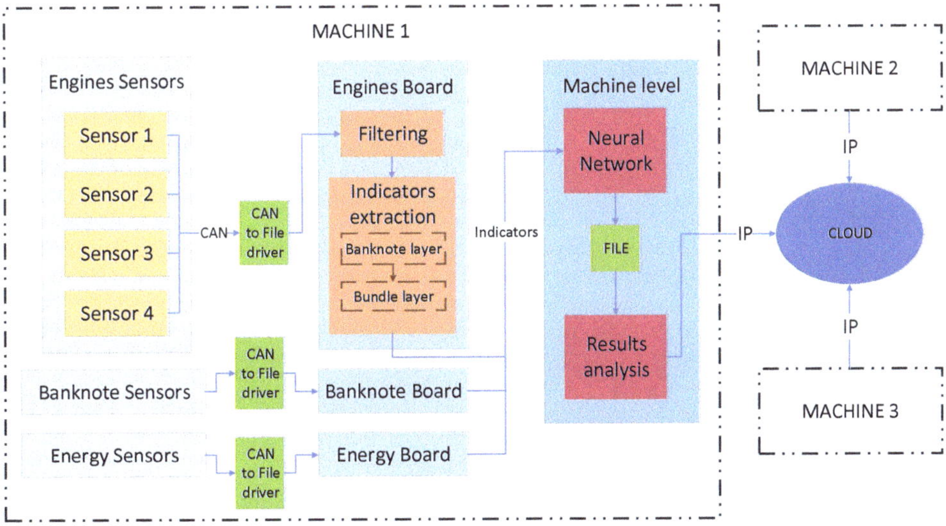

Figure 11. Scheme of application of the methodology to the specific case.

Table 1. Summary of the variables monitored.

Variable	Board	Abreviation	Unit	Bits
Transport engine current	Engines	I_trans	mA	12
Feeding engine current	Engines	I_feed		12
Transport engine encoder ticks	Engines	N_pul_trans	Number of counter ticks between two encoder pulses	12
Feeding engine encoder ticks	Engines	N_pul_feed		12
Infrared sensor 1a	Engines	IR1a	0 no obstacle, 1 obstacle	1
Infrared sensor 1b	Engines	IR1b		1
Infrared sensor 2	Engines	IR2		1
Infrared sensor 3a	Engines	IR3a		1
Infrared sensor 3b	Engines	IR3b		1
FFT of microphone measures	Engines	FFT	-	1024
Doubles sensor 1	Banknotes	Doubles1	Measure proprtional to banknote's thickness	32
Doubles sensor 2	Banknotes	Doubles2		32
Temperature	Energy	Temp	Celsius degrees	16
Internal voltage	Energy	Vint	V	16
Auxiliary voltage	Energy	Vaux		16

The tests will consist of passing bundles of 50 banknotes of the same denomination (5 €, 10 €, 20 € and 50 €) through the machine. In addition, each bundle will pass through the machine four times, placing all the banknotes in every possible orientation: front, back, reverse front, and reverse back (Figure 12). This results in 16 samples for each tested case, making a total of 800 banknotes analyzed per case.

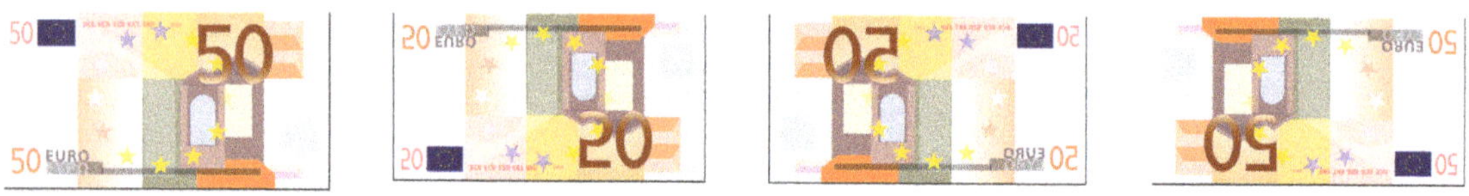

Figure 12. Disposition of a €50 banknote in the different orientations analyzed.

Regarding the failures analyzed, 13 defects (Table 2) were forced into the machine through variations in eccentricities in axles (4) and wheels (3), use of defective components such as springs (2), dented bearings (2), and deteriorated pulleys and worn belts (2):

This means that the whole dataset consists of 11,200 banknotes records. The proposed strategy will focus on detecting failures that could be called permanent, this means that they will appear throughout all the data collection and not sporadically, something that could also happen under real operating conditions.

Table 2. Summary of the tested cases.

Identifier		Name of Failure
0		Normal operation case
1		Effect of eccentricity in axle 2
	A	Concentrity deviation of 0.2 mm
	B	Concentrity deviation of 0.5 mm
2		Effect of eccentricity in axle 4
	A	Concentrity deviation of 0.2 mm.
	B	Concentrity deviation of 0.5 mm.
3		Effect of dented bearings:
	A	Dented bearing in axle 2.
	B	Dented bearing in axle 3.
4		Effect of defective springs:
	A	Spring without screw at BNF.
	B	Spring without screw at the entrance of the safe.
5		Effect of defective doubles sensors:
	A	Perforated doubles wheel.
	B	Eccentricity of 0.04 mm of the outer wheel.
	C	Eccentricity of 0.08 mm of the outer wheel.
6		Deteriorated pulleys and worn belts:
	A	Deteriorated 32 z pulley.
	B	Worn S2M 180 belt and deteriorated exit pulley.

3.2. Data Analysis

3.2.1. Layer 1: Sensor Data

Sensor data layer capture is accomplished by three boards, which gather the variables listed in Table 1. Once the data of the different failure cases have been obtained, they are analyzed and compared with those of the normal operation case. It is important to know the shapes and values of the data distributions in order to better understand the indicators to be extracted in subsequent layers, since this allows us to assess the best strategies to analyze them and perform a more efficient maintenance.

We will begin by analyzing the data obtained by the energy management board. Some of the measurements taken are voltage measurements at different points or temperature measurements, among others. When comparing the voltage data of various banknotes according to their orientation and obtained in different situations, differences can be appreciated. In Figure 13, it can be seen how the failure case shows higher values for *Vaux* than the case of normal operation. Although these are differences of a very small order of magnitude, given that the values present a very small variation, they must be taken into account.

The engines board has sensors that take measurements related to the mechanical operation of the machine, such as consumption of the machine's motors or an FFT for the vibrations analysis. Figure 14 shows the current consumption during the passage of different banknotes. We can see that when introducing the modification in the machine, the current figures have been altered. Although there is still a significant overlap in the ranges, the values of the failure case present values below what would be considered normal.

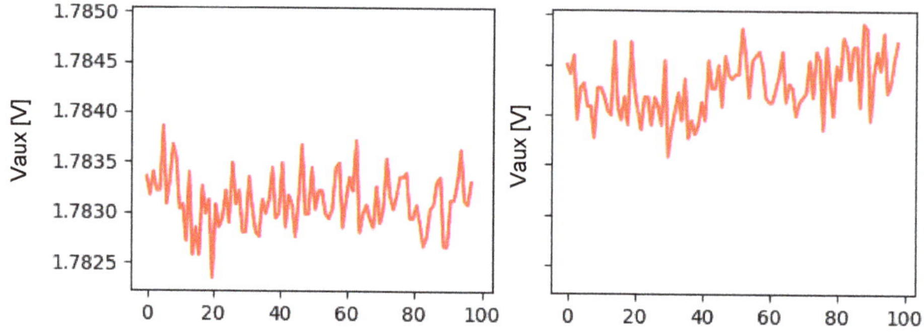

Figure 13. Comparison of the Vaux tension measurements of a 5 € banknote on reverse back orientation in the case of normal operation (**left**) and a failure case (**right**).

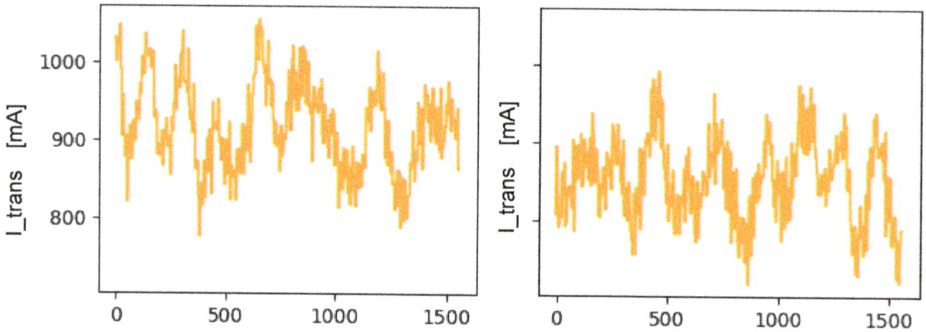

Figure 14. Comparison of the measurements of the transport motor current of a 50 € banknote on reverse front orientation in the case of normal operation (**left**) and that of a failure case (**right**).

The last of the boards included in the complex machine analyzed is responsible for monitoring the condition of the banknotes through various measurements. One of the sensors used provides values that are proportional to the thickness of the banknote passing through the machine, called the doubles sensor. Figure 15 shows the measurements obtained in the normal case versus one of the failure cases analyzed, presenting clear differences that would allow the identification of such operation as erroneous.

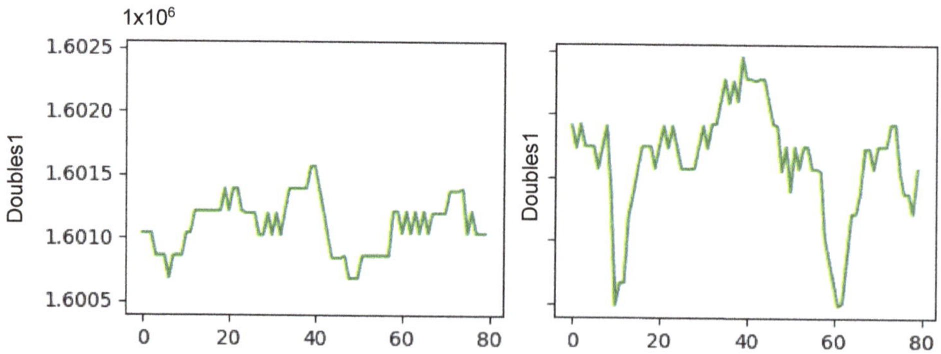

Figure 15. Comparison of the measurements of the double sensor 1 of a 20 € banknote on back orientation in the case of normal operation (**left**) and one case of failure (**right**).

3.2.2. Layer 2: Board Data

The amount of data provided by the machine is very high, since for each banknote (a banknote takes 610 ms to pass through the machine) 33,000 bytes would be received from the engines board, 9150 from the banknotes board and 7320 from the energy management board. Therefore, the need to reduce this number through filtering and extraction of indicators becomes evident.

In order to perform an initial filtering to reduce the number of data to be processed, it is decided to use only the data relating to the passage of a banknote through the machine. In this way, all data taken between banknotes are discarded. In addition, since the measurements of some sensors are only of interest when the banknote passes through them, it is necessary to generate a specific window for each of them. For this purpose, position sensors are used, which allow us to know the position of the banknote in the machine, being able to select the data only for those moments. Just through this filtering, we reduce to 8808 bytes per banknote from the engines board, 640 from the banknotes board and 394 from the power management board, a reduction of an order of magnitude.

When proposing the indicators to be extracted, a layered data analysis was chosen, as shown in Figure 16. The first layer would be the sensorization layer, the output of which is the raw data. After the filtering process that would follow the sensorization layer, the next layer would be that of the indicators per banknote, in which various indicators corresponding to each note are obtained. Finally, the last one would be that of the indicators per bundle, which aggregates the indicators per banknote into groups of a given number.

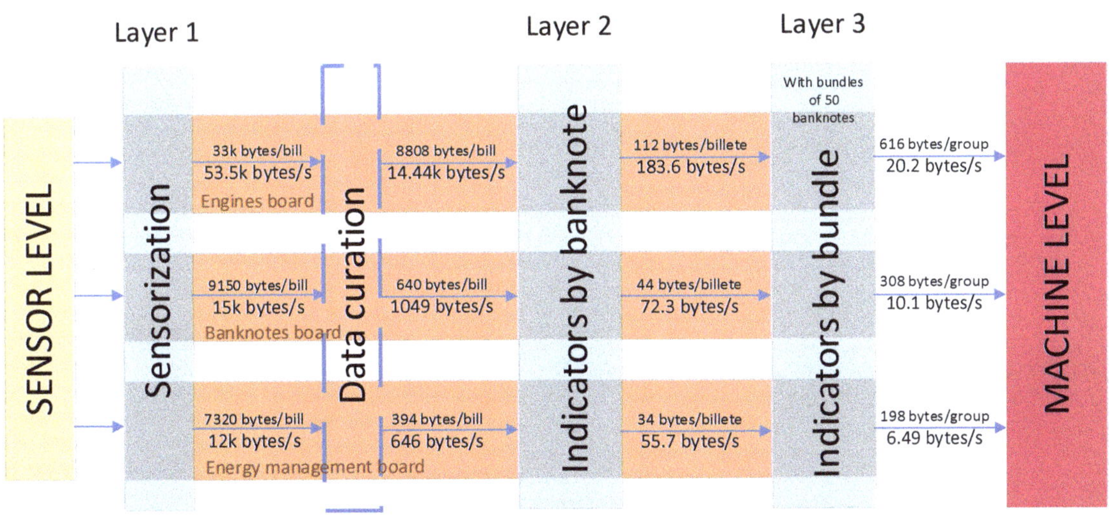

Figure 16. Diagram of the designed monitoring infrastructure.

Observing the output data rates of the last of the layers, it can be seen that a reduction of three orders of magnitude in the bytes per second that are obtained from each board has been achieved. With this extraction of indicators, by comparing the values of the training phase with the values obtained in subsequent measurements, it will be possible to detect deviations that will allow the identification of possible failures.

The indicators per banknote to be used are, in general, the means, medians, maximums, minimums, standard deviations, asymmetries and kurtosis of the different measurements available in the frames, adding the effective value in the case of currents.

Regarding the FFTs with a Hanning window (Figure 6) obtained for the vibration data, a more complex analysis is conducted. To obtain the indicators extracted in the layer of indicators per banknote, the areas under the curve of different parts of the FFT will be obtained. In order to discover the more interesting parts, we will begin by identifying the existing peaks. This identification consists of two phases: a first one in which the base noise is eliminated, leaving a flatter FFT in which the peaks stand out more; and a second one in which the peaks higher than half the maximum value are marked. Having identified the most interesting parts as those that concentrate the majority of the peaks, the areas under these parts of the FFTs will be used as the indicators of the vibration data (the limits of the areas are defined based on observation). In a preliminary analysis of the FFTs, it has been seen that in most of them, there are two areas of interest in which most of the peaks are concentrated (see Figure 17). Therefore, it is decided to work with these two areas for subsequent analysis.

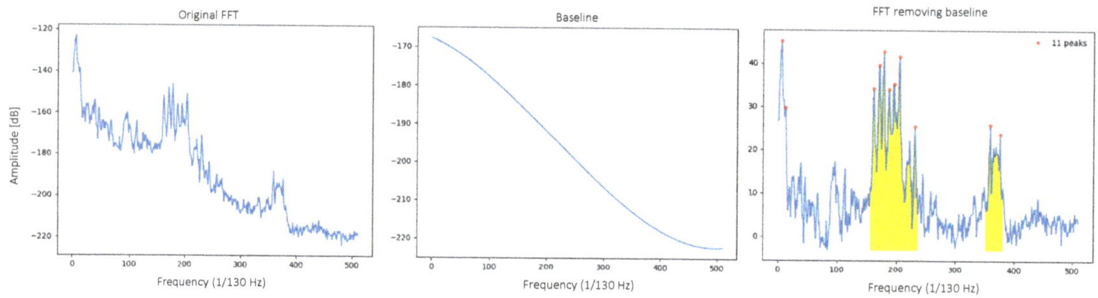

Figure 17. Graphs of the different phases of the identification process, from the original FFT, to the FFT without the base noise, with the detected peaks and the colored areas of interest.

Once the indicators per banknote have been extracted, they are passed to the layer of the indicators per bundle. Since the objective is the data reduction for a fault detection application, it is not sought to have an instantaneous view of the machine operation. A broader vision that allows observing the variations in a larger temporal space is more interesting. Therefore, the integration of the banknote level indicators will be done in groups of the same number of banknotes, from which indicators will be extracted per bundle. The indicators extracted from the indicators of the previous layer will be the same seven previous statistical values as above, the means, medians, maximums, minimums, the standard deviations, the skewness and the kurtosis.

3.2.3. Layer 3: Machine Data

Finally, after obtaining the indicators from the bundle layer, the indicators reach the machine level. In this layer, conclusions will be drawn from the indicators extracted in the previous processes. For this purpose, this analysis seeks to identify the most relevant indicators for each failure case, as well as the type of variation that should be expected based on their probability distributions. Next, we will comment on the results obtained by comparing the distributions of the indicators of the respective failure case with those of the normal case.

The objective is to indicate whether the indicators of the failure case have higher or lower values than those of the normal case, as well as the degree of discordance between the distributions of these indicators. The indicators analyzed will be the indicators per-bundle-mean. If no specific indicator is mentioned (AVG, Med, MAX, MIN, DES, SK or KUR), the mean values are assumed to be the ones mentioned.

To assess the direction of the variation, the median of the distributions is used. On the other hand, to assess the degree of discordance, the Kullback–Leibler divergence is used. This is a unitless measure that compares the probability densities of two distributions. It provides values close to zero with two similar distributions and it grows as the difference between both distributions increases. It is not symmetric, so two calculus are made, considering first the normal case and then the failure one $(P\|Q)$ and then viceversa $(Q\|P)$. From these two values, the larger one is the one considered. The choice of the

Kullback–Leibler divergence as a measure of comparison of the data distributions obtained in each failure case is based on the fact that the final model for failure classification will be implemented by neural networks trained with the cross-entropy cost function, which is directly related to the divergence measure. Thus, the nomenclature used is the one shown in Table 3 (limits used are based on experimental observation) and the results obtained can be seen in Table 4.

Table 3. Nomenclature used to classify the differences between distributions.

Divergence	KL > 4	KL > 5	KL > 10	KL > 15
Higher	LS	S	SS	SSS
Lower	LI	I	II	III

It is convenient to take the data in the summary table with caution, since there are distributions that, although they do not present divergences greater than the minimum, they do show variations with respect to the normal case.

Table 4. Summary of the information of the indicators of interest in each case.

Failures		1		2		3		4		5			6	
Indicators		A	B	A	B	A	B	A	B	A	B	C	A	B
Current	I_trans	SSS	=	II	II	I	=	III	III	III	II	III	III	III
	I_feed	=	=	=	=	=	=	=	=	=	=	=	=	=
Time between IR	T_IR11	=	=	S	S	=	=	=	=	=	=	=	=	=
	T_IR31	=	=	=	=	=	=	=	=	=	=	=	=	=
	T_IR33	=	=	=	=	=	=	=	=	=	=	=	=	=
N_pul_trans		=	=	=	=	=	=	=	=	=	=	=	=	=
N_pul_feed		=	=	=	=	=	=	=	=	=	=	=	=	=
Doubles sensors	Doubles 1	III	III	S	=	S	SSS	SSS	S	AVG SSS DES S	AVG SSS DES SSS SK III KUR SSS	AVG SSS DES SSS SK III KUR SSS	SS	AVG SSS DES S
	Doubles 2	AVG I DES II	AVG III DES II	S	LS	SS	SSS	SSS	S	AVG SSS	AVG III DES SSS	AVG III DES SSS SK I	SS	AVG SSS DES SSS SK SS
Voltages	Vint	=	=	=	=	=	=	=	=	=	=	LS	LS	=
	Vaux	=	=	=	=	=	S	=	=	=	=	=	=	=
FFTs	Energy 1	=	=	=	=	=	=	=	I	=	=	=	LI	=
	Energy 2	=	=	=	=	=	=	=	I	II	=	=	I	I

Next, the indicators of interest for the three failure cases in group five, associated with defects in the doubles sensor (case 5), will be shown. This case of failure has been chosen because it presents deviations in a great variety of indicators.

In the data of the **transport motor current** (Figure 18), it can be seen how in all cases of failure, the values obtained are reduced, the most notable being that of the first failure. In the following cases, in Table 5 it can be seen how the medians decrease in the distributions and the divergences increase as the eccentricity increases.

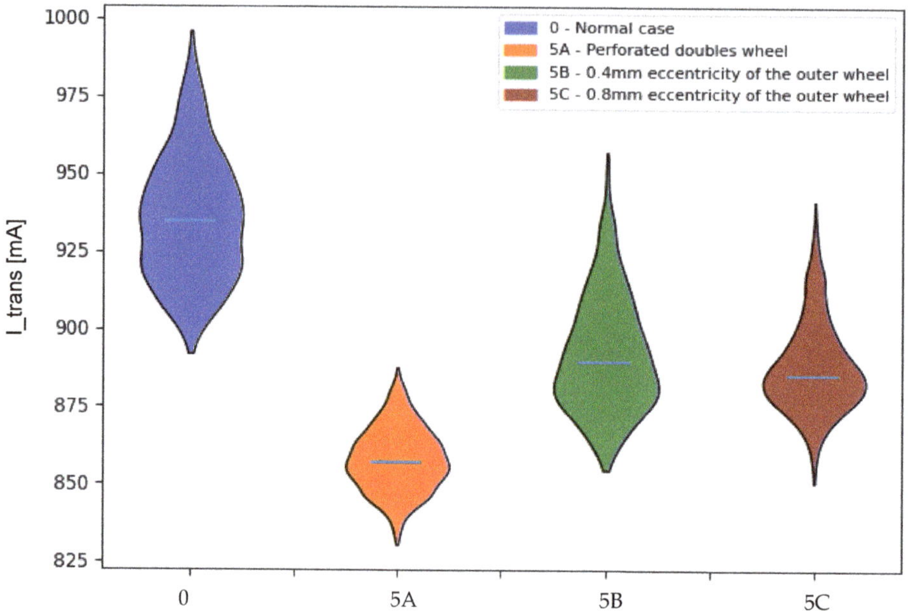

Figure 18. Comparison of the probability distributions obtained for the mean values of the transport motor current in the case of faults associated with the double sensors. Indicator: mean.

Table 5. Kullback–Leibler divergence values associated with the mean values of the transport motor current for the identification of the faults associated with the double sensors.

I_Trans	Kullback–Leibler					
	0-5A	**5A-0**	**0-5B**	**5B-0**	**0-5C**	**5C-0**
Mean	25.322	26.229	6.836	14.215	12.087	17.932

The data from the **doubles sensor 1** (Figure 19) shows deviations in all three cases analyzed, something that might be expected, as the defects are introduced in the sensor itself. Regarding the average values, it is seen that the one that suffers the most divergence is the first failure of the chopped roller, while those associated with eccentricities show smaller variations, but that vary in a linear way with increasing eccentricity. Analyzing the shape statistics, it can be seen how the first failure shows values quite similar to the normal case of standard deviations and coefficients of skewness and kurtosis. However, the failures associated with eccentricities show much larger differences, highlighting the standard deviation in the case of greater eccentricity, which is more than five times higher than that of the normal case. All this can be supported by the divergence values obtained in Table 6.

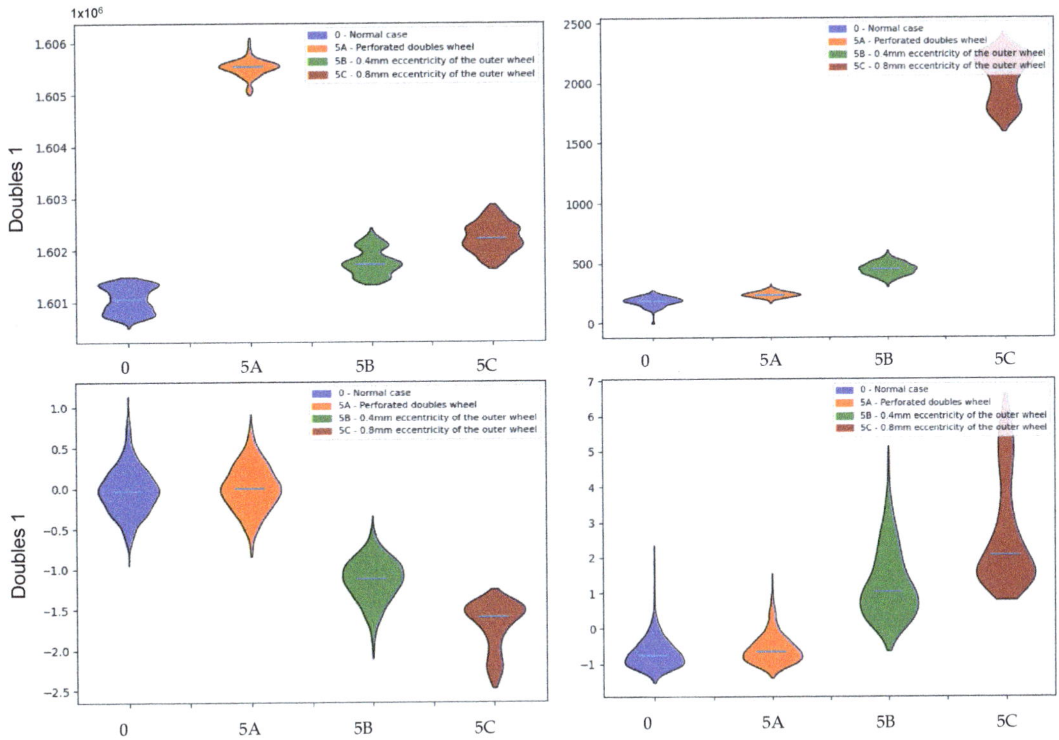

Figure 19. Comparison of the probability distributions obtained for the mean values of the doubles sensor 1 measurements in the case of failures associated with the doubles sensors. Indicators: (**top left**) mean, (**top right**) standard deviation, (**bottom left**) skewness and (**bottom right**) kurtosis.

Table 6. Kullback–Leibler divergence values associated with the means, standard deviations, skewness and kurtosis associated with the mean values of the double 1 sensor measurements for the identification of the failures associated with the double sensors.

Doubles 1	Kullback–Leibler					
	0-5A	5A-0	0-5B	5B-0	0-5C	5C-0
Mean	26.985	27.402	19.838	21.173	25.869	25.379
Std.Dev.	6.509	3.008	26.270	25.759	28.101	26.075
Skewness	0.355	0.202	22.922	22.689	25.899	26.216
Kurtosis	0.297	0.469	15.573	10.018	26.375	16.953

Regarding **the doubles sensor 2** (Figure 20), distributions of the mean values are very similar to those of the previous sensor. However, in the shape statistics, there are differences with respect to the previous sensor in the cases of failures associated with eccentricities. The standard deviations are no longer so far apart, although they still show considerable divergences (Table 7). The asymmetries no longer present values so far away from the normal ones, and only in the case of higher eccentricity is a significant divergence observed.

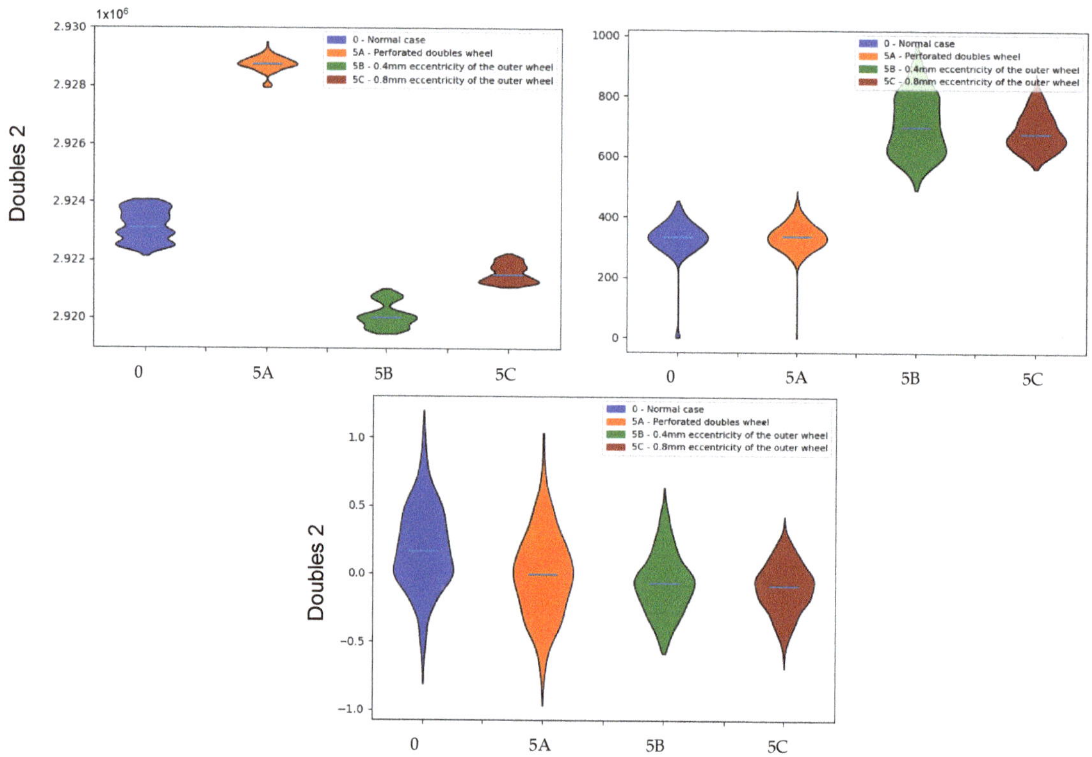

Figure 20. Comparison of the probability distributions obtained for the mean values of the doubles sensor 2 measurements in the case of failures associated with the doubles sensors. Indicators: (**top left**) mean, (**top right**) standard deviation, (**bottom**) skewness.

Table 7. Kullback–Leibler divergence values associated with the means, standard deviations and asymmetries associated with the mean values of the double 1 sensor measurements for the identification of the failures associated with the double sensors.

Doubles 2	Kullback–Leibler					
	0-5A	**5A-0**	**0-5B**	**5B-0**	**0-5C**	**5C-0**
Mean	26.516	27.226	25.925	26.189	25.012	24.793
Std. Dev	0.101	0.231	26.762	25.599	26.639	26.132
Skewness	0.588	0.663	2.222	0.761	5.800	0.983

Analyzing the values obtained for the Vint voltage (Figure 21), it can be seen that the second fault shows hardly any variations with respect to those of the normal case, while the other two show values higher than the usual ones. These effects are reflected in the divergences in Table 8, obtaining the highest value in the fault with the highest eccentricity.

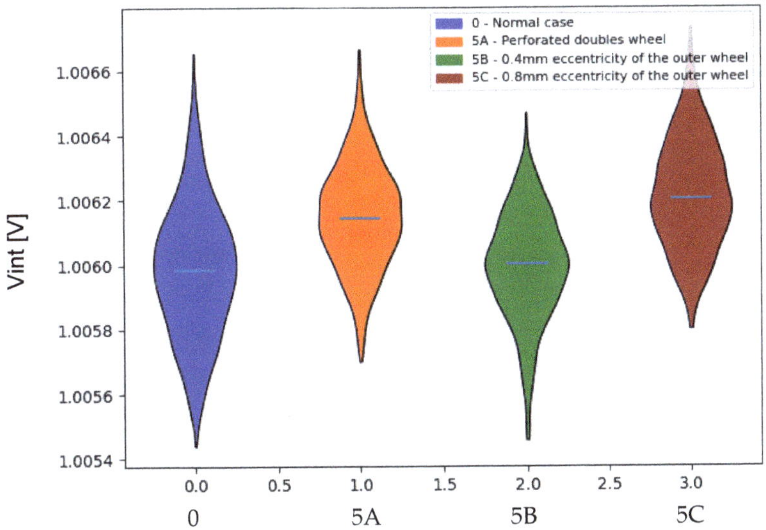

Figure 21. Comparison of the probability distributions obtained for the mean values of the voltage Vint in the case of failures associated with the double sensors. Indicator: mean.

Table 8. Kullback–Leibler divergence values associated with the means associated with the average values of the Vint voltage for the identification of the failures associated with the double sensors.

Vint	Kullback–Leibler					
	0-5A	**5A-0**	**0-5B**	**5B-0**	**0-5C**	**5C-0**
Mean	2.196	0.669	0.516	0.42	4.760	1.414

Observing the results of this first analysis (Table 4), it is clear that some measurements such as the feed motor currents (I_feed), some infrared pass-times or the intervals between encoder pulses do not provide much information. In the case of the supply currents, it may be due to the fact that none of the altered elements affected it directly, which means that there are hardly any changes from one case to another. As for the T_IR31 and T_IR33 times, since they are associated with short distances compared to the T_IR11 time, and with sections far away from the middle zone where the defects are located, this means that they are not so easily altered. Finally, the intervals between encoder pulses present a multimodal distribution that makes the extraction of information more difficult.

By detecting these differences with the naked eye on a measure related to the cross-entropy cost function used to train the network, it is expected that the neural model in charge of processing these indicators will also be able to identify the failures. To this end, a feedforward multilayer neural network with one hidden layer was explored and, after testing several architectures, a 39:128:14 architecture was chosen. The 39 input neurons correspond to the most relevant indicators, while the 14 output neurons are associated with the 13 failure cases analyzed and the case of normal operation.

Due to the large variety of ranges present in these input variables, a normalization is performed to equalize the contributions of each variable to the multilayer perceptron. This normalization is performed before the data enters the network. Between the two processing layers that compose it, another batch normalization layer has been added with the same objective.

The processing layers that compose the model are two dense layers of 128 and 14 neurons, respectively. The first layer uses the RELU function as the activation function. In the case of the second one, there is one neuron for each class and the activation function chosen is softmax. Thus, the output of each neuron will be between 0 and 1, summing all of them 1 (generates a probability distribution), assuming that the output with the maximum value is the correct failure/normal case.

The supervision labels follow the one-hot encoding (13 labels to zero and the corresponding class to one) and the optimization of the training hyperparameters was performed with the Adam optimization algorithm [28]. Figure 22 shows the summary of the implemented MLP architecture. Cross entropy was used as a cost function to perform the training over 40 epochs.

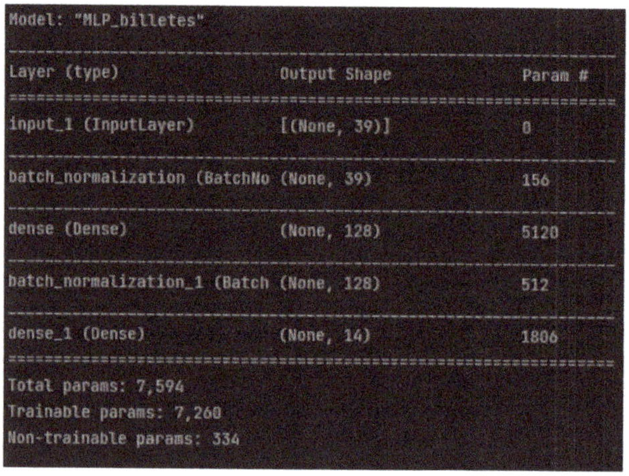

```
Model: "MLP_billetes"
_________________________________________________________________
Layer (type)                 Output Shape              Param #
=================================================================
input_1 (InputLayer)         [(None, 39)]              0
_________________________________________________________________
batch_normalization (BatchNo (None, 39)                156
_________________________________________________________________
dense (Dense)                (None, 128)               5120
_________________________________________________________________
batch_normalization_1 (Batch (None, 128)               512
_________________________________________________________________
dense_1 (Dense)              (None, 14)                1806
=================================================================
Total params: 7,594
Trainable params: 7,260
Non-trainable params: 334
```

Figure 22. Summary of the implemented MLP.

A cross-validation methodology was implemented to perform the training phase of several networks to generate the full model. To reduce the possibility of the data sets in the cross-validations becoming unbalanced due to sparse data, cross-validation with a low k-fold of 5 was used. In addition, to maintain data representativeness, these subsets were randomly generated using a stratified split to ensure that each test subset and training subset has statistical values of the label distribution equivalent to those of the whole dataset. In this way, five MLPs are generated, with the final model prediction being the mean or vote of the five different independent results.

At the end of the training, five MLPs were obtained with accuracies around 90%. However, as the classification result will be the result of the vote of the five networks, the reliability of the final model is even higher. The possibility of three networks being wrong, generating a bad prediction, would be close to 1%, which considerably increases the confidence of the classification. Figure 23 shows the confusion matrices of two of the five networks, which confirm the good performance of the methodology when applied to a specific case of fault detection on complex machinery.

Despite the good results obtained, it must be taken into account that this model has been made with data that were not isolated for validation. This has been the case because, as there is not a large amount of data available, we have worked with the available data, extracted manually by the manufacturers themselves. That is why the neural model, although showing good results and could be developed at some point, must currently be taken with caution. However, the fact that manual rule-based analysis of the results shows perceptible differences proves that the effort put into preprocessing the data to reduce the number of bytes to be sent has been successful. In this way, the objective of our proposal is reached, being able to identify the failures with a smaller amount of information and also being able to implement this methodology in complex machines with limited capacities.

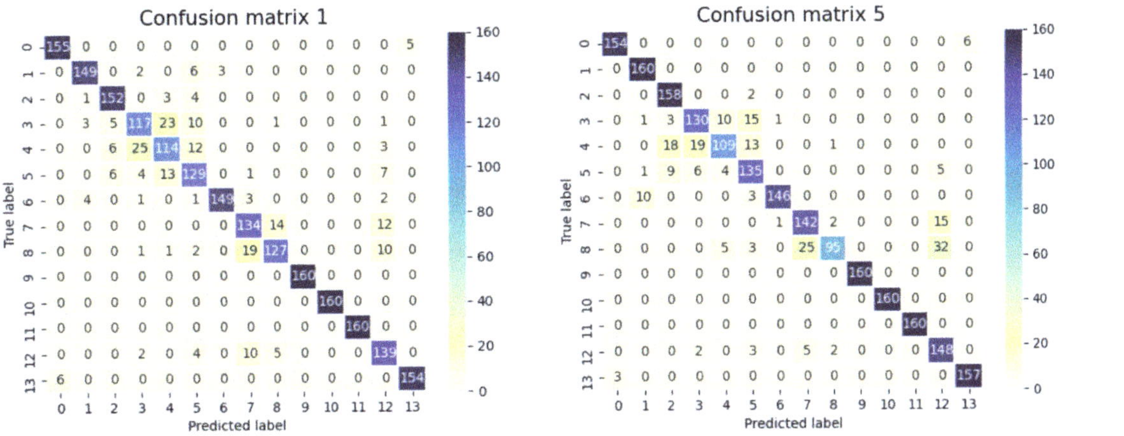

Figure 23. Confusion matrices of two neural networks designed to recognize the 14 failure cases.

4. Conclusions

In this work, a study of the state of the art of predictive maintenance techniques, focused on industrial environments, has been carried out. The variables to be monitored, the most relevant indicators and the most common treatment techniques have been analyzed to generate a basis on which to build the fault detection system to be carried out.

A monitoring methodology applicable to complex machines has been presented, in which work is carried out at different levels with the aim of extracting the most significant features from each piece of data. This methodology has been tested on a prototype cash counting machine, which meets the description of a complex machine.

By analyzing the results obtained in each of the levels in which we have worked, we have identified the most relevant data. In this way, it has been possible to significantly reduce the amount of data to be used, being able to continue identifying the corresponding failure in each case. All this allows a considerable reduction in the computational demand of the system.

As future work, the use of a cloud computing environment is proposed. Communication via IP with the cloud of numerous machines would allow the generation of a global neural model that reaches a higher degree of abstraction than can be achieved locally on the machine.

In short, a fault dectection system capable of identifying the failure that is occurring from the data extracted from a complex machine has been developed. This makes it posible to warn the operator who can correct the defect or it also can be used as a manufacturing quality control system.

Author Contributions: Conceptualization, R.C. and A.G.; methodology, R.C., A.G., D.B. and Á.M.; hardware, R.C. and A.G.; software, A.G. and Á.M.; validation, D.B. and Á.M.; formal analysis, R.C., A.G., D.B. and Á.M.; investigation, R.C. and A.G.; data curation, A.G.; writing, R.C., A.G., D.B. and Á.M. All authors have read and agreed to the published version of the manuscript.

Funding: This work has been partially supported by the Aragon Regional Government thought the program for R&D group (T59_20R) and by Sallén Tech SL. The work of Alvaro Marco has been partially supported by the Spanish Government, program Torres Quevedo (PTQ2017-09481).

Acknowledgments: We acknowledge the great work of Luis Lax, Jorge Lax, Carles Coll and Eduardo Salamero, who took part in the whole project and especially in the data collection.

Conflicts of Interest: The authors declare no conflict of interest.

References

1. Why Predictive Maintenance Is Driving Industry 4.0. Available online: https://files.solidworks.com/partners/pdfs/why-predictive-maintenance-is-driving-industry-4.0405.pdf (accessed on 3 December 2021).
2. Henriquez, P.; Alonso, J.B.; Ferrer, M.A.; Travieso, C.M. Review of Automatic Fault Diagnosis Systems Using Audio and Vibration Signals. *IEEE Trans. Syst. Man Cybern. Syst.* **2014**, *44*, 642–652. [CrossRef]
3. Choudhary, A.; Goyal, D.; Shimi, S.L.; Akula, A. Condition Monitoring and Fault Diagnosis of Induction Motors: A Review. *Arch. Comput. Methods Eng.* **2019**, *26*, 1221–1238. [CrossRef]
4. Castelli, M.; Andrade, M. Metodología de Monitoreo, Detección y Diagnóstico de Fallos En Motores Asíncronos de Inducción. *Mem. Investig. Ing.* **2007**, *5*, 65–76.
5. Mardacany, E. Smart cities characteristics: Importance of buit environments components. In Proceedings of the IET Conference on Future Intelligent Cities, London, UK, 4–5 December 2014; pp. 1–6.
6. Purnachand, K.; Shabbeer, M.; Rao, P.N.V.S.M.; Babu, C.M. Predictive maintenance of machines and industrial equipment. In Proceedings of the 2021 10th IEEE International Conference on Communication Systems and Network Technologies (CSNT), Bhopal, India, 18–19 June 2021; pp. 318–324.
7. Ayvaz, S.; Alpay, K. Predictive Maintenance System for Production Lines in Manufacturing: A Machine Learning Approach Using IoT Data in Real-Time. *Expert Syst. Appl.* **2021**, *173*, 114598. [CrossRef]
8. Mi, S.; Feng, Y.; Zheng, H.; Li, Z.; Gao, Y.; Tan, J. Integrated Intelligent Green Scheduling of Predictive Maintenance for Complex Equipment Based on Information Services. *IEEE Access* **2020**, *8*, 45797–45812. [CrossRef]
9. Nangia, S.; Makkar, S.; Hassan, R. IoT Based Predictive Maintenance in Manufacturing Sector. *SSRN Electron. J.* **2020**, 1–7. [CrossRef]
10. Dalzochio, J.; Kunst, R.; Pignaton, E.; Binotto, A.; Sanyal, S.; Favilla, J.; Barbosa, J. Machine Learning and Reasoning for Predictive Maintenance in Industry 4.0: Current Status and Challenges. *Comput. Ind.* **2020**, *123*, 103298. [CrossRef]
11. Liang, B.; Iwnicki, S.D.; Zhao, Y. Application of Power Spectrum, Cepstrum, Higher Order Spectrum and Neural Network Analyses for Induction Motor Fault Diagnosis. *Mech. Syst. Signal Process.* **2013**, *39*, 342–360. [CrossRef]
12. Bendjama, H.; Bouhouche, S.; Boucherit, M.S. Application of Wavelet Transform for Fault Diagnosis in Rotating Machinery. *Int. J. Mach. Learn. Comput.* **2012**, *2*, 82–87. [CrossRef]
13. Kiangala, K.S.; Wang, Z. An Effective Predictive Maintenance Framework for Conveyor Motors Using Dual Time-Series Imaging and Convolutional Neural Network in an Industry 4.0 Environment. *IEEE Access* **2020**, *8*, 121033–121049. [CrossRef]
14. Strauss, P.; Schmitz, M.; Wostmann, R.; Deuse, J. Enabling of predictive maintenance in the brownfield through low-cost Sensors, an IIoT-architecture and machine learning. In Proceedings of the 2018 IEEE International Conference on Big Data (Big Data), Seattle, WA, USA, 10–13 December 2018; pp. 1474–1483.
15. Saponara, S.; Fanucci, L.; Bernardo, F.; Falciani, A. A Network of Vibration Measuring Nodes with Integrated Signal Processing for Predictive Maintenance of High Power Transformers. In Proceedings of the 2015 IEEE 9th International Symposium on Intelligent Signal Processing (WISP), Sienna, Italy, 15–17 May 2015; pp. 1–4.
16. Pereira, R.R.; Diniz da Silva, V.A.; Brito, J.N.; Daniel Nolasco, J. On-line monitoring induction motors by fuzzy logic: A study for predictive maintenance operators. In Proceedings of the 2016 12th International Conference on Natural Computation, Fuzzy Systems and Knowledge Discovery (ICNC-FSKD), Changsha, China, 13–15 August 2016; pp. 1341–1346.
17. Novoa, C.G.; Berríos, G.A.G.; Söderberg, R.A. Predictive maintenance for motors based on vibration analysis with compact rio. In Proceedings of the 2017 IEEE Central America and Panama Student Conference (CONESCAPAN), Panama City, Panama, 20–22 September 2017; pp. 1–6.
18. Markiewicz, M.; Wielgosz, M.; Bochenski, M.; Tabaczynski, W.; Konieczny, T.; Kowalczyk, L. Predictive Maintenance of Induction Motors Using Ultra-Low Power Wireless Sensors and Compressed Recurrent Neural Networks. *IEEE Access* **2019**, *7*, 178891–178902. [CrossRef]
19. Patil, S.S.; Gaikwad, J.A. Vibration analysis of electrical rotating machines using FFT: A method of predictive maintenance. In Proceedings of the 2013 Fourth International Conference on Computing, Communications and Networking Technologies (ICCCNT), Tiruchengode, India, 4–6 July 2013; pp. 1–6.
20. Kavana, V.; Neethi, M. Fault analysis and predictive maintenance of induction motor using machine learning. In Proceedings of the 2018 International Conference on Electrical, Electronics, Communication, Computer, and Optimization Techniques (ICEECCOT), Msyuru, India, 14–15 December 2018; pp. 963–966.
21. Rosli, N.S.B.; Ibrahim, R.B.; Ismail, I. Optimized neural network of predictive maintenance for Air Booster Compressor (ABC) motor failure. In Proceedings of the 2019 IEEE International Instrumentation and Measurement Technology Conference (I2MTC), Auckland, New Zealand, 20–23 May 2019; pp. 1–6.
22. Godoy, W.F.; Silva, I.N.; Goedtel, A.; Palácios, R.H.C.; Lopes, T.D. Application of Intelligent Tools to Detect and Classify Broken Rotor Bars in Three-phase Induction Motors Fed by an Inverter. *IET Electr. Power Appl.* **2016**, *10*, 430–439. [CrossRef]
23. Amarnath, M.; Sugumaran, V.; Kumar, H. Exploiting Sound Signals for Fault Diagnosis of Bearings Using Decision Tree. *Measurement* **2013**, *46*, 1250–1256. [CrossRef]
24. Patidar, S.; Soni, P.K. An Overview on Vibration Analysis Techniques for the Diagnosis of Rolling Element Bearing Faults. *Int. J. Eng. Trends Technol. IJETT* **2013**, *4*, 1804–1809.

25. My NTi Audio. Available online: https://www.nti-audio.com/es/servicio/conocimientos/transformacion-rapida-de-fourier-fft (accessed on 1 December 2021).
26. Vilela, R.M.; Metrglho, J.C.; Cardoso, J.C. Machine and industrial monitorization system by analysis of acoustic signatures. In Proceedings of the 12th IEEE Mediterranean Electrotechnical Conference (IEEE Cat. No. 04CH37521), Dubrovnik, Croatia, 12–15 May 2004.
27. Zhang, W.; Jia, M.-P.; Zhu, L.; Yan, X.-A. Comprehensive Overview on Computational Intelligence Techniques for Machinery Condition Monitoring and Fault Diagnosis. *Chin. J. Mech. Eng.* **2017**, *30*, 782–795. [CrossRef]
28. Kingma, D.P.; Ba, J. Adam: A Method for Stochastic Optimization. *arXiv* **2014**, arXiv:1412.6980.

sensors

MDPI

Article

An Architecture for Service Integration to Fully Support Novel Personalized Smart Tourism Offerings

Andrea Sabbioni *, Thomas Villano and Antonio Corradi

Department of Computer Science and Engineering, University of Bologna, 40126 Bologna, Italy; thomas.villano@studio.unibo.it (T.V.); antonio.corradi@unibo.it (A.C.)
* Correspondence: andrea.sabbioni5@unibo.it

Abstract: The continuous evolution of IT (information technology) technologies is radically transforming many technical areas and social aspects, also reshaping the way we behave and looking for entertainment and leisure services. In that context, tourism experiences request to enhance the level of user involvement and integration and to create an ever more personalized and connected experience, by leveraging on the differentiated tourist services and information locally present in the territory, by pushing active participation of customers, and by taking advantage of the ever-increasing presence of sensors and IoT (Internet of Things) devices deployed in many realities. However, the deep fragmentation of services and technologies adopted in tourism context characterizes the whole information provided also by customer sensing and IoTs (Internet of Things) heterogeneity and deep clashes with an effective organization of smart tourism. This article presents APERTO5.0 (an Architecture for Personalization and Elaboration of services and data to Reshape Tourism Offers 5.0), an innovative architecture aiming at a whole integration and deep facilitation of tourism service and information organization and blending, to enable the re-provisioning of novel services as advanced aggregates or re-elaborated ones. The proposed solution will demonstrate its effectiveness in the context of Smart Tourism by choosing the real use case of the "Francigena way" (a pilgrim historical path), the Italian part.

Keywords: smart tourism; social sensing; sensors; e-tourism; distributed-architecture; Spark; Zenoh; FaaS; Elasticsearch

Citation: Sabbioni, A.; Villano, T.; Corradi, A. An Architecture for Service Integration to Fully Support Novel Personalized Smart Tourism Offerings. *Sensors* **2022**, *22*, 1619. https://doi.org/10.3390/s22041619

Academic Editors: Suparna De and Klaus Moessner

Received: 24 December 2021
Accepted: 15 February 2022
Published: 18 February 2022

1. Introduction

The rapid and pervasive evolution of digitalization covering all aspects of life has drastically changed the field of tourism [1] so as to propose a new pervasive experience, more and more based on online services and information; moreover, that trend is expected to further grow in acceptance and offerings. In fact, we already see a wide variety of services, datasets, and platforms concerning and supporting tourism in many of these new different forms. In the past decades, tourism acquired a key role in the development and economic growth of many countries, and so it is in Italy. As stated by Eurostat [2], in the EU (European Union) area tourism is EU's third largest socio-economic activity, representing around 10% of the EU's GDP (gross domestic product). Moreover, five EU Member States are among the world's top ten tourist destinations worldwide.

In Italy, more significantly than in other countries, tourism is in continuous growth and represents a vital contribution to the wealth of the nation. In fact, the total contribution of tourism to the Italian economy in 2017 was 223.2 billion euros, equal to 13% of Italian GDP and Italy was ranked the second destination for outbound trips made by EU residents within the EU, in terms of nights spent [2].

The fast increase of the tourism market is raising the need for a "Smarter" Tourism (Smart Tourism, or ST for short) more able to personalize and adapt customer experiences while creating a more culturally rich and even more sustainable offer.

To highlight the importance of ST in the sustainable development of a country, the European Commission launched the European Capital of Smart Tourism to stimulate the development and sharing of ST good practices. The European Capital of ST is an initiative to promote the integrated offers of innovative, inclusive, culturally diverse, and sustainable practices to tourism development by European cities [3]. Within that project, the European Commission has defined the concept of Smart Tourism as the combination of properties:

- **Accessibility**: To enable barrier-free destinations and enable access, regardless of age, cultural background [4], and physical disability.
- **Sustainability**: To protect natural resources of a city, reduce seasonality, and include local communities.
- **Digitalization**: To use digital technologies to enhance all aspects of the whole tourism experience.
- **Cultural heritage and creativity**: To protect and capitalize on the local heritage for the benefit of all stakeholders: the destination actors, the industries, and tourists.

The always-enlarging availability of information accessible through the network has modified the approach of visitors to the experience from an even structured and well-planned tourism offering to more dynamic and by-need ones. The development of ST will require not only a more *personalized experience* for tourists but also a more *dynamic service* proposal as key factors of the whole experience. Examples of dynamicity are modern apps that exploit geo-localization to retrieve more suitable local services in the locality of tourists and in a by need fashion [5]. That dynamicity further promotes a more personalized experience by using intelligent system recommenders: today the whole information about previous historical data and profiling plays a key role in ST decision-making processes [6]. Moreover, in the last years, the recent global pandemic has stressed further the importance of smart and dynamic tourism services based on geographic positions [7].

The pervasive diffusion of IoT and smart devices, connected with social sensing, represents an important accelerator to drive information retrieval toward service quality. In Social Sensing, the final customer can be actively and deeply involved in many ways, from contributing with her knowledge and sharing data gathered with smartphone and personal wearables, up to asking him to complete simple tasks while moving with her phones. Social sensing extends an already widespread and well-established series of techniques called crowdsensing [8] and has already been proposed to involve users in the process of data gathering [9]. Since initiatives of social sensing, crowdsensing, and crowdsourcing can play an essential role in the development of smarter tourism services, those initiatives are also coupled with incentivizing user participation via some forms of competition among users and via strategies of gamification. Tourism gamification extends strategies from game design and involvement strategies in non-game contexts [10], so to influence consumer engagement, customer loyalty, brand awareness, and user experience in tourism areas [11]. Examples of these initiatives include the usage of a scoring system related to customer action undertaken also rewarded with forms digital or material incentives and rewards.

As an example, in recent years many studies have proposed the use of social sensing to collect geo-tagged information and exploit them to identify tourism areas of interest [12], map tourist behaviors [13,14], compare and differentiate clusters of tourists [15], and discover and propose again noteworthy new places [16].

From the user perspective, the employment of ST combined with social sensing can provide better tailored information to tourists, about the quality and accessibility of a place, suited to their peculiar interests, either long-term or defined on the spot [17]. Examples are many, from the presence of barriers to the support of different languages in the service, from the current weather situation to the current mood of the entire group. In the city of Bologna, for example, there is an application to cancel barriers both in access to services and in mobility as demonstrated by projects like mPASS [18] or Kimap [19].

Additionally, we add that the tourism area itself asks for deep integration with many other fields, such as smart cities [20], smart transport, smart wealth, and relative services and data sources [21] as a few examples of connected areas.

To summarize, the integration and combination of ST information and services are expected to create great business value [22] and to enable the development of smarter tourism services and experiences, but the heterogeneity of formats and interactions protocols slows down the integration of multiple platforms, making impossible the acceptance of a unique and comprehensive standard, because of the different stakeholders and organizations proposing ST services and the lack of cogent regulation [23].

To solve these problems, the authors propose APERTO5.0, a reference Architecture for Personalization and Elaboration of services and data to Reshape Tourism Offerings 5.0, based on human-centric interactions and data gathering favoring a strong personalization with interested tourists (either single, in groups of interests, or in other aggregation forms). The aim of APERTO5.0 is not only to exploit better existing resources for a greater business value but also to stimulate the creation of novel tourist services toward the whole potential of aggregation and augmentation of the platform. APERTO5.0 can present local services per se, but also present new offerings, composing together local services chosen depending on specific tourist profiles and current needs. In fact, the APERTO5.0 aims both to enable the creation of a platform that can present a unified view of services and information to tourists and to offer a single access point for advanced and augmented information and facilities composition to third parties private and public organizations, so to encourage and enhance the development of positive and sustainable tourism offerings.

In collaboration with some local companies, the authors have implemented the first prototype of APERTO5.0 to handle a specific scenario of tourism paths ("cammini" in Italian or ways in English). "Tourism paths", also called pilgrim ways, are recognized routes connecting some historically relevant sites, and traditionally followed by pilgrims in the past: examples can be the "way of Saint James" to reach Santiago de Compostela, or the "Francigena way", an itinerary going to Rome by foot, both started since the Middle Age. Those tourism experiences are gaining more and more attraction as able to offer a healthy traditional and sustainable experience, in line with tradition and connected with the territories traversed by the path.

To validate the capabilities of APERTO5.0 architecture, the authors have extensively tested the proposed solution under different loading conditions, simulating the "Francigena way" behavior in the Italian final part, by testing the possible occurrences in the different conditions of service request typically experienced in tourism real scenarios. The paper compares the results for the presented solution in gathering and requesting information with the solutions that represent the de facto standard of the market. APERTO5.0 showed a very good capacity of distributing and requesting information under different loading conditions and diverse network and infrastructural topologies complexity.

In summary, this paper significantly advances the state-of-the-art literature in the field with the original contributions listed below:

(i) A novel architecture based on planes (see Section 3: Business, Cross Cutting Concerns, Data and Service) and Layers (Monitoring, Auditing Authorization Authentication, Presentation, Data, Analytics and Processing, Blending and Integration), to abstract the aspects related to the technologies used and group the different tasks.
(ii) A practical approach to address the problem of heterogeneity and dissemination of information and services in the context of tourism services.
(iii) An original implementation of the proposal based on open-source projects combining well established platforms with cutting-edge technologies.
(iv) An application of the proposed architecture to the real-use case of "tourism paths".
(v) Some in-the-field experimental results for simple deployment cases to show the feasibility of the proposed approach and the efficiency of the implemented architecture.

2. The Proposal

APERTO5.0 is based on an organization that put together on the one hand all possible information sources, and services toward a better integration, on the other hand, the best proposition possible for the differentiated needs of all tourists, either single or in different composition groups in number and interests. The proposed architecture has been designed driven in the middle of existing tourism services and information providers and the possible requests and needs of customers (Figure 1). APERTO5.0 aims at becoming the reference for the development of new smarter tourism services and platforms while adding value for both producers and consumers based on its integration and augmentation capacity.

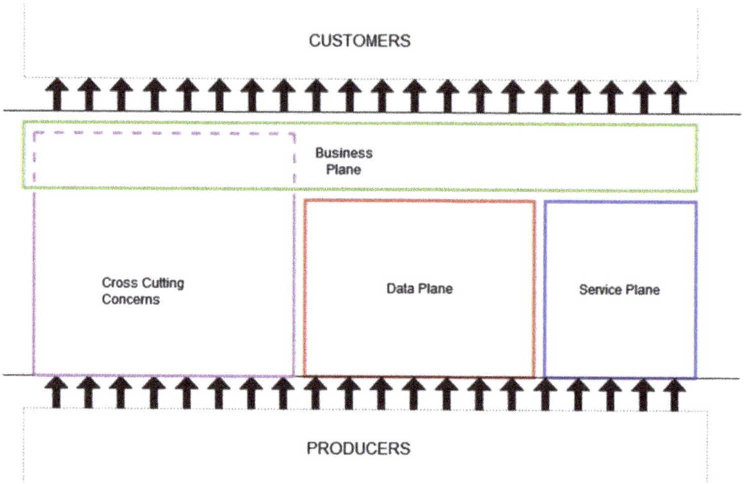

Figure 1. High-level vision of the APERTO5.0 architecture in terms of the layers connecting providers to customers.

The architecture considers a producer every entity that provides information or services potentially appealing for the tourism field and its development, by including partners belonging to the public and private sector, open data, and any connected things spread in the interested region, like connected transports, sensors, and user wearable devices. The absence of a common agreement and regulatory organs leads to a wide specter of interaction modes and formats, as well to many heterogeneous types of agreement and relationships proposed by different producers. On the other side, customers are users of the platform that can consume services and data resulting from the processes of augmentation, elaboration, and orchestration of information and services coming from Providers. The authors target does not consider only tourists as customers of the platform, but also municipalities, destination managers, third parties tourism services, and all the other actors possibly present in the territory and interested in participating in the formation of a smart tourism offer. Such heterogeneity in customers leads to the requirement of a high level of modularity and adaptability to suit the different needs of customers.

We must stress that a provider can also play the role of customer and vice-versa, by creating a circular Prod-Cons pattern, where providers can interact with the proposed platform as customers to grow their services, and customers can improve their experience through personal contributions to APERTO5.0. This positive evolution can be opportunistically encouraged through initiatives, such as crowdsourcing campaigns and the creation of local relationship networks. Authors claim that the introduction of the proposed digital platform can encourage the creation of a network of partners that can also increase, monitor, and guarantee the value of data and services. These types of relationships can also enable the development of opportunities by creating a mutual value, in the sense of social results in the tourism field, such as the creation of an agreement of multiple municipalities crossed by a path of cycle tourism. The network constituted with the different partners in the territory supported by the proposed digital platform constitutes the target supply chain for customers.

The architecture of APERTO5.0 is based on three well-defined layers called planes: the higher-layer business plane is responsible for the interaction with customers; the other two lower-layer planes are responsible for all possible services (service plane) and information (data plane) arriving from interested providers. It is important to stress that APERTO5.0 can expose new services based on the available existing ones; another lower layer component is the crosscutting one, in charge of all managing and monitoring functions of the entire architecture.

We are now expanding the details for the above planes. The business plane is the functional plane that addresses the complexity derived from interactions with customers with the main goal of providing a unique point of interaction, and, at the same time, of hiding from the customers the complexity of distributed datasets and services. The business plane has the main goal of uniform access to the heterogeneity of services, protocols, and interactions arriving from the underline planes to compose a solution offer. This plane drives the composition, coordination, orchestration, and exposition of services and information coming from the data plane and the service plane. This plane is capable of creating new synthetic tourism proposals, starting from existing services and information, such as in the creation of new packet experience, by booking public transportation, and by creating a path across most visited places. These features demand a high level of modularity and composability to adapt to the continuously evolving needs and interaction methods of customers, via tools like Dashboards, Apps, and APIs.

The data plane is the component of APERTO5.0 responsible for collecting, managing, and analyzing all the datasets and information collected from third parties providers, realizing the augmentation and conformation of data. This process is an essential step in the creation of new services and platforms, so that uniformity and standardization can reduce considerably the effort related to the management of different formats. The data plan can handle and process both data in motion and at rest (very static and very dynamic data, as extremes), enabling the exploitation of both historical data and fresh real-time information. To achieve a good value from data both stored and processed, it is necessary both the use blending techniques, over the data coming from multiple and diverse sources to merge them and the consolidation of a network of partnerships in the territory that can provide feedback and support, so as to specifically verify the information. The data plane supports multiple types of representation and analytics, in order to handle the different needs of customers, including geographical and time-based queries, up to computationally intensive processing like graph algorithms and machine learning techniques.

These planes, part of the proposed architecture, enable the representation, integration, and orchestration of provider services. Since the proposed solution does not replace or force migration of existing services but, on the contrary, focuses on empowering existing ones, the service plane has the goal of matching to each existing service one or more *synthetic* services representing it internally to the platform. Each *synthetic service* handles all the specificity of the target producer service like protocol, billings, and authentications taking charge of all the necessary coordination with other planes present in the proposed solution, specifically the *cross cutting concern plane*. Moreover, the services can be further composed or decomposed to create new *synthesis* services, so to create offerings at different granularities, e.g., a transportation service can be composed starting from a sharing mobility service and a public transport one. To enable these advanced techniques, the service plane also introduces a series of composing categories applicable to services that can not only enable a simpler composition of services but also simplify the suggestion of alternatives to the final customer, e.g., to suggest alternative places to visit, transportation to reach a point, or nearby hosting structures. Furthermore, these types of abstractions can enable smarter behaviors exploited by customers, leading them to a better tourism experience capable of reacting dynamically to events and information received, e.g., a signal of breakdown of a bus can trigger the automatic call of an alternative transportation partner or the proposal of remediation proposing like free hosting structures.

Finally, APERTO5.0 defines a *cross cutting concern plane* (CCCP) consisting of a series of components to implement all cross-cutting concerns and support other planes, in the whole management and interaction with internal and external services. Authors expect a continuous evolution of the CCCP while dealing with new challenges and new scenarios, especially distributed across a heterogeneous territory like the one covered by a pervasive platform of tourism. The CCCP supports the resolution of problems not only addressed internally to proposed infrastructure but also directly in relation with customers or providers, e.g., health checks of provider services and endpoints. Some services belongings to the CCCP can not only support other services but can become themselves part of the offer of other planes, e.g., the authentication services that can be provided as a service directly to customers behaving as part of the business plane. The components realized in the CCCP layer aim at operating transparently to the other component of the infrastructure. In this way, the evolution and introduction of new features in the CCCP plane can benefit, with little effort, multiple components belonging to other planes of the proposed solution.

3. APERTO5.0 Architecture Full Component Description

Going deeper into a more detailed description of APERTO5.0, we magnify the presented solution that is fully partitioned into detailed layers, each one corresponding to a single business process (Figure 2). We describe here: the presentation layer that implements and proposes a unique view of all services within the business plane; the blending and the data layers inside the data plane that allow the input of all information needed by ST by polishing and presenting, and also storing within the second component; the analytic and processing layer together with the *integration* layer constitute the service plane, where the former is capable of extracting any possible interesting service from the proposed available ones, while the latter is capable of getting to all available services available for ST. In addition, the *auditing, authentication, and authorization* (AAA) and the *monitoring* layers constitute the first two proposed modules that realize the cross cutting concerns plane.

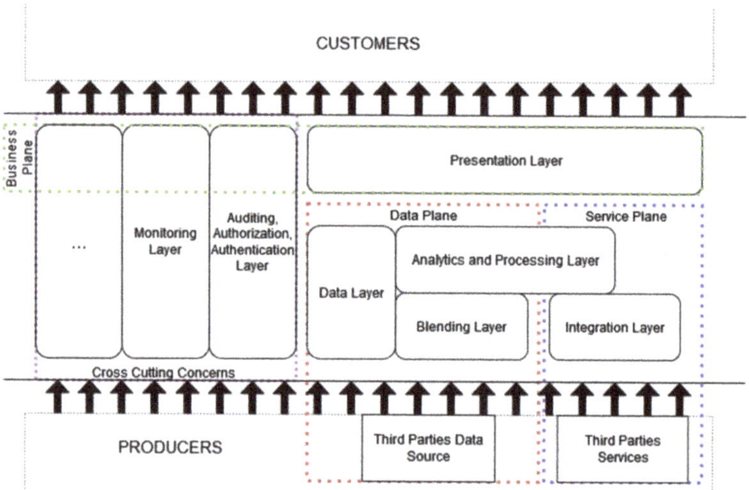

Figure 2. APERTO5.0 layers more detailed view. All components are put together for a more comprehensive effort of integration and synergic coordination.

The presentation layer constitutes the main component of the business plane to create and provide a unified view for tourism and third actors, by combining information and services coming from data plane and service plane to provide new smarter tourism services, typically the creation of a travel experience blending information on events and places in the destination with services of ticket purchase for transport. Supported customer interaction can employ heterogeneous technical protocols, e.g., pub-sub, client/server, fire and forget, and advanced query languages.

The data plane is subdivided into two horizontal layers: the blending layer and the analytics and processing layer (AP Layer) and one vertical layer: the data one.

The *blending layer* is responsible for gathering, cleaning, and adapting to a convenient format the information coming from third-party services, open data, and custom ad-hoc services. These data sets are then stored, according to predefined or dynamic policies on the different storage services composing the data layer. This lower layer implements the integration logic with the different forms of interactions and queries exposed by external data sources. The interaction methods are managed and adapted to require no changes on the provider side and can support reactive interaction, such as event-based traffic information systems and scheduled/polling-based ones like information about shows, fairs, and festivals periodically published on a site. The blending layer widely exploits principles of modularity and composability, and any introduction of new data sources or data manipulations follows the plug-in logic with a minimum effort of development and instantiation, enabling in such a way a sustainable growth of handled producers. This component implements a gathering approach direct from producer sources, in this way enhancing diversity and customized experience. This approach differs significantly from more traditional ones such as in booking.com [24], where the hosting infrastructure must register and constantly update its own data in a third-party portal, so requiring a standard format a-priori.

As part of the data plane, the *analytics and processing layer* aggregates and analyzes the different data sources stored in the data layer and exposed by the integration layer. This layer realizes the process of adding value to the information coming from local tourism offerings, via an internal creation of new aggregate datasets and the discovery of new insights through advanced elaboration techniques, such as big data processing, data mining, and AI (artificial intelligence) algorithms. This layer can also support real-time event-based and continuous-stream processing to enable advanced real-time queries and subscription mechanisms, as well batch operations for heavier time-consuming analytics.

The *data layer* is a vertical layer inside its plane, since it cooperates with all other layers, by storing elaborated data, schemes, and metadata and by exposing them through advanced indexing and query languages. The data layer handles the storing of fresh and past collected datasets and metadata, by enabling fast and advanced analytics and interactions through the exploitation of the data locality principle and advanced indexing, and by proposing customer-adapted viewing techniques. To support an effective memorization and query system the data layer exploits the most convenient storage strategy, so supporting any different data format and memorization technology.

The *integration layer* cooperates with the *analytics and processing layer* inside the service plane. The integration layer, in particular, is responsible for re-exposing in a convenient and optimized way the external services provided by third parties. The exploitation of many different categories enables the possibility to combine properly and substitute service calls to create smarter and reactive services, e.g., buy theater, cinemas, or public transportation tickets. This wrapping mechanism enables to hide and abstract from the peculiarities of each service, such as internal protocol, service call sequence, or rate-limiting, and facilitates the coordination and combination realized in a analytics and processing layer. Moreover, this layer interacts with all external services and datasets to obtain data not directly available, since they are filtered away, such as real-time number of available tickets or current position and updated time of arrival of a transport. The integration layer to support integration with different heterogeneous providers implements many types of interaction including periodically and reactively.

The *auditing, authentication, and authorization* (AAA) and the *monitoring* layers constitute the core of the cross cutting concerns plane, available to all other components, to operate in conjunction with all layers in the proposed architecture. These two layers also form an important part of the business plane as they provide important services to the final customer e.g., authentication service or metric.

In fact, the AAA layer is not only responsible for guaranteeing a proficient level of security to the infrastructure layer but also to provide a unique point of access, for final users, to the services covered and integrated into the platform. This allows to preventing registration and policy adaptation to any tourism service provider and enables a unique view for tourists and third-party organizations.

The *monitoring layer* provides useful insights on the service usage and the overall state of the platform to enable both elastic management of the infrastructure and significant added business value. In fact, from the monitoring layer it is possible to extract and underline trends in services usage with a geographical and temporal connotation and exploit them internally at the platform or supply "as they are" to external organizations, think to the trends in ticket buyout of a public transport localized in a determined time or region. Moreover, this layer can control malfunctions and unavailability of services and information provided by producers, by generating alerts by need to request automatic execution of recovery action.

The complexity and stratification of the proposed architecture derives from the exigence of addressing heterogeneity in tourism information and services as well as in customer needs. Moreover, this complexity is expected to further increase with many additional layers and components while increasingly addressing more and more ST use cases with their intrinsic exigences. Authors claim, however, the validity of the proposed architecture as a base for the structuration of a solution able to satisfy and integrate many ST vertical scenarios.

4. The Case Study of Tourism Paths

One of the most challenging scenarios in the context of smart tourism is the business of tourism paths (or ways or itineraries, sometimes pilgrim's ways), typically established very long ago to suggest routes to religious pilgrims in the middle age and to give advice in their ways toward their final destination. This paper takes the "Francigena way" as a use case, in particular, the Italian part passing through territory. This novel and more requested type of tourism offers are characterized by the requirement of extreme personalization and dynamicity of target user experience; tourists can choose via the information given by their smart devices how to continue the experience, which is always driven by current information derived from their connection on demand. Of course, the same always-connected feature can apply to many other areas apart from ST, so presented architecture can be crucial in those too.

In tourism paths, users intend to use ICT as an essential part of the experience and tend both to be driven by information they need to get either personally or as part of a group, and to feed information over the community depending on their current experience. We consider that tourists can become prosumers (consumers and producers at the same time) of the experience of tourism paths. As an example, via ICT tools, the user can interact more dynamically and satisfactorily, by choosing to read personalized paths and calibrating languages and contents based on specific levels of learning [25]. It becomes essential to gather as much information as possible, so as to provide customers with the necessary details and to provide the customers with the best experience possible. That high dynamicity constitutes a challenge for providers of smart tourism services characterized by huge distances covered, with a multitude of information gathered, with a high fluctuation in number of users requesting those services, and with an important level of heterogeneity in partners to be involved.

In collaboration with Imola Informatica, authors then decided to create a first prototype of APERTO 5.0 to support the creation of a platform of smart tourism for tourism paths. This platform aims to support customers in all the phases of the experience: from the *planning of the trip*, also *during the experience*, and finally in the *post-experience phase*.

From the user point of view, a basic service example should make it possible to (1) plan the trip in any plan details, such as choosing also where to sleep and eat; (2) contacting local shops and realities for the discovery and purchase of typical products; (3) explore the path you want to take, discovering what to visit. That is normal in the planning phase, but those services should be available also during the experience itself, and at the same time, must occur in the post experience, both to document the events and compare with possible different expectations for future planning.

Particular attention is devoted to the process of information gathering from partners, sensors, and user contributions. Users are continually encouraged to contribute during the whole experience with images, gathering data from sensors, expressing opinions, and sharing their GPS tracks.

This process of social sensing opens the APERTO5.0 platform to a continuous enrichment of the experience introducing in the ST offerings new variations to the route, points of interest, and information about places acquired by other users during similar experiences. Moreover, the entire information allows the platform to dynamically adapt customer experiences, to make the offer more accessible and enjoyable. In the scenario of tourism path, a parameter to be continuously monitored is the *difficulty* associated with a stretch of the route that depends on many environmental current factors, such as humidity in the air and the soil, condition of the track, and wind force.

To stimulate user contribution, the platform exploits a gamification approach, where each user contribution assigns some score points according to the relevance of the intervention. As an example, the signaling of a critical problem during the tourist path (particularly important for other users) can be estimated as a relevant contribution, while the review of a restaurant can be associated with fewer points. At the overpassing of some predefined thresholds, the user is rewarded with some prizes of many diverse types and values, such as showing the user status and coupons offering discounts on partners related to the experience.

5. Materials and Methods

To evaluate the potentiality of the presented architecture proposal the authors have developed a first implementation prototype based on some widely open source affirmed platforms, with the goal of not only helping us to evaluate the effectiveness of the proposal in some real use case scenarios but also discovering whether current technological solutions can fulfill the use case needs. As shown in Figure 3 we base the first prototype on 5 main technologies: (1) Eclipse Zenoh [26], (2) OpenFaaS [27], (3) Apache Kafka [28], (4) Apache Spark [29], (5) Elastic Stack (Elasticsearch and Kibana) [30], necessary to satisfy the needs of the "Francigena way" use case and presented from a bottom-up perspective from the data gathering to the final service and data representation to customers.

Potential technological gaps discovered during the execution of the testbed will constitute further directions of research to address and solve the problems of a general architecture and infrastructure to support the development of smart tourism.

One of the most challenging tasks demanded of the proposed infrastructure and demanded to the blending layer is the gathering and dynamic querying of information from different sensors, customers, and providers present in the territory. The difficulty of this task is further exacerbated by the use case scenarios that must cover a large territory with very heterogeneous characteristics, like connection quality, coverage, and density of devices. To overcome these challenges, we have introduced Eclipse Zenoh as the preferred interaction medium between the infrastructure and data sources present along pilgrim's paths. Zenoh is an open project developed by Eclipse Foundation born to fulfill the need for an efficient adaptation of fog-centric business. In the past decades, the cloud-centric model has taken hold in several fields and has been applied pervasively to countless business cases. Nevertheless, the cloud-centric model has shown some limitations under some application conditions and fields (e.g., limited connection bandwidth, low latency, negligible connection cost, etc.). Zenoh tries to fulfill those needs, by implementing a distributed pub/sub model with a limited footprint and overhead to reduce to the minimum the impact over the business resources. At the same time, Zenoh aims to keep as low as possible the latency and to increase to the maximum throughput.

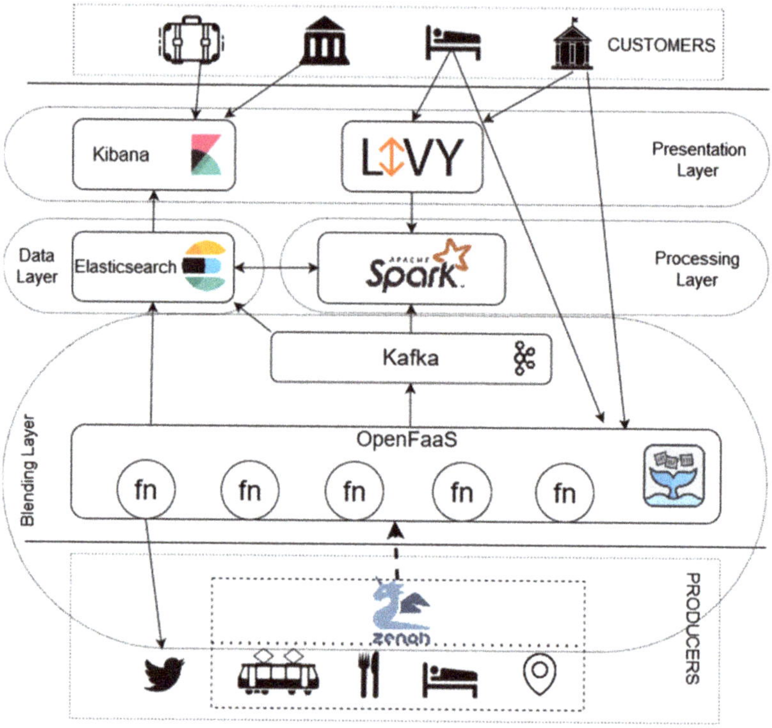

Figure 3. First prototype of APERTO 5.0 realized to satisfy pilgrims path needs.

Each node that uses the Zenoh APIs is declared to belong to one unit type: peer, client, and router, since these three units are the building blocks for any application. In fact, the application unit declared as *peer node* could connect to another node in the network allowing a complete graph topology or could select the nodes to connect to, to form a connected graph topology. The role of the *client node* is to connect to a single router or peer to communicate to the rest of the system. The *router node* is a software process able to perform level 7 routing (OSI model) and connect to different nodes to connect different topologies to each other to form any kind of topology.

Zenoh also provides and implements a configurable discovery mechanism that employs a gossip protocol over multicast, to auto-connect to the other nodes with the aim of using a so-called "path" to perform a request. The "path" is a string that works as a key thus acting as a symbolic location that expresses the logical function rather than IP addresses so that this organization allows the application to hide the exact position of the data enabling a geo-distributed storage implementation. To enable the gathering and dynamic querying of data across the distributed infrastructure, Zenoh provides a set of high-level APIs that hides the underline infrastructure complexity, while enabling advanced and well-established communication patterns, like publish/subscribe or request/reply. Those APIs together with the advanced infrastructure of Zenoh make possible for APERTO 5.0 the collecting of information from the action of user sensing and the contributions of partners, as well as the dynamic querying of data at rest present in the wide territory covered.

To address the great variability of sources in a Prosumer context, the authors introduced a layer of adaptation and abstraction at the forefront of the proposed architecture and realized by OpenFaaS an Open-Source Function as a Service (FaaS) platform. FaaS computing is a novel model of cloud computing that promotes the absence of customer control over the infrastructure by specifying only the creation and upload of the desired business logic. This model can enable an unprecedented speedup in the development of new components, such as the creation of ad-hoc components to manage and adapt data coming from various sources.

Moreover, the FaaS model is characterized by a fine-grained scaling of resources associated with each service. In fact, in FaaS platforms, the code representing the business logic is not always active and in execution but is dynamically executed by the platform only at the arrival of the user-associated events. That enables a good consolidation of different services deployed on the same hardware, potentially achieving an important cost-saving.

Finally, recent developments in FaaS, such as function chaining or map reduction, aim to introduce even more capabilities and paradigms support that can promise the standing of FaaS platforms as a crucial component to forefront complexity of tourism cases. The FaaS model paves its interaction model and scalability on asynchronous communication and stateless computation, while modern FaaS platforms, such as OpenFaaS, still lack mechanisms of synchronization and aggregation. Therefore, we had to recur to one well-spread message-oriented middleware, Apache Kafka, as a layer of interaction between functions and other functions or integrated platforms. Apache Kafka is a highly scalable, open-source streaming platform that provides the *Pub/Sub protocols*. We opted for Apache Kafka, among many open-source MoM present on the market, for its capacity of excellent scaling in handling a massive number of concurrent messages with multiple configurable qualities.

In fact, the FaaS layer can potentially create a large amount of information and as the importance of information can significantly vary, so a differentiation of QoS support in the service is essential. Data coming from the integration and blending layers are then convoyed as a stream in specific Kafka topics each one characterized not only by a business means but also by a specific QoS. The continuous stream of differentiated information can be either kept in persistent storage or streamed to the analytics and processing layer for further processing. Kafka enables not only to handle a huge amount of concurrent information coming from different providers but also to handle diverse sources with diverse levels of QoS, spanning all common policies, from best effort to atomic delivery [31].

To address the continuously evolving needs of customers we introduce in the processing layer Apache Spark, an open-source distributed analytics engine memory computation large-scale data processing that provides multiple modules like MLlib, GraphX, Structured Streaming to enable the parallel and in-memory computation of data at rest and streamed from Kafka with advanced techniques, including graph processing, machine learning, incremental computation, and stream processing [32]. Inside the proposed solutions, Spark is the essential component in charge of processing and interpolating data coming from sensors, social networks, and social sensing. For specific partner requirements, we can opportunistically open the possibility to directly exploit the advanced processing capabilities of Spark. The process of submitting a task to Spark, however, is not trivial and highly depends on the characteristics of the deployed environment. For this reason, we decided to provide a layer of abstraction through Apache Livy [33] to enable the direct submission of Spark Jobs via REST API. Furthermore, Livy also takes care of implementing politics and mechanisms to guarantee fault-tolerant submissions and concurrent request support in a multi-tenant environment.

All the information elaborated by Spark and the FaaS platform are eventually stored in Elasticsearch, a persistent, distributed, and fault-tolerant NoSQL database. Elasticsearch is a distributed and scalable open-source project part of the ELK Stack. Through its REST APIs Elasticsearch enables the store and advanced analytics and query of big volume of data in near-real time fashion.

In the data layer, in fact, two main needs to handle are the management of a huge amount of data and the creation of a query engine service with advanced capabilities to support visualizations and customer queries.

The introduction of Elasticsearch in proposed technological stack enables us to maintain relevant historical information while creating efficient and practical indexes and views to facilitate the retrieval of data. Data so memorized can be queried by customer through API or presented as advanced graphs, dashboards, dynamic maps in Kibana an open frontend service part of the elastic stack providing search and data visualization capabilities for data indexed in Elasticsearch.

The proposed architecture paves its strength on the modularity of the solutions composing the different layers. The different components proposed, in fact, while suiting the needs of the "Francigena way" cannot match the requirements of other ST use cases or result in overkilling and can be replaced accordingly. We expect, indeed, a great and continuous evolution in components of the single layers of the architecture presented, according to the increasing number of addressed use cases.

6. Results

The prototype of APERTO5.0 constitutes the backbone of services where the different partners constellating the tourism path and tourists receive and provide information also exploiting ad-hoc developed mobile-oriented applications.

Given the huge and ever-increasing number of data sources that APERTO5.0 is required to handle we are interested to present here two main quite common and potentially critical processes: (1) The action of gathering and querying data across complex and distributed topologies of sources; and (2) the processing and integration of cases to demonstrate the ability of proposed platform of gathering and process data and services in order to be re-exposed. We concentrate on these two specific functions, since the huge quantity of data produced by geographically distant places in the "Francigena way" requires a powerful and fast gathering processing phases, so as to maximize the value of the obtained information.

The tests are organized along the two major problems highlighted: with the first test-bed section aiming in proving the ability of Zenoh in supporting presented architecture in the phases of receiving and querying of data and with the second section with the goal of confirming the capacity of the chosen platform in adapting and processing incoming information. Afterward, we confirm the introduction of a FaaS platform as an essential part able to address the extreme variability in tourism service load while also facilitating the fast development of connectors to take on the Prod-Con heterogeneity.

The tests of the first testbed section are deployed on an infrastructure composed of 5 nodes equipped with an Intel(R) Core (TM) i5-3470 CPU running at 3.20 GHz and 8 GB of RAM each. All the tests of the second testbed section have been conducted on a physical infrastructure composed of five virtual machines each equipped with 8 vCores, 32 GB vRAM, and 150 GB data SSD.

6.1. Social Sensing Data Gathering

Infrastructure topologies interconnecting prosumers in the context of pilgrims paths can assume diverse levels of complexity by traversing many networks and applicative endpoints. This complexity is further worsened by the fact that a user will consume in their experiences not only local data but potentially data distributed far away, many stages before or after the current location. For this reason, this section aims at testing the capabilities of Zenoh in supporting APERTO5.0 in both processes of receiving data through a standard Pub/Sub protocol and querying data at rest present in separate locations, with varying interconnection topologies and load. We then compare the obtained results with two often adopted technologies for data gathering, namely RabbitMQ [34] and a cascade of Envoy HTTP Reverse proxy under the same conditions [35].

In the use case of "Francigena way", one of the most challenging tasks that proposed infrastructure requested to realize is the ability to gather and retrieve data traversing arbitrary complex networks and infrastructural topologies. This happens often during social sensing action where the data coming from sensors are requested to traverse different devices like user smartphone or Wi-Fi Access Points compared to the ability of Zenoh, RabbitMQ, and HTTP of traversing multiple hops to interrogate sources of data in the territory.

We have triggered an increasing number of requests starting from a frequency of 1 request per second and reaching 1000 requests per second. Results shown in Figure 4 highlight that in a request/reply interaction HTTP performs better when the number of hops traversed increases.

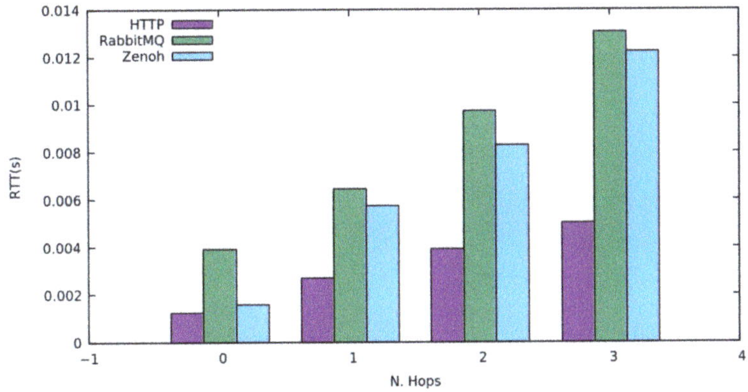

Figure 4. Round trip time of requests when traversing an increasing number of network hops.

We then tested the ability of the different solutions to satisfy higher rates of requests experienceable when multiple users and devices send contemporary a lot of data. We then repeat the precedent experiment in the case of 2 hops to be traversed and we sent an increasing rate of requests starting from 0 to 10,000 requests per second.

As Figure 5 shows, while HTTP forwarding realized through proxies performs better at low request rates when the rate exceeds 4000 requests per second the performance of this protocol starts to degrade with a fluctuating behavior and an RTT measured higher than once of Zenoh and RabbitMQ.

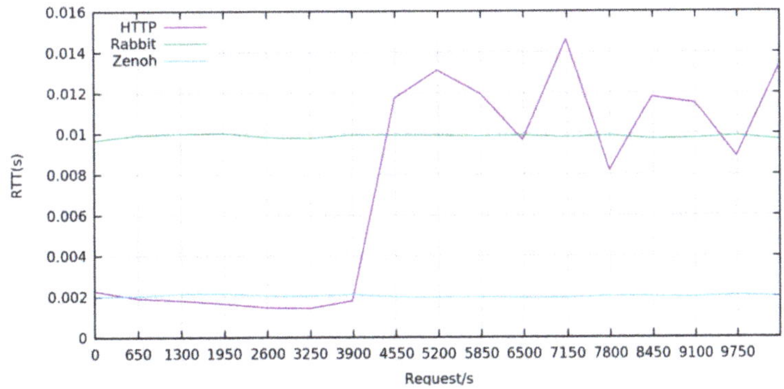

Figure 5. Round trip time of requests when traversing 2 hops with an increasing number of requests per second, starting from 0 up to 10,000.

We then tested the ability of RabbitMQ and Zenoh of distributing and collecting information through a typical Publish/Subscribe mechanism. This test foresees the exclusion of the HTTP Proxy and HTTP protocol does not support such type of interaction.

In particular, the third test follows the same guidelines as above to compare and verify the routing capacity and the delay introduced by the two message-oriented middlewares in distributing a constantly increasing amount of information among multiple subscribers, when traversing infrastructural and network topologies with different complexity.

The outcome of the test shows (Figure 6) that Zenoh introduces a delay in the delivery of information lower than 3 milliseconds, even when traversing multiple intermediaries. Moreover, the pub/sub protocol implementation realized by these two solutions seems to be less influenced by the complexity of traversed topologies, with a delay variation lower than 1 millisecond.

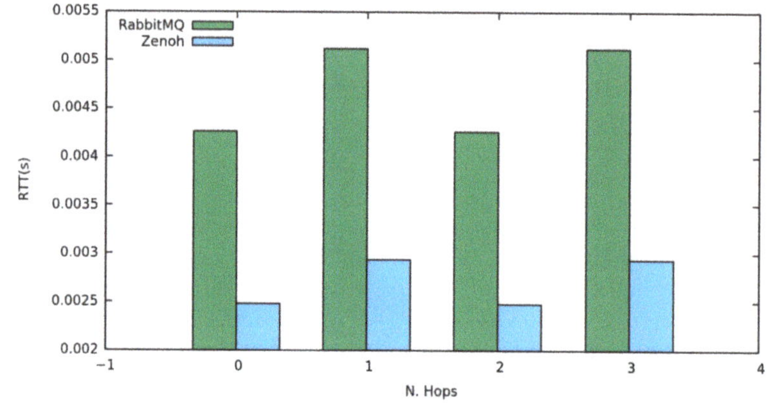

Figure 6. Registered delay in publish/subscribe mechanisms when traversing an increasing number of network hops starting from direct point to point communication to three intermediaries.

In conclusion, the tests of this section justify the choice of the authors of introducing Zenoh as a middleware, to retrieve and collect information in arbitrary complex scenarios not only because of its good and more stable performance, even at elevate request rates, but also for its flexibility in supporting multiple interactions paradigms, such as request-reply and pub/sub.

6.2. Data Integration

The whole information gathered from actions of Social Sensing, sensors, Providers, and user contributions must pass through a process of adaptation and elaboration so to be accessible and significant for tourists and partners of the pilgrim path. Through the processing phase it is possible to elaborate data coming from actions of social sensing and the different partners to create a customized experience and adapt to variating conditions such as weather, activation of new promotions, or unexpected events.

In this first test of this test-bed section, we aim at showing the effectiveness of the platform components chosen to elaborate and integrate data and services coming from producers. We created a simple function receiving data from either an HTTP endpoint or from a Zenoh Resource and applying a filtering operation and a conversion in JSON format. Transformed data are forwarded through a Kafka topic to Spark where a Spark Streaming will execute some computationally intensive operations on the data received such as square root of the number of occurrences of a character in the data received. The data so processed are then stored into Elasticsearch by the Spark-Elasticsearch connector. To measure the impact of each part on the latency before the data can be stored in the Elasticsearch, each service adds a timestamp in the payload at the receival of any message. We then sent a constant rate of 1000 requests per second and logged the timestamps so generated in Elasticsearch to be visualized through Kibana.

As mentioned before the fast elaboration of gathered data in order to obtain a small end-to-end latency presenting updates in a near real-time fashion to customers is one of the main requirements that emerged from the "Francigena way" use case.

Results shown in Figure 7 show that the platform proposed by authors is able to process and memorize a huge quantity of data and re-exposing them and to provide an extremely limited to provide a very limited total latency of less than 1 s to the customer. The decomposition of the end-to-end latency shows that the sawtooth behavior is due to Spark when processing data with its batch streaming approach.

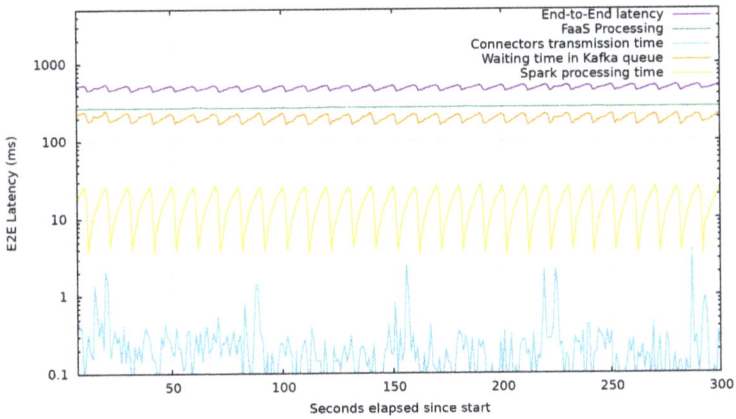

Figure 7. Average latency introduced by each component compared with the end-to-end latency when under a constant load of 1000 message/s (Log Scale).

We can also observe that while the processing delay carried on by the FaaS platform presents less jitter than the processing task by Spark, the many Spark optimizations carried, such as the pre-allocation of computational resources, lead to a processing time one order of magnitude faster. The two major components responsible for the resulting end-to-end latency are the message wait time and FaaS processing time. Messages stored in Kafka queues wait until the Spark adaptor is not ready to receive them in order to be computed and this latency introduced is mainly due to the batch streaming behavior of Spark. The latency introduced by the processing in the FaaS platforms is a known problem of these platforms caused by many factors derived from the dynamic creation of function at each request and is a highly active topic of research [36].

Those results also suggest that while the aim of authors of achieving a less than 1 s delivery time is achieved, when in need of faster end-to-end processing, the tuning of Spark Streaming is the applicable point of intervention. There are, in fact, several options to tune the performance of a Streaming process in Spark, such as increasing the level of parallelism in data receiving and serialization and by setting the right batch interval. Finding the right batch interval (BI), for a Spark Streaming application running on a cluster is an essential condition for its stability and requires that the system is able to process data as soon as it is being received [37]. We have sent a constantly increasing number of requests, starting from 0 to 10,000, during a time-lapse of 5 min and increased the batch interval to test how the infrastructure behaves, at the load increasing.

The variation of BI shows that pipelines configured with a higher batch interval present a higher jitter and latency, while they are more unaffected by variations of loads (Figure 8). In fact, the pipeline configured with a 2 s BI does not show a degraded behavior until the 150 second approximately (with 5000 message/s), while the one configured with a BI of 250 milliseconds shows already a degradation after 75 s (with 2500 message/s) from the test start. On the contrary, if the requirement is a low end-to-end latency, the best configuration is with a BI of 250 milliseconds that can process messages with several orders of magnitude less than other BI configurations.

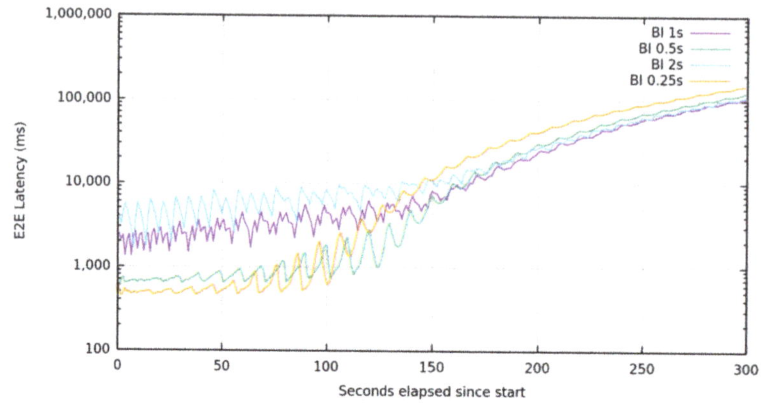

Figure 8. Average end-to-end latency in logarithmic scale when variating the batch interval (BI) in Spark Streaming and submitting an increasing number of messages from 0 to 10,000 in a time-lapse of 5 min.

Zooming in on the test with the BI of 250 milliseconds, we can see that the latency introduced by each part compared with the end-to-end latency is influenced by the processing of the Spark Streaming (Figure 9). These results confirm the necessity of introducing both the FaaS platform and Kafka that guarantee the best flexibility of the infrastructure. In fact, the FaaS platform with its fine-grained scalability has not only exhibited the best adaptation to different connection protocols and message formats, but also the flexibility of performing a preprocessing, before entering the "hard" processing of Spark. Apache Kafka, on its side, is capable of storing information until it is effectively requested by Spark, so as to enable the recovery of transient situations, where the load is excessive to be computed by resources assigned to Spark without losing data.

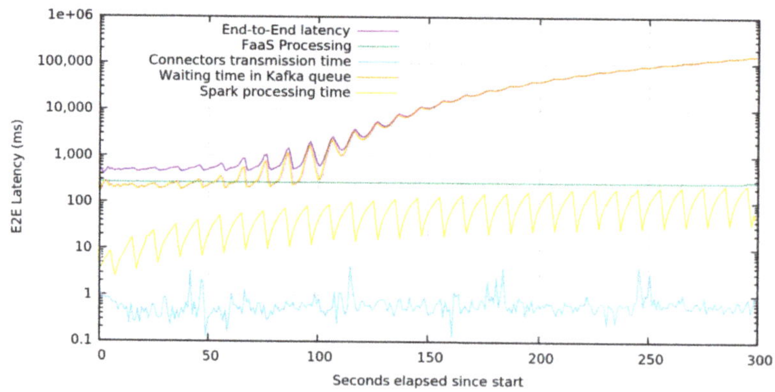

Figure 9. Average latency introduced by each part compared with the tail latency when under an increasing load of messages starting from 0 to 10,000 during a time-lapse of 5 min.

So, we can claim that a platform, like Apache Spark, can provide a fast, complete, and efficient solution to parallel processing massive amounts of data coming from social sensing, social networks, sensors, and tourism partners.

Altogether, even a tailored tuning of the Spark platform cannot satisfy all different requirements of ST while achieving an optimal usage of computational resources available [38,39]. We can state that Spark is not a solution to manage the heterogeneous fluctuating information of ST scenarios, so the only way to address ST challenging tasks requires to *ad-hoc setup* the Spark infrastructure dimensioning to the worst-case scenario with an obvious waste of resources [40].

On the contrary, the FaaS platform, even with introducing a greater latency in computation, has shown much greater flexibility in response to load variance (Figure 9). Future developments of these FaaS platforms introducing missing computation constructs (such as Map-Reduce) and enabling better performances can shed light to the offloading of the Spark computed part. In this way, it will be possible to leverage on the finer granularity and zero-scaling capabilities of FaaS platforms to grant the best dynamic adaptation to the continuous variations typical of the ST scenarios [36].

From the comparison with partners in the territory, the authors can claim that the proposed first prototype of APERTO5.0 is able to address the challenges in terms of data collecting and fast processing, presented by the creation of a unified platform to support ST services on the "Francigena way".

7. Related Work

To the best of our knowledge general-purpose architecture aiming to integrate and augment different tourist data-sources and services, has already been proposed, however neither as a proposal in literature nor as a business product. However, some work in the literature has formerly proposed infrastructures to support the development of smart tourism services.

RADON is an EU-funded project aiming to develop a model-driven DevOps framework capable of managing the entire lifecycle of services developed and deployed on top of FaaS platforms. The scope of this project also includes the porting of ADAMO, an application that combines tourist preferences with a city mobility network and points of interest in order to generate customized routes [41].

In [42], authors describe a model architecture for smart tourism systems (STS) tailored for cultural heritage and territorial data. The proposed architecture prompts a rethinking of the key paradigms: interactive travel, tourist gaze, hospitality, authenticity, and social networking data owing to the exploitation of software as a service platform tailored on the development of ST services.

However, the proposed platforms, while providing valid support for the development of new services, do not address the problem of heterogeneity of source in data and services nor propose an integration pattern to integrate with existing tourism offer.

In [43], the authors proposed a suite of small applications in tourism, by using a recommendation approach and supported in a microservice pattern, via a set of independent deployable services. The authors define the following functions: (1) to suggest routes and point of interest to users with respect to the choice of tourism activities; (2) the mutual supporting tourist and tourism services suppliers with location services; (3) to help communities in the preservation and valorization of cultural heritages; (4) to enable tourist to share their travel experiences to help other travelers in their decision-making process.

The proposed architecture of APERTO5.0 instead does not focus only on single tourism services or vertical tourism experiences but aims to integrate and involve as many data sources as possible to create a richer, more integrated, and unified tourism experience.

8. Discussion and Conclusions

In conclusion, the novel development opportunity of smart tourism via a pervasive redesign due to modern IT technologies constitutes a unique occasion toward more sustainable, inclusive, and culturally rich tourism tailored according to customer preferences. In this context, the cooperation with all stakeholders of the territory and the stimulation of local action of social and sensor sensing are necessary to consider and provide an unprecedented big data amount to enable and stimulate even more the creation of smarter and more connected tourism services. However, the deep fragmentation of services and technologies adopted by different actors in tourism that characterize also the whole information provided by customer sensing and IoTs heterogeneity deeply clash with an effective organization of smart tourism.

It is the authors' opinion that the introduction of APERTO5.0 can provide a significant contribution to Academia and Tourism business development and management. This paper has proposed APERTO5.0 as an architecture aiming to address the problem of heterogeneity by providing a unifying view in which any tourist item (data, service, and agents) can become part of an integration mosaic capable of accommodating any new possible element. Such an unifying environment is the authors' main design goal and represents a major innovation in academic research for its novel model and the innovative technological solutions employed in its first proposed prototype.

In fact, the introduction of APERTO5.0 as support to the structuration of ST services over the "Francigena way" allowed us to underline the great adaptability in providing a unifying support over the huge amount of information available. In this context, this solution introduces an innovative mechanism to query and gather data coming from complex scenarios in an efficient and scalable way enabling the introduction of actions of social sensing and partner involvement. A FaaS layer handles gathered data, to cope with the variety and availability fluctuations of information and then processed with Apache Spark. Results showed that the proposed platform is able to collect and process information in parallel, also traversing complex infrastructure topologies, with a resulting end-to-end latency lower than 1 s.

Considering the contribution of APERTO5.0 to the tourism management field, owing to the cooperation with realities of the territory and the explored use-case of tourism paths, authors already demonstrated the effectiveness of proposed solutions. However, a better exploration of the potentiality of a pervasive application of concepts and possibilities opened by Aperto5.0 in the many possible tourism facets and in particular in tourism management deserves better exploration by field experts.

As future directions, authors will work both in the use case of pilgrim's paths and in new technical widening directions.

Along the first line, through collaboration with tourism realities in the territory, we aim to further develop APERTO5.0 to support novel innovative and integrated forms of tourism. The integration of also other forms and use cases of tourism can increase furthermore the value of the interconnection of services and information toward a unified and pervasive support to ST development.

To follow novel technical directions, we plan to exploit better the potential of serverless infrastructure in providing fast fine-grained scaling of resources and in reducing the time-to-market development of new services. To pursue these capabilities, future research work will focus on optimizations and the introduction of new capabilities in the FaaS platform, so as to easily meet the needs of ST use cases. The main idea is to explore a more pervasive integration of this model of cloud computing with decentralized deployments over the so-called cloud continuum [41] and exploit data and service locality to achieve lower latencies and finer-grained customization of the platform. Moreover, we stress that a fast diffusion of the FaaS is highly likely, so many other features will be available very soon for even better performances.

Author Contributions: Conceptualization, A.S. and A.C.; Data curation, A.S., T.V. and A.C.; Formal analysis, A.C.; Funding acquisition, A.C.; Investigation, A.S. and T.V.; Methodology, A.S. and A.C.; Project administration, A.C.; Software, A.S. and T.V.; Supervision, A.S. and A.C.; Validation, T.V. and A.C.; Writing—original draft, A.S., T.V. and A.C. All authors have read and agreed to the published version of the manuscript.

Funding: This research received no external funding.

Institutional Review Board Statement: Not applicable.

Informed Consent Statement: Not applicable.

Data Availability Statement: Data available on request due to restrictions.

Acknowledgments: We would like to thank Stefano Maggiore and Imola Informatica for their collaboration in the definition and refinement of the Francigena way use case and for the providing of the computational resources employed during tests.

Conflicts of Interest: The authors declare no conflict of interest.

References

1. Cimbaljević, M.; Stankov, U.; Pavluković, V. Going beyond the traditional destination competitiveness–reflections on a smart destination in the current research. *Curr. Issues Tour.* **2019**, *22*, 2472–2477. [CrossRef]
2. Eurostat. Tourism Statistics. December 2018. Available online: https://ec.europa.eu/eurostat/statistics-explained/index.php?title=Tourism_statistics_ (accessed on 20 December 2021).
3. European Capital of Smart Tourism. Available online: https://smarttourismcapital.eu/cities-2020-winners/ (accessed on 27 October 2021).
4. Cuesta-Valiño, P.; Bolifa, F.; Núñez-Barriopedro, E. Sustainable, Smart and Muslim-Friendly Tourist Destinations. *Sustainability* **2020**, *12*, 1778. [CrossRef]
5. Gretzel, U.; Sigala, M.; Xiang, Z.; Koo, C. Smart tourism: Foundations and developments. *Electron. Mark.* **2015**, *25*, 179–188. [CrossRef]
6. Wen, I. Factors affecting the online travel buying decision: A review. *Int. J. Contemp. Hosp. Manag.* **2009**, *21*, 752–765. [CrossRef]
7. Radojević, B.; Lazić, L.; Cimbaljević, M. Rescaling smart destinations: The growing importance of smart geospatial services during and after COVID-19 pandemic. *Geogr. Pannonica* **2020**, *24*, 221–228. [CrossRef]
8. Cardone, G.; Foschini, L.; Bellavista, P.; Corradi, A.; Borcea, C.; Talasila, M.; Curtmola, R. Fostering participaction in smart cities: A geo-social crowdsensing platform. *IEEE Commun. Mag.* **2013**, *51*, 112–119. [CrossRef]
9. Cardone, G.; Cirri, A.; Corradi, A.; Foschini, L.; Ianniello, R.; Montanari, R. Crowdsensing in Urban Areas for City-Scale Mass Gathering Management: Geofencing and Activity Recognition. *IEEE Sens. J.* **2014**, *14*, 4185–4195. [CrossRef]
10. Deterding, S.; Dixon, D.; Khaled, R.; Nacke, L. From Game Design Elements to Gamefulness: Defining. In Proceedings of the International Academic MindTrek Conference: Envisioning Future Media Environments, MindTrek, Tampere, Finland, 28–30 September 2011; pp. 9–15. [CrossRef]
11. Xu, F.; Weber, J.; Buhalis, D. Gamification in Tourism. In *Information and Communication Technologies in Tourism 2014*; Springer Science and Business Media LLC: Cham, Switzerland, 2013; pp. 525–537.
12. Devkota, B.; Miyazaki, H.; Witayangkurn, A.; Kim, S.M. Using Volunteered Geographic Information and Nighttime Light Remote Sensing Data to Identify Tourism Areas of Interest. *Sustainability* **2019**, *11*, 4718. [CrossRef]
13. Lee, J.Y.; Tsou, M.-H. Mapping Spatiotemporal Tourist Behaviors and Hotspots through Location-Based Photo-Sharing Service (Flickr) Data. In Proceedings of the LBS 2018: 14th International Conference on Location Based Services, Zurich, Switzerland, 15–17 January 2017; pp. 315–334. [CrossRef]
14. Vu, H.Q.; Li, G.; Law, C.H.R.; Ye, B.H. Exploring the travel behaviors of inbound tourists to Hong Kong using geotagged photos. *Tour. Manag.* **2015**, *46*, 222–232. [CrossRef]
15. Maeda, T.N.; Yoshida, M.; Toriumi, F.; Ohashi, H. Extraction of Tourist Destinations and Comparative Analysis of Preferences between Foreign Tourists and Domestic Tourists on the Basis of Geotagged Social Media Data. *ISPRS Int. J. Geo-Inf.* **2018**, *7*, 99. [CrossRef]
16. Zhuang, C.; Ma, Q.; Liang, X.; Yoshikawa, M. Discovering Obscure Sightseeing Spots by Analysis of Geo-tagged Social Images. In Proceedings of the 2015 IEEE/ACM International Conference on Advances in Social Networks Analysis and Mining 2015—SONAM'15, Paris, France, 25–28 August 2015; pp. 590–595.
17. Corrêa, S.C.H.; Gosling, M.D.S. Travelers' Perception of Smart Tourism Experiences in Smart Tourism Destinations. *Tour. Plan. Dev.* **2021**, *18*, 415–434. [CrossRef]
18. Prandi, C.; Salomoni, P.; Mirri, S. mPASS: Integrating people sensing and crowdsourcing to map urban accessibility. In Proceedings of the 2014 IEEE 11th Consumer Communications and Networking Conference (CCNC), Las Vegas, NV, USA, 10–13 January 2014; pp. 591–595.
19. BOLOGNA–Prime Esperienze di Mappatura Dell'Accessibilità Cittadina. Dopo Firenze Kimappers in Emilia-Romagna. Available online: https://www.kimap.it/bologna-mappatura-accessibilita-cittadina/ (accessed on 20 December 2021).
20. Chung, N.; Lee, H.; Ham, J.; Koo, C. Smart Tourism Cities' Competitiveness Index: A Conceptual Model. In *Information and Communication Technologies in Tourism 2021*; Springer International Publishing: Cham, Switzerland, 2021; pp. 433–438.
21. Sagiroglu, S.; Sinanc, D. Big data: A review. In Proceedings of the 2013 International Conference on Collaboration Technologies and Systems (CTS), San Diego, CA, USA, 20–24 May 2013; pp. 42–47.
22. Hamid, R.A.; Albahri, A.; Alwan, J.K.; Al-Qaysi, Z.; Zaidan, A.; Alnoor, A.; Alamoodi, A.; Zaidan, B. How smart is e-tourism? A systematic review of smart tourism recommendation system applying data management. *Comput. Sci. Rev.* **2021**, *39*, 100337. [CrossRef]
23. Baggio, R.; Micera, R.; Del Chiappa, G. Smart tourism destinations: A critical reflection. *J. Hosp. Tour. Technol.* **2020**, *11*, 407–423. [CrossRef]

24. Programma Affiliati Booking.com Per Hotel–Guadagna dal tuo Sito. Available online: https://www.booking.com/affiliate-program/v2/index.html (accessed on 17 May 2021).
25. Varfolomeyev, A.; Korzun, D.; Ivanovs, A.; Soms, H.; Petrina, O. Smart Space based Recommendation Service for Historical Tourism. *Procedia Comput. Sci.* **2015**, *77*, 85–91. [CrossRef]
26. Baldoni, G.; Loudet, J.; Cominardi, L.; Corsaro, A.; He, Y. Facilitating distributed data-flow programming with Eclipse Zenoh. In *1st Workshop on Serverless Mobile Networking for 6G Communications*; ACM Press: New York, NY, USA, 2021; pp. 13–14.
27. Home OpenFaaS—Serverless Functions Made Simple. Available online: https://www.openfaas.com/ (accessed on 22 December 2021).
28. Kreps, J.; Corp, L.; Narkhede, N.; Rao, J. Kafka: A Distributed Messaging System for Log Processing. *Comput. Sci.* **2011**, *11*, 1–7.
29. Zaharia, M.; Chowdhury, M.; Franklin, M.J.; Shenker, S. *Spark: Cluster Computing with Working Sets*; ACM Press: New York, NY, USA, 2010.
30. Elastic Stack: Elasticsearch, Kibana, Beats & Logstash | Elastic. Available online: https://www.elastic.co/elastic-stack/ (accessed on 24 December 2021).
31. Kafka Inside Keystone Pipeline. The Second Story in Our Keystone ... | by Netflix Technology Blog | Netflix TechBlog. Available online: https://netflixtechblog.com/kafka-inside-keystone-pipeline-dd5aeabaf6bb (accessed on 20 July 2021).
32. Salloum, S.; Dautov, R.; Chen, X.; Peng, P.X.; Huang, J.Z. Big data analytics on Apache Spark. *Int. J. Data Sci. Anal.* **2016**, *1*, 145–164. [CrossRef]
33. Apache Livy. Available online: https://livy.incubator.apache.org/ (accessed on 22 December 2021).
34. Documentation: Table of Contents—RabbitMQ. Available online: https://www.rabbitmq.com/documentation.html (accessed on 26 December 2021).
35. Klein, M.; Lyft's Envoy: Experiences Operating a Large Service Mesh. In SREcon17 Americas (SREcon17 Americas). 2017. Available online: https://www.usenix.org/conference/srecon17americas/program/presentation/klein (accessed on 20 December 2021).
36. Jonas, E.; Schleier-Smith, J.; Sreekanti, V.; Tsai, C.; Khandelwal, A.; Pu, Q.; Shankar, V.; Carreira, J.M.; Krauth, K.; Yadwadkar, N.; et al. Cloud Programming Simplified: A Berkeley View on Serverless Computing. 2019. Available online: http://arxiv.org/abs/1902.03383 (accessed on 20 December 2021).
37. Spark Streaming—Spark 3.2.0 Documentation. Available online: https://spark.apache.org/docs/latest/streaming-programming-guide.html#setting-the-right-batch-interval (accessed on 30 November 2021).
38. Cheng, D.; Chen, Y.; Zhou, X.; Gmach, D.; Milojicic, D. Adaptive scheduling of parallel jobs in spark streaming. In Proceedings of the IEEE INFOCOM 2017–IEEE Conference on Computer Communications, Atlanta, GA, USA, 1–4 May 2017; pp. 1–9.
39. Liao, X.; Gao, Z.; Ji, W.; Wang, Y. An enforcement of real time scheduling in Spark Streaming. In Proceedings of the 2015 Sixth International Green and Sustainable Computing Conference (IGSC), Las Vegas, NV, USA, 14–16 December 2015; pp. 1–6.
40. HoseinyFarahabady, M.; Taheri, J.; Zomaya, A.Y.; Tari, Z. Spark-Tuner: An Elastic Auto-Tuner for Apache Spark Streaming. In Proceedings of the 2020 IEEE 13th International Conference on Cloud Computing (CLOUD), Beijing, China, 19–23 October 2020; pp. 544–548.
41. Casale, G.; Artač, M.; van den Heuvel, W.J.; van Hoorn, A.; Jakovits, P.; Leymann, F.; Long, M.; Papanikolaou, V.; Presenza, D.; Russo, A.; et al. RADON: Rational decomposition and orchestration for serverless computing. *Sics Softw. Intensive Cyber-Phys. Syst.* **2019**, *35*, 77–87. [CrossRef]
42. Angelaccio, M.; Basili, A.; Buttarazzi, B. Using Geo-business Intelligence and Social Integration for Smart Tourism Cultural Heritage Platforms. In Proceedings of the 2013 Workshops on Enabling Technologies: Infrastructure for Collaborative Enterprises, Hammamet, Tunisia, 17–20 June 2013; pp. 196–199.
43. Garcia, L.M.; Aciar, S.; Mendoza, R.; Puello, J.J. Smart Tourism Platform Based on Microservice Architecture and Recommender Services. In *Hybrid Learning and Education*; Springer Science and Business Media LLC: Berlin/Heidelberg, Germany, 2018; Volume 10995, pp. 167–180.

sensors

MDPI

Article

Investigating the Efficient Use of Word Embedding with Neural-Topic Models for Interpretable Topics from Short Texts

Riki Murakami [1] and Basabi Chakraborty [2,*]

[1] Graduate School of Software and Information Science, Iwate Prefectural University, 152-52 Sugo, Takizawa 020-0693, Iwate, Japan; g236r005@s.iwate-pu.ac.jp
[2] Faculty of Software and Information Science, Iwate Prefectural University, 152-52 Sugo, Takizawa 020-0693, Iwate, Japan
* Correspondence: basabi@iwate-pu.ac.jp

Abstract: With the rapid proliferation of social networking sites (SNS), automatic topic extraction from various text messages posted on SNS are becoming an important source of information for understanding current social trends or needs. Latent Dirichlet Allocation (LDA), a probabilistic generative model, is one of the popular topic models in the area of Natural Language Processing (NLP) and has been widely used in information retrieval, topic extraction, and document analysis. Unlike long texts from formal documents, messages on SNS are generally short. Traditional topic models such as LDA or pLSA (probabilistic latent semantic analysis) suffer performance degradation for short-text analysis due to a lack of word co-occurrence information in each short text. To cope with this problem, various techniques are evolving for interpretable topic modeling for short texts, pretrained word embedding with an external corpus combined with topic models is one of them. Due to recent developments of deep neural networks (DNN) and deep generative models, neural-topic models (NTM) are emerging to achieve flexibility and high performance in topic modeling. However, there are very few research works on neural-topic models with pretrained word embedding for generating high-quality topics from short texts. In this work, in addition to pretrained word embedding, a fine-tuning stage with an original corpus is proposed for training neural-topic models in order to generate semantically coherent, corpus-specific topics. An extensive study with eight neural-topic models has been completed to check the effectiveness of additional fine-tuning and pretrained word embedding in generating interpretable topics by simulation experiments with several benchmark datasets. The extracted topics are evaluated by different metrics of topic coherence and topic diversity. We have also studied the performance of the models in classification and clustering tasks. Our study concludes that though auxiliary word embedding with a large external corpus improves the topic coherency of short texts, an additional fine-tuning stage is needed for generating more corpus-specific topics from short-text data.

Keywords: short-text data; neural-topic model; pretrained word embedding; coherent topic; fine-tuning

Citation: Murakami, R.; Chakraborty, B. Investigating the Efficient Use of Word Embedding with Neural-Topic Models for Interpretable Topics from Short Texts. *Sensors* **2022**, *22*, 852. https://doi.org/10.3390/s22030852

Academic Editors: Suparna De and Klaus Moessner

Received: 26 December 2021
Accepted: 16 January 2022
Published: 23 January 2022

1. Introduction

Due to the rapid developments of computing and communication technologies and the widespread use of internet, people are gradually becoming accustomed to communicating through various online social platforms, such as microblogs, Twitter, webpages, Facebook, etc. These messages over web and social networking sites contain important information regarding current social situations and trends, people's opinions on different products and services, advertisements, and announcements of government policies, etc. An efficient text-processing technique is needed to automatically analyze these huge amounts of messages for extracting information. In the area of traditional natural language processing, a topic-modeling algorithm is considered an effective technique for the semantic understanding of text documents. Conventional topic models, such as pLSA [1] or LDA [2] and their various variants, are considerably good at extracting latent semantic structures from a text corpus without prior annotations and are widely used in emerging topic detection, document classification, comment summarizing, or event tracking. In these models, documents are viewed as a mixture of topics, while each topic is viewed as a particular distribution over all the words. Statistical tools are used to determine the latent topic distribution of each document, while higher-order word co-occurrence patterns are used to characterize each topic [3]. The efficient capture of document-level word co-occurrence patterns leads to the success of topic modeling.

The messages posted on various social network sites are generally short compared to the length of relatively formal documents such as newspapers or scientific articles. The main characteristics of these short texts are: (1) a limited number of words in one document, (2) the use of new and informal words, (3) meanings and usages of words that may change greatly depending on the posting, (4) spam posts, and (5) the restricted length of posts, such as API restrictions on Twitter. The direct application of traditional topic models for short-text analysis results in poor performance due to lack of word co-occurrence information in each short text document, originating from the above characteristics of short texts [4]. Earlier research on topic-modeling for short texts with traditional topic models used external, large-scale datasets such as Wikipedia, or related long-text datasets for a better estimation of word co-occurrences across short texts [5,6]. However, these methods work well only when the external dataset closely matches the original short-text data.

To cope with the problems of short-text topic-modeling by traditional topic models, three main categories of algorithms are found in the literature [7]. A simple solution is to aggregate a number of short texts into a long pseudo-document before training a standard topic model to improve word co-occurrence information. In [8], the tweets of an individual user are aggregated in one document. In [9,10], a short text is viewed as sampled from unobserved, long pseudo-documents, and topics are inferred from them. However, the performance of these methods depends on efficient aggregation and data type. When short texts of different semantic contents are aggregated to a long document, non-semantic word co-occurrence information can produce incoherent topics. In the second category, each short text document is assumed to consist of only a single topic. Based on this assumption, Direchlet Multinomial Mixture (DMM) model-based topic-modeling methods have been developed for short texts in [11–13]. Although this simple assumption eliminates data-sparsity problems to some extent, they fail to capture multiple topic elements in a document, which makes the model prone to over-fitting. Moreover, "shortness" is subjective and data-dependent; a single-topic assumption might be too strong for some datasets. A poisson-based DMM model (PDMM) [14] considers a small number of topics associated with each short text instead of only one. The third category of algorithms consider global word co-occurrence patterns for inferring latent topics. According to the usage, two types of models are developed. In [15], global word co-occurrence is directly used, while in [16], a word co-occurrence network is first constructed using global word co-occurrence, and then latent topics are inferred from this network. In the present work, we explored methods of exploiting this category for further improvement in the development of algorithms for extracting interpretable topics from short texts.

Another limitation of the above models for short-text analysis is that the context or background information is not used, resulting in the generation of not-so-coherent topics. The statistical information of words in the text document cannot fully capture words that are semantically correlated but that rarely co-occur. Recent advances in word embedding [17] provides an effective way of learning semantic word relations from a large corpus, which can help to develop models for generating more interpretable and coherent topics. Word embedding uses one-hot representation of words with vocabulary length vectors of zeroes with a single one, and words that are similar in semantics are close in a lower-dimensional vector space. An embedded topic model (ETM), a combination of LDA and word embedding which enjoys both the advantages of topic model and word embedding, has been proposed in [18]. Traditional topic models with word embedding for documents are explored in several other research works, cited in [19]. In [20], word embedding is combined with LDA to accelerate the inference process, resulting in the enhanced interpretability of topics. For short texts, models incorporating word embedding into DMM are proposed in [21,22]. In [23,24], short texts are merged into long pseudo-documents using word embedding. Word embedding in conjunction with conventional topic models seems to be a better technique for generating coherent topics.

The increasing complexity of inference processes in conventional topic models on large-text data, along with the recent developments of deep neural networks, has led to the emergence of neural-topic models (NTM). These models combine the performance, efficacy, scalability, and ease of leveraging parallel computing facilities, such as GPU, to probabilistic topic-modeling [25]. Neural-topic models are considered to be computationally simpler and easier for implementation, compared to traditional LDA models, and are increasingly used in various natural language processing tasks in which conventional topic models are difficult to use. A systematic study on the performances of several neural-topic models has been reported in [26]. Although various neural-topic models have been proposed, and although the reported experimental results on topic generation seem to be better than conventional topic models for long and formal texts, little research has been conducted on neural-topic models for effective analysis of short texts [27]. Most of the research works of topic modeling on short texts are based on extensions of Bayesian probabilistic topic models (BPTM) such as LDA.

The objective of the present research is to explore computationally easy and efficient techniques for improving the interpretability of generated topics from real-world short texts using neural-topic models. However, learning context information is the most challenging issue of topic-modeling for short texts, and incorporating pretrained word embedding into a topic model seems to be one of the most efficient ways of explicitly enriching the content information. Neural-topic models with pretrained word embedding for short-text analysis has not been extensively explored yet, compared to its long-text counterparts. In [28], we presented our preliminary analysis of short-text data (benchmark and real-world) with neural-topic models using pretrained word embedding. We found that although pretrained word embedding enhances the topic coherence of short texts that are similar to long and formal texts, the generated topics were often comprised of words having common meanings (which are found in the large external corpus used for pretraining) instead of the particular short-text-specific semantics of the word, which is especially important for real-world datasets. In other words, the learning of topic centroid vectors is influenced by pretraining the text corpus and fails to discover the important words of the particular short text. Our proposal is that this gap can be filled by adding a fine-tuning stage to the training of the topic model with the particular short-text corpus to be analysed. In this work, we have completed an extensive study to investigate the performance of recent neural-topic models with and without word embedding, and also with the proposed fine-tuning stage, for generating interpretable topics from short texts in terms of a number of performance metrics by simulation experiments on several datasets. We have also studied the performance of the NTM with pretrained word embedding added, with a fine-tuning stage for classification and clustering tasks. As a result of our experiments, we can confirm that the addition of a fine-tuning stage indeed enhances the topic quality of short texts in general, and generates topics with corpus-specific semantics.

In summary, our contributions in this paper are as follows:

- A proposal for fine-tuning with the original short-text corpus, along with the pre-trained word embedding with the large external corpus, for generating more interpretable and coherent corpus-specific topics from short texts;
- An extensive evaluation of the performance of several neural-topic models, with and without pretrained word embedding and with an added fine-tuning stage, in terms of topic quality and measured by several metrics of topic coherence and topic diversity;

- A performance evaluation of the proposed fine-tuned neural-topic models for classification and clustering tasks.

In the next section, neural-topic models are introduced in brief, followed by a short description of related works on neural-topic models (NTM), especially NTMs for short texts. The following sections contain our proposal, followed by simulation experiments and results. The final section presents the conclusion.

2. Neural-Topic Models and Related Works

The most popular neural-topic models (NTMs) are based on a variational autoencoder (VAE) [29], a deep generative model, and amortised variational inferences (AVI) [30]. The basic framework of VAE-based NTMs is described in the next section, in which generative and inference processes are modeled by a neural network-based decoder and encoder, respectively. Compared to the traditional Bayesian probabilistic topic models (BPTM), inference in neural-topic models is computationally simpler, their implementation is easier due to many existing deep learning frameworks, and NTMs are easy to be integrated with pretrained word embeddings for prior-knowledge acquisition. Several categories of VAE-based NTMs have been proposed. To name a few, there is the Neural Variational Document Model (NVDM) [31], Neural Variational Latent Dirichlet Allocation (NVLDA) [32], the Dirichlet Variational Autoencoder topic model (DVAE) [33], the Dirichlet Variational Autoencoder (DirVAE) [34], the Gaussian Softmax Model (GSM) [35], and iTM-VAE [36]. This list is not exhaustive and is still growing.

In addition to VAE-based NTMs, there are a few other frameworks for NTMs. In [37], an autoregressive NTM, named DocNade, has been proposed. Consequently, some extensions of DocNADE are found in the literature. Recently, some attempts have been made to use a GAN (Generative Adversarial Network) framework for topic-modeling [38,39]. Instead of considering a document as a sequence or a bag of words, a graph representation of a corpus of documents can be considered. In [40], a bipartite graph, with documents and words as two separate partitions and connected by word occurrences in documents as the weights, is used. Ref. [41] uses the framework of Wasserstein auto-encoders (WAEs), which minimizes the Wasserstein distance between reconstructed documents from the decoder and the real documents, similar to a VAE-based NTM. In [42], a NTM based on optimal transport that directly minimizes the optimal transport distance between the topic distribution learned by an encoder and the word distribution of a document has been introduced.

Neural-Topic Models for Short-Text Analysis

In order to generate coherent, meaningful, and interpretable topics from short texts by incorporating semantic and contextual information, a few researchers used NTMs in lieu of conventional topic models. In [43,44], a combination of NTM and either a recurrent neural network (RNN) or a memory network has been used, in which topics learned by the NTM are utilized for classification by a RNN or a memory network. In both works, the NTM shows better performance than conventional topic models in terms of topic coherence and a classification task. To enhance the discreteness of multiple topic distributions in a short text, in [27], the authors used Archimedean copulas. In [45], the authors introduced a new NTM with a topic-distribution quantization approach, producing peakier distributions, and also proposed a negative sampling decode, learning to minimize repetitive topics. As a result, the proposed model outperforms conventional topic models. In [46], the authors aggregated short texts into long documents and incorporated document embedding to provide word co-occurrence information. In [47], a variational autoencoder topic model (VAETM) and a supervised version (SVAETM) of it have been proposed by combining embedded representations of words and entities by employing an external corpus. To enhance contextual information, the authors in [48] proposed a graph neural network as the encoder of NTM, which accepts a bi-term graph of the words as inputs and produces the topic distribution of the corpus as the output. Ref. [49] proposed a context-reinforced neural-topic model with the assumption of a few salient topics for each short text, informing the word distributions of the topics using pretrained word embedding.

3. Proposal for Fine-Tuning of Neural-Topic Models for Short-Text Analysis

From the analysis of present research works on neural-topic models on short-text analysis, it seems that incorporating auxiliary information from an external corpus is one of the most popular and effective techniques for dealing with sparsity in short texts. As mentioned in the introduction, in our previous work [28], we found that although pretrained word embedding with a large external corpus helps with generating coherent topics from a short-text corpus, the generated topics lack the semantics expressed by the corpus-specific meaning of words. If the domain of the short-text corpus and the external corpus vary too much, the topic semantics become poor. This fact is also noted by other researchers [25].

In this work, we propose an additional fine-tuning stage, using the original short-text corpus, along with the pretrained word embedding and a large external corpus. For pretrained word embedding, we decided to use GloVe [50] after some preliminary experiments with two other techniques, namely, Word2Vec and Fast Text, as GloVe provided consistent results. Here, we have completed an extensive comparative study to evaluate the effect of pretrained embedding with our proposed additional fine-tuning stage using several short-text corpora and neural-topic models. The proposed study setting is represented in Figure 1.

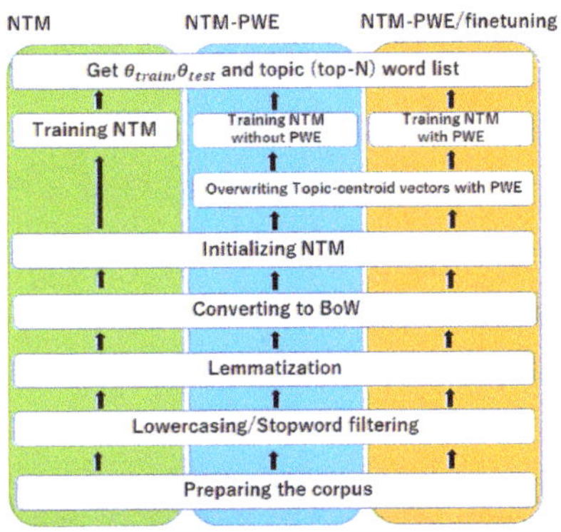

Figure 1. Proposed Study.

Here, pretrained word embedding is denoted as PWE. We have performed three sets of experiments for topic extraction, using only neural-topic models (NTM), neural-topic models with pretrained word embedding (NTM-PWE), and neural-topic models with pretrained embedding and the proposed fine-tuning step (NTM-PWE/fine-tuning). In all the cases, the data corpus is first pre-processed, and in NTM-PWE, word embedding vectors are replaced by PWE after the model parameters are initialized; the weights of PWE are not updated during the training step. In our proposed PWE/fine-tuning or simple fine-tuning (as mentioned in the text), the weights are gradually updated in the training step after replacing the word embedding vectors, as in PWE. In this case, it is possible to update the parameters at the same learning rate that is set to update the entire model, but experiments have shown that updating the PWE values at a large learning rate can easily over-fit the training data. Therefore, in the simulation experiments, we have set the learning rate of the word embedding vectors to a smaller value than the learning rate of the entire model.

We have used popular VAE-based neural-topic models with a few similar WAE (Wasserstein autoencoder)-based models, and ten popular benchmark datasets, for our simulation experiments. The performance of each neural-topic model with no word embedding, pretrained word embedding, and additional fine-tuning has been evaluated by the generated topic quality using different evaluation metrics of topic coherence and topic diversity. The neural-topic models, datasets, and evaluation metrics used in this study are described below.

3.1. Neural-Topic Models for Evaluation

In this section, the neural-topic models used in this study are described briefly. Table 1 describes the meaning of the notations used for description of the models.

Table 1. Table of notations used.

Indices:	
K	Number of topics $k \in \{1, \ldots, K\}$
L	Word embedding vectors dimension $l \in \{1, \ldots, L\}$
C	Number of classes $c \in \{1, \ldots, C\}$
Decision Variables:	
D	Set of documents
V	Set of lexicons, vocabularies
X	BoW matrix of all documents, $X \in \mathbb{R}_{+}^{\lvert V \rvert \times \lvert D \rvert}$
x_d	d's BoW representation vector, $x_d \in \mathbb{R}_{+}^{\lvert V \rvert}$
N	Number of words that occurred in document d
w_n	n-th word
Random Variables:	
$h^{(i)}$	i-th hidden layer's outputs
h	Gaussian random variables, $h \in \mathbb{R}^K$
z_n	latent topic for the n-th word
θ	Topic proportion vector, $\theta \in \mathbb{R}_{+}^{K}$
β	Topic–word distribution $\beta \in \mathbb{R}^{\lvert V \rvert \times K}$
α	Topic centroid vectors $\alpha \in \mathbb{R}^{L \times K}$
ρ	Word embedding vectors $\rho \in \mathbb{R}^{L \times \lvert V \rvert}$

Figures 2 and 3 describe the generalized architecture of the Variational autoencoder (VAE)- and Wasserstein autoencoder (WAE)-based neural-topic models, respectively. In both the models, the part of the network that generates θ is known as the encoder, which maps the input bag-of-words (BoW) to a latent document–topic vector, and the part that receives θ and outputs $p(x)$ is called the decoder, which maps the document–topic vector to a discrete distribution over the words in the vocabulary. They are called autoencoders because the decoder aims to reconstruct the word distribution of the input. In VAE, h is sampled by Gaussian distribution, and θ is created by performing some transformation on it. WAE, on the other hand, uses the Softmax function directly to create θ, so no sampling is required. Evidence lower bound (ELBO), the objective function of VAE, is defined below [29]:

$$\begin{aligned}\mathcal{L}_d = {} & \mathbb{E}_{q(\theta|d)}\left[\sum_{n=1}^{N_d} \log \sum_{z_n} [p(w_n \mid \beta_{z_n}) p(z_n \mid \theta)]\right] \\ & - D_{KL}\left[q(\theta \mid x_d) \| p\left(\theta \mid \mu_0, \sigma_0^2\right)\right]\end{aligned} \tag{1}$$

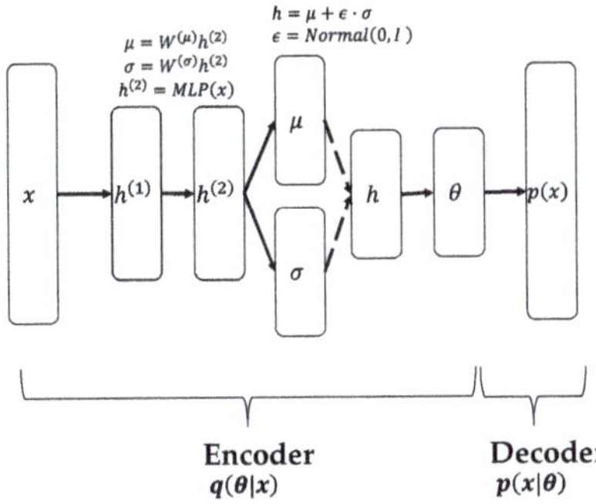

Figure 2. VAE-based model.

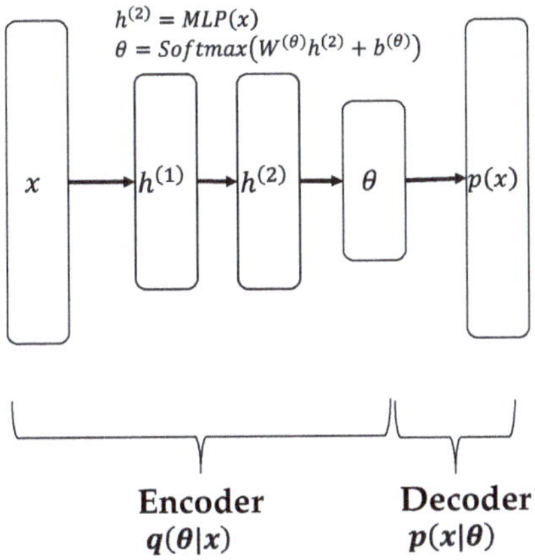

Figure 3. WAE-based model.

It is empirically known that maximizing this ELBO alone will result in smaller (worse) topic diversity. In order to solve this problem, some NTMs use a regularization term to increase the topic diversity [51]:

$$
\begin{aligned}
a(\alpha_i, \alpha_j) &= \arccos\left(\frac{|\alpha_i \cdot \alpha_j|}{||\alpha_i|| \cdot ||\alpha_j||}\right) \\
\zeta &= \frac{1}{K^2}\sum_i^K\sum_j^K a(\alpha_i, \alpha_j) \\
\nu &= \frac{1}{K^2}\sum_i^K\sum_j^K \left(a(\alpha_i, \alpha_j) - \zeta\right)^2 \\
\mathcal{J} &= \mathcal{L} + \lambda(\zeta - \nu)
\end{aligned}
\tag{2}
$$

where λ is a hyper-parameter that manipulates the influence of the regularization term; 10 was adopted here. This value was determined empirically. The VAE-based models in this paper use this regularization term.

The particular NTMs used in our study are mentioned in the next subsections.

3.1.1. Neural Variational Document Model (NVDM)

NVDM [31] is, to our knowledge, the first VAE-based document model proposed with the encoder implemented by a multilayer perceptron. This model uses the sample h from the Gaussian distribution as an input for the decoder, and variational inference is based on minimizing KL divergence. While most of the NTMs proposed after this one transform h to treat θ as a topic proportion vector, NVDM is a general VAE.

3.1.2. Neural Variational Latent Dirichlet Allocation (NVLDA)

NVLDA [32], another variant of NVDM, is a model that uses Neural Variational Inference to reproduce LDA. Here, the Softmax function is used to convert z to θ. The probability distribution that maps samples from a Gaussian distribution to the Softmax basis is called the Logisitic–Normal distribution, which is used as a surrogate for the Dirichlet distribution. Additionally, the decoder is $p(x) = softmax(\beta) \cdot \theta$. Unlike the NVDM, where both the topic proportions and the topic–word distribution are in the form of probability distributions, this model is a topic model. Logisitic–Normal distribution is defined as follows:

$$h \sim Normal(\mu, \sigma^2) \tag{3}$$
$$\theta = softmax(h) \tag{4}$$

3.1.3. Product-of-Experts Latent Dirichlet Allocation (ProdLDA)

ProdLDA [32] is an extension of NVLDA in which the decoder is designed by following the product of the expert model, and the topics-word distribution are not normalized.

3.1.4. Gaussian Softmax Model (GSM)

GSM [35] converts h to θ using Gaussian Softmax, as defined below:

$$h \sim Normal(\mu, \sigma^2) \tag{5}$$
$$\theta = \mathrm{softmax}(W_1^T h) \tag{6}$$

where $W_1 \in \mathbb{R}^{K \times K}$, a linear transformation, is the trainable parameters used as the connection weights.

3.1.5. Gaussian Stick-Breaking Model (GSB)

GSB [35] converts h to θ by Gaussian Stick-Breaking construction, which is defined as follows:

$$h \sim Normal(\mu, \sigma^2) \tag{7}$$
$$\eta = sigmoid(W_2^T h) \tag{8}$$
$$\theta = f_{SB}(\eta) \tag{9}$$

where $W_2 \in \mathbb{R}^{K \times K-1}$ is the trainable parameters used as connection weights, and the stick-breaking function f_{SB} is described by Algorithm 1:

Algorithm 1 Stick-Breaking Process (f_{SB})

Input: Return value from sigmoid function η
$\eta \in \mathbb{R}_{+}^{K}$, where $\forall \eta_k \in [0, 1]$
Output: Topic proportion vector θ
$\theta \in \mathbb{R}_{+}^{K}$, where $\sum_k(\theta_k) = 1$
1: Assign η_1 to the first element of the topic proportion vector θ_1.
$\theta_1 = \eta_1$
2: **for** $k = 2, \ldots K-1$ **do**

$$\theta_k = \eta_k \prod_{i=1}^{k-1}(1 - \eta_i)$$

3: **end for**
4:

$$\theta_K = \prod_{i=1}^{K-1}(1 - \eta_i)$$

3.1.6. Recurrent Stick-Breaking Model (RSB)

RSB [35] converts h to θ by recurrent Stick-Breaking construction, as defined below. Here, the stick-breaking construction is considered as a sequential draw from a recurrent neural network (RNN).

$$h \sim Normal(\mu, \sigma^2) \tag{10}$$

$$\eta = \mathrm{f}_{\mathrm{RNN}}(h) \tag{11}$$

$$\theta = \mathrm{f}_{\mathrm{SB}}(\eta) \tag{12}$$

where f_{RNN} is decomposed as:

$$h_k = \mathrm{RNN}_{\mathrm{SB}}(h_{k-1}) \tag{13}$$

$$\eta_k = \mathrm{sigmoid}(h_{k-1}^T z) \tag{14}$$

and $f_{SB}(\eta)$ is the same as in GSB.

3.1.7. Wasserstein Latent Dirichlet Allocation (WLDA)

WLDA [41] is a topic model based on a Wasserstein autoencoder (WAE). Though various probability distributions can be used for the prior distribution of θ, in this paper, we use the Dirichlet distribution, which we believe is the most basic. In WAE, two training methods are available, GAN (Generative Adversarial Network)-based training and MMD (Maximum Mean Discrepancy)-based training, but in WLDA, MMD is used because of the ease of convergence of training loss. In VAE, the loss function is composed of the KL Divergence used as the regularization term for θ and the reconstruction error, while in WLDA, MMD is used as the regularization term.

If P_Θ is a θ's prior distribution, and Q_Θ is a fake samples's prior distribution, maximum mean discrepancy (MMD) is defined as:

$$\mathrm{MMD}_{\mathbf{k}}(Q_\Theta, P_\Theta) = \left\| \int_\Theta \mathbf{k}(\theta, \cdot) dP_\Theta(\theta) - \int_\Theta \mathbf{k}(\theta, \cdot) dQ_\Theta(\theta) \right\|_{\mathcal{H}_{\mathbf{k}}} \tag{15}$$

where $\mathcal{H}$ means the reproducing kernel Hilbert space (RKHS) of real-valued functions mapping Θ to $\mathbb{R}$, and k is the kernel function.

$$\mathbf{k}(\theta, \theta') = \exp\left(-\arccos^2\left(\sum_{k=1}^{K} \sqrt{\theta_k \theta'_k} \right) \right) \tag{16}$$

3.1.8. Neural Sinkhorn Topic Model (NSTM)

NSTM [52] is trained using optimal transport [42], as in WLDA. Since we assume that θ encodes x into a low-dimensional latent space while preserving sufficient information about x, the optimal transport distance between θ and x is calculated by the Sinkhorn Algorithm. The sum of this optimal transport distance and the negative log likelihood is used as the loss function.

3.2. Datasets

Table 2 presents the details of the benchmark datasets used in this work. The first column represents the name of the dataset, followed by the number of documents ($|D|$), vocabulary size ($|V|$), the total number of tokens ($\sum X$), average document length (ave dL), maximum document length (max dL), sparsity, number of classes (C,) and the source of the data in the respective columns. In the source column, 1 and 2 represent OCCITS and STTM, respectively.

1. OCTIS: https://aclanthology.org/2021.eacl-demos.31/ (accessed on 14 January 2022).
2. STTM: https://arxiv.org/pdf/1701.00185.pdf (accessed on 14 January 2022).

The first two datasets fall in the category of long documents, and the other eight datasets can be considered as the short-text corpus, as the average document length is quite short compared to the long documents.

Table 2. Details of Data set.

Name	$\|D\|$	$\|V\|$	$\sum X$	ave dL	max dL	Sparsity	C	Source
BBC_news	2225	2949	267,259	120.12	1176	0.027811	5	1
20NewsGroup	16,309	1612	783,151	48.02	3033	0.018855	20	1
SearchSnippets	12,295	4720	177,338	14.42	37	0.002167	8	2
TrecTweet	2472	5098	21,148	8.56	20	0.001561	89	2
Biomedical	19,448	3892	144,683	7.44	28	0.001870	20	2
GoogleNews	11,108	8110	69,229	6.23	14	0.000764	152	2
M10	8355	1696	49,385	5.91	21	0.003411	10	1
DBLP	54,595	1513	294,757	5.40	21	0.003527	4	1
PascalFlicker	4834	3431	25,980	5.37	19	0.001548	20	2
StackOverflow	16,407	2303	82,342	5.02	17	0.002145	20	2

The datasets shown in the table are pre-processed. HTML tags and other symbols have been removed from each dataset, and all words have been lowercased. Then, the stop-words were removed and lemmatized. From these datasets, 80% of the total documents was used as the training data and the rest as the test data. These pre-processed corpora are then converted into a BoW (Bag-of-Words), which basically has word frequency as an element, to be used as input data for the NTM. However, for the NSTM, the vector corresponding to each document in the BoW is divided by the total value of the vectors, as in the original paper.

3.3. Evaluation of Topic Quality

It is quite challenging to evaluate the performance of topic models, including NTMs, according to the quality of the generated topics. Topics generated by topic models can be considered as soft clusters of words. Under the constraints of the topic model, this is a probability distribution that collects the probability of word generation for each topic; the same is true for NTM, but this may not be in the form of a probability distribution for document models that impose even weaker constraints than the topic model. Either way, a topic here is a topic–word distribution, and each distribution has as many dimensions as the number of lexemes that occur in the corpus. It is very difficult to understand the goodness of a topic by directly comparing them with human topics. Therefore, in practice, analysts check a list of N words characteristic of a topic based on the values of the word distributions. In most cases, the list of the top-N words in terms of the large probability values in the word distribution is used.

Various metrics have been proposed to evaluate the quality of the top-N words with two main directions. One is to check whether the meaning of words belonging to the top-N words are consistent with each other, defined as topic coherence (TC). The other is to measure the diversity of the top-N words of each pair of topics, defined as topic diversity (TD) or topic uniqueness. Topics with high TC may have low TD. In this case, the top-N words of most topics will be nearly the same, which is not desirable. So, to evaluate the quality of a topic for human-like interpretability, it should have high TC as well as high TD.

3.3.1. Topic Coherence (TC)

For computing TC, general coherence between two sets of words are estimated based on word co-occurrence counts in a reference corpus [53]. The choices are (1) the training corpus for topic modeling; (2) a large external corpus (e.g., Wikipedia); (3) word embedding vectors trained on a large external corpus (e.g., Wikipedia). The scores may differ according to different computations. Choice 1 is easy, but the results are affected by the size of the training corpus. Choices 2 and 3 are more popular, although choice 2 is computationally costly. However, if the domain gap of the training corpus and the external corpus is high, the evaluation is not proper. In this work, we have used the following metrics for computation of topic coherence:

- Normalized Point-Wise Mutual Information (NPMI) [54]: NPMI is a measure of the semantic coherence of a group of words. It is considered to have the largest correlations with human ratings, and is defined by the following equation:

$$NPMI(w) = \frac{1}{N(N-1)} \sum_{j=2}^{N} \sum_{i=1}^{j-1} \frac{log \frac{P(w_i, w_j)}{P(w_i)P(w_j)}}{-log P(w_i, w_j)} \tag{17}$$

  where w is the list of the top-N words for a topic. N is usually set to 10. For K topics, averages of NPMI over all topics are used for evaluation;
- Word Embeddings Topic Coherence (WETC) [55]: WETC represents word embedding-based topic coherence, and pair-wise WETC for a particular topic is defined as:

$$\text{WETC}_{PW}(E^{(k)}) = \frac{1}{N(N-1)} \sum_{j=2}^{N} \sum_{i=1}^{j-1} \left\langle E^{(k)}_{i,:}, E^{(k)}_{j,:} \right\rangle \tag{18}$$

  where $\langle .,. \rangle$ denotes the inner product. For the calculation of the WETC score, pre-trained weights of GloVe [50] have been used, and $E^{(k)}$ is the word embedding vector sequence of GloVe corresponding to the top-N words for topic k; $E^{(k)}_i$ means and all vectors are normalized as follows: $||E^{(k)}_{i,:}|| = 1$, N is taken as 10.
  WETC_c (centroid WETC) is defined as follows:

$$\begin{aligned} \text{WETC}_c(E^{(k)}) &= \frac{1}{N} \sum_{n=1}^{N} E^{(k)} t \\ t &= \frac{\alpha_{:,k}}{||\alpha_{:,k}||} \end{aligned} \tag{19}$$

3.3.2. Topic Diversity

Topic diversity is defined here as the percentage of unique words in the top 25 words of all topics, according to [18]. A diversity close to 0 represents a redundant topic, and those close to 1 indicate more varied topics. Here, we have also used two other metrics, inverted rank-biased overlap (InvertedRBO) [56] and mean squared cosine deviation among topics

(MSCD) [57], as a measure of diversity of the generated topics. InvertedRBO is a measure of disjointedness between topics weighted on word rankings, based on the top-N words. The higher these metrics are, the better. MSCD is the cosine similarity of the word distribution of each topic, so it should be lower for better topics. In general, NTM training updates parameters to maximize ELBO, but such a naive implementation can easily lead to poor TD. However, in this case, since we use the topic centroid vectors as trainable parameters, we regularize parameters of NTM to increase the angle formed by each topic centroid vector in order to increase the TD.

4. Simulation Experiments and Results

The simulation experiments have been performed with several benchmark datasets, and the performance of the topic models are evaluated by topic coherence and topic diversity measures.

4.1. Experimental Configuration

For the purpose of comparison and evaluation, the experimental setting should be similar for all the neural-topic models and all the datasets. At the beginning, we completed some trial experiments, and determined that the optimum topic size parameter should be set at $K = 50$, based on topic coherence and perplexity, so that there are a sufficient number of topics without becoming very large, considering the length of short text. This value is also in accordance with the value used for related experiments in similar research works. The number of dimensions of the word embeddings was fixed at $L = 300$. This is in accordance with the GloVe's Common Crawl-based trained word embedding vectors, publicly available at https://nlp.stanford.edu/projects/glove/ (accessed on 14 January 2022), which cover largest number of vocabularies.

The other experimental parameters are set as follows: number of units of the encoder's hidden layers: $H^{(1)} = 500$, $H^{(2)} = 500$; Dropout rate: $p_{\text{dropout}} = 0.2$; Minibatch size: 256; Max epochs: 200; Learning rate for the encoder network: 0.005; Learning rate for the decoder network: 0.001. We employ Adam as the optimizer and Softplus as the activation function of the encoder networks.

On WLDA, we employ Dirichlet as the prior distribution for topic proportion-generation, using MMD for this model's training. On NSTM, Sinkhorn algorithm's max number of updates is 2000, and the threshold value for updating termination condition is 0.05, constant value $\alpha_{\text{Sinkhorn}} = 20$.

4.2. Results for Topic Coherence

Tables 3–12 represent the detail results of different topic coherence metrics (NPMI and WETC) for different neural models and different datasets, respectively. Values in bold faces indicate the best results. We have used two versions of GloVe, differing in terms of the size of the corpus.

For many datasets, NVLDA-PWE/fine-tuning has the highest TC. One of the challenges of using PWE without fine-tuning is that the high domain gap between the PWE training corpus and the corpus for topic modeling has a negative impact. In many cases, our proposal produces better results, but not for all the datasets or for all the models. The dataset "GoogleNews" often has the best TC with PWE, and does not show better performance with additional fine-tuning. This is probably because this corpus has a similar domain as the training data for PWE. In a few datasets, the best performance is noticed when no pretraining word embedding is used. It is verified that for those datasets, the original corpus contains sufficient word co-occurrence information.

However, we noted that the TC value changes significantly depending on the type of word embedding. This result suggests that the quality of the word embeddings may have a significant impact on the training of the topic model. In particular, whether or not the unique words in the training corpus are included in the unique words in the PWE has a significant impact. If the coverage of this word dictionary is large, the PWE can be used for evaluation, but if there are many missing words, the reliability of the evaluation value will be greatly compromised.

Figure 4 presents the summary of topic coherence over all the neural-topic models for the long-text corpus (2 datasets) and the short-text corpus (8 datasets), which shows the overall trend. In the case of long texts, the scores of the PWE/fine-tuning metrics for TC are either a little worse or the same as the others. One of the reasons for this is that the long-text corpus used in this study is composed of relatively formal documents, which is close to the domain of PWE. In contrast, the short-text corpus shows better performance in all metrics. The overall trend is none < PWE < PWE/finetuning.

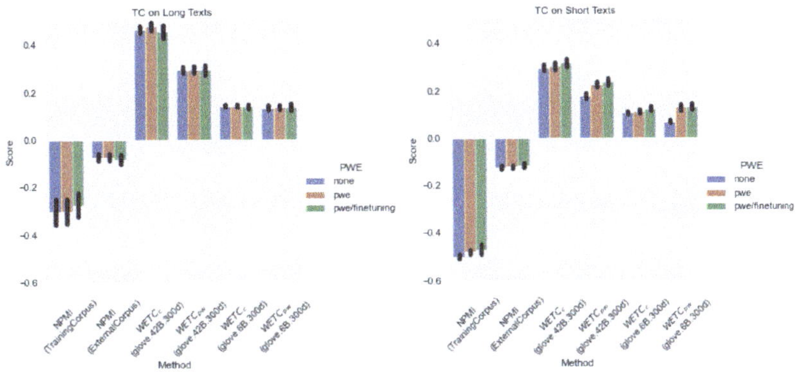

TC on Long Texts TC on Short Texts

Figure 4. Summary of TC results.

Table 3. Topic Coherence on 20NewsGroups.

Model	Resources Method PWE	TrainingCorpus npmi	ExternalCorpus npmi	glove.42B.300d WETC$_c$	 WETC$_{pw}$	glove.6B.300d WETC$_c$	 WETC$_{pw}$
GSB	none	−0.16	−0.04	0.47	0.33	0.16	0.15
	pwe	−0.21	−0.05	0.50	0.32	0.16	0.15
	pwe/fine-tuning	−0.12	**−0.03**	0.48	0.34	**0.17**	0.17
GSM	none	−0.18	−0.05	0.50	0.32	0.15	0.15
	pwe	−0.21	−0.05	0.50	0.32	0.16	0.15
	pwe/fine-tuning	−0.21	−0.09	0.43	0.29	0.12	0.13
NSTM	none	−0.17	−0.05	0.53	0.33	0.13	0.16
	pwe	−0.19	−0.05	0.53	0.33	0.15	0.15
	pwe/fine-tuning	−0.16	−0.04	0.53	0.33	0.14	0.16
NVDM	none	−0.27	−0.05	0.47	0.30	0.14	0.14
	pwe	−0.22	−0.05	0.51	0.32	0.15	0.15
	pwe/fine-tuning	−0.27	−0.08	0.45	0.28	0.12	0.12
NVLDA	none	−0.18	−0.04	0.52	0.34	0.14	0.16
	pwe	−0.19	−0.04	0.52	0.33	0.14	0.15
	pwe/fine-tuning	**−0.10**	−0.04	**0.55**	**0.36**	0.16	**0.18**
ProdLDA	none	−0.21	−0.05	0.48	0.31	0.16	0.14
	pwe	−0.22	−0.05	0.50	0.32	0.15	0.15
	pwe/fine-tuning	−0.21	−0.09	0.46	0.29	0.12	0.12
RSB	none	−0.19	−0.06	0.48	0.31	0.15	0.14
	pwe	−0.18	−0.04	0.52	0.33	0.15	0.16
	pwe/fine-tuning	−0.26	−0.09	0.42	0.26	0.09	0.11
WLDA	none	−0.28	-	0.44	0.29	0.13	0.12
	pwe	−0.17	−0.04	0.52	0.33	0.15	0.16
	pwe/fine-tuning	−0.32	−0.07	0.51	0.32	0.15	0.14

Table 4. Topic Coherence on BBCNews.

Model	Resources Method PWE	TrainingCorpus npmi	ExternalCorpus npmi	glove.42B.300d WETC$_c$	glove.42B.300d WETC$_{pw}$	glove.6B.300d WETC$_c$	glove.6B.300d WETC$_{pw}$
GSB	none	−0.36	−0.09	0.44	0.28	0.13	0.13
	pwe	−0.42	−0.11	0.43	0.26	0.14	0.11
	pwe/fine-tuning	−0.35	−0.10	0.41	0.27	0.15	0.12
GSM	none	−0.40	−0.10	0.43	0.27	0.13	0.12
	pwe	−0.41	−0.10	0.44	0.27	0.14	0.12
	pwe/fine-tuning	−0.32	−0.09	0.45	0.29	**0.16**	0.14
NSTM	none	−0.40	−0.10	0.44	0.27	0.14	0.12
	pwe	−0.41	−0.10	0.46	0.27	0.13	0.12
	pwe/fine-tuning	−0.41	−0.09	0.45	0.27	0.14	0.12
NVDM	none	−0.44	−0.12	0.42	0.25	0.13	0.10
	pwe	−0.42	−0.11	0.43	0.26	0.12	0.11
	pwe/fine-tuning	−0.45	−0.17	0.38	0.22	0.12	0.08
NVLDA	none	−0.38	−0.09	0.46	0.28	0.15	0.13
	pwe	−0.39	−0.09	0.45	0.27	0.13	0.12
	pwe/fine-tuning	**−0.18**	**−0.03**	**0.53**	**0.36**	**0.16**	**0.20**
ProdLDA	none	−0.43	−0.13	0.41	0.25	0.12	0.10
	pwe	−0.42	−0.10	0.43	0.26	0.13	0.11
	pwe/fine-tuning	−0.27	−0.06	0.46	0.31	0.15	0.16
RSB	none	−0.40	−0.09	0.44	0.27	0.13	0.12
	pwe	−0.40	−0.09	0.45	0.27	0.13	0.12
	pwe/fine-tuning	−0.43	−0.18	0.34	0.21	0.14	0.09
WLDA	none	−0.41	-	0.44	0.27	0.14	0.12
	pwe	−0.43	−0.11	0.43	0.26	0.13	0.11
	pwe/fine-tuning	−0.39	−0.10	0.43	0.27	0.15	0.12

Table 5. Topic Coherence on Biomedical.

Model	Resources Method PWE	TrainingCorpus npmi	ExternalCorpus npmi	glove.42B.300d WETC$_c$	glove.42B.300d WETC$_{pw}$	glove.6B.300d WETC$_c$	glove.6B.300d WETC$_{pw}$
GSB	none	−0.52	−0.13	0.19	0.11	0.09	0.06
	pwe	−0.46	−0.12	0.22	0.24	0.12	0.19
	pwe/fine-tuning	−0.51	−0.14	0.19	0.23	0.09	0.18
GSM	none	−0.52	−0.12	0.19	0.10	0.09	0.06
	pwe	−0.47	−0.12	0.21	0.24	0.11	0.19
	pwe/fine-tuning	−0.49	−0.13	0.19	0.23	0.10	0.19
NSTM	none	−0.52	−0.13	0.21	0.12	0.10	0.06
	pwe	−0.49	−0.13	0.21	0.12	0.10	0.06
	pwe/fine-tuning	−0.52	−0.13	0.21	0.13	0.10	0.06
NVDM	none	−0.53	−0.12	0.18	0.09	0.08	0.05
	pwe	−0.51	−0.13	0.20	0.12	0.09	0.06
	pwe/fine-tuning	−0.52	−0.14	0.24	0.13	0.10	0.06
NVLDA	none	−0.47	−0.13	0.25	0.16	0.11	0.08
	pwe	−0.49	−0.13	0.21	0.23	0.10	0.18
	pwe/fine-tuning	**−0.15**	**−0.01**	**0.29**	**0.36**	**0.33**	**0.27**
ProdLDA	none	−0.50	−0.11	0.16	0.09	0.09	0.06
	pwe	−0.51	−0.13	0.21	0.23	0.09	0.18
	pwe/fine-tuning	−0.44	−0.12	0.25	0.25	0.14	0.19
RSB	none	−0.49	−0.14	0.19	0.12	0.10	0.07
	pwe	−0.48	−0.13	0.22	0.24	0.11	0.18
	pwe/fine-tuning	−0.47	−0.13	0.18	0.25	0.11	0.20
WLDA	none	−0.51	-	0.25	0.14	0.10	0.07
	pwe	−0.48	−0.13	0.20	0.13	0.11	0.07
	pwe/fine-tuning	−0.55	−0.11	0.15	0.08	0.09	0.06

Table 6. Topic Coherence on DBLP.

Model	Resources Method PWE	TrainingCorpus npmi	ExternalCorpus npmi	glove.42B.300d $WETC_c$	glove.42B.300d $WETC_{pw}$	glove.6B.300d $WETC_c$	glove.6B.300d $WETC_{pw}$
GSB	none	−0.52	−0.14	0.34	0.21	0.11	0.09
	pwe	−0.35	−0.10	0.37	0.25	0.13	0.11
	pwe/fine-tuning	−0.42	−0.13	0.35	0.23	0.12	0.10
GSM	none	−0.52	−0.14	0.33	0.21	0.12	0.09
	pwe	−0.38	−0.11	0.38	0.24	0.13	0.11
	pwe/fine-tuning	−0.43	−0.13	0.36	0.22	0.12	0.10
NSTM	none	−0.33	−0.08	0.37	0.26	0.13	0.13
	pwe	−0.42	−0.13	0.37	0.23	0.11	0.10
	pwe/fine-tuning	−0.31	−0.07	**0.40**	**0.27**	**0.15**	**0.14**
NVDM	none	−0.48	−0.14	0.34	0.22	0.11	0.09
	pwe	−0.47	−0.13	0.34	0.22	0.11	0.09
	pwe/fine-tuning	−0.45	−0.13	0.35	0.23	0.13	0.10
NVLDA	none	−0.41	−0.11	0.37	0.24	0.13	0.11
	pwe	−0.41	−0.12	0.37	0.24	0.13	0.10
	pwe/fine-tuning	**−0.29**	−0.09	0.39	**0.27**	**0.15**	0.13
ProdLDA	none	−0.43	−0.14	0.34	0.21	0.12	0.09
	pwe	−0.41	−0.12	0.37	0.24	0.13	0.11
	pwe/fine-tuning	−0.45	−0.14	0.35	0.21	0.12	0.09
RSB	none	−0.41	−0.13	0.36	0.23	0.13	0.10
	pwe	−0.37	−0.11	0.38	0.24	0.12	0.11
	pwe/fine-tuning	−0.37	−0.13	0.33	0.22	0.13	0.10
WLDA	none	−0.43	-	0.37	0.24	0.13	0.11
	pwe	−0.39	−0.11	0.38	0.25	0.13	0.11
	pwe/fine-tuning	−0.35	**−0.06**	0.38	0.26	0.13	0.12

Table 7. Topic Coherence on GoogleNews.

Model	Resources Method PWE	TrainingCorpus npmi	ExternalCorpus npmi	glove.42B.300d $WETC_c$	glove.42B.300d $WETC_{pw}$	glove.6B.300d $WETC_c$	glove.6B.300d $WETC_{pw}$
GSB	none	−0.52	−0.15	0.30	0.14	0.09	0.03
	pwe	**−0.48**	−0.16	0.27	**0.29**	0.09	**0.25**
	pwe/fine-tuning	−0.51	−0.17	0.27	0.22	0.08	0.15
GSM	none	−0.53	−0.16	0.28	0.13	0.09	0.03
	pwe	**−0.48**	−0.16	0.28	0.27	0.09	0.21
	pwe/fine-tuning	−0.51	−0.15	0.28	0.24	0.08	0.18
NSTM	none	−0.54	−0.16	0.32	0.17	0.10	0.05
	pwe	**−0.48**	−0.16	0.29	0.13	0.08	0.03
	pwe/fine-tuning	−0.53	−0.16	0.31	0.16	0.09	0.04
NVDM	none	−0.53	−0.15	0.25	0.11	0.07	0.02
	pwe	−0.50	−0.16	0.28	0.13	0.08	0.03
	pwe/fine-tuning	−0.51	−0.15	0.29	0.15	0.08	0.04
NVLDA	none	−0.50	−0.15	0.34	0.19	0.10	0.06
	pwe	−0.49	−0.16	0.28	0.24	0.08	0.17
	pwe/fine-tuning	−0.55	−0.14	0.39	0.27	0.12	0.19
ProdLDA	none	−0.51	−0.15	0.28	0.12	0.08	0.03
	pwe	−0.49	−0.16	0.27	0.22	0.08	0.15
	pwe/fine-tuning	−0.53	−0.18	0.29	0.22	0.09	0.14
RSB	none	−0.53	−0.18	0.30	0.15	0.09	0.04
	pwe	**−0.48**	−0.16	0.29	0.25	0.09	0.18
	pwe/fine-tuning	−0.58	−0.18	0.30	0.24	0.10	0.17
WLDA	none	−0.53	-	0.25	0.11	0.08	0.02
	pwe	−0.50	−0.16	0.28	0.13	0.08	0.03
	pwe/fine-tuning	−0.55	**−0.12**	**0.41**	0.24	**0.13**	0.10

Table 8. Topic Coherence on M10.

Model	Resources Method PWE	TrainingCorpus npmi	ExternalCorpus npmi	glove.42B.300d WETC$_c$	glove.42B.300d WETC$_{pw}$	glove.6B.300d WETC$_c$	glove.6B.300d WETC$_{pw}$
GSB	none	−0.54	−0.13	0.37	0.23	0.11	0.10
	pwe	−0.51	−0.11	0.41	0.25	0.13	0.11
	pwe/fine-tuning	−0.54	−0.13	0.38	0.22	0.11	0.10
GSM	none	−0.55	−0.14	0.37	0.22	0.11	0.10
	pwe	−0.51	−0.12	0.39	0.24	0.13	0.11
	pwe/fine-tuning	−0.54	−0.15	0.35	0.20	0.10	0.08
NSTM	none	−0.43	−0.08	0.42	0.27	0.13	0.13
	pwe	−0.52	−0.12	0.42	0.24	0.13	0.11
	pwe/fine-tuning	−0.45	−0.08	0.42	0.27	0.13	0.13
NVDM	none	−0.55	−0.14	0.36	0.22	0.12	0.09
	pwe	−0.54	−0.12	0.37	0.23	0.12	0.10
	pwe/fine-tuning	−0.53	−0.13	0.38	0.24	0.12	0.11
NVLDA	none	−0.51	−0.11	0.39	0.24	0.13	0.11
	pwe	−0.51	−0.12	0.39	0.24	0.12	0.11
	pwe/fine-tuning	**−0.41**	**−0.03**	**0.45**	**0.31**	**0.17**	**0.16**
ProdLDA	none	−0.52	−0.14	0.37	0.22	0.12	0.09
	pwe	−0.53	−0.12	0.38	0.24	0.12	0.10
	pwe/fine-tuning	−0.52	−0.13	0.35	0.22	0.11	0.10
RSB	none	−0.52	−0.13	0.37	0.23	0.11	0.10
	pwe	−0.50	−0.10	0.40	0.25	0.14	0.12
	pwe/fine-tuning	−0.53	−0.14	0.36	0.22	0.12	0.10
WLDA	none	−0.55	-	0.36	0.21	0.11	0.09
	pwe	−0.55	−0.13	0.38	0.23	0.12	0.10
	pwe/fine-tuning	−0.54	−0.11	0.40	0.25	0.13	0.11

Table 9. Topic Coherence on PascalFlicker.

Model	Resources Method PWE	TrainingCorpus npmi	ExternalCorpus npmi	glove.42B.300d WETC$_c$	glove.42B.300d WETC$_{pw}$	glove.6B.300d WETC$_c$	glove.6B.300d WETC$_{pw}$
GSB	none	−0.51	−0.10	0.26	0.16	0.08	0.06
	pwe	−0.46	−0.10	0.27	0.24	0.08	0.17
	pwe/fine-tuning	−0.49	−0.09	0.27	0.23	0.10	0.15
GSM	none	−0.52	−0.10	0.26	0.16	0.08	0.06
	pwe	−0.46	−0.10	0.26	0.24	0.09	0.17
	pwe/fine-tuning	−0.47	−0.09	0.26	0.21	0.09	0.14
NSTM	none	−0.52	−0.09	0.24	0.15	0.08	0.05
	pwe	−0.48	−0.09	0.24	0.15	0.09	0.05
	pwe/fine-tuning	−0.51	−0.09	0.25	0.15	0.08	0.05
NVDM	none	−0.52	−0.09	0.25	0.14	0.09	0.05
	pwe	−0.51	−0.09	0.25	0.14	0.08	0.05
	pwe/fine-tuning	−0.50	−0.10	0.27	0.15	0.08	0.06
NVLDA	none	−0.48	−0.10	0.28	0.18	0.09	0.07
	pwe	−0.45	−0.10	0.27	0.23	0.09	0.17
	pwe/fine-tuning	**−0.40**	**−0.07**	**0.45**	**0.34**	**0.16**	**0.25**
ProdLDA	none	−0.48	−0.09	0.25	0.14	0.08	0.05
	pwe	−0.49	−0.10	0.26	0.22	0.08	0.15
	pwe/fine-tuning	**−0.40**	**−0.07**	0.39	0.30	0.12	0.21
RSB	none	−0.48	−0.09	0.31	0.20	0.09	0.08
	pwe	−0.45	−0.10	0.26	0.24	0.08	0.18
	pwe/fine-tuning	−0.48	−0.10	0.32	0.27	0.11	0.19
WLDA	none	−0.49	-	0.26	0.15	0.08	0.05
	pwe	−0.52	−0.09	0.26	0.14	0.08	0.05
	pwe/fine-tuning	−0.54	−0.11	0.30	0.19	0.10	0.08

Table 10. Topic Coherence on SearchSnippets.

Model	Resources Method PWE	TrainingCorpus npmi	ExternalCorpus npmi	glove.42B.300d WETC$_c$	WETC$_{pw}$	glove.6B.300d WETC$_c$	WETC$_{pw}$
GSB	none	−0.49	−0.15	0.27	0.14	0.09	0.04
	pwe	−0.46	−0.14	0.32	0.28	0.12	0.20
	pwe/fine-tuning	−0.49	−0.17	0.26	0.22	0.10	0.14
GSM	none	−0.50	−0.16	0.25	0.13	0.09	0.03
	pwe	−0.46	−0.14	0.31	0.27	0.12	0.18
	pwe/fine-tuning	−0.48	−0.16	0.27	0.21	0.10	0.13
NSTM	none	−0.49	−0.13	0.32	0.19	0.11	0.06
	pwe	−0.47	−0.14	0.30	0.18	0.11	0.06
	pwe/fine-tuning	−0.49	−0.13	0.31	0.18	0.10	0.06
NVDM	none	−0.50	−0.15	0.26	0.13	0.09	0.03
	pwe	−0.48	−0.14	0.31	0.17	0.10	0.05
	pwe/fine-tuning	−0.48	−0.13	0.32	0.20	0.11	0.06
NVLDA	none	−0.46	−0.12	0.36	0.22	0.13	0.09
	pwe	−0.45	−0.13	0.30	0.26	0.11	0.17
	pwe/fine-tuning	**−0.26**	**−0.06**	**0.48**	**0.36**	**0.18**	**0.24**
ProdLDA	none	−0.48	−0.15	0.28	0.16	0.11	0.05
	pwe	−0.48	−0.14	0.31	0.24	0.11	0.14
	pwe/fine-tuning	−0.42	−0.14	0.30	0.24	0.14	0.15
RSB	none	−0.47	−0.15	0.27	0.15	0.10	0.05
	pwe	−0.46	−0.13	0.31	0.28	0.12	0.19
	pwe/fine-tuning	−0.48	−0.14	0.24	0.20	0.09	0.13
WLDA	none	−0.49	-	0.29	0.17	0.11	0.05
	pwe	−0.47	−0.13	0.32	0.18	0.11	0.06
	pwe/fine-tuning	−0.31	−0.07	0.44	0.33	0.17	0.17

Table 11. Topic Coherence on StackOverflow.

Model	Resources Method PWE	TrainingCorpus npmi	ExternalCorpus npmi	glove.42B.300d WETC$_c$	WETC$_{pw}$	glove.6B.300d WETC$_c$	WETC$_{pw}$
GSB	none	−0.53	−0.10	0.31	0.22	0.15	0.08
	pwe	−0.48	−0.10	0.30	0.31	0.17	0.21
	pwe/fine-tuning	−0.50	−0.12	0.27	0.29	0.14	0.17
GSM	none	−0.53	−0.10	0.30	0.21	0.14	0.08
	pwe	−0.48	−0.11	0.30	0.31	0.16	0.21
	pwe/fine-tuning	−0.49	−0.10	0.24	0.26	0.14	0.15
NSTM	none	−0.48	−0.10	0.28	0.23	0.16	0.09
	pwe	−0.50	−0.10	0.31	0.23	0.14	0.09
	pwe/fine-tuning	−0.47	−0.10	0.30	0.24	0.15	0.09
NVDM	none	−0.55	−0.10	0.25	0.17	0.12	0.07
	pwe	−0.52	−0.10	0.30	0.22	0.14	0.08
	pwe/fine-tuning	−0.53	−0.09	0.30	0.22	0.14	0.08
NVLDA	none	−0.48	−0.09	0.32	0.25	0.16	0.10
	pwe	−0.50	−0.10	0.29	0.30	0.15	0.20
	pwe/fine-tuning	**−0.26**	**−0.05**	**0.37**	**0.39**	**0.25**	**0.24**
ProdLDA	none	−0.51	−0.10	0.25	0.19	0.12	0.07
	pwe	−0.52	−0.10	0.29	0.28	0.15	0.17
	pwe/fine-tuning	−0.44	−0.12	0.26	0.29	0.17	0.16
RSB	none	−0.49	−0.11	0.26	0.22	0.17	0.08
	pwe	−0.48	−0.09	0.30	0.31	0.15	0.21
	pwe/fine-tuning	−0.48	−0.13	0.19	0.28	0.17	0.16
WLDA	none	−0.53	-	0.31	0.23	0.15	0.09
	pwe	−0.53	−0.10	0.29	0.22	0.15	0.08
	pwe/fine-tuning	−0.51	−0.09	0.34	0.28	0.18	0.11

Table 12. Topic Coherence on TrecTweet.

Model	Resources Method PWE	TrainingCorpus npmi	ExternalCorpus npmi	glove.42B.300d WETC$_c$	glove.42B.300d WETC$_{pw}$	glove.6B.300d WETC$_c$	glove.6B.300d WETC$_{pw}$
GSB	none	−0.53	−0.14	0.29	0.14	0.08	0.04
	pwe	**−0.50**	−0.14	0.29	0.24	0.09	**0.17**
	pwe/fine-tuning	−0.54	−0.15	0.29	0.23	0.08	0.15
GSM	none	−0.53	−0.14	0.28	0.14	0.08	0.03
	pwe	**−0.50**	−0.14	0.27	0.24	0.08	**0.17**
	pwe/fine-tuning	−0.53	−0.14	0.28	0.22	0.09	0.14
NSTM	none	−0.53	−0.13	0.29	0.15	0.09	0.04
	pwe	−0.51	−0.14	0.29	0.14	0.08	0.03
	pwe/fine-tuning	−0.53	−0.14	0.29	0.15	0.09	0.04
NVDM	none	−0.51	−0.15	0.28	0.14	0.08	0.04
	pwe	−0.51	−0.14	0.28	0.14	0.08	0.03
	pwe/fine-tuning	−0.52	−0.14	0.29	0.15	0.09	0.04
NVLDA	none	−0.51	−0.14	0.32	0.18	0.09	0.05
	pwe	**−0.50**	−0.14	0.29	0.24	0.09	**0.17**
	pwe/fine-tuning	−0.55	**−0.12**	**0.35**	**0.27**	**0.11**	**0.17**
ProdLDA	none	−0.53	−0.14	0.28	0.13	0.08	0.03
	pwe	−0.51	−0.14	0.29	0.21	0.09	0.13
	pwe/fine-tuning	−0.55	−0.14	0.32	0.25	0.10	0.16
RSB	none	−0.53	−0.15	0.30	0.16	0.09	0.04
	pwe	**−0.50**	−0.13	0.30	0.25	0.09	**0.17**
	pwe/fine-tuning	−0.57	−0.15	0.28	**0.27**	0.10	**0.17**
WLDA	none	**−0.50**	-	0.29	0.14	0.09	0.04
	pwe	−0.51	−0.14	0.27	0.14	0.08	0.03
	pwe/fine-tuning	−0.54	−0.15	0.34	0.18	0.10	0.06

4.3. Results for Topic Diversity

Tables 13–22 represent the detailed results of different metrics (TopicDiversity, Inverted RBO, and MSCD) expressing a diversity of topics for different neural-topic models and different datasets, respectively. Values in bold indicate the best results. InvertedRBO focuses on the weight of the top-N words. It shows the highest values in almost all of the cases, from which it can be inferred that we were able to construct the topics with high diversity. This result shows that it was useful to add a regularization term that maximizes the distance between topic-centroid vectors, resulting in highly diverse topics.

Furthermore, WLDA and NSTM show similar results without this regularization term, indicating that these models are able to learn without compromising topic diversity in their raw form. To check if this regularization term is working well, we have added TopicCentroidDistance (TCD) in the tables. The larger this metric is, the better, but the values are almost the same for all cases. This metric was evaluated based on two PWEs, and since the values varied, we can infer that the quality of the embedding has a significant impact on the evaluation of the topic model.

Although the results of TopicDiversity varied greatly depending on model and dataset, when checked individually, the scores were sufficiently better in many cases. However, as in the case of Biomedical's NVLDA-pwe/fine-tuning results, there were cases where the TC showed good scores but the TD showed bad scores. In this respect, InvertedRBO also shows a good score, but MSCD, which is an evaluation using the entire topic–word distribution, shows a relatively large value (i.e., a bad score), indicating that the topics are relatively tangled. Metrics such as TopicDiversity and InvertedRBO, which are based on the top-N words, are useful for evaluating topic diversity, but it is also important to evaluate the entire topic–word distribution.

Figure 5 presents the summary of topic diversity results over all the neural-topic models for the long-text corpus (2 datasets) and the short-text corpus (8 datasets), which shows the overall trend. Among the metrics related to TD, the InvertedRBO score is almost the highest in all cases. This indicates that there is sufficient diversity in all conditions. However, for the other scores, the performance is slightly worse for PWE and PWE/fine-tuning.

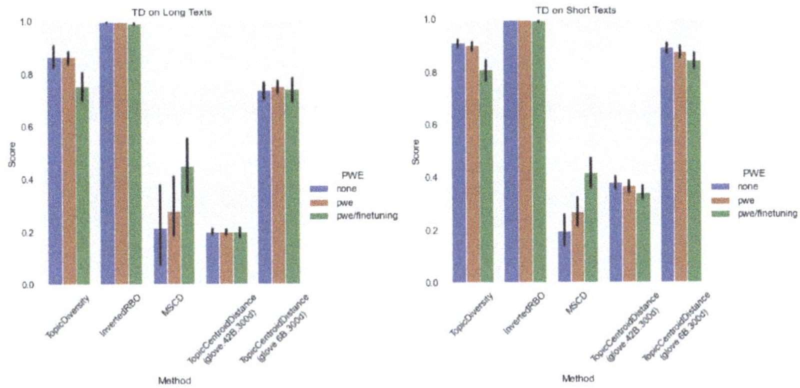

TD on Long Texts　　　TD on Short Texts

Figure 5. Summary of TD Results.

Table 13. Topic Diversity on 20NewsGroups.

Model	PWE	Topic Diversity	Inverted RBO	MSCD	TCD (42B)	TCD (6B)
GSB	none	0.78	0.99	0.08	0.17	0.60
	pwe	0.79	0.99	0.20	−0.18	0.67
	pwe/fine-tuning	0.68	0.98	0.31	0.15	0.52
GSM	none	0.84	**1.00**	0.09	0.17	0.66
	pwe	0.85	**1.00**	0.21	0.18	0.71
	pwe/fine-tuning	0.72	0.99	0.39	0.19	0.77
NSTM	none	0.83	0.99	0.93	0.17	0.74
	pwe	0.85	**1.00**	0.91	0.17	0.68
	pwe/fine-tuning	0.82	0.99	0.91	0.17	0.75
NVDM	none	0.70	0.99	0.18	0.19	0.73
	pwe	0.85	**1.00**	0.21	0.18	0.71
	pwe/fine-tuning	0.83	**1.00**	0.55	0.20	0.84
NVLDA	none	**0.91**	**1.00**	0.07	0.17	0.70
	pwe	0.82	0.99	0.19	0.17	0.83
	pwe/fine-tuning	0.64	0.99	0.35	0.15	0.65
ProdLDA	none	0.77	0.99	0.11	0.18	0.68
	pwe	0.84	**1.00**	0.22	0.18	0.73
	pwe/fine-tuning	0.64	0.99	0.39	0.20	0.84
RSB	none	0.88	**1.00**	0.07	0.18	0.65
	pwe	0.78	0.99	0.21	0.17	0.69
	pwe/fine-tuning	0.62	0.99	0.44	**0.21**	**0.94**
WLDA	none	0.78	0.99	0.55	0.20	0.77
	0pwe	0.83	**1.00**	0.26	0.17	0.72
	pwe/fine-tuning	0.56	0.98	**0.00**	0.18	0.73

Table 14. Topic Diversity on BBCNews.

Model	PWE	Topic Diversity	Inverted RBO	MSCD	TCD (42B)	TCD (6B)
GSB	none	0.85	0.99	0.07	0.21	0.80
	pwe	0.90	**1.00**	0.15	0.22	0.75
	pwe/fine-tuning	0.87	**1.00**	0.25	0.20	0.71
GSM	none	0.92	**1.00**	0.07	0.22	0.79
	pwe	0.91	**1.00**	0.15	0.22	0.80
	pwe/fine-tuning	0.84	**1.00**	0.30	0.19	0.68
NSTM	none	**1.00**	**1.00**	0.97	0.21	0.74
	pwe	0.90	**1.00**	0.86	0.22	0.80
	pwe/fine-tuning	0.95	**1.00**	0.76	0.21	0.74
NVDM	none	0.92	**1.00**	0.05	0.23	0.80
	pwe	0.92	**1.00**	0.18	0.23	0.78
	pwe/fine-tuning	0.88	**1.00**	0.65	**0.27**	0.81
NVLDA	none	0.97	**1.00**	0.06	0.21	0.74
	pwe	0.91	**1.00**	0.17	0.22	0.78
	pwe/fine-tuning	0.70	0.99	0.43	0.15	0.69
ProdLDA	none	0.81	0.99	0.05	0.24	**0.85**
	pwe	0.91	**1.00**	0.16	0.23	0.75
	pwe/fine-tuning	0.69	0.99	0.39	0.18	0.71
RSB	none	0.95	**1.00**	0.06	0.21	0.78
	pwe	0.88	**1.00**	0.19	0.21	0.80
	pwe/fine-tuning	0.73	0.99	0.40	**0.27**	0.72
WLDA	none	0.94	**1.00**	**0.04**	0.22	0.75
	pwe	0.90	**1.00**	0.18	0.22	0.79
	pwe/fine-tuning	0.85	**1.00**	0.65	0.21	0.71

Table 15. Topic Diversity on Biomedical.

Model	PWE	Topic Diversity	Inverted RBO	MSCD	TCD (42B)	TCD (6B)
GSB	none	0.96	**1.00**	0.08	0.56	0.93
	pwe	0.92	**1.00**	0.19	0.49	0.85
	pwe/fine-tuning	0.95	**1.00**	0.35	0.54	0.89
GSM	none	0.95	**1.00**	0.08	0.55	0.93
	pwe	0.93	**1.00**	0.17	0.50	0.86
	pwe/fine-tuning	0.91	**1.00**	0.30	0.53	0.89
NSTM	none	**1.00**	**1.00**	0.79	0.50	0.90
	pwe	0.95	**1.00**	0.84	0.49	0.86
	pwe/fine-tuning	**1.00**	**1.00**	0.80	0.53	0.91
NVDM	none	0.85	**1.00**	0.19	0.59	**0.97**
	pwe	0.93	**1.00**	0.11	0.51	0.94
	pwe/fine-tuning	0.95	**1.00**	0.17	0.46	0.90
NVLDA	none	0.97	**1.00**	0.06	0.45	0.85
	pwe	0.92	**1.00**	0.13	0.54	0.89
	pwe/fine-tuning	0.35	0.96	0.49	0.26	0.37
ProdLDA	none	0.78	0.99	0.12	0.67	0.92
	pwe	0.94	**1.00**	0.14	0.49	0.90
	pwe/fine-tuning	0.85	0.99	0.43	0.39	0.78
RSB	none	0.91	**1.00**	0.08	0.56	0.89
	pwe	0.88	**1.00**	0.19	0.49	0.85
	pwe/fine-tuning	0.62	0.98	0.44	0.62	0.87
WLDA	none	0.91	**1.00**	0.26	0.45	0.88
	pwe	0.92	**1.00**	0.22	0.52	0.87
	pwe/fine-tuning	0.73	0.99	**0.00**	**0.71**	**0.97**

Table 16. Topic Diversity on DBLP.

Model	PWE	Topic Diversity	Inverted RBO	MSCD	TCD (42B)	TCD (6B)
GSB	none	0.86	**1.00**	0.09	0.27	0.88
	pwe	0.83	**1.00**	0.29	0.24	0.81
	pwe/fine-tuning	0.89	**1.00**	0.27	0.27	0.83
GSM	none	0.86	**1.00**	0.09	0.28	0.86
	pwe	0.86	**1.00**	0.28	0.25	0.85
	pwe/fine-tuning	0.80	0.99	0.18	0.28	0.85
NSTM	none	0.71	0.98	0.76	0.27	**0.89**
	pwe	0.87	**1.00**	0.91	0.26	**0.89**
	pwe/fine-tuning	0.70	0.98	0.77	0.26	0.82
NVDM	none	0.88	**1.00**	0.58	0.27	0.87
	pwe	0.87	**1.00**	0.21	0.26	0.85
	pwe/fine-tuning	**0.90**	**1.00**	0.12	0.27	0.85
NVLDA	none	**0.90**	**1.00**	0.06	0.25	0.81
	pwe	0.87	**1.00**	0.29	0.25	0.80
	pwe/fine-tuning	0.74	0.99	0.11	0.22	0.72
ProdLDA	none	0.82	0.99	0.12	**0.31**	0.85
	pwe	0.87	**1.00**	0.26	0.25	0.79
	pwe/fine-tuning	0.82	**1.00**	0.10	0.29	0.85
RSB	none	0.86	**1.00**	0.18	0.27	0.81
	pwe	0.65	0.98	0.34	0.25	0.84
	pwe/fine-tuning	0.68	0.99	0.40	0.27	0.81
WLDA	none	0.85	**1.00**	0.38	0.25	0.79
	pwe	0.84	**1.00**	0.26	0.24	0.81
	pwe/fine-tuning	0.78	0.99	**0.00**	0.24	0.80

Table 17. Topic Diversity on GoogleNews.

Model	PWE	Topic Diversity	Inverted RBO	MSCD	TCD (42B)	TCD (6B)
GSB	none	0.98	**1.00**	0.09	0.38	0.96
	pwe	0.96	**1.00**	0.13	0.43	0.95
	pwe/fine-tuning	0.96	**1.00**	0.34	0.45	0.98
GSM	none	0.98	**1.00**	0.09	0.41	0.97
	pwe	0.97	**1.00**	0.12	0.42	0.97
	pwe/fine-tuning	0.91	**1.00**	0.43	0.41	0.96
NSTM	none	**1.00**	**1.00**	0.84	0.34	0.93
	pwe	0.99	**1.00**	0.69	0.40	0.98
	pwe/fine-tuning	**1.00**	**1.00**	0.81	0.36	0.94
NVDM	none	0.96	**1.00**	0.07	**0.50**	**0.99**
	pwe	0.97	**1.00**	**0.06**	0.43	**0.99**
	pwe/fine-tuning	0.96	**1.00**	0.10	0.39	0.95
NVLDA	none	0.99	**1.00**	**0.06**	0.32	0.90
	pwe	0.97	**1.00**	0.10	0.44	0.97
	pwe/fine-tuning	0.72	0.99	0.45	0.27	0.86
ProdLDA	none	0.90	**1.00**	0.11	0.43	0.97
	pwe	0.97	**1.00**	0.09	0.45	0.98
	pwe/fine-tuning	0.91	**1.00**	0.49	0.38	0.97
RSB	none	0.89	**1.00**	0.14	0.39	0.93
	pwe	0.87	**1.00**	0.20	0.43	0.96
	pwe/fine-tuning	0.71	0.99	0.40	0.37	0.92
WLDA	none	0.93	**1.00**	0.21	0.47	0.98
	pwe	0.92	**1.00**	0.11	0.41	0.97
	pwe/fine-tuning	0.82	0.99	0.24	0.25	0.82

Table 18. Topic Diversity on M10.

Model	PWE	Topic Diversity	Inverted RBO	MSCD	TCD (42B)	TCD (6B)
GSB	none	**0.91**	**1.00**	0.08	0.26	0.84
	pwe	0.86	**1.00**	0.22	0.23	0.81
	pwe/fine-tuning	0.89	**1.00**	0.28	0.25	0.88
GSM	none	0.89	**1.00**	0.08	0.27	0.85
	pwe	0.86	**1.00**	0.20	0.25	0.83
	pwe/fine-tuning	0.77	0.99	0.32	**0.29**	**0.90**
NSTM	none	0.75	0.98	0.77	0.23	0.87
	pwe	0.84	0.99	0.90	0.24	0.82
	pwe/fine-tuning	0.82	0.99	0.75	0.24	0.86
NVDM	none	0.85	**1.00**	**0.04**	0.28	0.86
	pwe	0.87	**1.00**	0.21	0.26	0.84
	pwe/fine-tuning	0.89	**1.00**	0.32	0.25	0.86
NVLDA	none	**0.91**	**1.00**	0.07	0.25	0.83
	pwe	0.87	**1.00**	0.22	0.24	0.83
	pwe/fine-tuning	0.64	0.99	0.20	0.19	0.68
ProdLDA	none	0.79	0.99	0.05	0.28	0.88
	pwe	0.87	**1.00**	0.19	0.26	0.84
	pwe/fine-tuning	0.81	0.99	0.20	0.27	0.87
RSB	none	0.87	**1.00**	0.15	0.25	0.83
	pwe	0.67	0.98	0.24	0.24	0.80
	pwe/fine-tuning	0.69	0.99	0.41	0.26	0.82
WLDA	none	0.87	**1.00**	0.16	**0.29**	**0.90**
	pwe	0.79	0.99	0.33	0.26	0.82
	pwe/fine-tuning	0.85	**1.00**	0.44	0.23	0.80

Table 19. Topic Diversity on PascalFlicker.

Model	PWE	Topic Diversity	Inverted RBO	MSCD	TCD (42B)	TCD (6B)
GSB	none	0.95	**1.00**	0.07	0.41	0.98
	pwe	0.91	**1.00**	0.17	0.40	0.97
	pwe/fine-tuning	0.93	**1.00**	0.55	0.38	0.95
GSM	none	0.94	**1.00**	0.07	0.42	0.97
	pwe	0.93	**1.00**	0.17	0.42	0.96
	pwe/fine-tuning	0.89	**1.00**	0.61	0.39	0.97
NSTM	none	**1.00**	**1.00**	0.79	**0.48**	**0.99**
	pwe	0.95	**1.00**	0.85	0.44	0.95
	pwe/fine-tuning	**1.00**	**1.00**	0.78	0.45	0.98
NVDM	none	0.93	**1.00**	0.05	0.46	0.98
	pwe	0.93	**1.00**	0.17	0.44	**0.99**
	pwe/fine-tuning	0.92	**1.00**	0.24	0.40	0.98
NVLDA	none	0.95	**1.00**	0.06	0.38	0.96
	pwe	0.91	**1.00**	0.18	0.40	0.97
	pwe/fine-tuning	0.49	0.98	0.75	0.18	0.71
ProdLDA	none	0.77	0.99	0.07	0.47	0.98
	pwe	0.93	**1.00**	0.16	0.43	0.98
	pwe/fine-tuning	0.59	0.97	0.82	0.22	0.84
RSB	none	0.91	**1.00**	0.09	0.32	0.94
	pwe	0.86	**1.00**	0.18	0.42	0.98
	pwe/fine-tuning	0.62	0.99	0.53	0.29	0.90
WLDA	none	0.94	**1.00**	**0.03**	0.43	0.98
	pwe	0.90	**1.00**	0.17	0.43	**0.99**
	pwe/fine-tuning	0.83	0.99	0.40	0.34	0.95

Table 20. Topic Diversity on SearchSnippets.

Model	PWE	Topic Diversity	Inverted RBO	MSCD	TCD (42B)	TCD (6B)
GSB	none	0.97	**1.00**	0.07	0.41	0.94
	pwe	0.94	**1.00**	0.20	0.32	0.81
	pwe/fine-tuning	0.96	**1.00**	0.34	0.42	0.91
GSM	none	0.95	**1.00**	0.07	0.45	0.93
	pwe	0.94	**1.00**	0.16	0.34	0.83
	pwe/fine-tuning	0.93	**1.00**	0.33	0.41	0.92
NSTM	none	**1.00**	**1.00**	0.86	0.32	0.87
	pwe	0.96	**1.00**	0.81	0.34	0.85
	pwe/fine-tuning	**1.00**	**1.00**	0.83	0.33	0.89
NVDM	none	0.90	**1.00**	0.07	0.45	**0.95**
	pwe	0.94	**1.00**	0.10	0.34	0.87
	pwe/fine-tuning	0.96	**1.00**	0.14	0.32	0.86
NVLDA	none	0.98	**1.00**	**0.06**	0.28	0.78
	pwe	0.95	**1.00**	0.13	0.33	0.85
	pwe/fine-tuning	0.51	0.97	0.64	0.17	0.68
ProdLDA	none	0.88	**1.00**	0.07	0.37	0.87
	pwe	0.95	**1.00**	0.12	0.34	0.89
	pwe/fine-tuning	0.78	0.99	0.56	0.32	0.82
RSB	none	0.92	**1.00**	0.08	0.43	0.91
	pwe	0.86	0.99	0.23	0.33	0.82
	pwe/fine-tuning	0.78	0.99	0.39	**0.48**	0.94
WLDA	none	0.92	**1.00**	0.25	0.36	0.89
	pwe	0.93	**1.00**	0.15	0.34	0.86
	pwe/fine-tuning	0.34	0.96	0.72	0.18	0.64

Table 21. Topic Diversity on StackOverflow.

Model	PWE	Topic Diversity	Inverted RBO	MSCD	TCD (42B)	TCD (6B)
GSB	none	0.94	**1.00**	0.09	0.29	0.70
	pwe	0.89	**1.00**	0.19	0.28	0.69
	pwe/fine-tuning	0.93	**1.00**	0.34	0.34	0.71
GSM	none	0.93	**1.00**	0.09	0.32	0.74
	pwe	0.90	**1.00**	0.20	0.34	0.68
	pwe/fine-tuning	0.79	0.99	0.38	0.37	0.69
NSTM	none	0.95	**1.00**	0.77	0.32	0.71
	pwe	0.90	**1.00**	0.89	0.28	0.74
	pwe/fine-tuning	**0.96**	**1.00**	0.78	0.32	0.71
NVDM	none	0.82	0.99	0.12	0.43	**0.82**
	pwe	0.89	**1.00**	0.18	0.34	0.72
	pwe/fine-tuning	0.92	**1.00**	0.25	0.32	0.75
NVLDA	none	0.94	**1.00**	**0.06**	0.29	0.69
	pwe	0.88	**1.00**	0.20	0.34	0.69
	pwe/fine-tuning	0.46	0.97	0.23	0.22	0.47
ProdLDA	none	0.75	0.99	**0.06**	0.38	0.79
	pwe	0.90	**1.00**	0.19	0.30	0.70
	pwe/fine-tuning	0.77	0.99	0.32	0.34	0.63
RSB	none	0.88	**1.00**	0.13	0.37	0.64
	pwe	0.71	0.99	0.25	0.33	0.68
	pwe/fine-tuning	0.65	0.99	0.39	**0.49**	0.70
WLDA	none	0.88	**1.00**	0.40	0.29	0.71
	pwe	0.84	**1.00**	0.30	0.34	0.71
	pwe/fine-tuning	0.83	**1.00**	0.16	0.24	0.59

Table 22. Topic Diversity on TrecTweet.

Model	PWE	Topic Diversity	Inverted RBO	MSCD	TCD (42B)	TCD (6B)
GSB	none	0.97	**1.00**	0.06	0.41	0.97
	pwe	0.95	**1.00**	0.15	0.42	0.96
	pwe/fine-tuning	0.94	**1.00**	0.43	0.36	0.96
GSM	none	0.97	**1.00**	0.06	0.42	0.96
	pwe	0.96	**1.00**	0.14	0.43	0.96
	pwe/fine-tuning	0.95	**1.00**	0.53	0.40	0.96
NSTM	none	**1.00**	**1.00**	0.55	0.42	0.96
	pwe	0.97	**1.00**	0.79	0.41	0.97
	pwe/fine-tuning	**1.00**	**1.00**	0.77	0.43	0.96
NVDM	none	0.96	**1.00**	**0.05**	0.41	0.96
	pwe	0.95	**1.00**	0.14	**0.44**	**0.98**
	pwe/fine-tuning	0.97	**1.00**	0.25	0.39	0.96
NVLDA	none	0.98	**1.00**	0.06	0.36	0.94
	pwe	0.95	**1.00**	0.14	0.40	0.96
	pwe/fine-tuning	0.83	0.99	0.41	0.26	0.87
ProdLDA	none	0.92	**1.00**	**0.05**	0.42	0.97
	pwe	0.96	**1.00**	0.14	0.43	0.96
	pwe/fine-tuning	0.76	0.99	0.64	0.31	0.92
RSB	none	0.92	**1.00**	0.08	0.36	0.95
	pwe	0.88	**1.00**	0.19	0.39	0.95
	pwe/fine-tuning	0.52	0.98	0.57	0.35	0.96
WLDA	none	0.96	**1.00**	**0.05**	0.40	0.96
	pwe	0.94	**1.00**	0.12	**0.44**	0.97
	pwe/fine-tuning	0.87	**1.00**	0.44	0.33	0.94

4.4. Classification and Clustering Performance

Tables 23–32 represent the classification and clustering performance of all models and all datasets, respectively. Values in the bold face represent best results. For the TrecTweet dataset, the classification results could not be obtained, possibly due to some technical problem. Figure 6 presents the average classification and clustering performance of the models over long- and short-text datasets. Classification has been performed by a SVM (Support Vector Machine) with linear and rbf kernels. Classification accuracy, precision, recall, and F1 scores have been used for performance assessment and for supervised classification, and NMI (Normalized Mutual Information) and Purity have been used for unsupervised classification.

For classification, VAE-based models, such as NVDM and GSM, exhibit good performance, while WAE-based models, such as WLDA and NSTM, show relatively poor performance. NSTM shows good performance in TC and TD, especially in TD, without adding any regularization term. However, the application to downstream tasks using WAE variants remains a challenge. Considering the overall trend, for long texts, PWE with fine-tuning improves all the scores, but for short texts, the performance is the best for the cases without embedding. Although, after fine-tuning, the scores got better than those obtained with pretrained embedding only.

For clustering results, the large NMI and Purity scores for all models and all datasets for both long and short texts indicate that there is a concentration of documents with the same label around the topic-centroid vector, which proves that the proposal of PWE/fine-tuning improves topic cohesion. Therefore, we can see that our proposal of PWE/fine-tuning contributes to narrowing the domain gap between the training corpus and PWE.

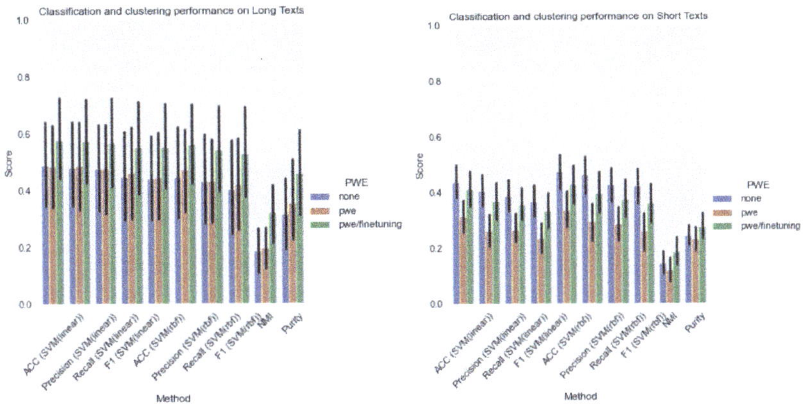

For Long Texts For Short Texts

Figure 6. Classification and clustering performance.

Table 23. Classification and Clustering performance on 20NewsGroups.

Model	Classifier Method PWE	SVM (Linear) ACC	Precision	Recall	F1	SVM (rbm) ACC	Precision	Recall	F1	- NMI	Purity
GSB	none	0.34	0.32	0.32	0.29	0.30	0.29	0.28	0.27	0.18	0.14
	pwe	0.34	0.33	0.33	0.31	0.34	0.34	0.32	0.31	0.26	0.25
	pwe/fine-tuning	0.48	0.46	0.47	0.45	0.49	0.48	0.48	0.47	0.33	**0.33**
GSM	none	0.30	0.28	0.28	0.25	0.25	0.23	0.24	0.21	0.20	0.14
	pwe	0.31	0.28	0.30	0.27	0.30	0.29	0.28	0.27	0.24	0.21
	pwe/fine-tuning	0.45	0.42	0.44	0.41	0.45	0.43	0.43	0.42	**0.35**	**0.33**
NSTM	none	0.22	0.23	0.21	0.18	0.07	0.15	0.07	0.03	0.02	0.07
	pwe	0.14	0.18	0.13	0.11	0.08	0.29	0.08	0.06	0.04	0.08
	pwe/fine-tuning	0.19	0.19	0.18	0.15	0.07	0.05	0.07	0.02	0.01	0.06
NVDM	none	0.36	0.34	0.35	0.32	0.23	0.22	0.22	0.20	0.02	0.07
	pwe	0.38	0.36	0.37	0.34	0.33	0.33	0.32	0.32	0.11	0.13
	pwe/fine-tuning	**0.54**	**0.53**	**0.52**	**0.50**	**0.55**	**0.53**	**0.53**	**0.53**	0.21	0.20
NVLDA	none	0.31	0.30	0.30	0.27	0.26	0.29	0.25	0.25	0.06	0.11
	pwe	0.20	0.20	0.19	0.15	0.12	0.15	0.11	0.08	0.07	0.09
	pwe/fine-tuning	0.36	0.34	0.35	0.31	0.27	0.34	0.26	0.23	0.33	0.29
ProdLDA	none	0.18	0.15	0.17	0.12	0.11	0.08	0.10	0.06	0.05	0.07
	pwe	0.18	0.17	0.17	0.13	0.10	0.14	0.10	0.06	0.05	0.08
	pwe/fine-tuning	0.36	0.34	0.34	0.30	0.27	0.33	0.26	0.23	0.32	0.27
RSB	none	0.15	0.09	0.14	0.07	0.14	0.09	0.13	0.06	0.18	0.13
	pwe	0.11	0.05	0.10	0.05	0.10	0.03	0.09	0.03	0.08	0.09
	pwe/fine-tuning	0.30	0.26	0.29	0.24	0.30	0.29	0.29	0.25	0.21	0.15
WLDA	none	0.07	0.10	0.07	0.03	0.07	0.13	0.07	0.03	0.00	0.06
	pwe	0.07	0.11	0.07	0.04	0.07	0.12	0.07	0.04	0.00	0.06
	pwe/fine-tuning	0.07	0.10	0.07	0.04	0.07	0.10	0.07	0.04	0.00	0.06

Table 24. Classification and Clustering performance on BBCNews.

Model	Classifier Method PWE	SVM (Linear) ACC	Precision	Recall	F1	SVM (rbm) ACC	Precision	Recall	F1	- NMI	Purity
GSB	none	0.94	0.94	0.94	0.94	**0.95**	**0.95**	0.94	0.94	0.49	0.79
	pwe	0.89	0.89	0.89	0.89	0.89	0.89	0.88	0.88	0.42	0.81
	pwe/fine-tuning	0.92	0.92	0.92	0.92	0.91	0.92	0.91	0.91	0.51	0.86
GSM	none	0.94	0.94	0.94	0.94	0.94	0.94	0.94	0.94	0.51	0.78
	pwe	0.89	0.89	0.88	0.88	0.88	0.88	0.88	0.88	0.43	0.81
	pwe/fine-tuning	0.94	0.94	0.93	0.93	0.92	0.92	0.92	0.92	0.56	0.91
NSTM	none	0.57	0.69	0.56	0.57	0.52	0.66	0.51	0.51	0.16	0.51
	pwe	0.63	0.69	0.62	0.63	0.51	0.65	0.50	0.50	0.17	0.46
	pwe/fine-tuning	0.60	0.63	0.58	0.58	0.52	0.60	0.50	0.49	0.16	0.42
NVDM	none	0.92	0.92	0.91	0.91	0.83	0.83	0.83	0.83	0.21	0.41
	pwe	0.92	0.93	0.92	0.92	0.88	0.88	0.88	0.88	0.32	0.57
	pwe/fine-tuning	**0.95**	**0.95**	**0.95**	**0.95**	**0.95**	**0.95**	**0.95**	**0.95**	0.44	0.78
NVLDA	none	0.81	0.81	0.80	0.80	0.83	0.83	0.82	0.82	0.14	0.47
	pwe	0.87	0.88	0.87	0.87	0.78	0.82	0.77	0.78	0.40	0.72
	pwe/fine-tuning	0.92	0.93	0.92	0.92	0.90	0.91	0.90	0.90	0.61	**0.92**
ProdLDA	none	0.92	0.93	0.92	0.92	0.88	0.88	0.87	0.87	0.44	0.62
	pwe	0.85	0.87	0.85	0.85	0.74	0.81	0.73	0.74	0.33	0.59
	pwe/fine-tuning	0.93	0.93	0.93	0.93	0.91	0.91	0.91	0.91	**0.64**	**0.92**
RSB	none	0.43	0.29	0.39	0.27	0.43	0.28	0.39	0.27	0.30	0.41
	pwe	0.66	0.66	0.65	0.64	0.66	0.66	0.65	0.64	0.22	0.46
	pwe/fine-tuning	0.93	0.93	0.93	0.93	0.93	0.93	0.93	0.93	0.42	0.58
WLDA	none	0.34	0.36	0.32	0.27	0.28	0.27	0.25	0.16	0.01	0.24
	pwe	0.30	0.29	0.28	0.25	0.29	0.27	0.26	0.21	0.00	0.23
	pwe/fine-tuning	0.31	0.33	0.29	0.26	0.29	0.31	0.27	0.22	0.00	0.23

Table 25. Classification and Clustering performance on Biomedical.

Model	Classifier Method PWE	SVM (Linear) ACC	Precision	Recall	F1	SVM (rbm) ACC	Precision	Recall	F1	- NMI	Purity
GSB	none	0.39	0.39	0.39	0.35	0.49	0.51	0.49	0.49	0.11	0.19
	pwe	0.22	0.20	0.22	0.18	0.25	0.25	0.25	0.23	0.07	0.14
	pwe/fine-tuning	0.31	0.30	0.31	0.28	0.36	0.36	0.36	0.35	0.13	0.21
GSM	none	0.40	0.38	0.40	0.36	0.49	0.51	0.49	0.49	0.11	0.20
	pwe	0.25	0.23	0.25	0.22	0.26	0.27	0.26	0.25	0.10	0.18
	pwe/fine-tuning	0.32	0.30	0.32	0.28	0.35	0.35	0.35	0.34	**0.15**	**0.23**
NSTM	none	0.25	0.23	0.24	0.21	0.29	0.28	0.29	0.28	0.00	0.05
	pwe	0.28	0.26	0.28	0.24	0.32	0.32	0.32	0.31	0.01	0.06
	pwe/fine-tuning	0.23	0.22	0.23	0.19	0.28	0.27	0.27	0.26	0.00	0.05
NVDM	none	0.42	0.41	0.42	0.38	0.55	0.57	0.55	0.55	0.06	0.14
	pwe	0.38	0.36	0.38	0.34	0.51	0.52	0.51	0.51	0.05	0.12
	pwe/fine-tuning	**0.45**	**0.44**	**0.44**	**0.41**	**0.57**	**0.59**	**0.57**	**0.57**	0.05	0.12
NVLDA	none	0.28	0.25	0.28	0.23	0.26	0.25	0.26	0.24	0.06	0.15
	pwe	0.06	0.02	0.06	0.01	0.06	0.02	0.06	0.01	0.00	0.05
	pwe/fine-tuning	0.10	0.07	0.10	0.06	0.07	0.04	0.07	0.03	0.03	0.08
ProdLDA	none	0.14	0.13	0.14	0.10	0.09	0.10	0.09	0.06	0.05	0.10
	pwe	0.06	0.01	0.05	0.01	0.06	0.01	0.05	0.01	0.00	0.05
	pwe/fine-tuning	0.11	0.08	0.11	0.07	0.07	0.05	0.07	0.03	0.03	0.08
RSB	none	0.18	0.14	0.18	0.13	0.20	0.17	0.19	0.16	0.04	0.09
	pwe	0.08	0.03	0.08	0.03	0.08	0.03	0.08	0.03	0.01	0.06
	pwe/fine-tuning	0.16	0.13	0.16	0.13	0.17	0.15	0.17	0.14	0.04	0.10
WLDA	none	0.08	0.06	0.08	0.04	0.08	0.06	0.08	0.04	0.00	0.05
	pwe	0.08	0.05	0.08	0.04	0.08	0.05	0.08	0.04	0.00	0.05
	pwe/fine-tuning	0.07	0.05	0.07	0.03	0.08	0.06	0.07	0.04	0.00	0.05

Table 26. Classification and Clustering performance on DBLP.

Model	Classifier Method PWE	SVM (Linear) ACC	Precision	Recall	F1	SVM (rbm) ACC	Precision	Recall	F1	- NMI	Purity
GSB	none	0.69	0.65	0.58	0.58	0.74	0.70	0.65	0.67	0.04	0.46
	pwe	0.62	0.60	0.52	0.53	0.65	0.63	0.55	0.57	0.09	0.55
	pwe/fine-tuning	**0.72**	0.68	**0.62**	0.63	0.75	0.71	0.66	0.68	0.11	0.56
GSM	none	0.68	0.64	0.57	0.57	0.74	0.70	0.65	0.67	0.04	0.44
	pwe	0.64	0.61	0.55	0.56	0.66	0.64	0.56	0.58	0.11	0.59
	pwe/fine-tuning	**0.72**	**0.69**	**0.62**	**0.64**	0.74	0.71	0.66	0.67	0.14	0.61
NSTM	none	0.48	0.46	0.36	0.35	0.52	0.51	0.41	0.41	0.01	0.39
	pwe	0.50	0.47	0.40	0.39	0.56	0.55	0.46	0.47	0.00	0.38
	pwe/fine-tuning	0.46	0.46	0.34	0.32	0.50	0.50	0.40	0.40	0.00	0.38
NVDM	none	0.71	0.67	0.61	0.62	**0.76**	**0.72**	**0.68**	**0.70**	0.04	0.46
	pwe	0.64	0.62	0.54	0.54	0.72	0.69	0.63	0.65	0.02	0.42
	pwe/fine-tuning	0.71	0.67	0.61	0.61	**0.76**	**0.72**	**0.68**	**0.70**	0.04	0.45
NVLDA	none	0.53	0.50	0.42	0.41	0.54	0.53	0.43	0.43	0.03	0.45
	pwe	0.38	0.15	0.25	0.15	0.38	0.13	0.25	0.15	0.00	0.38
	pwe/fine-tuning	0.57	0.54	0.45	0.44	0.54	0.52	0.42	0.40	0.15	0.55
ProdLDA	none	0.66	0.65	0.54	0.56	0.63	0.66	0.52	0.54	**0.20**	**0.64**
	pwe	0.38	0.13	0.25	0.14	0.38	0.11	0.25	0.14	0.00	0.38
	pwe/fine-tuning	0.57	0.48	0.44	0.42	0.54	0.47	0.41	0.38	0.14	0.53
RSB	none	0.71	0.67	0.61	0.61	0.75	0.71	0.66	0.68	0.04	0.43
	pwe	0.41	0.23	0.31	0.24	0.40	0.19	0.30	0.22	0.02	0.39
	pwe/fine-tuning	0.69	0.64	0.58	0.58	0.70	0.66	0.61	0.62	0.11	0.50
WLDA	none	0.39	0.23	0.26	0.18	0.39	0.24	0.26	0.18	0.00	0.38
	pwe	0.39	0.25	0.26	0.18	0.39	0.28	0.26	0.18	0.00	0.38
	pwe/fine-tuning	0.39	0.24	0.26	0.17	0.39	0.27	0.26	0.17	0.00	0.38

Table 27. Classification and Clustering performance on GoogleNews.

Model	Classifier Method PWE	SVM (Linear) ACC	Precision	Recall	F1	SVM (rbm) ACC	Precision	Recall	F1	- NMI	Purity
GSB	none	0.85	0.81	0.78	0.78	0.85	0.82	0.74	0.77	0.46	0.25
	pwe	0.55	0.37	0.36	0.33	0.52	0.30	0.30	0.27	0.63	0.41
	pwe/fine-tuning	0.85	0.77	0.74	0.74	0.83	0.76	0.70	0.71	0.54	0.27
GSM	none	0.84	0.82	0.75	0.77	0.85	0.82	0.74	0.77	0.47	0.25
	pwe	0.53	0.30	0.31	0.27	0.51	0.24	0.27	0.23	**0.68**	**0.46**
	pwe/fine-tuning	0.80	0.65	0.60	0.60	0.82	0.68	0.64	0.65	0.67	0.40
NSTM	none	0.15	0.04	0.04	0.03	0.09	0.02	0.02	0.02	0.00	0.04
	pwe	0.52	0.46	0.40	0.41	0.49	0.44	0.32	0.35	0.01	0.04
	pwe/fine-tuning	0.15	0.04	0.05	0.03	0.11	0.03	0.03	0.02	0.01	0.04
NVDM	none	0.87	**0.85**	**0.83**	**0.83**	**0.89**	**0.87**	**0.82**	**0.84**	0.30	0.15
	pwe	0.85	0.83	0.81	0.81	0.88	**0.87**	0.80	0.83	0.32	0.17
	pwe/fine-tuning	**0.88**	0.83	**0.83**	0.82	**0.89**	**0.87**	0.81	0.83	0.27	0.13
NVLDA	none	0.56	0.45	0.41	0.40	0.48	0.37	0.30	0.30	0.33	0.19
	pwe	0.14	0.03	0.04	0.03	0.14	0.03	0.04	0.03	0.17	0.12
	pwe/fine-tuning	0.32	0.08	0.12	0.08	0.29	0.07	0.10	0.07	0.53	0.30
ProdLDA	none	0.41	0.12	0.17	0.12	0.38	0.12	0.15	0.11	0.61	0.36
	pwe	0.11	0.02	0.03	0.02	0.10	0.02	0.03	0.02	0.11	0.08
	pwe/fine-tuning	0.35	0.08	0.13	0.09	0.33	0.08	0.12	0.08	0.59	0.32
RSB	none	0.59	0.49	0.48	0.47	0.59	0.48	0.47	0.47	0.34	0.12
	pwe	0.07	0.00	0.01	0.00	0.07	0.00	0.01	0.00	0.06	0.05
	pwe/fine-tuning	0.65	0.43	0.44	0.41	0.64	0.39	0.41	0.38	0.37	0.14
WLDA	none	0.05	0.01	0.01	0.01	0.06	0.01	0.01	0.01	0.00	0.04
	pwe	0.06	0.01	0.02	0.01	0.06	0.01	0.02	0.01	0.00	0.04
	pwe/fine-tuning	0.06	0.01	0.02	0.01	0.06	0.01	0.02	0.01	0.00	0.04

Table 28. Classification and Clustering performance on M10.

Model	Classifier Method PWE	SVM (Linear) ACC	Precision	Recall	F1	SVM (rbm) ACC	Precision	Recall	F1	- NMI	Purity
GSB	none	0.66	0.62	0.59	0.59	0.73	0.74	0.67	0.68	0.14	0.34
	pwe	0.52	0.46	0.46	0.45	0.53	0.50	0.47	0.47	0.22	0.46
	pwe/fine-tuning	0.66	0.62	0.59	0.59	0.68	0.67	0.63	0.63	0.26	0.50
GSM	none	0.66	0.62	0.59	0.59	0.72	0.74	0.67	0.68	0.14	0.34
	pwe	0.54	0.49	0.48	0.48	0.54	0.50	0.48	0.48	0.24	0.50
	pwe/fine-tuning	0.66	0.63	0.60	0.60	0.68	0.68	0.62	0.63	0.30	0.54
NSTM	none	0.40	0.36	0.35	0.35	0.50	0.45	0.45	0.45	0.04	0.17
	pwe	0.41	0.39	0.36	0.36	0.49	0.49	0.44	0.44	0.00	0.14
	pwe/fine-tuning	0.36	0.33	0.32	0.31	0.44	0.42	0.40	0.40	0.01	0.14
NVDM	none	0.62	0.60	0.56	0.55	0.72	0.74	0.66	0.67	0.08	0.26
	pwe	0.57	0.53	0.50	0.50	0.70	0.73	0.64	0.65	0.07	0.26
	pwe/fine-tuning	**0.69**	**0.67**	**0.64**	**0.64**	**0.76**	**0.77**	**0.71**	**0.73**	0.10	0.28
NVLDA	none	0.41	0.37	0.36	0.35	0.43	0.40	0.38	0.38	0.08	0.28
	pwe	0.19	0.15	0.15	0.10	0.20	0.18	0.16	0.11	0.08	0.20
	pwe/fine-tuning	0.52	0.53	0.45	0.45	0.50	0.57	0.43	0.44	0.33	0.51
ProdLDA	none	0.64	0.58	0.57	0.56	0.56	0.63	0.50	0.52	**0.34**	**0.59**
	pwe	0.16	0.11	0.12	0.06	0.18	0.16	0.14	0.08	0.05	0.18
	pwe/fine-tuning	0.54	0.57	0.48	0.48	0.52	0.58	0.45	0.46	0.33	0.54
RSB	none	0.64	0.59	0.57	0.57	0.69	0.70	0.63	0.64	0.11	0.26
	pwe	0.22	0.13	0.18	0.12	0.21	0.13	0.18	0.11	0.05	0.19
	pwe/fine-tuning	0.64	0.57	0.58	0.57	0.66	0.59	0.59	0.59	0.21	0.32
WLDA	none	0.16	0.11	0.13	0.09	0.16	0.11	0.13	0.09	0.00	0.13
	pwe	0.16	0.11	0.13	0.08	0.16	0.13	0.13	0.09	0.00	0.13
	pwe/fine-tuning	0.16	0.12	0.13	0.08	0.16	0.12	0.13	0.08	0.00	0.13

Table 29. Classification and Clustering performance on PascalFlicker.

Model	Classifier Method PWE	SVM (Linear) ACC	Precision	Recall	F1	SVM (rbm) ACC	Precision	Recall	F1	- NMI	Purity
GSB	none	0.34	0.34	0.34	0.31	0.41	0.45	0.42	0.41	0.14	0.19
	pwe	0.22	0.21	0.22	0.19	0.23	0.22	0.23	0.21	0.14	0.18
	pwe/fine-tuning	0.30	0.28	0.30	0.27	0.32	0.33	0.32	0.31	0.18	**0.21**
GSM	none	0.34	0.32	0.34	0.31	0.40	0.44	0.40	0.40	0.14	0.20
	pwe	0.22	0.21	0.22	0.19	0.22	0.22	0.22	0.20	0.15	0.19
	pwe/fine-tuning	0.31	0.31	0.32	0.29	0.31	0.32	0.31	0.29	**0.19**	**0.21**
NSTM	none	0.19	0.23	0.19	0.16	0.11	0.12	0.11	0.09	0.00	0.05
	pwe	0.14	0.13	0.15	0.11	0.13	0.13	0.13	0.10	0.01	0.06
	pwe/fine-tuning	0.11	0.10	0.11	0.06	0.09	0.07	0.09	0.04	0.00	0.05
NVDM	none	0.34	0.32	0.34	0.31	0.46	0.49	0.46	0.46	0.06	0.12
	pwe	0.35	0.33	0.35	0.32	0.46	0.50	0.46	0.47	0.09	0.14
	pwe/fine-tuning	**0.40**	**0.38**	**0.40**	**0.37**	**0.50**	**0.53**	**0.50**	**0.50**	0.10	0.15
NVLDA	none	0.25	0.24	0.26	0.22	0.26	0.26	0.26	0.25	0.11	0.15
	pwe	0.06	0.02	0.06	0.02	0.06	0.04	0.06	0.02	0.00	0.06
	pwe/fine-tuning	0.08	0.05	0.08	0.04	0.08	0.05	0.08	0.04	0.07	0.10
ProdLDA	none	0.30	0.29	0.30	0.27	0.23	0.27	0.23	0.21	0.13	0.16
	pwe	0.05	0.01	0.05	0.01	0.05	0.01	0.05	0.01	0.00	0.05
	pwe/fine-tuning	0.09	0.06	0.09	0.05	0.09	0.06	0.08	0.05	0.08	0.11
RSB	none	0.23	0.20	0.23	0.19	0.26	0.25	0.26	0.25	0.07	0.11
	pwe	0.10	0.05	0.10	0.04	0.10	0.03	0.10	0.03	0.08	0.10
	pwe/fine-tuning	0.17	0.13	0.17	0.13	0.17	0.14	0.17	0.13	0.10	0.11
WLDA	none	0.07	0.06	0.07	0.04	0.07	0.05	0.07	0.04	0.00	0.05
	pwe	0.06	0.05	0.06	0.04	0.07	0.05	0.07	0.04	0.00	0.05
	pwe/fine-tuning	0.06	0.04	0.06	0.04	0.06	0.04	0.06	0.03	0.00	0.05

Table 30. Classification and Clustering performance on SearchSnippets.

Model	Classifier Method PWE	SVM (Linear) ACC	Precision	Recall	F1	SVM (rbm) ACC	Precision	Recall	F1	- NMI	Purity
GSB	none	0.51	0.54	0.48	0.49	0.78	0.80	0.76	0.78	0.08	0.33
	pwe	0.41	0.39	0.34	0.34	0.48	0.49	0.41	0.42	0.09	0.35
	pwe/fine-tuning	0.60	0.59	0.53	**0.54**	0.71	0.73	0.67	0.68	0.16	0.41
GSM	none	0.50	0.55	0.45	0.47	0.78	0.81	0.75	0.77	0.08	0.33
	pwe	0.45	0.42	0.38	0.38	0.48	0.53	0.41	0.42	0.13	0.40
	pwe/fine-tuning	**0.61**	**0.61**	**0.54**	**0.54**	0.70	0.72	0.65	0.67	**0.18**	**0.45**
NSTM	none	0.25	0.32	0.16	0.11	0.27	0.41	0.19	0.16	0.01	0.22
	pwe	0.29	0.43	0.20	0.16	0.26	0.59	0.17	0.13	0.02	0.23
	pwe/fine-tuning	0.24	0.29	0.15	0.10	0.25	0.32	0.17	0.13	0.01	0.22
NVDM	none	0.44	0.49	0.41	0.43	0.80	0.83	0.77	0.79	0.03	0.25
	pwe	0.44	0.50	0.40	0.41	0.78	0.82	0.76	0.78	0.03	0.25
	pwe/fine-tuning	0.52	0.55	0.51	0.52	**0.83**	**0.85**	**0.81**	**0.82**	0.04	0.26
NVLDA	none	0.37	0.36	0.31	0.31	0.40	0.42	0.33	0.34	0.05	0.28
	pwe	0.27	0.23	0.18	0.14	0.24	0.20	0.15	0.10	0.03	0.24
	pwe/fine-tuning	0.27	0.19	0.19	0.14	0.25	0.12	0.16	0.11	0.03	0.26
ProdLDA	none	0.33	0.36	0.25	0.24	0.46	0.56	0.38	0.40	0.06	0.30
	pwe	0.23	0.11	0.14	0.08	0.22	0.11	0.13	0.06	0.01	0.22
	pwe/fine-tuning	0.27	0.17	0.18	0.13	0.25	0.14	0.16	0.11	0.03	0.25
RSB	none	0.30	0.22	0.21	0.18	0.42	0.38	0.35	0.33	0.02	0.23
	pwe	0.22	0.09	0.13	0.07	0.22	0.05	0.13	0.06	0.01	0.22
	pwe/fine-tuning	0.32	0.28	0.24	0.21	0.38	0.35	0.30	0.30	0.03	0.25
WLDA	none	0.22	0.16	0.14	0.08	0.22	0.16	0.14	0.08	0.00	0.22
	pwe	0.22	0.14	0.14	0.08	0.23	0.17	0.14	0.08	0.00	0.22
	pwe/fine-tuning	0.23	0.15	0.14	0.09	0.23	0.16	0.14	0.09	0.00	0.22

Table 31. Classification and Clustering performance on StackOverflow.

Model	Classifier Method PWE	SVM (Linear) ACC	Precision	Recall	F1	SVM (rbm) ACC	Precision	Recall	F1	- NMI	Purity
GSB	none	0.71	0.68	0.70	0.68	0.72	0.72	0.71	0.71	0.16	0.23
	pwe	0.70	0.71	0.70	0.70	0.72	0.78	0.72	0.74	0.50	0.60
	pwe/fine-tuning	0.68	0.66	0.67	0.66	0.69	0.69	0.68	0.68	0.35	0.44
GSM	none	0.71	0.69	0.70	0.69	0.72	0.71	0.71	0.71	0.17	0.24
	pwe	0.73	0.76	0.73	0.73	0.74	**0.81**	0.73	0.76	**0.59**	**0.70**
	pwe/fine-tuning	0.71	0.70	0.71	0.70	0.71	0.72	0.71	0.71	0.47	0.59
NSTM	none	0.22	0.20	0.21	0.19	0.25	0.24	0.24	0.24	0.00	0.06
	pwe	0.45	0.42	0.44	0.41	0.47	0.45	0.46	0.45	0.00	0.06
	pwe/fine-tuning	0.21	0.19	0.20	0.18	0.22	0.22	0.22	0.22	0.00	0.06
NVDM	none	0.74	0.72	0.74	0.72	0.76	0.76	0.75	0.76	0.13	0.19
	pwe	0.71	0.68	0.70	0.68	0.75	0.75	0.74	0.75	0.10	0.19
	pwe/fine-tuning	**0.78**	**0.77**	**0.77**	**0.77**	**0.79**	0.80	**0.79**	**0.79**	0.15	0.22
NVLDA	none	0.50	0.47	0.49	0.46	0.48	0.47	0.47	0.46	0.19	0.30
	pwe	0.07	0.02	0.06	0.02	0.07	0.01	0.06	0.02	0.00	0.06
	pwe/fine-tuning	0.22	0.15	0.21	0.15	0.15	0.12	0.14	0.10	0.15	0.16
ProdLDA	none	0.44	0.43	0.43	0.39	0.32	0.38	0.31	0.30	0.35	0.34
	pwe	0.06	0.01	0.06	0.01	0.06	0.02	0.06	0.01	0.00	0.06
	pwe/fine-tuning	0.29	0.25	0.28	0.23	0.18	0.17	0.17	0.14	0.21	0.19
RSB	none	0.64	0.61	0.63	0.60	0.66	0.64	0.65	0.64	0.18	0.17
	pwe	0.10	0.04	0.10	0.04	0.10	0.04	0.09	0.03	0.05	0.08
	pwe/fine-tuning	0.66	0.63	0.65	0.63	0.67	0.66	0.66	0.65	0.29	0.22
WLDA	none	0.09	0.07	0.09	0.05	0.09	0.08	0.09	0.05	0.00	0.06
	pwe	0.10	0.08	0.10	0.06	0.10	0.08	0.10	0.06	0.00	0.06
	pwe/fine-tuning	0.11	0.08	0.10	0.07	0.10	0.09	0.10	0.07	0.00	0.06

Table 32. Classification and Clustering performance on TrecTweet.

Model	Classifier Method PWE	SVM (Linear) ACC	Precision	Recall	F1	SVM (rbm) ACC	Precision	Recall	F1	- NMI	Purity
GSB	none	-	-	-	-	-	-	-	-	0.50	0.39
	pwe	-	-	-	-	-	-	-	-	0.62	0.54
	pwe/fine-tuning	-	-	-	-	-	-	-	-	0.66	0.49
GSM	none	-	-	-	-	-	-	-	-	0.52	0.40
	pwe	-	-	-	-	-	-	-	-	0.65	0.58
	pwe/fine-tuning	-	-	-	-	-	-	-	-	**0.74**	0.56
NSTM	none	-	-	-	-	-	-	-	-	0.04	0.12
	pwe	-	-	-	-	-	-	-	-	0.03	0.13
	pwe/fine-tuning	-	-	-	-	-	-	-	-	0.00	0.10
NVDM	none	-	-	-	-	-	-	-	-	0.39	0.30
	pwe	-	-	-	-	-	-	-	-	0.37	0.27
	pwe/fine-tuning	-	-	-	-	-	-	-	-	0.43	0.33
NVLDA	none	-	-	-	-	-	-	-	-	0.44	0.36
	pwe	-	-	-	-	-	-	-	-	0.39	0.34
	pwe/fine-tuning	-	-	-	-	-	-	-	-	0.70	**0.59**
ProdLDA	none	-	-	-	-	-	-	-	-	0.60	0.50
	pwe	-	-	-	-	-	-	-	-	0.16	0.18
	pwe/fine-tuning	-	-	-	-	-	-	-	-	0.62	0.49
RSB	none	-	-	-	-	-	-	-	-	0.38	0.26
	pwe	-	-	-	-	-	-	-	-	0.17	0.18
	pwe/fine-tuning	-	-	-	-	-	-	-	-	0.41	0.28
WLDA	none	-	-	-	-	-	-	-	-	0.00	0.10
	pwe	-	-	-	-	-	-	-	-	0.01	0.10
	pwe/fine-tuning	-	-	-	-	-	-	-	-	0.00	0.10

5. Conclusions

Short-text data are now becoming ubiquitous in the real world through various social networking sites. The importance of analysing these short messages is also growing day by day. Unlike long texts or documents, short texts suffer from a lack of word co-occurrence information due to their restricted lengths, posing a difficulty in generating coherent and interpretable topics with popular topic-model techniques.

The use of pretrained word embedding in neural-topic models is a good choice to easily increase the generated topic quality as measured by topic coherence and topic diversity. This is effective for both long and short texts, and reduces the number of trainable parameters, thus shortening the training step time. However, to achieve better topic coherence, especially in short texts, or to make the top-N words of a topic more relevant to the real semantic contents of the training corpus, the additional fine-tuning stage proposed in this work is indeed necessary. The extensive study in this work with several neural-topic models and benchmark datasets justifies our proposal.

However, the use of pretrained word embedding (PWE) has its inherent limitations, which may affect the quality of the extracted topics from short texts. The short-text corpus to be analyzed may contain words that are not included in the vocabulary covered by the corpus used for pretrained word embedding. In this case, NTM-PWE uses a vector initialized with zero. As the vocabulary coverage increases, the performance is likely to deteriorate. Moreover, in the case of NTM-PWE/fine-tuning, there is a possibility that the number of parameter updates will increase until the loss function converges, resulting in an increase in training time. If the time difference between the corpus used for PWE training and the corpus to be analyzed is too large, the meanings of words may change with time, which may have a negative impact on the production of interpretable topics.

It is also seen that the improvement in topic quality after introducing a fine-tuning stage is not the same for all the datasets and all the models. It is difficult to define the correlation between the structure of neural-topic models and the inherent characteristics of the datasets, which poses a challenge to our study. In this work, we limited our study to benchmark datasets available on the internet. Currently, we are collecting data for the evaluation of our proposal with real-world datasets.

By incorporating the additional training with the original training corpus, along with pretrained word embedding with the external corpus, we can improve the purity and NMI of the topics evaluated using the class labels of the documents. Thus, we can construct topics that are more suitable for the training corpus. This method can also be expected to improve the performance of downstream tasks, such as classifications for long texts. Even for short texts, the performance of the downstream tasks is better than when using pretrained word embedding without fine-tuning.

Author Contributions: Conceptualization, R.M.; methodology, R.M.; software, R.M.; validation, R.M. and B.C.; investigation, R.M.; data curation, R.M.; writing—original draft preparation, R.M.; writing—review and editing, B.C.; visualization, R.M. and B.C.; supervision, B.C.; project administration, B.C.; funding acquisition, B.C. All authors have read and agreed to the published version of the manuscript.

Funding: This research received no external funding.

Institutional Review Board Statement: Not applicable.

Informed Consent Statement: Not applicable.

Acknowledgments: The authors acknowledge the technical support of the Pattern Recognition and Machine Learning Laboratory, Department of Software, Iwate Prefectural University, Japan.

Conflicts of Interest: The authors declare no conflict of interest.

References

1. Hofmann, T. Probabilistic latent semantic indexing. In Proceedings of the Twenty-Second Annual International SIGIR Conference, Berkeley, CA, USA, 15–19 August 1999; pp. 50–57.
2. Blei, D.M.; Ng, A.Y.; Jordan, M.I. Latent Dirichlet Allocation. *J. Mach. Learn. Res. JMLR* **2003**, *3*, 993–1022.
3. Blei, D.M. Probabilistic topic models. *Commun. ACM* **2012**, *55*, 77–84. [CrossRef]
4. Hong, L.; Davison, B. Empirical study of topic modeling in Twitter. In Proceedings of the First Workshop on Social Media Analytics, Washington, DC, USA, 25–28 July 2020; ACM: New York, NY, USA, 2010; pp. 80–88.
5. Phan, X.; Nguyen, L.; Horiguchi, S. Learning to classify short and sparse text & web with hidden topics from large scale data collections. In Proceedings of the 17th International Conference on World Wide Web, Beijing, China, 21–25 April 2008; pp. 91–100.
6. Jin, O.; Liu, N.; Zhao, K.; Yu, Y.; Yang, Q. Transferring topical knowledge from auxiliary long texts for short text clustering. In Proceedings of the 20th International Conference on Information and Knowledge Management, Scotland, UK, 24–28 October 2011; pp. 775–784.
7. Qiang, J.; Qian, Z.; Li, Y.; Yuan, Y.; Wu, X. Short Text Topic Modeling Techniques, Applications, and Performance: A Survey. *IEEE Trans. Knowl. Data Eng.* **2020**, early access. [CrossRef]
8. Weng, J.; Lim, E.; Jiang, J.; He, Q. Twitterrank: Finding topic-sensitive influential twitterers. In Proceedings of the Third ACM International Conference on Web Search and Data Mining WSDM, New York, NY, USA, 3–6 February 2010; pp. 261–270.
9. Quan, X.; Kit, C.; Ge, Y.; Pan, S.J. Short and sparse text topic modeling via self-aggregation. In Proceedings of the 24th International Conference on Artificial Intelligence, Buenos Aires, Argentina, 25–31 July 2015; pp. 2270–2276.
10. Zuo, Y.; Wu, J.; Zhang, H.; Lin, H.; Wang, F.; Xu, K.; Xiong, H. Topic modeling of short texts: A pseudo-document view. In Proceedings of the 22nd ACM SIGKDD International Conference on Knowledge Discovery and Data Mining, San Francisco, CA, USA, 13–17 August 2016; pp. 2105–2114.
11. Zhao, W.X.; Jiang, J.; Weng, J.; He, J.; Lim, E.P.; Yan, H.; Li, X. Comparing twitter and traditional media using topic models. In *Advances in Information Retrieval*; Springer: Berlin/Heidelberg, Germany, 2011; pp. 338–349.
12. Yin, J.; Wang, J. A dirichlet multinomial mixture model-based approach for short text clustering. In Proceedings of the 20th ACM SIGKDD International Conference on Knowledge Discovery and Data Mining, New York, NY, USA, 24–27 August 2014; pp. 233–242.
13. Nigam, K.; McCallum, A.K.; Thrun, S.; Mitchell, T. Text classification from labeled and unlabeled documents using EM. *Mach. Learn.* **2000**, *39*, 103–134. [CrossRef]
14. Li, C.; Duan, Y.; Wang, H.; Zhang, Z.; Sun, A.; Ma, Z. Enhancing topic modeling for short texts with auxiliary word embeddings. *ACM Trans. Inf. Syst. (TOIS)* **2017**, *36*, 11. [CrossRef]
15. Cheng, X.; Yan, X.; Lan, Y.; Guo, J. BTM: Topic modeling over short texts. *IEEE Trans. Knowl. Data Eng.* **2014**, *26*, 2928–2941. [CrossRef]
16. Zuo, Y.; Zhao, J.; Xu, K. Word network topic model: A simple but general solution for short and imbalanced texts. *Knowl. Inf. Syst.* **2016**, *48*, 379–398. [CrossRef]
17. Almeida, F.; Xexeo, G. Word Embeddings: A Survey. *arXiv* **2019**, arXiv:1901.09069.
18. Dieng, A.B.; Ruiz, F.J.R.; Blei, D.M. Topic Modeling in Embedding Spaces. *Trans. Assoc. Comput. Linguist.* **2020**, *8*, 439–453. [CrossRef]
19. Chauhan, U.; Shah, A. Topic Modeling Using Latent Dirichlet allocation: A Survey. *ACM Comput. Surv.* **2021**, *54*, 145. [CrossRef]

20. Bunk, S.; Krestel, R. WELDA: Enhancing topic models by incorporating local word context. In Proceedings of the 18th ACM/IEEE on Joint Conference on Digital Libraries, Fort Worth, TX, USA, 3–7 June 2018; pp. 293–302.
21. Nguyen, D.Q.; Billingsley, R.; Du, L.; Johnson, M. Improving topic models with latent feature word representations. *Trans. Assoc. Comput. Linguist.* **2015**, *3*, 299–313. [CrossRef]
22. Li, C.; Wang, H.; Zhang, Z.; Sun, A.; Ma, Z. Topic modeling for short texts with auxiliary word embeddings. In Proceedings of the 39th International ACM SIGIR Conference on Research and Development in Information Retrieval, Pisa, Italy, 17–21 July 2016; pp. 165–174.
23. Qiang, J.; Chen, P.; Wang, T.; Wu, X. Topic modeling over short texts by incorporating word embeddings. In Proceedings of the Pacific-Asia Conference on Knowledge Discovery and Data Mining, Jeju, Korea, 23–26 May 2017; pp. 363–374.
24. Bicalho, P.V.; Pita, M.; Pedrosa, G.; Lacerda, A.; Pappa, G.L. A general framework to expand short text for topic modeling. *Inf. Sci.* **2017**, *393*, 66–81. [CrossRef]
25. Zhao, H.; Phung, D.; Huynh, V.; Jin, Y.; Du, L.; Buntine, W. Topic Modelling Meets Deep Neural Networks: A Survey. In Proceedings of the Thirtieth International Joint Conference on Artificial Intelligence (IJCAI-21), Montreal, QC, Canada, 19–27 August 2021; pp. 4713–4720.
26. Doan, T.; Hoang, T. Benchmarking Neural Topic Models: An Empirical Study. In Proceedings of the Findings of the Association for Computational Linguistics: ACL-IJCNLP 2021, Online Event, 1 August 2021; pp. 4363–4368.
27. Lin, L.; Jiang, H.; Rao, Y. Copula Guided Neural Topic Modelling for Short Texts. In Proceedings of the 43rd International ACM SIGIR Conference on Research and Development in Information SIGIR 20, Xi'an, China, 25–30 July 2020; pp. 1773–1776. [CrossRef]
28. Murakami, R.; Chakraborty, B. Neural topic models for short text using pretrained embeddings and its application to real data. In Proceedings of the 2021 IEEE 4th International Conference on Knowledge Innovation and Invention (ICKII), Taichung, Taiwan, 23–25 July 2021; pp. 146–150. [CrossRef]
29. Kingma, D.P.; Welling, M. Auto-encoding variational Bayes. In Proceedings of the ICML, Beijing, China, 21–26 June 2014.
30. Rezende, D.J.; Mohamed, S.; Wierstra, D. Stochastic backpropagation and approximate inference in deep generative models. In Proceedings of the ICML, Beijing, China, 21–26 June 2014.
31. Miao, Y.; Yu, L.; Blunsom, P. Neural Variational Inference for text processing. In Proceedings of 33rd International Conference on Machine Learning, New York, NY, USA, 19–24 June 2016; pp. 1727–1736.
32. Srivastava, A.; Sutton, C.A. Autoencoding variational inference for topic models. In Proceedings of the International Conference on Learning Representations 2017, Toulon, France, 24–26 April 2017.
33. Burkhardt, S.; Kramer, S. Decoupling sparsity and smoothness in the dirichlet variational autoencoder topic model. *J. Mach. Learn. Res.* **2019**, *20*, 1–27.
34. Joo, W.; Lee, W.; Park, S.; Moon, I.C. Dirichlet Variational Autoencoder. *Pattern Recognit.* **2020**, *107*, 107514. [CrossRef]
35. Miao, Y.; Grefenstette, E.; Blunsom, P. Discovering discrete latent topics with neural variational inference. In Proceedings of the 34th International Conference on Machine Learning, Sydney, Australia, 6–11 August 2017; pp. 2410–2419.
36. Ning, X.; Zheng, Y.; Jiang, Z.; Wang, Y.; Yang, H.; Huang, J. Nonparametric Topic Modeling with Neural Inference. *Neurocomputing* **2020**, *399*, 296–306. [CrossRef]
37. Larochelle, H.; Lauly, S. A neural Autoregressive topic model. *Adv. Neural Inf. Process. Syst.* **2012**, *4*, 2708–2716.
38. Wang, R.; Zhou, D.; He, Y. ATM: Adversarial neural topic model. *Inf. Process. Manag.* **2019**, *56*, 102098. [CrossRef]
39. Wang, R.; Hu, X.; Zhou, D.; He, Y.; Xiong, Y.; Ye, C.; Xu, H. Neural Topic Modeling with Bidirectional Adversarial Training. In Proceedings of the 58th Annual Meeting of Association for Computational Linguistics, Online Event, 5–10 July 2020; pp. 340–350.
40. Yang, L.; Wu, F.; Gu, J.; Wang, C.; Cao, X.; Jin, D.; Guo, Y. Graph Attention Topic Modeling Network. In Proceedings of the WWW '20: Proceedings of The Web Conference 2020, Ljubljana, Slovenia, 19–23 April 2020; pp. 144–154. [CrossRef]
41. Nan, F.; Ding, R.; Nallapati, R.; Xiang, B. Topic Modeling with Wasserstein autoencoders. In Proceedings of the 2019 Meeting of the Association for Computational Linguistic, Florence, Italy, 28 July–2 August 2019; pp. 6345–6381. [CrossRef]
42. Zhao, H.; Phung, D.; Huynh, V.; Le, T.; Buntine, W. Neural topic model via optimal transport. In Proceedings of the ICLR 2021, Vienna, Austria, 4 May 2021.
43. Wang, X.; Yang, Y. Neural topic model with attention for supervised learning. In Proceedings of the 23rd International Conference on artificial Intelligence and Statistics (AISTSTS), Palermo, Italy, 26–28 August 2020; pp. 1147–1156.
44. Zeng, J.; Li, J.; Song, Y.; Gao, C.; Lyu, M.R.; King, I. Topic memory networks for short text classification. In Proceedings of the 2018 Conference on Empirical Methods in Natural Language Processing EMNLP, Brussels, Belgium, 31 October–4 November 2018; pp. 3120–3131.
45. Wu, X.; Li, C.; Zhu, Y.; Miao, Y. Short text topic modeling with topic distribution quantization and negative sampling decoder. In Proceedings of the 2020 Conference on Empirical Methods in Natural Language Processing EMNLP, Online Event, 13 November 2020; pp. 1772–1782.
46. Niu, Y.; Zhang, H.; Li, J. A Nested Chinese Restaurant Topic Model for Short Texts with Document Embeddings. *Appl. Sci.* **2021**, *11*, 8708. [CrossRef]
47. Zhao, X.; Wang, D.; Zhao, Z.; Liu, W.; Lu, C.; Zhuang, F. A neural topic model with word vectors and entity vectors for short texts. *Inf. Process. Manag.* **2021**, *58*, 102455. [CrossRef]

48. Zhu, Q.; Feng, Z.; Li, X. Graphbtm: Graph enhanced autoencoded variational inference for biterm topic model. In Proceedings of the 2018 Conference on Empirical Methods in Natural Language Processing EMNLP, Brussels, Belgium, 31 October–4 November 2018; pp. 4663–4672.
49. Feng, J.; Zhang, Z.; Ding, C.; Rao, Y.; Xie, H. Context reinforced neural topic modeling over short texts. *arXiv* **2020**, arXiv:2008.04545.
50. Pennington, J.; Socher, R.; Manning, C. Glove: Global vectors for word representation. In Proceedings of the 2014 Conference on Empirical Methods in Natural Language Processing (EMNLP), Doha, Qatar, 25–29 October 2014.
51. Xie, P.; Deng, Y.; Xing, E. Diversifying Restricted Boltzmann Machine for Document Modeling. In Proceedings of the 21th ACM SIGKDD International Conference on Knowledge Discovery and Data Mining, Sydney, Australia, 10–13 August 2015; pp. 1315–1324. [CrossRef]
52. Zhao, H.; Phung, D.Q.; Huynh, V.; Le, T.; Buntine, W.L. Neural Topic Model via Optimal Transport. *arXiv* **2020**, arXiv:2008.13537.
53. Roder, M.; Both, A.; Hinneburg, A. Exploring the Space of Topic Coherence Measures. In Proceedings of the Eighth ACM International Conference on Web Search and Data Mining, Shanghai, China, 31 January–6 February 2015; pp. 399–408.
54. Lau, J.H.; Newman, D.; Baldwin, T. Machine reading tea leaves: Automatically evaluating topic coherence and topic model quality. In Proceedings of the 14th Conference of the European Chapter of the Association for Computational Linguistics, Gothenburg, Sweden, 26–30 April 2014; pp. 530–539.
55. Ding, R.; Nallapati, R.; Xiang, B. Coherence-Aware neural topic modeling. In Proceedings of the 2018 Conference on Empirical Methods in Natural Language Processing, Brussels, Belgium, 31 October–4 November 2018; Association for Computational Linguistics: Stroudsburg, PA, USA, 2018; pp. 830–836.
56. Carbone, G.; Sarti, G. ETC-NLG: End-to-end Topic-Conditioned Natural Language Generation. *Ital. J. Comput. Linguist.* **2020**, *6*, 61–77. [CrossRef]
57. Chen, Y.; Zaki, M.J. KATE: K-Competitive Autoencoder for Text. In Proceedings of the 23rd ACM SIGKDD International Conference on Knowledge Discovery and Data Mining, Halifax, NS, USA, 13–17 August 2017; pp. 85–94.

sensors MDPI

Article

A Novel Privacy Preserving Scheme for Smart Grid-Based Home Area Networks

Wajahat Ali [1], Ikram Ud Din [1,*], Ahmad Almogren [2] and Byung-Seo Kim [3,*]

[1] Department of Information Technology, The University of Haripur, Haripur 22620, Pakistan; wajahat.haripur@gmail.com
[2] Department of Computer Science, College of Computer and Information Sciences, King Saud University, Riyadh 11633, Saudi Arabia; ahalmogren@ksu.edu.sa
[3] Department of Software and Communications Engineering, Hongik University, Sejong 30016, Korea
* Correspondence: ikramuddin205@yahoo.com (I.U.D.); jsnbs@hongik.ac.kr (B.-S.K.)

Abstract: Despite the benefits of smart grids, concerns about security and privacy arise when a large number of heterogeneous devices communicate via a public network. A novel privacy-preserving method for smart grid-based home area networks (HAN) is proposed in this research. To aggregate data from diverse household appliances, the proposed approach uses homomorphic Paillier encryption, Chinese remainder theorem, and one-way hash function. The privacy in Internet of things (IoT)-enabled smart homes is one of the major concerns of the research community. In the proposed scheme, the sink node not only aggregates the data but also enables the early detection of false data injection and replay attacks. According to the security analysis, the proposed approach offers adequate security. The smart grid distributes power and facilitates a two-way communications channel that leads to transparency and developing trust.

Keywords: aggregation; authentication; key management; privacy; smart home; smart meter

Citation: Ali, W.; Din, I.U.; Almogren, A.; Kim, B.-S. A Novel Privacy Preserving Scheme for Smart Grid-Based Home Area Networks. *Sensors* **2022**, *22*, 2269. https://doi.org/10.3390/s22062269

Academic Editor: Raffaele Bruno

Received: 3 December 2021
Accepted: 14 March 2022
Published: 15 March 2022

Publisher's Note: MDPI stays neutral with regard to jurisdictional claims in published maps and institutional affiliations.

1. Introduction

The term power grid is commonly referred to as an electricity distribution system that supplies energy to a territory. A power grid actually comprises power generation, distribution, and transmission [1]. The traditional power grid just supplies energy to the consumer which results in simplified management but at the cost of short falls when supply and demand do not catch up. There are certain limitations of the traditional grid such as the losses at transmission lines and lack of information, or we can say that lack of demand knowledge, which further leads to inefficient power management [2,3]. For example, the traditional grid feeds constant power during peak and off peak hours. To overcome the limitations of the traditional power grid, certain changes must be made in traditional grid [4]. The power sector needs to be revolutionized to meet the needs of modern living. A smart grid promises to replace the traditional grid with better performance and is also open enough to meet the upcoming revolution in the power sector. The SG consists of a power generation unit, power transmission and distribution units, smart meter (SM), smart homes, smart energy management systems, and smart appliances [5]. The communication among power generation, transmission, distribution, and customers is usually a two way communication managed by the CC and service provider enabling real time communications between consumers and the utility/service provider [6,7].

An SM senses the energy consumption of a home and sends it to a substation or the gateway or control room of that region. There may be a number of devices between the SM and service provider. The SM reports the energy consumption about every 15 min to the service provider [8]. The control room receives data from all the SMs in the neighborhood and transmits the combined energy usage report to the control center. The control center uses the consumption report to run load management and power distribution and uses the information for billing purposes [9]. Abbreviations contains all the terminologies and their definition used in the paper.

1.1. Architecture of Smart Grid

Smart grid refers to the electricity distribution network that uses communication channels to detect any change in local power usage and acts accordingly without external interference. It uses smart home appliances, SM, and green energy resources. The smart grid utilizes a two-way communication channel and allows consumer to interact with the grid. It facilitates the consumers, service provider, and government establishment by overcoming the drawbacks of traditional grid. It reduces the energy consumption and decreases the consumer's cost of electricity by smart means.

1.2. Working Architecture

The smart meter works inside a HAN as follows:

- The SMs are installed at the home, offices, and factory premises. The SM communicates to the local control center (CC), which is the nearest data gathering center. An SM can provide instantaneous consumption, cumulative energy, time of day energy data, and maximum demand (in kW).
- The local data center transmits information gathered from SMs in a locality to the data center at the utility provider or any third-party service provider using wired or wireless means.
- The data at the utility provider side can be accessed using a web portal. The utility providers gather data from local CC in real time and process it. It reports any tampering of meters, billing information, energy usage, load status, etc.

1.3. Our Contributions

The communication that makes up the grid smart is one of the obstacles that the smart grid deployment faces. To date, several schemes have been proposed and review surveys have been published, for example, [10–14]. Most of these schemes highlight either communication, technology standards, and infrastructure or home energy management and security. To the best of our knowledge, however, privacy is a big issue that still has to be thoroughly examined.

Since the smart grid is designed to facilitate its consumers, keeping the privacy of end users in a HAN is important. To date, a few schemes have been proposed, such as [15–23] to create a safe communication route over vulnerable public networks. These schemes are aimed at establishing a framework that can potentially protect end users' privacy. The majority of the plans cover the smart grid in basic terms, but they leave out smart homes and HANs. As it runs in the field, a HAN is the most vulnerable to cyber threats, theft, and data tampering. Because a consumer may be unaware of cyber security standards, it is critical to have built-in security to protect HANs from various cyber threats. In short, the issues associated with smart homes are rarely explored in published articles. This paper discusses smart homes and gathers the prominent articles in this domain, and proposes a novel privacy preserving scheme. Here, we highlight the following points:

- The development of smart grids is discussed along with the architecture of smart grids.
- Data aggregation, privacy preservation, key management, and user authentication are discussed within the scope of a smart home.
- A comprehensive literature review along with the pros and cons of existing schemes is discussed in addition to presenting some advanced literature to solve these issues.
- The paper tries to focus on smart homes and the privacy concerns of consumers along with discussing the future directions for faster transformation from traditional to smart grids.
- Finally, a privacy preserving data aggregation scheme is proposed for HANs that gathers the readings from all home appliances at the sink node, performs an early stage fault tolerance and aggregates the received reading data into one, and sends the result to the SM for further analysis.

The remainder of the paper is laid out as follows: The smart home is introduced in Section 2. Section 3 discusses the notion of privacy, its parameters, goals and attacks, and the threats to privacy. Section 4 presents a comparative analysis of advanced privacy-preserving techniques, including their benefits and drawbacks, as well as countermeasures. Section 5 describes a privacy-preserving data aggregation technique for fault-tolerant smart homes. Section 6 examines the proposed scheme's security measures, followed by a performance evaluation in Section 7. Future study directions are discussed in Section 8, and the work is concluded in Section 9.

2. Smart Home

A smart home consists of an SM and various appliances. Appliances may be a low voltage devices or high voltage devices which aggregate their energy consumption and send information to SMs, as shown in Figure 1. The SM receives energy consumption from appliances and forward to utilities for further processing [24,25]. Home energy management system (HEMS) is an automated system consist of hardware and software which controls and monitors the various devices and their operations. Users manually manage and control the electricity generation and production [26]. Different hourly block rates are offered for 24 h. Needless devices are automatically turn off with a short notification. Demand side management, demand response, direct load control, real time pricing, time of use, and real time peak pricing are recent examples of HEMS [27].

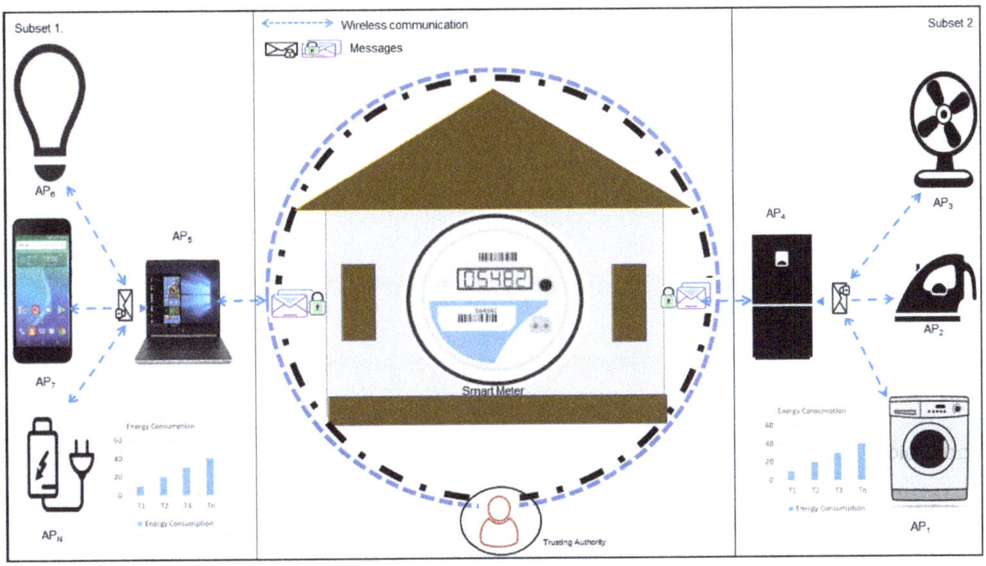

Figure 1. A home area network.

2.1. Smart Meter

An SM is an electric meter that performs the following functions: (i) measuring energy consumption, (ii) measuring energy consumption, (iii) report consumption data to meter management system, (iv) receiving electricity consumption cost or control signals, and (v) inform all the home appliances [28]. The SM data is one of the major sources of information similar to other actors, e.g., distribution system, transmission system, and generation sources for smooth running and construction of smart grid. As per European commission report around 80 percent of energy meters will be replaced with SM in 2020 [29].

2.2. Home Appliances

Home appliances are household devices installed in user home or apartments. These devices are connected to SM for monitoring and reporting [30]. Users can schedule appliances per use. In [16], the appliances are classified into four groups: Group 1 includes light load normal appliances, e.g., light bulbs and phone chargers; Group 2 consists of non-stoppable home appliances, e.g., microwave ovens, Group 3 comprises schedule-oriented appliances, e.g., washing machines and heaters; Group 4 includes electrical vehicles. Consumer can also schedule their appliances with different hourly changing rates to control their electricity cost [16]. Appliances monitor their energy readings and send to the SM after every time interval, which is usually 15 min duration [28].

2.3. Wireless Sensor

With the invention of new technologies, wireless sensors are being used in industries, health care, education, and utility grids. Due to their sensing capabilities, it makes them able to interact with machines, devices, and various appliances for controlling and monitoring [17,31]. Similarly, with the deployment of smart grids, wireless sensors are being deployed in smart homes and also at utility. Wireless sensor performs conversion of analog signals to digital, analog signal processing, transformation of information via bidirectional bus, manipulation of sensor derived signals, and addressing [32]. In a HAN, kitchen appliances, heating system, security frameworks, lighting system, theater setups, and water and sewage systems are totally instrumented with wireless sensors performing various operations. Access to these systems is through a home management system (HMS), which could be through the Internet or a cell phone application [32,33].

2.4. Consumer

Consumer is the main stakeholder for which the smart grid is designed. Consumers can schedule energy consumption, generate green energy for themselves, and store energy for future purposes [34]. Consumers can control their appliances' function and information flows, such as home automation, home energy management system, and industrial automation system [24].

2.5. Advanced Smart Home Applications

To upgrade the functionality of conventional grids, smart grids have introduced various new applications in smart homes, e.g., energy generation and storage and demand response. Smart grids have enabled the consumer to control their electricity bill. Other benefits that smart home applications have provided are scheduling power usage during the on/off peak hours, they can demand extra energy from the grid in advance, and, if required, they can also feed any green energy generated back to the national grid.

2.5.1. Two-Way Communications

A smart grid establishes a communications channel between consumers and service providers. A consumer can request peak and off-peak hours tariff rates to schedule the electricity usage appropriately. The SP can also obtain a future consumption forecast and can therefore control the energy production [35].

2.5.2. Renewable Energy Resources

With renewable energy resources an individual home can generate its own energy using mostly solar panels, but also with windmills or biogas. A consumer uses part of the energy for their own purpose and can feed the extra energy into the national grid. Hence, a consumer can also participate in the national grid and play a useful role for the national cause [36].

2.5.3. Energy Generation and Storage

With the invention of energy renewable resources such as solar, biogas, wind, and electricity storage sources such as electrical vehicles, smart transformers, and appliances, many home users generate electricity using photovoltaic panels for their daily use and sell extra electricity to the national power grid [35]. In each area, every application of SG is based on necessities such as voltage support, power quality, and service reliability [36]. However, smart grids have a serious issue with generating and storing electricity and, similarly, with the evaluation of the distribution system and integration of the evaluated grid components. Currently, gas and diesel generators, tides, and solar and wind are conventional energy generation resources, which provide power whenever natural resources are not available. The authors of [37] proposed optimization techniques for domestic users to control the operations in a HAN. Similarly, in [38], the integer-programming model is presented for electricity storage based on electric vehicles and photovoltaic panels.

2.5.4. Demand Response

With the rapid increase in population and wireless devices in HAN, demand for electricity has also increased. In a conventional grid system, it is hard to accommodate the electricity needs because conventional systems have various challenges, such as maintenance, erection, operation, and design [39]. This strategy is used to control or reduce the electricity consumption at peak hours. It does not only reduce the use of electricity consumption at peak hours but also reduces the electricity consumption cost. Different rates have already been offered for various hours [18,19].

3. Privacy

Smart grids are a promising technology describing electrical power infrastructure for transmission and distribution with integrated information and communications technologies [40]. The purpose of a smart grid is to bill the customers accurately and manage and distribute electrical energy in an efficient way. In a smart grid, an SM is the key entity. When an SM is deployed, the concern regarding meter tempering and consumer privacy is raised. There is a need for legislation of SMs. In compliance with privacy requirements, certain properties ensuring the privacy are confidentiality, integrity, authenticity, and availability [24,41]. An SM is prone to data tempering where an adversary can invade the SM. If an SM is compromised, it is then easy to access a cryptographic key. By exploiting a common vulnerability, a large number of SMs can be compromised and can result in manipulating real-time consumption. Therefore, a scalable access control is needed to prevent meter compromises and make sure that any stored information is used for the purpose of billing operations and other value-added services [42]. The major benefit of an SM is accurate billing, but the frequent sharing of consumption information with the utility might leak some private information. In order to protect billing information techniques—e.g., battery management—a zero knowledge homomorphic encryption technique is proposed in [43].

3.1. *Privacy Goals*

An SM collects the energy consumption from home appliances usually after a 15 min duration. An SM then generates a consumption report and sends it to the utility company. There might be an adversary peeking at energy consumption of a smart home that can further predict the lifestyle and routine of a homeowner. This poses a threat to security as well as privacy of the smart home. Therefore, to preserve privacy and security, energy consumption is encrypted either at the appliance level or before it leaves the HAN. The energy consumption sent by the HAN is further processed and the CC receives energy consumption details from various SMs. Each meter reports either its own consumption or an aggregation technique is used to sum-up the consumption reports from all meters in a sub-region/zone and send a bulk consumption report after the aggregation. Based on this, the CC generates monthly bills and maintains a profile picture of that region to show that how much electricity consumption is required [44]. The CC continuously collects consumption details from different devices. The received consumption is encrypted or aggregated because if an intruder changes the message, it could easily be identified [20]. This section describes various privacy goals.

3.1.1. Confidentiality

Home appliances send their energy detail to aggregator, SM or SP. This detail may reflect a users' personal profile. Privacy and confidentiality are interdependent of each other [45]. By securing messages from unauthorized access, privacy of home incumbents will be retained. If privacy of home users is compromised, then confidentiality is automatically violated. To ensure confidentiality, various schemes for HAN are employed, e.g., homomorphic encryption, blind signature, in-network aggregation, etc. [16,21].

3.1.2. Integrity

Smart appliances continuously send consumption patterns to an aggregator or SM after an interval. It may be possible that an aggregator or an SM are physically secure but vulnerable to different attacks such as man-in-middle attacks, alteration attacks, or replay attacks [46]. If integrity is compromised, precious information will be compromised and wrong decisions for managing and controlling the network might be made. Integrity refers to the message sent by sender; the same message without any modification is received by receiver. The following schemes are used for HAN data integrity: message digest, MAC, digital signature, and H-MAC [16,22].

3.1.3. Anonymity

Anonymity refers to the situation where the real identity of a person is kept secret. During sharing of secret control signals or reading, a device may protect their real identity from other appliances or devices [47]. Even an appliance or SM cannot recognize other devices communicating with them in a HAN. The purpose of anonymity is to hide one's identity from appliance to appliance, appliance to SM, SM to SP, and SP to appliance. Various techniques for anonymity include PALK, ASF, and TAI [48,49].

3.1.4. Availability

Availability indicates that data, applications, and systems are available to end-users when they are required. Availability is compromised if someone pretends to be an authorized user to access the system and make the network busy [50]. Distributed denial of service (DDoS) is the most basic availability attack. In DDoS attacks, the incoming traffic is originated from multiple sources; therefore, making it difficult for offensive measures to identify a single malfunctioning device. A DDoS attack in IoT devices happens due to lack of security measures. The alternating direction method of multipliers [51] and honeypot game models [52] are used to protect the systems from DDoS attacks.

3.2. *Cyber Attacks on Smart Homes*

In a HAN, various heterogeneous devices are connected to each other. The devices are interoperable and are managed remotely. An adversary is always searching to find an entry point to enter the network for different attacks.

3.2.1. Impersonation Attack

Each device's status—ON/OFF—is saved in the SM memory. Every 15 min the appliance sends its consumption to the SM. If an appliance is compromised and impersonates another device, this can result in a false reading for a time period unless it is detected and recovered, for example, if AC is switched ON and impersonates a fan or light bulb and vice versa, then it has a huge impact on power billing [53]. Even if an appliance impersonates itself as the SM and requests that the other appliances send consumption reports every 15 min, the result could be dangerous and can lead to some disaster or might lead to electricity theft [54].

3.2.2. Eavesdropping

Smart grids are meant not for the electricity supply from grid to home or home to grid but also as a communication channel between a smart home to the SG and also sends various control messages and forecasts the power demand in advance [53]. If an adversary eavesdrops or sneaks into someone's SM, he or she can easily know the homeowner's routine/lifestyle, living habits, and interests (tuned TV channel) as well as the time they go to work and when a person is at home or not. This information may result in compromising the customer's privacy and also can also be used to plan for theft and other activities.

3.2.3. Replay Attack

Smart homes and SG are continuously communicating and sharing information about electricity usage and forecasting the future power demands. If there is a compromised appliance or SM, an adversary can see the consumption report and can replay an old report in place of the current report and can also change the demand-to-supply report or even replay an old control message. For example, if extra power is demanded or an appliance asks to be scheduled for off peak hours, a replay attack could alter the demand to low power or the appliance can be switched ON at once and cause an inconvenience [55].

3.2.4. Alteration Attack

An alteration attack happens when the HAN, an appliance, or the SM is compromised and an adversary maliciously alters the consumption report or forges a message. The forged message or consumption report can lead to false execution, for example, if a message is sent to set the oven to 120 °C but when altered sets the water heating system to 120 °C then it might lead to injuring a person at home or can also lead to system failure or short circuit. Even if a consumption report is forged, it may cause the customer to pay for electricity that he has not consumed [56].

3.2.5. Message Modification Attack

Communication is a key way in which the SG differs from a traditional grid. If there is an adversary between SG and HAN, it can modify the messages sent to or received from the SG/HAN, which may result in a trust deficit between the working entities and thus leading to serious damage at either side [23].

3.2.6. Energy Import/Export Attack

The SG allows for distributed power generation, where a consumer can install the renewable power generation resources at the consumer's premises. It feeds the surpass energy into the national grid and can also demand extra energy resources from the grid when needed [23]. For example, an adversary demands the energy import from the grid, which is not needed, and exports the energy from home to the grid even when it is needed at home [56]. Similarly, if a plug-in electrical vehicle is charged and imports unnecessary energy from the grid, which is not needed at peak hours, it can lead to power shortfalls and load shedding.

4. Advanced Privacy Preserving Scheme and Its Countermeasures

In this section, we study the latest privacy preserving schemes related to user authentication, data aggregation, key management, and CIA triad.

In [32], the authors have reviewed security issues related to smart homes. The purpose is to portray the scenarios that pose a threat to smart homes, which are an essential part of the smart grid. The smart grid security objectives adopted in this paper are confidentiality, integrity, availability, authenticity, authorization, and nonrepudiation attacks. Furthermore, in [32], the authors examine potential cyber and physical security threats in terms of security objectives.

The communication infrastructure of the smart grid, while considering reliability and challenges to security in the smart grid, is provided in [57]. In [58], an emerging technology, i.e., software defined network (SDN), is discussed. A complete overview of the HEMS literature with reference to main principles, setups, and enabling technologies is offered in [59]. The scheme in [60] comprises existing architectures, applications, and prototypes of IoT-assisted SG systems, and provides an overview of IoT-assisted SG systems.

In the studied literature, the following points have been observed:

- The security challenges of the smart grid are discussed and threats are evaluated.

- The existing architectures, prototypes, and communications challenges are discussed.
- The challenges related to interoperability of various technologies at the hardware level of the smart grid are discussed.

In [61], big data collection and management is surveyed. The authors have used an analytical method to study big data and its applications associated with smart grids. The paper gives an insight about the sources of big data in smart grids and real-time processing to predict a pattern for decision making. In [62], a detailed survey on the future wireless communications systems is performed. The authors have reviewed the energy utilization, redistribution, and trading. The authors have performed a comprehensive study of the current literature and have observed the concern about security vulnerabilities of smart grids. In [63], with the use of new IoT technologies, an overview of smart grid security improvements and weaknesses is provided.

4.1. Data Aggregation

Data aggregation is the process of gathering data from several sources and combining it into a variable or report. In smart homes, various appliances are connected to an SM and send their demand/consumption report to the SM. It creates a communications overhead and privacy hole [64]. To avoid this issue, an aggregator is used that collects messages from various appliances and aggregates them into a single message. The following techniques are discussed for data aggregation while preserving the privacy.

In [22], a scheme has been presented, which is based on incremental hash operation. This scheme reports the cost to the operation center instead of energy consumption readings. After an interval of time, the SM calculates the cost of the recorded reading using hash function and sends it to an operation center. The operation center first receives all the consumption costs from different residential areas and then aggregates them for forwarding to utility providers for the verification of integrity. Utility providers sum up all the received values and compare it with the power distribution for that time interval to validate the integrity. If the value of cost and distribution is not equal, the entire consumption reading is discarded automatically.

A framework based on Shamir's secret sharing is proposed in [65] in order to effectively reduce computational overhead and dependency on a single dedicated aggregator. The scheme also prevents the electrical utility from linking its data to a single SM. The architecture describes that the area under the supply of one service provider is divided into subregions. Each SM divides its reading into shares and connects it to several aggregators. The scheme masks the SM form the utility by sending the aggregated reading and reduces the dependency on a single aggregator.

An in-network data aggregation scheme is proposed in [30], which aggregates the data hop by hop. Each appliance has its own chip code and spreads the energy consumption using these chip codes, which are sent to the SM after every time interval. These chip codes are unique among appliances. The SM can extract each appliance's consumption by knowing the chip codes. Since each appliance has its own chip code, any malfunctioning appliance cannot alter the consumption of other appliances.

In [66], a multidimensional aggregation scheme is used to save the communication bandwidth and increase the computational speed of the SM. There is a gateway between the CC and HAN, which receives the encrypted data from a large number of SMs and then aggregates the data before sending it to CC. A TTP is used to mask the gateway from HANs to avoid any mishandling. Any failure or attack on the TTP end can lead to a serious disturbance in communications between the CC and HAN.

Summary: Table 1 provides a detailed aggregation summary of the above analyzed techniques. It is perceived that the majority of the aggregation steps are performed by a separate third party device or CC [22,65–67]. Similarly, in [65,66], the selection of devices for aggregation and their group header nomination also increases computation overhead. The authors of [67] assume that all entities taking part in the communications are secure and resistant to tampering and modification attacks.

Table 1. Comparison of data aggregation schemes.

Schemes	Technique Used	Appliances Aggr:	SM Aggr:	Separate Device for Aggr:	CC Aggr:	Descriptions
[67]	MPC	No	No	No	Yes	**Pros:** Used multiparty computation scheme, universal composition, deals multiple recipients, create subsets of SMs, and fault tolerant. **Cons:** Cost of communication server and communication overhead.
[22]	PPCR	No	No	Yes	Yes	**Pros:** Incremental-hash function used and only cost of consumption is circulated. **Cons:** Aggregation performed by an outside device, i.e., operation center. Did not address various attacks, key generation, and authenticity.
[65]	Distributed aggregation	No	No	Yes	No	**Pros:** Grouping, each group has an aggregator, and slices send randomly to the multiple group aggregator. **Cons:** Creation of groups, selection of group header, reassembling of various slices at the CC end, and communication overhead.
[30]	In-network aggregation	Yes	No	No	No	**Pros:** Before installation devices are authenticated, hop-by-hop aggregation. **Cons:** Creation of chip codes, computation burden, and no procedure to update key.
[66]	MLTD	No	No	Yes	Yes	**Pros:** Blinding factor generated by TTP, provides unforgeability, resistant to MIMT, alteration, and spoofing. **Cons:** A separate device for aggregation is an issue, provides computation and communication overhead, and did not provide key generation and updating process.

4.2. CIA Triad and Anonymity

In this section, we present schemes that ensure CIA triad and anonymity while preserving the privacy of HAN. A scheme proposed in [68] divides the users of a residential area in subsets based on the energy consumption ranges over a period of time. The energy consumption is then summed up for each subset. The TTP and Paillier homomorphic schemes are used to ensure the privacy of data. However, a damaged SM may not report the data correctly and the malfunctioning or misuse of TTP can lead to serious concerns regarding the authenticity of aggregation reports.

In [69], a Q-learning technique, which is based on artificial intelligence, is proposed and presented. The structure is that there are three kinds of information shared between a HAN/BAN or SCC: control flow, data flow, and power flow. Smart appliances and SMs constitute the HAN. Different HANs that are in the same building constitute a BAN. The regional power supplier which manages multiple BANs is called NAN. The NAN sends information such as dispatch instruction, billing, real-time reporting, and uploads the data to the SCC. Before sending data to the control center, the data is distributed to uniformly random secret shares. SCC outsources information to professional cloud server operators to train the Q-Learning model using edge computing. The secret shares are randomly distributed so that cloud servers could not obtain the information. However, if the two servers collude, then it can be a very serious privacy breach. The scheme also has it own protocols for selection and addition and subtraction but, as we know, the honest but curious entities in the network can access the information from the secret shares anytime.

Similarly, in [16], a homomorphic scheme is proposed for smart homes, which consists of home appliances, SM, and a third-party aggregator. The third party aggregator assigns an ID to every appliance at the time of installation. All appliances in a home are similarly arranged in a sequence order as per given IDs. All appliances report their consumption report to an SM. Before sending their consumption, they add homomorphic features and forward data to the aggregator for the current round. The aggregator appliances sum up all received readings, encrypt it with SM's public key, and send to the SM. The SM authenticates the aggregator appliance using a private key.

The SM encrypts the consumption and gives the identity to the SS. After verifying the identity, SS generates the group blind signature and generates the tags for each data block which the CC acquires and matches with the corresponding data block. In this way CC verifies the data integrity. The author supports the scheme by following that if an adversary or SS somehow can obtain the encrypted consumption but could not obtain the CC's private key. This is because in order to guess the private key prime numbers must be used and exact prime numbers are difficult to match in a polynomial equation. Thus, the possibility of compromising the CC's private key is almost negligible. However, the CC is assumed to be honest in this scheme [21].

In many privacy preserving schemes, TTP is certification authority to generate public and private keys. To avoid TTP, Xiaoli et al. presented a secure privacy preserving scheme. At the time of physical configuration each SM is assigned an ID by the CC [70]. The same ID is also registered with the CC. Every time the CC sends a request message to the SM for sending the energy consumption pattern, the request message includes SM, CC ID, and the key material. Using ID and key material, the SM first generates a random number and then a secret key. The SM will encrypt the energy consumption report by using their secret key and current time stamp. The encrypted message is then forward to CC for identity verification and decryption. The CC first verifies the SM identity by its ID and then decrypts the message using the same secret key.

Summary: PPMA [68], LiPSG [69], and lattice-based homomorphic schemes [16] provide confidentiality and integrity, but not anonymity and availability (see Table 2). PPMA and lattice-based homomorphic schemes are resistant against passive and active attacks, but blind signature [21] fails to do this. Similarly, [16,21,68,70] do not update their encryption key.

4.3. User Authentication

Authentication is a process of associating the incoming activation requests with the already set authentication rights [71]. These authentication rights are stored in file systems or databases. When any device sends its consumption to the SM based on the designed schemes, the system allows or denies the request.

In [72], a scheme is designed, which is based on elliptic curve cryptography and consists of three phases, i.e., system-setup phase, registration phase, and key agreement and authentication phase. Initially, in the system setup phase, the trust anchor shares the system parameters using an elliptic curve and publishes these parameters. In the registration phase, the trust anchor generates the private key for both the SM and the SP using Schnorr's signature. After registration, the SM and SP communicate directly without the involvement of a trust anchor. In the last phase, the SM and SP automatically generate a session key and authenticate each other via session and private keys.

In [15], data source authentication and data aggregation are performed for a particular residential area over a defined time period while ensuring the privacy of each user's data aggregation and fault tolerance. This scheme provides a high level of control over data collection and the processing phase in addition to verifying the integrity of the data and validates the data source.

To eradicate computations and communication resources, a lightweight authentication scheme is presented in [73], which is based on a physically-unclonable function. Before any communications, the SM and neighborhood gateway authenticate each other. The SM sends the ID to the neighborhood gateway. The neighborhood gateway checks the SM ID in its database and creates two random numbers, concatenates these numbers with the time stamp, and the result is XoRed with R-response and sent to the SM. The SM authenticates the neighborhood gateway for further communications.

In [21], an SG is divided into three layers. The CC lies in the middle layer and is responsible for generating system parameters, user registrations, and the verification of data. The SM is placed at the lowest layer and monitors/sends real-time consumption; therefore, it is prone to data being tampered or manipulated.

Table 2. Comparative analysis of CIA models and anonymity.

Ref.	Method	Confidentiality	Integrity	Availability	Anonymity	Description
[68]	PPMA	Yes	Yes	No	No	**Pros:** Uses Paillier cryptography, provides individual privacy, and creates parameters for key generation. **Cons:** Shares key parameters on wireless media, uses external device for aggregation, and does not discuss availability and various attacks.
[69]	LiPSG	Yes	Yes	No	No	**Pros:** Uses Q-learning, good at computation, and splits energy consumption into two subsets. **Cons:** Depends on third party server, does not discuss attacks, and has no aggregation point.
[16]	Lattic based scheme	Yes	Yes	No	No	**Pros:** Aggregation performed by every device in rounds, every sender is authenticated by receiver, and except CC, no one can decrypt consumption. **Cons:** Outside network third party issues appliance ID and generates keys and the key remains the same.
[21]	Blind signature	Yes	Yes	Yes	No	**Pros:** Creation of blocks, shares key material instead of secret keys, and provides anonymity and traceability. **Cons:** Ignores smart appliances, has no mechanism for key updating, has no verification of authenticity inside homes, and computes overhead at CC end.
[70]	LWPPS	Yes	Yes	No	No	**Pros:** Shares key material instead of private key, does not involve a third party, and performs encryption and decryption at the CC end. **Cons:** CC sends request message, communications overhead, and does not explain threat model and authenticity.

Similarly, in [74], elliptic curve cryptography is used to authenticate the entities in the SG to preserve the communication between them over a public and insecure channel. First of all, TTP generates all system parameters and then authenticates the SG device and UC in an offline mode. The scheme is robust against certain attacks; however, the pre-loaded system information may affect the computation power of the smart devices.

In [75], the authors have proposed a scheme to achieve anonymity for the SM to avail all the services provided by the UC, without the involvement of TTP. TTP is only responsible for the registration phase, and its role is limited. The SM is supposed to send the consumption report and control signals to UC, which is an aggregator as well as controller for monitoring the energy consumption trends. Authentication will take place between the UC and the SM.

Summary: Table 3 contains a summary of the analyzed techniques for authentication. In [21,74,75], the SM and CC authenticate each other but the appliances are not authenticated. In [72], only the CC performs authentication; the SM and appliance are just relay nodes. Similarly, [74,75] are prone to cyber security attacks and require higher computational cost.

Table 3. Summary of authentication schemes.

Scheme	Appliances Auth:	SM Auth:	CC Auth:	Descriptions
Schnorr's signature [72]	No	No	Yes	**Pros:** Uses Schnorr's signature for authentication and key agreement, low communications and computation cost, and updating secret as well as session keys after intervals. **Cons:** The process of registration is performed by third party, TA. It does not discuss HAN and appliances. No mechanism is defined for aggregation.
EFFECT [15]	No	Yes	Yes	**Pros:** Introduces a threshold value for data aggregation. Provides integrity, authenticity, and availability. **Cons:** The TCA sets up the whole architecture, is responsible for key generation, aggregation, and selection of gateways. Increases computation by using secret sharing.
PUC [73]	No	Yes	No	**Pros:** Prevents inside and outside attacks. Detects any attack easily using PUF. **Cons:** The whole functioning of SM fails if anyone tries a physical attack. Has an expensive chip and lengthy process of initial registration.
Blind signature [21]	No	Yes	Yes	**Pros:** Creates blocks, shares key material instead of secret keys, and provides anonymity and traceability. **Cons:** Ignores smart appliances, no mechanism for key updating, authenticity is verified outside the home, and computation overhead at the CC end.
ECCAuth [74]	No	Yes	Yes	**Pros:** Proposes real model with mutual authentication of devices locally and globally. **Cons:** High in computation and does not address anonymity and denial of service attacks.
IBS [75]	No	Yes	Yes	**Pros:** Minimum role of TTP. TTP is only involved in initial registration, key generation, and parameter creation. Provides anonymity, authenticity, traceability, and confidentiality. **Cons:** Does not explain how to update the private key, aggregation, and DDos attacks. Requires higher computation resources and scalability.

4.4. Key Management

A key is a bit of code encrypting and decrypting the message. Each key has a specific length of code. A strong encryption process requires a high key size [76]. In cryptography, private keys, session keys, and public kesy are frequently used. Below, different HAN models are discussed to illustrate how they used the cryptographic techniques.

In [55], a HAN sends its consumption to the NAN gateway, which is a trusted service provider and an interface between the HAN and utility provider. A NAN is distributed over a village, city, and sometimes over a residential or commercial area. The communication between the utility provider and the SM takes place via a gateway. The gateway should communicate with the SM in an offline mode. However, the proposed scheme establishes a session key using mutual authentication between the SM and gateway.

In HAN, appliances are arranged in two groups [77]. The first group is for one-way communication devices such as light bulbs, chargers, etc., while the second one consists of two-way communication appliances, e.g., electric vehicles, AC, etc. Before deployment, every smart appliance is assigned with an ID and master key. On the basis of the master key, the group header assigns a unique key and group controller key to every smart appliance and SM. The appliances encrypt their consumption using a unique key and send it to the group controller, which forwards it to the SM for further processing and verification. This scheme prevents man-in-the-middle attacks, Sybil attacks, and replay attacks, but ignores key updating.

Similarly, a cloud-based security scheme is proposed in [33] for smart homes, where home appliances are categorized into two different groups. Appliances which performs simple basic functions are placed in group 1. Group 2 contains controllable and monitoring devices which have two-way communication. Both groups have a group header. In this architecture, the SMs are not considered as the part of the smart home. The SM is considered as part of the AMI smart grid. Group headers are responsible for communicating with a home management system or cloud server. HMS is placed in a local cloud, which is controlled by a remote or simple device. Before deployment, every appliance and group header is assigned an ID. Using this ID, HMS generates a group key and shares it with the appliances and the group controller. Appliances use groups to further generate a unique key for communication inside the group. Every appliance before sending consumption or control signals, encrypts the data with unique key that is automatically generated by HMS.

IEC 61850 standard transmits a message in the time limit of 4 ms, which was more suitable than the existing schemes. To overcome the time bounded activity and privacy issues in existing schemes some proposals have been outlined. In [78], an authentication scheme is proposed, which comprises two phases: registration and key agreement. In the registration phase, a secure channel is established between the substation and data center, while in the key agreement phase—on the basis of a secure channel—unique session keys are created for communication and authentication. In the key agreement phase, the substation and data center authenticate each other and then a unique session key is established on the basis of passed parameters, i.e., certificate, ID, random number, and time stamp.

Summary: Table 4 presents the summary of key generation, key updating, and key sharing in [33,55,77–79] schemes. Schemes in [77,78] update their secret keys, but in the schemes discussed in [33,55,79], the secret keys remain the same.

Table 4. Comparative analysis of key management schemes.

Scheme	Key Generation	Key Sharing	Key Updating	Description
COT [77]	Yes	No	Yes	**Pros:** IoMT-based HAN structure, based on mesh topology with a single source of energy collection. **Cons:** Selection of header for group and TA authenticate home appliances.
LAKA [55]	Yes	Yes	No	**Pros:** Lower communication overhead, provides mutual authentication, and key generation and anonymity. **Cons:** Ignores appliances operations and no key updating and traceability.
KMP [79]	Yes	No	No	**Pros:** Uses elliptic curve, provides session key agreement, and used less communication and computation overhead. **Cons:** Involves a third party for authentication, no key updating, traceability, and anonymity.
IEC [78]	Yes	No	Yes	**Pros:** Updates IEC 62,351 standard with respect to symmetric key encryption and creates session key with a transmission of message within 4 ms. **Cons:** No mechanism for data aggregation or communication overhead.

Table 4. *Cont.*

Scheme	Key Generation	Key Sharing	Key Updating	Description
COT-HAN [33]	Yes	Yes	No	**Pros:** The HAN architecture is based on HMS and cloud based infrastructure and a cell phone has the authority to access cloud services. **Cons:** SM is not considered part of the HMS; no proper structure for authentication and the symmetric key value remains the same; vulnerable to MIMA, DDoS, and alteration attacks; and no aggregation point.

4.5. Observations

This section highlights various shortcomings of the schemes discussed under data aggregation, CIA triad and anonymity, user authenticity, and key management. According to the findings, some issues are at hardware level and we need an international standard for the equipment used for the deployment of smart grids. Moreover, the hardware cost is another issue whereby the most important is that a majority of the proposed schemes discuss only one problem—e.g., authenticity or key management—and ignore other features, especially in the HAN context. If a proposed scheme does not consider the HAN model or some of the threads, then during the implementation phase of the smart grid, it can bring a number of severe issues such as cyber security concerns at the application level.

5. Proposed Scheme

In this section, we present a privacy preserving data aggregation scheme that employs the Chinese remainder theorem, one-way hash chain, and properties of modulo n^2 to aggregate the data [68,80,81].

5.1. System Model

In the proposed model, we consider smart home appliances, a dedicated sink node, a smart meter, and a third party trusted authority.

- Smart Home Appliances: Every smart appliance deployed in a HAN is equipped with sensing and communication equipment, which enables the sensing device to report its reading to the smart meter through a dedicated sink node. For simplicity, we can group home appliances based on their functionalities, e.g., lighting, fans, or kitchen appliances (microwave, refrigerator, toaster), etc. All devices with the same functionalities will be placed in the same group. There can be another grouping strategy such as by room with appliances as a separate group. Let us we have N number of appliances, AP. We can divide these appliances into k subsets such as $G_1, G_2, G_3, G_4 \ldots G_k$ where the size of subset G_i is $\mid G_i \mid = G_i$. Since the smart appliances have limited computational power, we do not apply any time consuming and computationally extensive encryption algorithms.Therefore, lightweight security mechanisms are desired for smart appliances.
- Sink Node: First, we will make it clear, that we have a dedicated aggregator device with computational ability installed in each home. This dedicated device will be called a sink node. The sink node is really important as it acts as a relay device between home appliances and the smart meter. In particular, the sink node will aggregate the reading data from all the home appliances and forward the aggregated data to the smart meter. The sink node also applies some rules that help smart meters to identify any external attacks.
- Smart Meter: The smart meter receives the data from the sink node and does some data analytics. Since the data comes from heterogeneous devices, it is not appropriate to directly operate on all data. Therefore, the smart meter first calculates the mean and

variance for each subset, which is a group of particular appliances. For the mean, we have the following equation:

$$M(G_k) = \sum_{G_i \epsilon G_k} x_i / N_k; \quad (1)$$

While for the variance, we can calculate it with:

$$Var(G_k) = \sum_{G_i \epsilon G_k} x_i^2 / N_k - M(G_k)^2; \quad (2)$$

- Trusted Authority: Trusted authority is a trusted third party which initializes the system and manages key generation and other parameters for each entity in the network and assign keys to all the entities in the network including home appliances, sink nodes, and smart meters. Trusted authority will only be active while initiating the system and adding new appliances. It will will be offline afterwards. The trusted party will not participate in the following actions.

5.2. Threat Model

We assume that the trusted authority is a trusted third party and it will not be involved in any misconduct that can compromise the privacy of the HAN while the smart meter and sink node are honest but curious. The smart meter and sink node may be affected by undetected malware and those malware might eavesdrop on the smart appliances. The smart meter and sink node are honest, meaning that they will follow the design protocols. They are also curious, that is, they are also curious about smart appliance's data privacy. They will not collude with each other. Smart appliances are not resourceful, so they are vulnerable to attacks. The attacks that might affect the smart appliances are false data injection by an external attacker or attacks may prevent a device from reporting readings or replay an old message. However, we have a resourceful smart meter and sink node, and the sink node will apply some techniques to check whether an appliance is malfunctioning or it is simply inactive at the time. If an appliance is inactive it will simply send a zero in its reading. The sink node can filter out the false data and will not include false data during the aggregation process.

5.3. Proposed Scheme

In this section, we present the proposed scheme, which consists of system initialization, appliance report, data aggregation, and analysis phases.

System Initialization

The trusted authority is a completely trustworthy party that starts the system. The trusted authority chooses two random prime numbers m and n, where $m = 2m' + 1$ and $n = 2n' + 1$ and $|m| = |n| = k_0$, compute $p = mn$, and $\lambda = lcm(m-1, n-1) = 2m'n'$; and defines a function, as in [82]

$$L(x) = x - 1/m \quad (3)$$

Then, consider that there are N home appliances inside a HAN. The trusted authority chooses $N + 2$ random numbers such that

$$\sum_{i=0}^{N+1} p_i = 0 mod \lambda \quad (4)$$

Suppose that there are k subgroups in the home area network and the maximum communications range for any group is $[0, X_j]$, then we can define the range of data sensing for an appliance as $X = max(X_1, X_2, X_3, \ldots, X_k)$. Note that, the range $[0, X_k]$ is a small message space as compared to Z_n. With this knowledge, the trusted authority chooses k + 1 prime numbers $\alpha_0, n_1, n_2, n_3, \ldots n_k$, and computes

$$\begin{cases} R = n_1 \times n_2 \times \cdots \times n_k \\ R_i = \dfrac{R}{n_i}, \quad y_i \equiv \dfrac{1}{R_i} \bmod n_i \\ \sigma_i = R_i \cdot y_i \end{cases} \tag{5}$$

where all the prime numbers are of the same length, i.e., $\mid n_i \mid = k_1$ for $1 \le i \le k$. The condition of parameters is as follows (as taken from [82])

$$\begin{cases} N \cdot X^2 \le \sigma_0, \quad N \cdot (X^2 + X \cdot \sigma_0) < n_i \\ k_1 \cdot (k+1) + \lg k < |n| \end{cases} \tag{6}$$

which enables us to gather all data in one cipher text. The trusted authority then chooses two secure hash functions h, H where h = $(0,1)^l$ and H = $(0,1)^* \epsilon Z_n^*$ and a random number $t_0 \epsilon (0,1)^l$ as the secret key. As the system initializes, the home appliance will report the power consumption periodically after a specific time. Thus, we divide the reporting time into w time slots for ease. At each time slot, the reporting appliance reports its reading and will send zero when a device is off or inactive. Thus, the trusted authority chooses a random number t_0 and generates a chain of N one way hash functions such as $HC_1, HC_2, HC_3, \ldots, HC_N$ where each chain contains $HC_N = h_{i1}, h_{i2}, \ldots, h_{iw}$ which is of length $(w+1)$, and $h_{iw} \epsilon (0,1)^l$ is a randomly chosen number.

$$h_{ij} = h(h_{i(j+1)} \parallel T_j) j = 0, 1, 2, \ldots, w-1 \tag{7}$$

For each h_{ij}, $1 \le j \le w$, the trusted authority will also compute its corresponding key

$$key_{ij} = h(h_{ij} \parallel t_0) \tag{8}$$

The proposed scheme utilizes the property of a one-time password for authentication and encryption. For that purpose, we have h_{ij} and key_{ij} in time slot T_j. The header of each hash chain $h_{10}, h_{20}, \ldots, h_{N0}$ will be signed with α by the third party trusted authority to ensure the validity of the hash chains for authentication. The scheme employs the AES algorithm for home appliances for encrypting the reading before sending it to the sink node. We have public parameters for the system, *parameter*: N, n_i: i = 1, 2, 3, ..., k, σ_j: (j = 1, 2, 3, ..., k), h, H, L(x), AES. Then, we have the public parameters for the system, so we will calculate key and assign it to the network entities.

- Every smart home appliance is assigned with a private key p_i, secret hash chain HC_i, corresponding keys $K = key_{i0}, key_{i1}, key_{i2}, \ldots, key_{iw}$, and public *parameter*, which are shared over a secure channel.
- Then, a random number is chosen as the *shared key sk* between the sink node and the smart meter, which we assign to the sink node along with signed hash chain heads $h_{10}, h_{20}, h_{30}, \ldots, h_{N0}, \alpha$, and secret key p_{N+1}, t_0 and corresponding public *parameters* to the sink node.
- The smart meter is assigned the same key that is shared between the sink node and the smart meter and a secret key p_0, β, along with the public *parameters*.

5.4. Home Appliance Reporting

At every time slot T_s, each appliance will report its reading to the sink node by calculating the following:

- Step 1: Appliance uses its secret key p_i and (σ_0, σ_j) to compute

$$c_{ip} = [1 + n \cdot \sigma_j \cdot (x_i \cdot \sigma_0 + x_i^2)] \cdot H(T_s)^{n \cdot p_i} \bmod n^2 \tag{9}$$

and uses the key key_{ip} to compute C_{ip} = $AES_{key_{ip}}(c_{ip})$. This method is used to prevent any external attacker from knowing the readings or to avoid the worst-case scenario that HAN might communicate with an unauthorized or compromised sink node.

- Step 2: AP_i uses the hash value h_{ip} from hash chain vector to compute

$$mac_{ip} = h(C_{ip}||h_{ip}) \quad (10)$$

- Step 3: The appliance then forwards the (C_{ip}, h_{ip}, mac_{ip}) to the sink node. The appliance can efficiently compute these parameters, especially if $H(T_s)^{n \cdot p_i}$ is computed in advance.

5.5. Sink Node Data Aggregation

Upon receiving the (C_{ip}, h_{ip}, mac_{ip}) in time slot T_s, the sink node checks whether the data is sent by an authenticated sender.

- Step 1: The sink node holds h_{i0} from α; thus, the authenticity of each h_{ij} on the hash chain HC_i is simple to check. The sink node compares the freshness of the received hash value with the previously received hash values. If h_{ip} has never been received before, h_{ip} is accepted, otherwise, it is refused.
- Step 2: If h_{ip} is valid, the sink node will compute mac'_{ip} by comparing it with the received value to check whether C_{ip} has been altered or not.

$$mac'_{ip} = h(C_{ip}||h_{ip}) \quad (11)$$

- Step 3: If C_{ip} is accepted, the sink node computes key_{ip} = h(h_{ip} | | t_0) and uses key_{ip} to reproduce c_{ip} from C_{ip} = $AES_{key_{ip}}(c_{ip})$.

After verifying the encrypted readings received from all home appliances, the sink node runs the following data aggregation operation and calculates a single cipher text C_p and sends (C_p, mac_p) to the smart meter.

$$\begin{cases} C_p = \left(\prod_{i=1}^{N} c_{ip}\right) \cdot H(T_s)^{n \cdot p_{N+1}} \bmod n^2 \\ mac_s = h(C_p||T_s||sk) \end{cases} \quad (12)$$

5.6. Smart Meter

Upon receiving the (C_p,mac_p) in time slot T_s, the smart meter verifies mac_p using the shared key between the sink node and smart meter, as in [82]. Then, it validates the C_p. If C_p is valid, the smart meter performs a report reading and analyses the received aggregated reading. Moreover, it calculates $H(T_s)^{n \cdot p_0}$ through its secret keys. Next, the smart meter computes using the following equations (as adapted from [82])

$$C'_p = C_p \cdot H(T_s)^{n \cdot p_0} \bmod n^2 \quad (13)$$

$$C'_p = \prod_{i=1}^{N} c_{ip} \cdot H(T_s)^{n \cdot (p_0 + p_{N+1})} \bmod n^2 \quad (14)$$

$$C'_p = \prod_{i=1}^{N} [1 + n \cdot \sigma_j \cdot (x_i \cdot \sigma_0 + x_i^2)] \cdot H(T_s)^{n \cdot p_i} \bmod n^2 \quad \cdot H(T_s)^{n \cdot (p_0 + p_{N+1})} \bmod n^2 \quad (15)$$

$$C'_p = \prod_{i=1}^{N} [1 + n \cdot \sigma *_j \cdot (x_i \cdot \sigma *_0 + x_i^2)] \cdot \prod_{i=1}^{N+1} H(T_s)^{n \cdot p_i} \bmod n^2 \quad (16)$$

$$C'_p = \prod_{i=1}^{N} [1 + n \cdot \sigma *_j \cdot (x_i \cdot \sigma_0 + x_i^2)] \cdot H(T_s)^{n \cdot \sum_{i=1}^{N+1} p_i} \bmod n^2 \quad (17)$$

$$C'_p = \prod_{i=1}^{N}[1+n \cdot \sigma *_j \cdot (x_i \cdot \sigma_0 + x_i^2)] \cdot H(T_s)^{n \cdot \lambda \cdot k} \bmod n^2 \tag{18}$$

$$C'_p = \prod_{i=1}^{N}[1+n \cdot \sigma *_j \cdot (x_i \cdot \sigma_0 + x_i^2)] \bmod n^2 \tag{19}$$

$$C'_p = 1+n \cdot \sum_{i=1}^{N}[\sigma *_j \cdot (x_i \cdot \sigma_0 + x_i^2)] \bmod n^2 \tag{20}$$

$$C'_p = 1+n \cdot \sum_{j=1}^{k} \sigma_j (\sum_{G_i \epsilon \mathcal{G}_j} (x_i \cdot \sigma_0 + x_i^2)) \bmod n^2 \tag{21}$$

The smart meter has the ability to calculate its mean and variance as

$$M_j = M \bmod n_j = \sum_{i=1}^{N_j} (x_i \cdot \sigma_0 + x_i^2) \tag{22}$$

$$E(\mathcal{G}_j) = \frac{M_j - (M_j \bmod \sigma_0)}{\sigma_0 \cdot N_j} \tag{23}$$

$$Var(\mathcal{G}_j) = \frac{M_j \bmod \sigma_0}{N_j} - E(\mathcal{G}_j)^2 \tag{24}$$

Fault Tolerance

In cases where an appliance is not reporting according to protocols, the sink node will aggregate the data received from other appliances and send the results to the smart meter and inform the smart meter that the appliance AP_a is malfunctioning. The smart meter then uses the following method to calculate the mean and variance for the malfunctioning device.

- Step 1: C_p^* is mathematically represented as

$$C_p^* = \left(1 + n \cdot \sum_{i=1, i \neq a}^{N} \sigma_j^* \cdot (x_i \cdot \sigma_0 + x_i^2)\right) \cdot \prod_{i=1, i \neq a}^{N+1} H(T_s)^{n \cdot p_i} \bmod n^2 \tag{25}$$

Therefore, the smart meter computes

$$\begin{aligned} M_s^* &= C_p^{*\lambda} \bmod n^2 \\ &\xrightarrow{(1+n \cdot x)^{\lambda} \equiv (1+n \cdot \lambda x) \bmod n^2, \quad x^{\lambda n} \equiv 1 \bmod n^2} \\ &= 1 + n \cdot \lambda \cdot \sum_{i=1, i \neq a}^{N} \sigma_j^* \cdot (x_i \cdot \sigma_0 + x_i^2) \bmod n^2 \end{aligned} \tag{26}$$

and M can be calculated as

$$M = \left(\frac{M_s^* - 1}{n \cdot \lambda} \bmod n\right) \bmod Q \tag{27}$$

- Step 2: Except for the subset containing malfunctioning devices, the smart meter calculates mean and variance using Equations (13) and (15).

- Step 3: For the subset containing malfunctioning devices, the smart meter computes M_b from Equation (13) and gains the mean and variance through

$$E(\mathcal{G}_b) = \frac{M_b - (M_b \bmod \sigma_0)}{\sigma_0 \cdot (N_b - 1)} \tag{28}$$

$$Var(\mathcal{G}_b) = \frac{M_b \bmod \sigma_0}{N_b - 1} - E(\mathcal{G}_b)^2 \tag{29}$$

Hence, the proposed approach is still workable even if some devices are malfunctioning. As a result, the proposed approach satisfies the need for fault tolerance.

6. Security Analysis

Here, we will determine that how the proposed approach can achieve the prevention of false data injection attacks and privacy preserving data aggregation.

6.1. Prevent False Data Injection

In the proposed scheme, the trusted authority uses one-way hash function, generates hash chains, and assigns a hash chain to each device for every time slot T_s. For every home appliance and for each time slot T_{s-1}, we have a hash value $h_{i(p-1)}$. From $h_{i(p-1)} = h(h_{ip}||T_s)$, we can authenticate h_{ip} from T_s. However, because of the one-way nature of the hash function, we cannot obtain the h_{ip} from $h_{i(p-1)}$. Moreover, since every device reports its reading directly to the sink node—and only if a device reports the correct data—we will receive the fresh h_{ip} as we have assumed that the device will not act abnormally. If the h_{ip} is not fresh in the time slot T_s, it means that it has been attacked and the false data is injected externally or the device is compromised and a replay attack has been launched externally. Therefore, the sink node will reject the false data. Thus, it is ensured that the proposed scheme is resistant to false data injection attacks.

6.2. Privacy Preserving

In the proposed scheme, if we consider the encrypted text by a home appliance

$$c_{ip} = [1 + n \cdot \sigma_j \cdot (x_i \cdot \sigma_0 + x_i^2)] \cdot H(T_s)^{n \cdot p_i} \bmod n^2 \tag{30}$$

and that of the aggregated cipher text [82]

$$C_p^* = \left(1 + n \cdot \sum_{i=1, i \neq a}^{N} \sigma_j^* \cdot (x_i \cdot \sigma_0 + x_i^2)\right) \cdot \prod_{i=1, i \neq a}^{N+1} H(T_s)^{n \cdot p_i} \bmod n^2$$

and take $\sigma_j \cdot (x_i \cdot \sigma_0 + x_i^2)$ as a $\overline{message}$ and $\cdot H(T_s)^{\cdot p_i}$ as a random number $\overline{rand}$ and $\sum_{i=1,i\neq a}^{N} \sigma_j^* \cdot (x_i \cdot \sigma_0 + x_i^2)$ as $\overline{Message}$ and $\prod_{i=1,i\neq a}^{N+1} H(T_s)^{\cdot p_i}$ as $\overline{Rand}$ then

$$c_{ip} = [1 + n \cdot \overline{message}] \cdot \overline{rand}^n \bmod n^2 \tag{31}$$

and

$$C_p^* = (1 + n \cdot \overline{Message}) \cdot \overline{Rand}^n \bmod n^2$$

are valid Paillier ciphers. Under the chosen plaintext attack, Paillier encryption is indistinguishable, and an external attacker cannot obtain the exact message. The sink node may be curious about the exact message. However, without the knowledge of secret keys p_0 and λ, the sink node has no knowledge of the information. Whereas the smart meter can recover $\overline{Message}$ and may want to recover the message sent by individual devices. For that purpose, it has to collude with the sink node, which is not possible under the threat model. Hence, the suggested technique protects aggregated data privacy.

7. Performance Evaluation

In this section, we analyze the communication and processing overhead of household appliances, sink nodes, and smart meters.

7.1. Communication Overhead

The proposed privacy preserving scheme aggregates the data from different subsets into one and the smart meter can recover the mean and variance of the individual subset. To demonstrate the efficiency of the proposed technique, we compare it to the basic Paillier encryption [80] in which the bit length of n^2 is 2048 and that of n is 1024. Therefore, the communication overhead from N devices to sink node is 2048 × N bits, because each home appliance encrypts both x_i and x_i^2 into one cipher text, as shown in Equation (9). However, because the data in the BPE is encrypted into two cipher texts, the transmission cost is doubled, and the overhead is 4096 × N bits.

In the proposed scheme, the sink node and smart meter are independent of the number of devices as the aggregation is undertaken at the sink node. In BPE, the communications cost is dependent on the number of subsets. If there are k subsets, then the communications cost is 4096 × k bits, while in the proposed scheme, all data have been aggregated in one cipher text. Therefore, the communications overhead from the sink node to the SM is only 2048 bits. Figure 2 plots a graph for the communications overhead from the home appliances to the sink node, and Figure 3 shows the communications overhead pattern of the sink node to the SM. As a result, it is obvious that the suggested method is superior to BPE in terms of communication costs. We use the Chinese remainder theorem that enables a careful parameter choice and, hence, in real-time scenarios the message size is small.

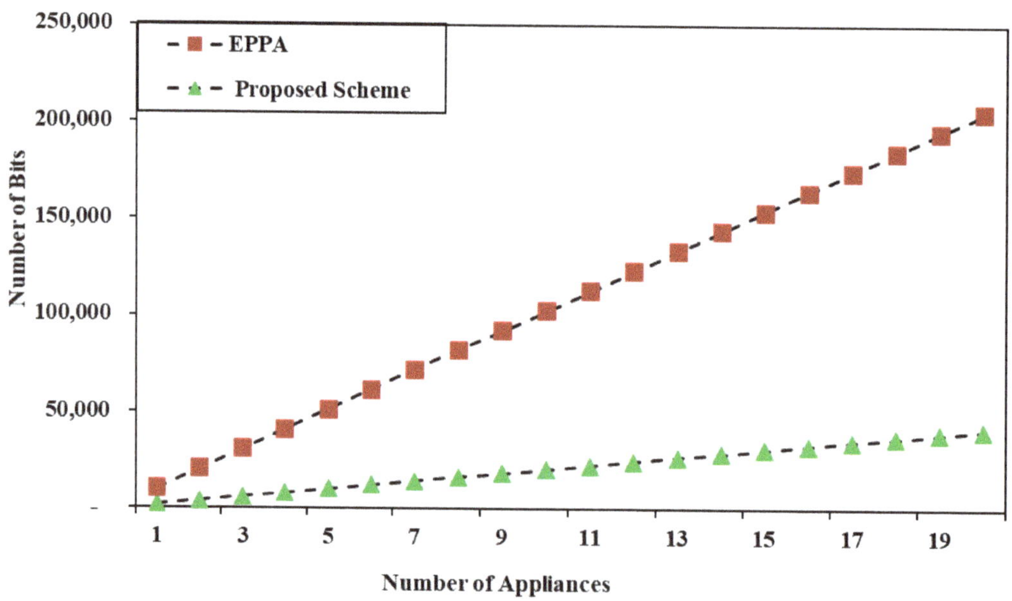

Figure 2. Communication overhead from home appliances to sink node.

7.2. Computation Overhead

The suggested system is lightweight in terms of communication costs since we only use the time-consuming modulo operation, and each node in the network, such as a sink node, a smart meter, and a home appliance, has at least one modulo operation. The proposed approach will become more efficient if the modulo exponent is computed ahead of time.

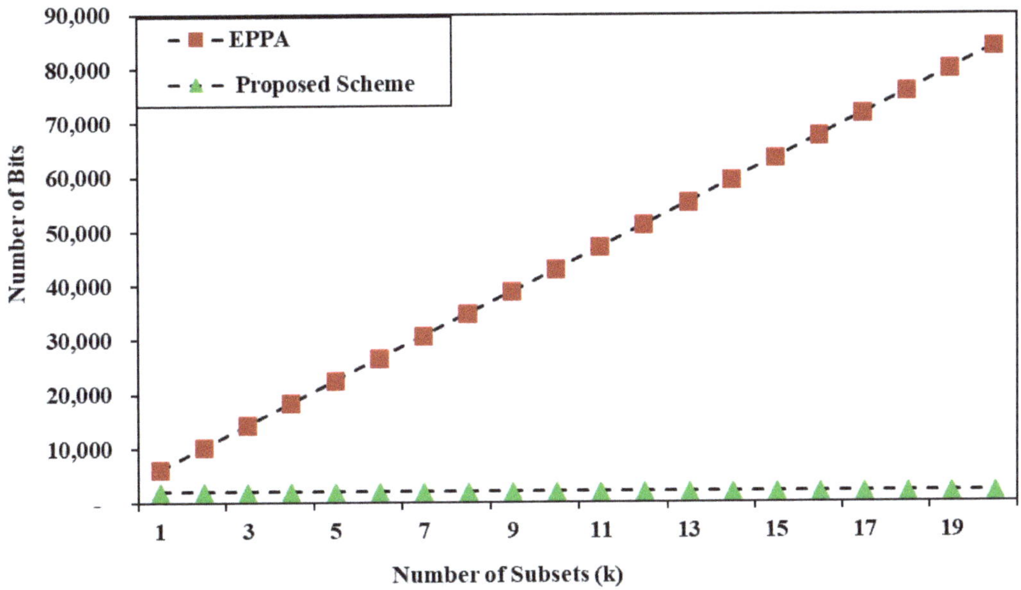

Figure 3. Communication overhead from sink node to smart meter.

8. Future Directions and Challenges

In this paper, we have discussed smart grids and the relation among different entities of the smart grid and how they interact with each other. After that, we reviewed the literature regarding countermeasures to the posed threats and discussed some promising solutions that are suggested in order to overcome privacy related issues and threats. Lastly, we proposed a novel privacy preservation scheme for HANs. To complete the effort, we have devised a direction for future research and challenges, which are discussed below.

8.1. A Safe and Secure Trust Mechanism for Home Incumbents in Smart Grid

As discussed in Section 1, smart grids work through the coordination of different entities. The smart gird is a network of different entities and different subnetworks working together. Each entity and subnetwork has its own requirements. The continuous communication is essential to ensure the smart grid remains active. Iinteroperable and uninterrupted communication between the different subnetworks is an intimidating task. Therefore, a universal standardized trusted framework is essential for any communication.

8.2. Government Authorities to Regulate and Maintain Smart Grids

Most of the research work performed is voluntary and in order to make smart grids a success, it is necessary to have a government authority to evaluate the standard and conformity of the research undertaken on smart grids. Thus, the authorities can make the necessary decisions and improvements needed to regulate smart grids.

8.3. New Goals and Standards to Evaluate Privacy Preserving Mechanism and Solution

The authorities should set standards and new metrics to evaluate any new research or protocols. Each new research idea should be evaluated on common standards and then a decision should be made whether to make it a standard or revise an old standard.

8.4. Legal Code for Preserving the Privacy

A legal framework through the contributions of both governments and business authorities should be made to protect the privacy of consumers and other network entities. The legal framework would help to set the standard to what extent a user's data can be collected and how it can be manipulated to further increase the efficiency of the smart grid. In cases of breach of this contract, the legal framework should outline what the consequences would be faced by the entity breaching the contract.

8.5. A Framework for Aggregation without Third Party Involvement

The third-party involvement in the aggregation can be compromised anytime. A scheme should be devised that could aggregate the data without knowing the meaning of the aggregated data that could not harm or lead the smart grid to instability.

9. Conclusions

We proposed a novel privacy-preserving data aggregation approach for HANs in a smart grid in this paper. The proposed approach deploys a sink node between the household appliances and the smart meter, which not only filters false data injection attacks but also provides for early fault tolerance. The technology also combines data from several home appliances, which may belong to distinct subsets, into a single stream and sends it to the smart meter. The suggested technique is secure, and the performance evaluation reveals that it is more efficient in terms of communications and computation overhead than aggregation using basic Paillier encryption. Detecting and avoiding new threats, IDS architectures for smart grid privacy, IoT-driven smart grids, new privacy metrics, and privacy for IoT are all demanding research fields that need to be further explored in the future.

Author Contributions: Conceptualization, W.A., I.U.D., and A.A.; methodology, I.U.D. and A.A.; software, W.A.; validation, A.A.; formal analysis, I.U.D. and B.-S.K.; investigation, A.A. and B.-S.K.; resources, B.-S.K.; data curation, W.A.; writing—original draft preparation, W.A.; writing—review and editing, I.U.D. and A.A.; visualization, B.-S.K. and A.A.; supervision, I.U.D.; project administration, A.A.; funding acquisition, B.-S.K. All authors have read and agreed to the published version of the manuscript.

Funding: This work was supported in part by the National Research Foundation (NRF), Korea, under Project BK21 FOUR (F21YY8102068) and in part by King Saud University, Riyadh, Saudi Arabia, through Researchers Supporting Project Number RSP-2021/184.

Acknowledgments: This work was supported in part by the National Research Foundation (NRF), Korea, under Project BK21 FOUR (F21YY8102068) and in part by King Saud University, Riyadh, Saudi Arabia, through Researchers Supporting Project Number RSP-2021/184.

Conflicts of Interest: The authors declare no conflict of interest.

Abbreviations

Abbreviation	Definition
ADDM	Distributed State Estimation
Aggr:	Aggregator
Auth:	Authentication
BAN	Building area network
BPE	Basic Paillier Encryption
CC	Control center
CIA	Confidentiality, integrity, and availability
COT	Clouds-of-things
COT-HAN	COT home-area-network
DDOS	Denial-of-service attack
DR	Demand respond
EPPA	Efficient Privacy-Preserving Aggregation

Abbreviation	Definition
HAN	Home-area-network
H-MAC	Hashed based MAC
HMS	Home management system
IBS	Identity based signature
ID	Identification
KMP	Key-Management protocol
LAKA	Lightweight Authentication key agreement
LPPS	Lightweight privacy preservation
LWPPS	Lightweight-privacy-preserving
MAC	Message authentication code
MLAN	Metropolitan LAN
MLTD	Multidimensional aggregation
MPC	Multiparty computations
NAN	Neighbour area network
PALK	Password-based anonymous lightweight key
PK	Private key
PPCR	Privacy-preserving cheat resilient
PPMA	Privacy-preserving multi-subset
PS	Power supplier
PUC	Physical unclonable function
SA	Smart Appliance
SG	Smart grid
SM	Smart meter
SP	Service provider
TAI	Threshold-Based Anonymous Identification
TTP	Trusted third party
WAN	Wide area network

References

1. Bagri, D.; Rathore, S.K. Research Issues Based on Comparative Work Related to Data Security and Privacy Preservation in Smart Grid. In Proceedings of the 2018 4th IEEE International Conference on Computing Sciences (ICCS), Phagwara, India, 30–31 August 2018; pp. 88–91.
2. Dileep, G. A survey on smart grid technologies and applications. *Renew. Energy* **2020**, *146*, 2589–2625. [CrossRef]
3. Lopez, G.; Matanza, J.; De La Vega, D.; Castro, M.; Arrinda, A.; Moreno, J.I.; Sendin, A. The role of power line communications in the smart grid revisited: applications, challenges, and research initiatives. *IEEE Access* **2019**, *7*, 117346–117368. [CrossRef]
4. Saxena, N.; Choi, B.J. Integrated distributed authentication protocol for smart grid communications. *IEEE Syst. J.* **2016**, *12*, 2545–2556. [CrossRef]
5. Boyapally, H.; Mathew, P.; Patranabis, S.; Chatterjee, U.; Agarwal, U.; Maheshwari, M.; Dey, S.; Mukhopadhyay, D. Safe is the new Smart: PUF-based Authentication for Load Modification-Resistant Smart Meters. *IEEE Trans. Dependable Secur. Comput.* **2020**, *19*, 663–680.
6. Greer, C.; Wollman, D.; Prochaska, D.; Boynton, P.; Mazer, J.; Nguyen, C.; FitzPatrick, G.; Nelson, T.; Koepke, G.; , Hefner, A., Jr.; et al. NIST Framework and Roadmap for Smart Grid Interoperability Standards, Release 3.0, Special Publication (NIST SP), National Institute of Standards and Technology, Gaithersburg, MD, 2014. Available online: https://tsapps.nist.gov/publication/get_pdf.cfm?pub_id=916755 (accessed on 2 December 2021). [CrossRef]
7. Kumar, A.; Agarwal, A. Research issues related to cryptography algorithms and key generation for smart grid: A survey. In Proceedings of the 2016 7th IEEE India International Conference on Power Electronics (IICPE), Patiala, India, 17–19 November 2016; pp. 1–5.
8. Alahakoon, D.; Yu, X. Smart electricity meter data intelligence for future energy systems: A survey. *IEEE Trans. Ind. Inform.* **2015**, *12*, 425–436. [CrossRef]
9. Alamatsaz, N.; Boustani, A.; Jadliwala, M.; Namboodiri, V. Agsec: Secure and efficient cdma-based aggregation for smart metering systems. In Proceedings of the 2014 IEEE 11th Consumer Communications and Networking Conference (CCNC), Las Vegas, NV, USA, 10–13 January 2014; pp. 489–494.
10. Kumar, P.; Lin, Y.; Bai, G.; Paverd, A.; Dong, J.S.; Martin, A. Smart grid metering networks: A survey on security, privacy and open research issues. *IEEE Commun. Surv. Tutorials* **2019**, *21*, 2886–2927. [CrossRef]
11. Desai, S.; Alhadad, R.; Chilamkurti, N.; Mahmood, A. A survey of privacy preserving schemes in IoE enabled smart grid advanced metering infrastructure. *Clust. Comput.* **2019**, *22*, 43–69. [CrossRef]
12. Llaria, A.; Dos Santos, J.; Terrasson, G.; Boussaada, Z.; Merlo, C.; Curea, O. Intelligent Buildings in Smart Grids: A Survey on Security and Privacy Issues Related to Energy Management. *Energies* **2021**, *14*, 2733. [CrossRef]

13. Tan, S.; De, D.; Song, W.Z.; Yang, J.; Das, S.K. Survey of security advances in smart grid: A data driven approach. *IEEE Commun. Surv. Tutorials* **2016**, *19*, 397–422. [CrossRef]
14. Ferrag, M.A.; Babaghayou, M.; Yazici, M.A. Cyber security for fog-based smart grid SCADA systems: Solutions and challenges. *J. Inf. Secur. Appl.* **2020**, *52*, 102500. [CrossRef]
15. Guan, Z.; Zhang, Y.; Zhu, L.; Wu, L.; Yu, S. EFFECT: An efficient flexible privacy-preserving data aggregation scheme with authentication in smart grid. *Sci. China Inf. Sci.* **2019**, *62*, 32103. [CrossRef]
16. Abdallah, A.; Shen, X.S. A lightweight lattice-based homomorphic privacy-preserving data aggregation scheme for smart grid. *IEEE Trans. Smart Grid* **2016**, *9*, 396–405. [CrossRef]
17. Akila, V.; Sheela, T. Preserving data and key privacy in Data Aggregation for Wireless Sensor Networks. In Proceedings of the 2017 2nd IEEE International Conference on Computing and Communications Technologies (ICCCT), Russia, Moscow, 20–22 September 2017; pp. 282–287.
18. Kement, C.E.; Gultekin, H.; Tavli, B. A Holistic Analysis of Privacy Aware Smart Grid Demand Response. *IEEE Trans. Ind. Electron.* **2020**, *68*, 7631–7641. [CrossRef]
19. Chen, Z.; Wu, L. Residential appliance DR energy management with electric privacy protection by online stochastic optimization. *IEEE Trans. Smart Grid* **2013**, *4*, 1861–1869. [CrossRef]
20. Liu, Y.; Guo, W.; Fan, C.I.; Chang, L.; Cheng, C. A practical privacy-preserving data aggregation (3PDA) scheme for smart grid. *IEEE Trans. Ind. Inform.* **2018**, *15*, 1767–1774. [CrossRef]
21. Kong, W.; Shen, J.; Vijayakumar, P.; Cho, Y.; Chang, V. A practical group blind signature scheme for privacy protection in smart grid. *J. Parallel Distrib. Comput.* **2020**, *136*, 29–39. [CrossRef]
22. Yip, S.C.; Wong, K.; Phan, R.C.W.; Tan, S.W.; Ku, I.; Hew, W.P. A Privacy-Preserving and Cheat-Resilient electricity consumption reporting Scheme for smart grids. In Proceedings of the 2014 IEEE International Conference on Computer, Information and Telecommunication Systems (CITS), Jeju, Korea, 7–9 July 2014; pp. 1–5.
23. Diao, F.; Zhang, F.; Cheng, X. A privacy-preserving smart metering scheme using linkable anonymous credential. *IEEE Trans. Smart Grid* **2014**, *6*, 461–467. [CrossRef]
24. Asghar, M.R.; Dán, G.; Miorandi, D.; Chlamtac, I. Smart meter data privacy: A survey. *IEEE Commun. Surv. Tutorials* **2017**, *19*, 2820–2835. [CrossRef]
25. Bedi, G.; Venayagamoorthy, G.K.; Singh, R. Internet of Things (IoT) sensors for smart home electric energy usage management. In Proceedings of the 2016 IEEE International Conference on Information and Automation for Sustainability (ICIAfS), St. Gallen, Switzerland, 16–19 December 2016; pp. 1–6.
26. Gholinejad, H.R.; Loni, A.; Adabi, J.; Marzband, M. A hierarchical energy management system for multiple home energy hubs in neighborhood grids. *J. Build. Eng.* **2020**, *28*, 101028. [CrossRef]
27. Son, Y.S.; Pulkkinen, T.; Moon, K.D.; Kim, C. Home energy management system based on power line communication. *IEEE Trans. Consum. Electron.* **2010**, *56*, 1380–1386. [CrossRef]
28. Chen, Y.; Martínez-Ortega, J.F.; Castillejo, P.; López, L. A homomorphic-based multiple data aggregation scheme for smart grid. *IEEE Sens. J.* **2019**, *19*, 3921–3929. [CrossRef]
29. Agarkar, A.; Agrawal, H. R-LWE based lightweight privacy preserving scheme for Smart Grid. In Proceedings of the 2016 IEEE International Conference on Computing, Analytics and Security Trends (CAST), Pune, India, 19–21 December 2016; pp. 410–415.
30. Yan, Y.; Qian, Y.; Sharif, H. A secure data aggregation and dispatch scheme for home area networks in smart grid. In Proceedings of the 2011 IEEE Global Telecommunications Conference-GLOBECOM 2011, Houston, TX, USA, 5–9 December 2011; pp. 1–6.
31. Tizazu, G.A.; Hussen, H.R.; Kim, K.H. Secure session key exchange scheme for smart grid home area networks. In Proceedings of the 2013 IEEE International Conference on ICT Convergence (ICTC), Jeju Island, Korea, 20–22 October 2013; pp. 1116–1120.
32. Komninos, N.; Philippou, E.; Pitsillides, A. Survey in smart grid and smart home security: Issues, challenges and countermeasures. *IEEE Commun. Surv. Tutorials* **2014**, *16*, 1933–1954. [CrossRef]
33. Alohali, B.; Merabti, M.; Kifayat, K. A secure scheme for a smart house based on Cloud of Things (CoT). In Proceedings of the 2014 6th IEEE Computer Science and Electronic Engineering Conference (CEEC), Colchester, UK, 25–26 September 2014; pp. 115–120.
34. Li, T.; Ren, J.; Tang, X. Secure wireless monitoring and control systems for smart grid and smart home. *IEEE Wirel. Commun.* **2012**, *19*, 66–73.
35. Gong, H.; Ionel, D.M. Optimization of Aggregated EV Power in Residential Communities with Smart Homes. In Proceedings of the 2020 IEEE Transportation Electrification Conference & Expo (ITEC), Chicago, IL, USA, 23–26 June 2020; pp. 779–782.
36. Martins, P.E.T.; Oleskovicz, M.; da Silva Pessoa, A.L. A Survey on Smart Grids: concerns, advances, and trends. In Proceedings of the 2019 IEEE PES Innovative Smart Grid Technologies Conference-Latin America (ISGT Latin America), Gramado City, Brazil, 15–18 September 2019; pp. 1–6.
37. Molderink, A.; Bakker, V.; Bosman, M.G.; Hurink, J.L.; Smit, G.J. Domestic energy management methodology for optimizing efficiency in smart grids. In Proceedings of the 2009 IEEE Bucharest PowerTech, Bucharest, Romania, 28 June–2 July 2009; pp. 1–7.
38. Thomas, D.; Deblecker, O.; Ioakimidis, C.S. Optimal operation of an energy management system for a grid-connected smart building considering photovoltaics' uncertainty and stochastic electric vehicles' driving schedule. *Appl. Energy* **2018**, *210*, 1188–1206. [CrossRef]

39. Shakeri, M.; Pasupuleti, J.; Amin, N.; Rokonuzzaman, M.; Low, F.W.; Yaw, C.T.; Asim, N.; Samsudin, N.A.; Tiong, S.K.; Hen, C.K.; et al. An Overview of the Building Energy Management System Considering the Demand Response Programs, Smart Strategies and Smart Grid. *Energies* **2020**, *13*, 3299. [CrossRef]
40. Pindoriya, N.M.; Dasgupta, D.; Srinivasan, D.; Carvalho, M. Infrastructure security for smart electric grids: A survey. In *Optimization and Security Challenges in Smart Power Grids*; Springer: Berlin/Heidelberg, Germany, 2013; pp. 161–180.
41. Walgama, S.; Hasinthara, U.; Herath, A.; Daranagama, K.; Kumarawadu, S. An Optimal Electrical Energy Management Scheme for Future Smart Homes. In Proceedings of the 2020 IEEE 8th International Conference on Smart Energy Grid Engineering (SEGE), Oshawa, ON, Canada, 12–14 August 2020; pp. 137–141.
42. Paukstadt, U. *A Survey of Smart Energy Services for Private Households*. In Proceedings of the 14th International Conference on Wirtschaftsinformatik, Siegen, Germany, 24–27 February, 2019; pp. 1448–1462.
43. Li, H.; Lu, R.; Zhou, L.; Yang, B.; Shen, X. An efficient merkle-tree-based authentication scheme for smart grid. *IEEE Syst. J.* **2013**, *8*, 655–663. [CrossRef]
44. Shen, H.; Liu, Y.; Xia, Z.; Zhang, M. An Efficient Aggregation Scheme Resisting on Malicious Data Mining Attacks for Smart Grid. *Inf. Sci.* **2020**, *526*, 289–300. [CrossRef]
45. Fan, X.; Gong, G. Security challenges in smart-grid metering and control systems. *Technol. Innov. Manag. Rev.* **2013**, *3*, 42–49. [CrossRef]
46. Ge, L.; Yu, W.; Moulema, P.; Xu, G.; Griffith, D.; Golmie, N. Detecting Data Integrity Attacks in Smart Grid. *Secur. Priv. Cyber-Phys. Syst. Found. Princ. Appl.* **2017**, 281–303.
47. Sui, Z.; Niedermeier, M. TAI: A threshold-based anonymous identification scheme for demand-response in smart grids. *IEEE Trans. Smart Grid* **2016**, *9*, 3496–3506. [CrossRef]
48. Kumar, P.; Braeken, A.; Gurtov, A.; Iinatti, J.; Ha, P.H. Anonymous secure framework in connected smart home environments. *IEEE Trans. Inf. Forensics Secur.* **2017**, *12*, 968–979. [CrossRef]
49. Khan, A.A.; Kumar, V.; Ahmad, M.; Rana, S.; Mishra, D. PALK: Password-based anonymous lightweight key agreement framework for smart grid. *Int. J. Electr. Power Energy Syst.* **2020**, *121*, 106121. [CrossRef]
50. Kim, M. A survey on guaranteeing availability in smart grid communications. In Proceedings of the 2012 14th IEEE International Conference on Advanced Communication Technology (ICACT), PyeongChang, Korea, 19–22 February 2012; pp. 314–317.
51. Du, D.; Li, X.; Li, W.; Chen, R.; Fei, M.; Wu, L. ADMM-based distributed state estimation of smart grid under data deception and denial of service attacks. *IEEE Trans. Syst. Man Cybern. Syst.* **2019**, *49*, 1698–1711. [CrossRef]
52. Wang, K.; Du, M.; Maharjan, S.; Sun, Y. Strategic honeypot game model for distributed denial of service attacks in the smart grid. *IEEE Trans. Smart Grid* **2017**, *8*, 2474–2482. [CrossRef]
53. Mahmud, R.; Vallakati, R.; Mukherjee, A.; Ranganathan, P.; Nejadpak, A. A survey on smart grid metering infrastructures: Threats and solutions. In Proceedings of the 2015 IEEE International Conference on Electro/Information Technology (EIT), DeKalb, IL, USA, 21–23 May 2015; pp. 386–391.
54. Peng, C.; Sun, H.; Yang, M.; Wang, Y.L. A survey on security communication and control for smart grids under malicious cyber attacks. *IEEE Trans. Syst. Man Cybern. Syst.* **2019**, *49*, 1554–1569. [CrossRef]
55. Kumar, P.; Gurtov, A.; Sain, M.; Martin, A.; Ha, P.H. Lightweight authentication and key agreement for smart metering in smart energy networks. *IEEE Trans. Smart Grid* **2018**, *10*, 4349–4359. [CrossRef]
56. Depuru, S.S.S.R.; Wang, L.; Devabhaktuni, V.; Gudi, N. Smart meters for power grid—Challenges, issues, advantages and status. In Proceedings of the 2011 IEEE/PES Power Systems Conference and Exposition, Phoenix, AZ, USA, 20–23 March 2011; pp. 1–7.
57. Yan, Y.; Qian, Y.; Sharif, H.; Tipper, D. A survey on smart grid communication infrastructures: Motivations, requirements and challenges. *IEEE Commun. Surv. Tutorials* **2012**, *15*, 5–20. [CrossRef]
58. Rehmani, M.H.; Davy, A.; Jennings, B.; Assi, C. Software defined networks-based smart grid communication: A comprehensive survey. *IEEE Commun. Surv. Tutorials* **2019**, *21*, 2637–2670. [CrossRef]
59. Zafar, U.; Bayhan, S.; Sanfilippo, A. Home energy management system concepts, configurations, and technologies for the smart grid. *IEEE Access* **2020**, *8*, 119271–119286. [CrossRef]
60. Saleem, Y.; Crespi, N.; Rehmani, M.H.; Copeland, R. Internet of things-aided smart grid: Technologies, architectures, applications, prototypes, and future research directions. *IEEE Access* **2019**, *7*, 62962–63003. [CrossRef]
61. Syed, D.; Zainab, A.; Ghrayeb, A.; Refaat, S.S.; Abu-Rub, H.; Bouhali, O. Smart grid big data analytics: Survey of technologies, techniques, and applications. *IEEE Access* **2020**, *9*, 59564–59585. [CrossRef]
62. Hu, S.; Chen, X.; Ni, W.; Wang, X.; Hossain, E. Modeling and analysis of energy harvesting and smart grid-powered wireless communication networks: A contemporary survey. *IEEE Trans. Green Commun. Netw.* **2020**, *4*, 461–496. [CrossRef]
63. Abir, S.A.A.; Anwar, A.; Choi, J.; Kayes, A. IoT-Enabled Smart Energy Grid: Applications and Challenges. *IEEE Access* **2021**, *9*, 50961–50981. [CrossRef]
64. Hanganu, C.; Chcnaru, O.; Ichim, L.; Popescu, D. Efficient Solution for Smart Home Applications. In Proceedings of the 2018 26th IEEE Telecommunications Forum (TELFOR), Belgrade, Serbia, 20–21 November 2018; pp. 1–4.
65. Wagh, G.S.; Gupta, S.; Mishra, S. A distributed privacy preserving framework for the Smart Grid. In Proceedings of the 2020 IEEE Power & Energy Society Innovative Smart Grid Technologies Conference (ISGT), Delft, The Netherlands, 17–20 February 2020; pp. 1–5.
66. Ming, Y.; Zhang, X.; Shen, X. Efficient privacy-preserving multi-dimensional data aggregation scheme in smart grid. *IEEE Access* **2019**, *7*, 32907–32921. [CrossRef]

67. Mustafa, M.A.; Cleemput, S.; Aly, A.; Abidin, A. A secure and privacy-preserving protocol for smart metering operational data collection. *IEEE Trans. Smart Grid* **2019**, *10*, 6481–6490. [CrossRef]
68. Li, S.; Xue, K.; Yang, Q.; Hong, P. PPMA: Privacy-preserving multisubset data aggregation in smart grid. *IEEE Trans. Ind. Inform.* **2017**, *14*, 462–471. [CrossRef]
69. Wang, Z.; Liu, Y.; Ma, Z.; Liu, X.; Ma, J. LiPSG: Lightweight Privacy-Preserving Q-Learning-Based Energy Management for the IoT-Enabled Smart Grid. *IEEE Internet Things J.* **2020**, *7*, 3935–3947. [CrossRef]
70. Zeng, X.; Liu, Q.; Huang, H.; Jia, X. A lightweight privacy-preserving scheme for metering data collection in smart grid. In Proceedings of the 2017 IEEE 18th International Symposium on A World of Wireless, Mobile and Multimedia Networks (WoWMoM), Macau, China, 12–15 June 2017; pp. 1–6.
71. Alfakeeh, A.S.; Khan, S.; Al-Bayatti, A.H. A Multi-User, Single-Authentication Protocol for Smart Grid Architectures. *Sensors* **2020**, *20*, 1581. [CrossRef]
72. Jo, M.; Jangirala, S.; Das, A.K.; Li, X.; Khan, M.K. Designing Anonymous Signature-Based Authenticated Key Exchange Scheme for IoT-Enabled Smart Grid Systems. *IEEE Trans. Ind. Inform.* **2020**, *17*, 4425–4436.
73. Kaveh, M.; Mosavi, M.R. A Lightweight Mutual Authentication for Smart Grid Neighborhood Area Network Communications Based on Physically Unclonable Function. *IEEE Syst. J.* **2020**, *14*, 4535–4544. [CrossRef]
74. Kumar, N.; Aujla, G.S.; Das, A.K.; Conti, M. ECCAuth: A secure authentication protocol for demand response management in a smart grid system. *IEEE Trans. Ind. Inform.* **2019**, *15*, 6572–6582. [CrossRef]
75. Mahmood, K.; Li, X.; Chaudhry, S.A.; Naqvi, H.; Kumari, S.; Sangaiah, A.K.; Rodrigues, J.J. Pairing based anonymous and secure key agreement protocol for smart grid edge computing infrastructure. *Future Gener. Comput. Syst.* **2018**, *88*, 491–500. [CrossRef]
76. Ghosal, A.; Conti, M. Key management systems for smart grid advanced metering infrastructure: A survey. *IEEE Commun. Surv. Tutorials* **2019**, *21*, 2831–2848. [CrossRef]
77. Alohali, B.; Merabti, M.; Kifayat, K. A cloud of things (cot) based security for home area network (han) in the smart grid. In Proceedings of the 2014 Eighth IEEE International Conference on Next Generation Mobile Apps, Services and Technologies, Oxford, UK, 10–12 September 2014; pp. 326–330.
78. Moghadam, M.F.; Nikooghadam, M.; Mohajerzadeh, A.H.; Movali, B. A lightweight key management protocol for secure communication in smart grids. *Electr. Power Syst. Res.* **2020**, *178*, 106024. [CrossRef]
79. Qi, M.; Chen, J. Two-Pass Privacy Preserving Authenticated Key Agreement Scheme for Smart Grid. *IEEE Syst. J.* **2020**, *15*, 3201–3207. [CrossRef]
80. Lu, R.; Liang, X.; Li, X.; Lin, X.; Shen, X. EPPA: An efficient and privacy-preserving aggregation scheme for secure smart grid communications. *IEEE Trans. Parallel Distrib. Syst.* **2012**, *23*, 1621–1631.
81. Shen, H.; Zhang, M.; Shen, J. Efficient privacy-preserving cube-data aggregation scheme for smart grids. *IEEE Trans. Inf. Forensics Secur.* **2017**, *12*, 1369–1381. [CrossRef]
82. Lu, R.; Heung, K.; Lashkari, A.H.; Ghorbani, A.A. A lightweight privacy-preserving data aggregation scheme for fog computing-enhanced IoT. *IEEE Access* **2017**, *5*, 3302–3312. [CrossRef]

sensors

Review

Applications of Wireless Sensor Networks and Internet of Things Frameworks in the Industry Revolution 4.0: A Systematic Literature Review

Mamoona Majid [1], Shaista Habib [1], Abdul Rehman Javed [2], Muhammad Rizwan [3], Gautam Srivastava [4,5], Thippa Reddy Gadekallu [6] and Jerry Chun-Wei Lin [7,*]

1 School of System and Technology, University of Management and Technology, Lahore 54782, Pakistan; mamoona.majid@umt.edu.pk (M.M.); shaista.habib@umt.edu.pk (S.H.)
2 Department of Cyber Security, PAF Complex, E-9, Air University, Islamabad 44000, Pakistan; abdulrehman.cs@au.edu.pk
3 Department of Computer Science, Kinnaird College for Women, Lahore 54000, Pakistan; muhammad.rizwan@kinnaird.edu.pk
4 Department of Mathematics and Computer Science, Brandon University, Brandon, MB R7A 6A9, Canada; srivastavag@brandonu.ca
5 Research Center for Interneural Computing, China Medical University, Taichung 406040, Taiwan
6 School of Information Technology and Engineering, Vellore Institute of Technology, Vellore 632014, Tamil Nadu, India; thippareddy.g@vit.ac.in
7 Department of Computer Science, Electrical Engineering and Mathematical Sciences, Western Norway University of Applied Sciences, 5063 Bergen, Norway
* Correspondence: jerrylin@ieee.org

Citation: Majid, M.; Habib, S.; Javed, A.R.; Rizwan, M.; Srivastava, G.; Gadekallu, T.R.; Lin, J.C.-W. Applications of Wireless Sensor Networks and Internet of Things Frameworks in the Industry Revolution 4.0: A Systematic Literature Review. *Sensors* **2022**, *22*, 2087. https://doi.org/10.3390/s22062087

Academic Editors: Suparna De and Klaus Moessner

Received: 9 February 2022
Accepted: 2 March 2022
Published: 8 March 2022

Abstract: The 21st century has seen rapid changes in technology, industry, and social patterns. Most industries have moved towards automation, and human intervention has decreased, which has led to a revolution in industries, named the fourth industrial revolution (Industry 4.0). Industry 4.0 or the fourth industrial revolution (IR 4.0) relies heavily on the Internet of Things (IoT) and wireless sensor networks (WSN). IoT and WSN are used in various control systems, including environmental monitoring, home automation, and chemical/biological attack detection. IoT devices and applications are used to process extracted data from WSN devices and transmit them to remote locations. This systematic literature review offers a wide range of information on Industry 4.0, finds research gaps, and recommends future directions. Seven research questions are addressed in this article: (i) What are the contributions of WSN in IR 4.0? (ii) What are the contributions of IoT in IR 4.0? (iii) What are the types of WSN coverage areas for IR 4.0? (iv) What are the major types of network intruders in WSN and IoT systems? (v) What are the prominent network security attacks in WSN and IoT? (vi) What are the significant issues in IoT and WSN frameworks? and (vii) What are the limitations and research gaps in the existing work? This study mainly focuses on research solutions and new techniques to automate Industry 4.0. In this research, we analyzed over 130 articles from 2014 until 2021. This paper covers several aspects of Industry 4.0, from the designing phase to security needs, from the deployment stage to the classification of the network, the difficulties, challenges, and future directions.

Keywords: Internet of Things (IoT); industrial revolution 4.0 (IR 4.0); computer networks; network security; wireless sensor networks (WSN); systematic literature review (SLR); state-of-the-art

1. Introduction

Smart technologies play a crucial role in sustainable economic growth. They transform houses, offices, factories, and even cities into autonomic, self-controlled systems without human intervention [1]. This modern automation trend and ever-increasing use of cutting-edge technologies are boosting the world's economy [2]. The Internet of Things (IoT) and Wireless Sensor Networks (WSN) both play vital roles in this modernization [3]. IoT is a branch of engineering primarily concerned with offering thousands of miniature, physical connected objects, which may collaborate to achieve a shared goal. IoT has gained much importance due to the abundant usage of these tiny networked devices. These are smart, yet basic things that can sense and communicate wirelessly [4]. WSN is a collection of sensor and routing nodes, as shown in Figure 1, which may be put together in the environment to predict physical conditions, such as wind, temperature, and many others. These networks collect and process data from tiny nodes and then transfer it to the operators. Figure 2 illustrates that sensor networks are used in a variety of control systems, including environmental monitoring, home automation, chemical and biological assault detection, smart grid deployment [5], surveillance, and many more. WSN also plays a significant role in aquaculture and the oil industry, including data collection, offshore exploration, disaster prevention, tactical surveillance, and pollution monitoring [6–8].

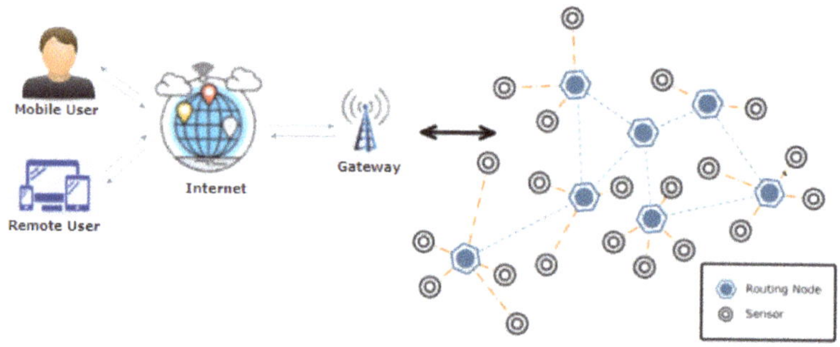

Figure 1. Architecture of wireless sensor network (WSN).

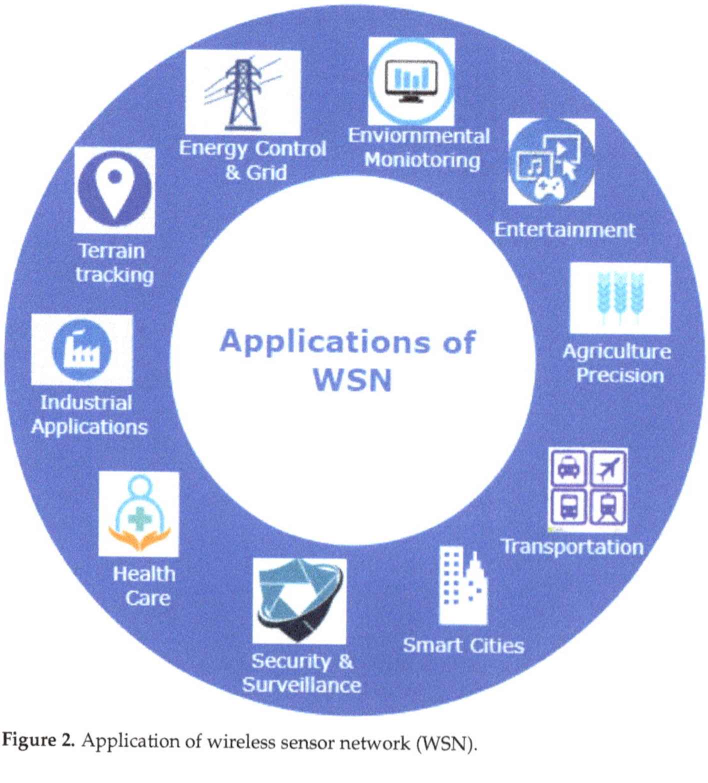

Figure 2. Application of wireless sensor network (WSN).

WSN are often deployed in remote areas where human intervention is not possible for post-deployment maintenance. Therefore, efforts are being made to enhance their efficiency and durability [9]. There are many barriers to WSN deployment, such as power consumption–long-distance deployment. Due to automation trends and applications developed, these barriers are no longer barriers for large-scale remote deployment. In general, WSN follows a star topology to decrease the network failure probability by connecting all systems to a central node. While ad-hoc networks follow mesh topology where each node is human-driven [10].

In physical production systems, grid and energy-saving applications minimize the energy resources and noise pollution. In the last few decades, transportation has improved a lot with the usage of smart IoT devices, such as signals and high-resolution cameras on roads, which has led to an increase in traffic flow. RFID readers are deployed at toll booths that automatically deduct toll amounts after reading RFID tags on vehicles. In the transportation sector, smart vehicles reduce the travelling time and also fuel consumption with low cost of mobility and reduced human efforts [11], atmospheric monitoring reduces pollution, and surveillance applications reduce crime. Nowadays, WSN also plays a role in precision agriculture. On the other hand, WSN applications facilitate our day to day lives, making them more comfortable, such as healthcare applications that improve our health and longevity.

Besides WSN, IoT has also played an important role in human life. IoT and the digital age play essential roles in overcoming social and physical barriers and providing ease and mobility to people, resulting in improved and equal opportunities, and access to information [12,13]. IoT also has many application areas such as agribusiness, climate, clinical care, education, transportation, and finance, as shown in Figure 2.

In regard to information and communication technology, researchers are attracted to IoT [14]. By adopting this essential technology, companies have become smarter, more competitive, automated, and sustainable in the global supply chain. In today's competitive marketplace, supply chains are struggling as they compete with each other. Therefore, IoT devices are an effective way to authenticate, monitor, and track products using GPS and many technologies [15,16]. Industry 4.0 stands for the fourth industrial revolution in the digital age, it is associated with virtualizing real-world scenarios of production and processing without human intervention. This virtual world is linked to IoT devices, allowing the creation of cyber–physical systems to communicate and cooperate [17,18]. This fully connected manufacturing system—operating without human intervention by generating, transferring, receiving, and processing necessary data to conduct all required tasks for producing all kinds of goods—is one of Industry 4.0's key "constructs". The concept of Industry 4.0 is based on the combination of three main elements: people, things, and business [19]. A complete cyber–physical production system created by the integration of IoT devices, things and objects (IoT), sensor nodes (WSN), and people, is shown in Figure 3. CPS is a typical example of Industry 4.0. IoT is the connection of smart devices, objects, or machines to the internet and with each other. In WSN systems, there is no direct connection of these devices to the Internet. These systems can send their data to the Internet by connecting several sensor nodes to a central routing node. While CPS systems involve the integration of IoT devices, computation, networking, and physical process, IoT is an essential component of CPS. CPS systems are key elements in the implementation of IR 4.0 [20]. Industry 4.0 is the network-enabled entity that automates the whole process of manufacturing, connecting business and processes. Market demands and the advancements in new technologies are transforming manufacturing firms' business operations into smart factories and warehouses. Due to this automation, IoT devices are producing a massive amount of data daily, known as big data [21,22]. Statistics show that, at the end of 2021, there were more than 10 billion active IoT devices globally [23]. By 2030, the number of active IoT devices is expected to exceed 10 billion to 25.4 billion. By 2025, the data created by IoT devices will reach 73.1 ZB (zeta bytes) [24]. In 2020, the IoT industry was predicted to generate more than USD 450 billion, including hardware, software, systems integration, and data services. By the end of 2021, it reached USD 520 billion. The global amount expected to be spent on the IoT in 2022 is USD 1 trillion. The IoT industry is predicted to grow to more than USD 2 trillion by 2027 [25,26]. The increasing number of devices and the usage by humans shows the importance of IoT devices; moreover, the industry is growing and gaining revenue.

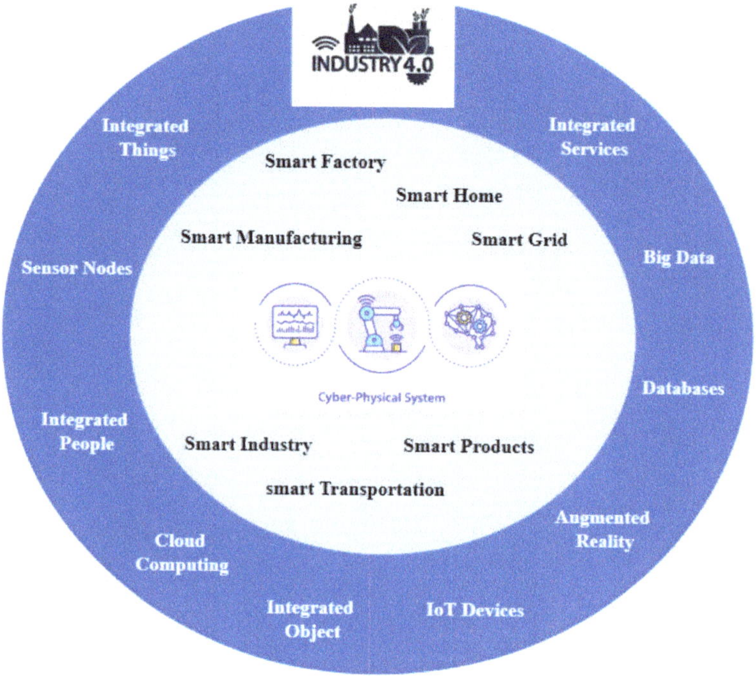

Figure 3. CPS system–integration of IoT, wireless devices, and people in Industry 4.0.

In this paper, we conduct a detailed systematic literature review on the applications and contributions of IoT and WSN in Industry 4.0. We had a large corpus of data to analyze papers using systematic approaches. Among the selected articles, we obtained 22 articles describing the detailed review of existing security techniques, applications used, advantages/disadvantages, and limitations of IoT and WSN. Most of them reviewed the articles in terms of application of IoT and attacks only. The paper mainly focuses on research challenges, issues, limitations, and the future direction of IoT and WSN frameworks in Industry 4.0.

1.1. Motivation

The primary motivation for performing SLR is the ever-increasing trend of automation in Industry 4.0. Industry 4.0 is made up of both WSN technology and IoT to digitize work. Over time, we see how digitization and technology are replacing people in the workplace, and dramatically changing the global workforce. Technology has brought ease to human life and the efficiency of things by making them eco-friendly, more agile, and productive. From smart cities to smart industries, a drastic change has happened due to the intelligent usage of IoT and WSN and IR 4.0. For example, a smart factory integrates virtual and physical systems and calibrating instruments to record their readings immediately. In short, the integration of IoT and WSN with Industry 4.0 has reduced labour needs, freed humans from low-level skilled work, and improved the quantity and quality of work. Therefore, To achieve better analysis results, we systematically analyzed almost all research data related to IoT and WSN domains in Industry 4.0.

1.2. Contribution

From smart factories to individual lives, IoT and WSN play significant roles. There are many advantages, but security problems have arisen as these devices generate considerable amounts of data daily. These papers amalgamated both sides: IoT and WSN contributions and their security risks. This paper focuses more on the contributions of IoT and WSN in Industry 4.0 and presents an in-depth review and analysis of IoT and WSN. The paper's main contribution involves formulating research questions from filtered data and building a strong work foundation using SLR methods. We discuss various applications and security attacks in IR 4.0. Furthermore, new findings of the paper pertain to the key challenges and open issues of WSN integrated with IoT in Industry 4.0, to optimize different criteria.

1.3. Paper Organization

The remainder of the paper is organized into the following sections. Table 1 presents the notation used in the entire paper. Section 2 provides the related works. Comparative analyses of these review papers are given in Table 2. Section 3 presents a detailed systematic literature survey (SLR). Research questions presented in Table 3 are addressed in Section 4. Section 5 presents the challenges and issues. Future directions are elaborated in Section 6. Finally, the conclusion is presented in Section 7.

Table 1. List of Abbreviations.

Abbreviation	Description
5G	fifth generation
6G	sixth generation
AR	augmented reality
CPS	cyber physical system
DDoS	distributed denial of service
DNS	domain name system
DoS	denial of service
DT	digital twin
FoI	field of interest
GUI	graphical user interface
ID 4.0	Industry 4.0
IIoT	industrial internet of things
IoT	internet of things
IR 4.0	industry revolution 4.0
IWSN	industrial wireless sensor network
IWSAN	industrial wireless sensor and actuator network
PREQ	request packet
QoS	quality of service
RFID	radio frequency identification
SEPTIC	self protecting databases from attacks

Table 1. *Cont.*

Abbreviation	Description
SG	smart grid
SIRP	self-optimized smart routing protocol
SLR	systematic literature review
UASN	underwater acoustic sensor networks
WSN	wireless sensor network
WSAN	wireless sensor area network

2. Related Studies

In this section, we analyze the state-of-the-art research studies on IoT and WSN. With the fourth industrial revolution, it is observed that communication, computation, and storage costs have remarkably decreased, which make integration of IoT and WSN possible and cost-effective globally. We studied many review articles and original research. Existing review articles lack, in many aspects, research challenges, issues, limitations, and future directions of IoT and WSN, but the systematic literature review (SLR) given in this work is precise enough to deal with the aspects of IoT and WSN area. Related work of review papers is provided next. Moreover, Table 2 is presented which shows the comparison between the proposed research work and the existing state-of-the-art analysis.

Table 2. Comparative analysis of the existing review papers. Key: deployment category—DC, literature review—LR, security overview—SO, bibliometric literature review—BLR, systematic literature review—SLR.

Reference	Year	Review Type	DC	Application Types	IoT and WSN Architecture Used	Challenges and Issues	R. Q.
[1]	2015	LR	Industry	802.11 (WiFi) technology in smart cities	WSN	×	×
[4]	2020	LR	Science	IoT sensing applications discussed using sensing technology	WSN using RFID	Energy harvesting, communication interference, fault tolerance, higher capacities to handling data processing, cost feasibility.	×
[5]	2016	SO	Industry	IoT application in smart grids	IoT	Challenges discussed along with solutions to cope with	×
[6]	2017	SO	Industry	Deployment techniques discussed using sensor network	WSN	Communication cost, coverage time, accuracy, etc.	×
[10]	2016	LR	WSN applications in urban areas	Urban areas	WSN	Problems and solution of each WSN application	×
[17]	2014	SO	Industry	Network security protocols discussed in industrial applications	WSN	Challenges of stack protocol and their solutions	×
[19]	2020	SLR	Smart factories	Scope and conceptualization of IoT in Industry 4.0	IoT	×	✓
[27]	2020	LR	Smart IoT devices	Detailed survey on security threat models applicable for IoT and WSN. They also discussed communication attacks and taxonomy of IoT and WSN	Both	✓	×
[28]	2019	LR	-	Discussed technical and social perspective of IoT for future technology enhancement	IoT	✓	×

Table 2. Cont.

Reference	Year	Review Type	DC	Application Types	IoT and WSN Architecture Used	Challenges & Issues	R. Q.
[29]	2017	SLR	Smart cities	Applications, security, and taxonomy in IoT	IoT	×	×
[30]	2019	LR	Industrial	Applications and usage of actuators and sensor networks using MAC protocol.	IWSN	Security challenges on different layers of the stack, also discussed their solutions	×
[31]	2016	LR	–	Technologies, innovations, and applications of IoT discussed.	IoT	✓	×
[32]	2014	LR	Industrial	Coverage areas of WSN are discussed	WSN	Challenges they face were: Node type, depth type, communication range, etc.	×
[33]	2016	SUR	Industrial	Applications of intrusion detection system in IoT	IoT	×	×
[34]	2015	SUR	Industrial	Only explore and analyze existing solution to detect sinkhole attack	WSN	×	×
[35]	2020	LR	–	×	Both	Discuss attacks in IoT ans WSN with their solutions, advantages, and limitation	×
[36]	2021	SLR	Smart mobiles	Routing attacks and security measures in mobile network are discussed	WSN	×	×
[37]	2021	LR	Industrial	Detection of wormhole in both domains	Both	×	✓
[38]	2017	SUR	IoT systems	Software board and chips, crypto algorithms, security of IoT systems, and network protocols are discussed	IoT	–	×
[39]	2015	SUR	–	Existing security approaches of IoT system are described	IoT	✓	×
[40]	2016	SUR	–	Deployment models for sensor network to achieve coverage, their classification and working was discussed	WSN	×	×
[41]	2018	BLR	Smart factory & Industry	Discuss 12 approaches of Industry 4.0. in business and account management fields	IoT	×	×
This Paper	2021	SLR	Smart industry and Factory	Applications and contribution of both IoT and WSN are discussed in detail	IoT and WSN (both)	Key challenges and open issues of both IoT and WSN in Industry 4.0 are discussed	✓

The authors in [1] discussed the very novel technology "WiFi". They discussed how this technology helps IoT devices–used in various applications, such as smart cities, healthcare systems, and smart houses—communicate effectively. Landaluce et al. [4] discussed RFID and WSN technology in detail. They discussed how RFIDs are used to trace devices while WSN gathers information about them from interconnected devices. The authors also discussed the obstacles and challenges, such as energy consumption, fault-tolerant, communication interference, and cost feasibility, along with detailed surveys. They provided the advantages and limitations of wearable sensor devices. Energy consumption is increasing each day. Therefore, Dailipi [4] explored how IoT evolution has managed the electricity consumption process and provided many benefits to grid stations, consumers, and suppliers. They also discussed the security issues and challenges after introducing IoT devices in smart grids.

In [6], the authors discussed how WSN advancement had played a significant role in UASN. They traced the location of sensor nodes deployed underwater in the ocean using localization algorithms. They reviewed many applications of UASN, their advantages and their disadvantages. They also discussed the challenges they faced during deployment and presented future directions in the acoustic area. In [10], researchers discussed the applications and advantages of WSN being deployed everywhere due to their flexible and dynamic nature. They discussed each application of WSN in urban areas and their solutions. They analyzed how WSN deployment in urban areas demanded much more care and attention due to harsh weather and perverse channel conditions.

WSN is applicable in many domains, such as industrial automation, and the requirement elicitation of the industrial process is different from general WSN requirement gatherings. In [17], the authors presented some standard protocols that were used to measure the requirements of industrial applications. They also provide solutions to WSN protocols by discussing MAC, routing, and transport in detail. They also discussed the security issues in detail and identified the unsolved challenges encountered during designing standard protocols. In [19], the authors conducted SLR, which is mainly focused on scope definition, concept, literature review, analysis, synthesis, and future research directions. Their selected study has contributed to eight thematic perspectives: intelligence factories, CPS, data handling, IT infrastructure, digital transformation, HCI, IoT, and cloud [18].

Due to the rapid evolution in IoT and WSN, technology is becoming more vulnerable to security threats [42–44]. Therefore authors in [27] presented threat models for the security of WSN and IoT devices communication. In [28], the authors discussed IoT applications, advantages, challenges, and security issues from both technological and social perspectives. Researchers have provided detailed architectures of IoT and WSN and discussions of IDS system protocols. They also discussed the security challenges and attacks on IoT and WSN communication devices. Moreover, in [29], the authors conducted extensive research related to smart homes, applications, and IoT. They collected 229 articles, analyzed them thoroughly, and divided them into four categories. They discussed smart home IoT applications in the first category. The second category concerned with IoT applications in smart home technology. In the third category, they developed a framework to operate further. In the fourth category, they developed smart IoT home applications.

IoT has dramatically changed human life, especially regarding communication devices integrating technologies. Traditional industry is changing in the digital industry, and WSN and wireless sensor and actuator networks (WSANs) are the core parts of Industry 4.0. In the article [30], the authors discussed the industrial wireless sensor network (IWSN) and industrial wireless sensor and actuator network (IWSAN) in detail. They discussed IWSAN requirements, applications, challenges, solutions, and future directions in detail. IWSN/IWSAN are compelling technologies due to their promising benefits, such as low-cost deployment, less complexity, and mobility support.

In [31], the authors discussed how IoT plays a vital role in bringing the physical world close to the digital world. They discussed technologies, various challenges, future directions, and various Internet of Things (IoT) applications.

In [32], Sharma et al. have described the sensor nodes according to coverage point of view. They analyzed the full coverage issues by considering node type, deployment type, communication and sensing range, and positioning-based independent algorithms. They also discussed the research challenges of WSN.

Andrey et al. [33] described a detailed survey on IDS systems and presented the methods proposed for IoT. Using a cross-platform distributed approach, they analyzed the IDS system, their platform differences, and current research trends in IDS. In [34], the authors analyzed and discussed the solutions to identify and detect sinkhole attacks in the WSN domain. They discussed the advantages and limitations of the proposed solution as well. In [35], the authors presented a detailed review on security attacks of WSN and IoT along with their preventive measures, mitigations, and detection mechanisms. They stated that the integration of IoT and WSN has raised new challenges and open security issues. Although technology has increased, it has become prone to external attacks.

In [36], the authors presented a review on the security of mobile networks. They discussed the integration of WSN with IoT via the Internet and how the inter-connected devices have guarded networks against external attacks, keeping the router in a secure and protected environment. They discussed the attacks and their detection mechanisms over the

Internet. Similarly, the authors in [37] have discussed the wormhole attack and its solution in IoT and WSN domains. They stated that the detection algorithm performed much better for IoT (70%) than WSN (20%). In [45], the authors discussed side-channel attacks in smartphones. Similarly, the authors in [38] discussed the security threats, challenges, and solutions in the IoT domain. While in [39], the authors analyzed existing protocols for secure communication between IoT devices. They also discussed open issues and challenges raised during the communication of IoT devices and future directions in IoT. The authors in [40] presented a detailed review regarding deployment schemes, classification, working, and comparative analyses of sensor nodes. This growing technology trend has converged the "world sense" from traditional systems to CPS—this transition is called Industry 4.0. The authors in [41] conducted a bibliometric review of 12 different approaches of critical aspects of Industry 4.0.

From the above-detailed literature review, we noticed that authors and researchers have worked on IoT and WSN, but the integration of both IoT & WSN with Industry 4.0 is benign. They discussed their applications, security attacks, advantages, and limitations at each level.

The proposed paper is more oriented towards the applications and contributions of IoT and WSN in Industry 4.0, along with the security attacks, their challenges, and open issues in each domain. This paper also provides the limitations and future directions for IoT and WSN in Industry 4.0.

3. Research Methodology

The SLR followed in this work is based on the template of IEEE SLR; all research steps and guidelines were followed using the same template. The research process and results were manually evaluated by comparing different research methods and techniques presented by different authors based on efficiency, security, limitations, and performance. First, we formulated our research questions based on the reviewed articles and then searched for these keywords in various databases. Using avoidance and consideration criteria, only relevant papers were considered. Then, we evaluated the quality of the papers, and then extracted the relevant features.

3.1. Planning Review

The planning of the review was based on the research questions and their objectives. Analyzes of IoT and WSN were conducted to show their importance in daily life. The use of smart and appropriate devices has been studied in this field but still needs critical analysis. Therefore, it is important to make a systematic review article in this realm, especially related to IoT and WSN, to show future directions. Therefore, we established the search technique, search strings, inclusion/exclusion criteria, and quality assessment criteria for the papers collected from different repositories. Figure 4 shows the step of our proposed review planning process.

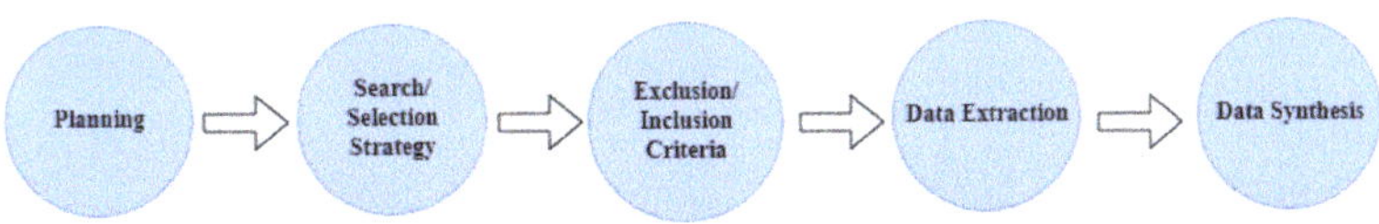

Figure 4. Planning process of systematic literature review (SLR).

3.2. Research Goals

The main objective of this study is to find out the major contributions, solved problems, and challenges of IoT and WSN in Industry 4.0. In addition, the research gap and limitations of current work in these areas helped us find room for improvement; we also explored future directions and possible outcomes in this area. We designed questionnaires on these domains to find high quality research papers. Some of them are listed in Table 3. In addition, we answer the research questions in Table 3 and in Section 4 .

Table 3. Research questions.

	Research Questions (RQ)
RQ1	What are the contributions of WSN in IR 4.0?
RQ2	What are the contributions of IoT in IR 4.0?
RQ3	What are the types of WSN coverage areas for IR 4.0?
RQ4	What are the major types of network intruders in WSN and IoT systems?
RQ5	What are the prominent network security attacks in WSN and IoT?
RQ6	What are the major issues in IoT and WSN frameworks?
RQ7	What are the limitations and research gaps in the existing work?

3.3. *Selection of Primary Studies*

It was a challenge to search the specific and limited computer science/engineering databases to get a "complete picture" of the research questions. After formulating the research questions, we collected research papers from various repositories, such as IEEE Xplore, ACM, Wiley Online Search, Elsevier, etc. The papers from electronic databases with the areas of IoT, WSN, and Industry 4.0 were efficiently evaluated. Figure 5 shows the names of the repositories where the research articles were collected from 2014 to June 2021.

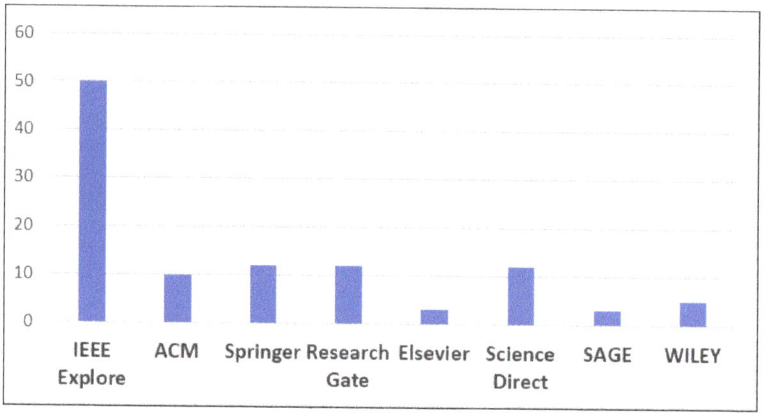

Figure 5. Repositories versus number of studies Used.

3.4. *Selection/Search Criteria*

This SLR combines three domains: IoT, WSN, and Industrial Revolution 4.0, so the search strings were related to each domain and its applications in Industry 4.0. Table 4 shows the subject search strings used to search for relevant articles. The search strings were divided into five groups to search for relevant articles from reputable journals and conferences.

Table 4. Search strings.

Sr. No.	Groups	Group Search Query
1	Group 1	Application of WSN in IR 4.0.
2	Group 2	Implementation of IoT infrastructure in IR 4.0.
3	Group 3	Industrial Revolution 4.0. for smart manufacturing
4	Group 4	Security attacks, issues, and challenges of IoT and WSN in IR 4.0.
5	Group 5	Role of WSN and IoT systems in IR 4.0.

3.5. Inclusion and Exclusion Criteria

Several research articles were found on the mentioned domains, IoT, WSN and IR 4.0. To extract the most relevant and concise data from these articles, the following criteria were chosen, as shown in Table 5. The publicly available papers written in English and related to the search strings and research objectives in IoT, WSN, and IR 4.0 published between 2014 and 2021 were considered for further SLR tracking.

Table 5. Inclusion and exclusion criteria.

	Inclusion Criteria
1	Include only those papers written in the English language.
2	Include papers that were published in 2014–2021.
3	Include papers that reflected enough knowledge about the search strings and search objectives.
4	Include papers whose titles, keywords, abstracts, and conclusions provided enough information related to WSN, IoT, and IR 4.0.
5	Include papers whose content focused on WSN, IoT, and IR 4.0 content and provided in depth insights.
	Exclusion Criteria
1	Exclude papers written in a language other than the English language.
2	Exclude gray papers.
3	Exclude papers that were not published within 2014–2021.
4	Exclude research papers containing less than three pages.
5	Exclude papers that failed to meet the inclusion criteria.

3.6. Selection Results

As mentioned earlier, the downloaded articles were based on the initial screening processes, which were based on inclusion criteria. The articles were screened using the initial quality assessment criteria (QAC). The main objective of using the QAC was to ensure that the primary studies selected were appropriate to address the concerns of previous studies. Nearly 300 articles were found on the above topics. After applying a duplication filter, 40 articles were discarded, leaving 260 articles that were reviewed based on the above exclusion criteria. In the next step, 60 articles were excluded based on the inclusion/exclusion criteria. A quality analysis was performed for the remaining 200 articles. After the quality analysis, only 120 articles were reviewed. Figure 6 shows the year-by-year distribution of articles reviewed that were used for the SLR study. There is no doubt that the eliminated articles contained valuable material, but they were not studied because they did not meet the screening criteria. The number of articles was initially limited to 120, after which they were stored in the citation manager software for information synthesis.

The selected papers were evaluated against the quality assessment criteria (QAC) listed in Table 6. Table 6 shows the quality assessment criteria (QAC) used to screen the articles for response grading. Studies selected for screening are described in Table 7. Studies with a score greater than or equal to 80 were selected according to the grading criteria shown in Figure 7.

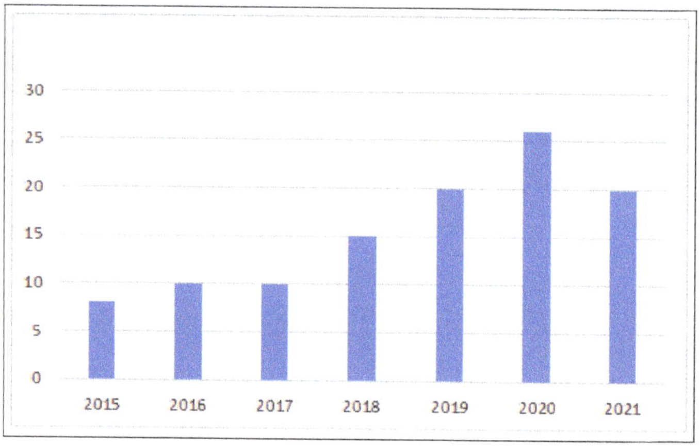

Figure 6. Year-wise distribution of articles.

Table 6. Selection criteria versus response graded.

Criteria	Selection Criteria	Graded Response
C1	Is the aim of research and context clearly defined?	1, 0.5, 0 (yes, nominally, no)
C2	Is the context of research well addressed?	1, 0.5, 0 (yes, nominally, no)
C3	Are the findings clearly stated?	1, 0.5, 0 (yes, nominally, no)
C4	Based on the findings, how valuable is the research?	>80% = 1 , <20% = 0 , in between = 0.5

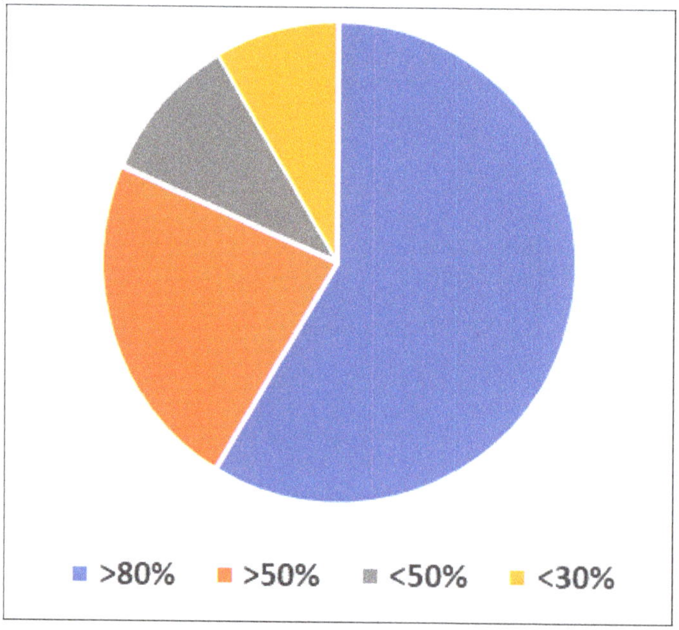

Figure 7. Graded percentage of selected studies.

Table 7. Selected studies used for SLR analysis.

Sr. No.	Title of Research	Authors	Year
1.	Cyber-Physical Systems Security: Analysis, Challenges, and Solutions	Y. Ashibani and Q. H. Mahmoud	2017
2.	A Review of IoT sensing applications and challenges using RFID and wireless sensor networks	H. Landaluce, L. Arjona, A. Perallos, F. Falcone, I. Angulo, and F. Muralter	2020
3.	Enhancement of relay nodes communication approach in WSN-IoT for underground coal mine	R. Sharma and S. Prakash	2020
4.	Applications of wireless sensor networks for urban areas: A survey	B. Rashid and M. H. Rehmani	2016
5.	An empirical study of application layer protocols for IoT	U. Tandale, B. Momin, and D. P. Seetharam	2017
6.	Digital twin technologies and smart cities.	M. Farsi, A. Daneshkhah, A. Hosseinian-Far, and H. Jahankhani	2020
7.	Internet of things (IoT) embedded future supply chains for Industry 4.0: An assessment from ERP-based fashion apparel and footwear industry	M. A. A. Majeed and T. D. Rupasinghe	2017
8.	Towards Industry 4.0 utilizing data-mining techniques: a case study on quality improvement	H. Oliff and Y. Liu	2017
9.	An industrial perspective on wireless sensor networks-a survey of requirements, protocols, and challenges	A. A. Kumar S., K. Ovsthus, and L. M. Kristensen.	2014
10.	The smart factory as a key construct of Industry 4.0: A systematic literature review	P. Osterrieder, L. Budde, and T. Friedli	2020
11.	Social expectations and market changes in the context of developing the Industry 4.0 concept	S. Saniuk, S. Grabowska, and B. Gajdzik	2020
12.	Key IoT Statistics	B. Jovanović	2021
13.	30 Internet of Things – IoT stats from reputable sources in 2021	A. Multiple	2021
14.	Wide-area and short-range IoT devices	S. O'Dea	2021
15.	The Future of Industrial Communication: Automation Networks in the Era of the Internet of Things and Industry 4.0	M. Wollschlaeger and T. Sauter and J. Jasperneite	2017
16.	Internet of things (IoT): a technological analysis and survey on vision, concepts, challenges, innovation directions, technologies, and applications	G. Misra, V. Kumar, A. Agarwal, and K. Agarwal	2016
17.	EDHRP: Energy-efficient event-driven hybrid routing protocol for densely deployed wireless sensor networks	Faheem M, Abbas MZ, Tuna G, Gungor VC.	2015
18.	A survey on deployment techniques, localization algorithms, and research challenges for underwater acoustic sensor networks.	Tuna G, Gungor VC	2017
19.	Lrp: Link quality-aware queue-based spectral clustering routing protocol for underwater acoustic sensor networks	Faheem M, Tuna G, Gungor VC	2017
20.	Design and deployment of a smart system for data gathering in aquaculture tanks using wireless sensor networks	Parra L, Sendra S, Lloret J, Rodrigues JJ.	2017

Table 7. *Cont.*

Sr. No.	Title of Research	Authors	Year
21.	WSN-and IoT-based smart homes and their extension to intelligent buildings. Sensors	Ghayvat H, Mukhopadhyay S, Gui X, Suryadevara N	2015
22.	Conceptual model for informing user with an innovative smart wearable device in Industry 4.0	M. Periša, T. M. Kuljanić, I. Cvitić, and P. Kolarovszki	2019
23.	Evolution of wireless sensor network for air quality measurements	Arroyo, P.; Lozano, J.; Suárez, J.	2018
24.	Industrial wireless sensor and actuator networks in Industry 4.0: Exploring requirements, protocols, and challenges—A MAC survey	S. Raza, M. Faheem, and M. Genes	2019
25.	Cause the Industry 4.0 in the automated industry to new requirements on the user interface	C. Wittenberg	2015
26.	Impact of 5G technologies on Industry 4.0	G. S. Rao and R. Prasad	2018
27.	Material efficiency in manufacturing: Swedish evidence on potential, barriers, and strategies	S. Shahbazi et al.	2016
28.	Organizational change, and industry 4.0 (id4). A perspective on possible future challenges for human resources management	J. Radel	2017
29.	Organizational culture as an indication of readiness to implement Industry 4.0	Z. Nafchi and M.Mohelská	2020
30.	Smart production planning and control: concept, use-cases, and sustainability implications	O.E, Oluyisola	2020
31.	Fortune favors the prepared: How SMEs approach business model innovations in Industry 4.0.	J. M. Müller et al.	2018
32.	Visual computing as a critical enabling technology for industries 4.0 and industrial Internet	J. Posada et al.	2015
33.	Digitalization and energy consumption. Does ICT reduce energy demand	S. Lange	2020
34.	Industry 4.0: adoption challenges and benefits for SMEs	T. Masood and P. Sonntag	2020
35.	Measurement and analysis of corporate operating vitality in the age of digital business models	J. Zhu et al.	2020
36.	Cyber security and the Internet of Things: vulnerabilities, threats, intruders and attacks	M. Abomhara and G. M. Køien	2015
37.	Sharing user IoT devices in the cloud	Y. Benazzouz, C. Munilla O. Gunalp, M. Gallissot, and L. Gurgen	2014
38.	Security in Internet of things: Challenges, solutions, and future directions	S. A. Kumar, T. Vealey, and H. Srivastava	2016
39.	Survey of intrusion detection system towards an end-to-end secure internet of things	A. A. Gendreau, M. Moorman	2016
40.	Recent advances and trends in predictive manufacturing systems in a big data environment	J. Lee et al.	2015
41.	A comprehensive dependability model for QOM-aware industrial WSN when performing visual area coverage in occluded scenarios	T. C. Jesus, P. Portugal, D. G. Costa, and F. Vasques	2020

Table 7. *Cont.*

Sr. No.	Title of Research	Authors	Year
42.	Security issues and challenges on wireless sensor networks	M. A. Elsadig, A. Altigani, and M. A. A. Baraka	2019
43.	Challenges of Wireless Sensor Networks and Issues associated with Time Synchronization	G. S. Karthik and A. A. Kumar	2015
44.	Design and analysis of intrusion detection protocols for hierarchical wireless sensor networks	M. Wazid	2017
45.	Intrusion detection protocols in wireless sensor networks integrated to the Internet of Things deployment: survey and future challenges	S. Pundir, M. Wazid, D. P. Singh, A. K. Das, J. J. P. C. Rodrigues, and Y. Park	2020
46.	Robust malware detection for Internet of (battlefield) Things devices using deep Eigenspace learning [46]	Azmoodeh, A. Dehghantanha, and K.-K.-R. Choo	2019
47.	LSDAR: A lightweight structure-based data aggregation routing protocol with secure IoT integrated next-generation sensor networks.	Haseeb K, Islam N, Saba T, Rehman A, Mehmood Z.	2020
48.	SEPTIC: Detecting injection attacks and vulnerabilities inside the DBMS.	Medeiros, M. Beatriz, N. Neves, and M. Correia	2019
49.	An efficient ECC-based provably secure three-factor user authentication and key agreement protocol for wireless healthcare sensor networks. Computers and Electrical Engineering	Challa S, Das AK, Odelu V, Kumar N, Kumari S, Khan MK, et al.	2018
50.	Internet of Things: vision, applications and challenges	Rishika Mehta, Jyoti Sahnib, Kavita Khannac	2018
51.	A roadmap for security challenges in the Internet of Things	Arabia Riahi Sfar, Enrico Natalizio, Yacine Challal, Zied Chtourou	2018
52.	A novel low-rate denial of service attack detection approach in ZigBee wireless sensor network by combining Hilbert-Huang transformation and trust evaluation	H. Chen, C. Meng, Z. Shan, Z. Fu, and B. K. Bhargava	2019
53.	Analysis of quantities influencing the performance of time synchronization based on linear regression in low-cost WSN	D. Capriglione, D. Casinelli, and L. Ferrigno	2016
54.	C–Sync: Counter-based synchronization for duty-cycled wireless sensor networks	K.-P. Ng, C. Tsimenidis, and W. L. Woo	2017
55.	Time synchronization in WSN with random bounded communication delays.	Y.-P. Tian	2017
56.	A novel model of Sybil attack in cluster-based wireless sensor networks and propose a distributed algorithm to defend It	M. Jamshidi, E. Zangeneh, M. Esnaashari, A. M. Darwesh, and A. J. Meybodi	2019
57.	Challenges, threats, security issues, and new trends of underwater wireless sensor networks	G. Yang, L. Dai, and Z. Wei	2018
58.	Industry 4.0 key research topics: A bibliometric review	D. Trotta and P. Garengo	2018
59.	Privacy in the Internet of Things: threats and challenges	J. H. Ziegeldorf, O. G. Morchon, and K. Wehrl	2015

Table 7. *Cont.*

Sr. No.	Title of Research	Authors	Year
60.	On the security and privacy of the Internet of Things architectures and systems.	E. Vasilomanolakis, J. Daubert, M. Luthra, V. Gazis, A. Wiesmaier and P. Kikiras	2015
61.	Cybersecurity issues in wireless sensor networks: current challenges and solutions	D. E. Boubiche, S. Athmani, S. Boubiche, and H. Toral-Cruz	2020
62.	A security model for IoT-based systems	Z. Safdar, S. Farid, M. Pasha, and K. Safdar	2017
63.	Security issues and challenges in IoT routing over wireless communication	G. Saibabu, A. Jain, and V. K. Sharma	2020
64.	Security and privacy consideration for Internet of Things in smart home environments	Desai, Drushti, and Hardik Upadhyay	2015
65.	E.D. Security and grand privacy challenges for the Internet of Things	Fink, G.A., Zarzhitsky, D. V., Carroll, T.E., and Farquhar	2015
66.	A comprehensive approach to privacy in the cloud-based Internet of Things.	Henze, M., Hermerschmidt, L., Kerpen, D., Häußling, R., Rumpe, B., and Wehrle, K.	2016
67.	Towards an analysis of security issues, challenges, and open problems on the internet of Things.	Hossain, A. J., Fotouhi, M., and Hasan, R.	2015
68.	An End-to-end view of IoT security and privacy	Zhen Ling, Kaizheng Liu, Yiling Xu, YierJin, XinwenFu	2017
69.	Security and privacy considerations for IoT application on smart grids: Survey and research challenges	Dalipi, F.; Yayilgan, S.Y.	2016
70.	Internet of Things security: A survey	Alaba, Fadele Ayotunde, et al.	2017
71.	Security for the Internet of things: a survey of existing protocols and open research issues	J. Granjal, E. Monteiro, J. Silva	2015
72.	Security, privacy and trust in Internet of things: the road ahead	S. Sicari, A. Rizzardi, L.A. Grieco, A. Coen-Porisini	2015
73.	Access control and authentication in the Internet of Things environment	A.K. Ranjan, G. Somani	2016
74.	Toward secure and provable authentication for the Internet of Things: realizing Industry 4.0	S. Garg, K. Kaur, G. Kaddoum, and K. K. R. Choo	2020
75.	Prediction of satellite shadowing in smart cities with application to IoT	S. Hornillo-Mellado, R. Martín-Clemente, and V. Baena-Lecuyer	2020
76.	Software-defined industrial Internet of Things in the context of Industry 4.0	J. Wan et al.	2016
77.	Residual energy-based cluster-head selection in WSN for IoT application.	T. M. Behera, G. S. Mohapatra, U. C. Samal, M. G. S. Han, M. Daneshmand, and A. H. Gandomi	2019
78.	DistB-SDoIndustry: enhancing security in Industry 4.0 services based on the distributed blockchain through software-defined networking-IoT enabled architecture,	A. Rahman et al.	2020
79.	Application of IoT-aided simulation to manufacturing systems in the cyber-physical system	Y. Tan, W. Yang, K. Yoshida, and S. Takakuwa	2019
80.	Convergence of blockchain and edge computing for secure and scalable IIoT critical infrastructures in Industry 4.0 [47]	Y. Wu, H.-N. Dai, and H. Wang	2020

Table 7. *Cont.*

Sr. No.	Title of Research	Authors	Year
81.	Comparative study of IoT-based topology maintenance protocol in a wireless sensor network for structural health monitoring	M. E. Haque, M. Asikuzzaman, I. U. Khan, I. H. Ra, M. S. Hossain, and S. B. Hussain Shah	2020
82.	Toward dynamic resources management for IoT-based manufacturing	J. Wan et al.	2018
83.	SENET: A novel architecture for IoT-based body sensor networks	Z. Arabi Bulaghi, A. Habibi Zad Navin, M. Hosseinzadeh, and A. Rezaee	2020
84.	Bio-inspired routing protocol for WSN-based smart grid applications in the context of Industry 4.0	M. Faheem et al.	2019
85.	IoT and wireless sensor network-based autonomous farming robot	A. Khan, S. Aziz, M. Bashir, and M. U. Khan	2020
86.	Efficient and secure three-party mutual authentication key agreement protocol for WSN in IoT environments	C. T. Chen, C. C. Lee, and I. C. Lin	2020
87.	Wireless sensor network combined with cloud computing for air quality monitoring	P. Arroyo, J. L. Herrero, J. I. Suárez, and J. Lozano	2019
88.	Edge computing-enabled wireless sensor networks for multiple data collection tasks in Smart Agriculture	X. Li, L. Zhu, X. Chu, and H. Fu	2020
89.	Cluster centroid-based energy-efficient routing protocol for WSN-Assisted IoT	N. Prophess, R. Kumar, and J. B. Gnanadhas	2020
90.	An energy-efficient and secure IoT-based WSN framework: an application to smart agriculture	K. Haseeb, I. U. Din, A. Almogren, and N. Islam	2020
91.	Deployment schemes in a wireless sensor network to achieve blanket coverage in large-scale open area	Vikrant Sharmaa, R.B. Patelb, H.S. Bhadauriaa, D. Prasadc	2016

3.7. Data Extraction and Synthesis Process

After collecting articles relevant to the research questions and objectives, we performed a SLR according to various characteristics, such as year of publication, limitations, and future work. The information or previously collected characteristics were integrated with the responses collected through questionnaires to summarize the information.

After an extensive and systematic review of the literature, the research questions are answered and described in the following sections. The contributions and types of WSN and IoT are respectively explained in the Sections 4.1–4.3. Before attacks network intrusions are briefly classified in Section 4.4. In contrast, network security attacks and WSN and IoT layer issues are discussed in Sections 4.5 and 4.6. In Section 4.7, we discuss the limitations and future work of the selected papers. Finally, we conclude the paper in Section 7.

4. Results

In this section, we have briefly discuss the results of the SLR work. We have formulated the research questions presented in Table 3 and divided the results section into seven subsections to answer them. The information about the contribution of WSN and IoT in IR 4.0, network security attacks and intruders in WSN and IoT, WSN coverage, issues in IoT and WSN framework, and limitations of existing reviews are explained in this section. The challenges section summarizes all the problems encountered in WSN and IoT usage.

4.1. RQ1: Contributions of WSN in IR 4.0

The use of WSN has attracted a lot of attention in industry. Because of their prevalence and use in industry, WSN have given rise to IWSN and IWSAN, respectively. These networks enable autonomous work without human intervention. The in-network transmission characteristics are fundamental properties of WSN. Sensor nodes do not transmit raw data, but integrate it to save communication costs. Due to their unique properties and wide range of applications, they are used in many systems, such as military, surveillance, home automation, smart cities, smart buildings, and healthcare monitoring [27]. WSN- and IoT-based devices are used to create reliable, realistic, efficient, flexible, and economical smart cities and buildings in heterogeneous environments [48].

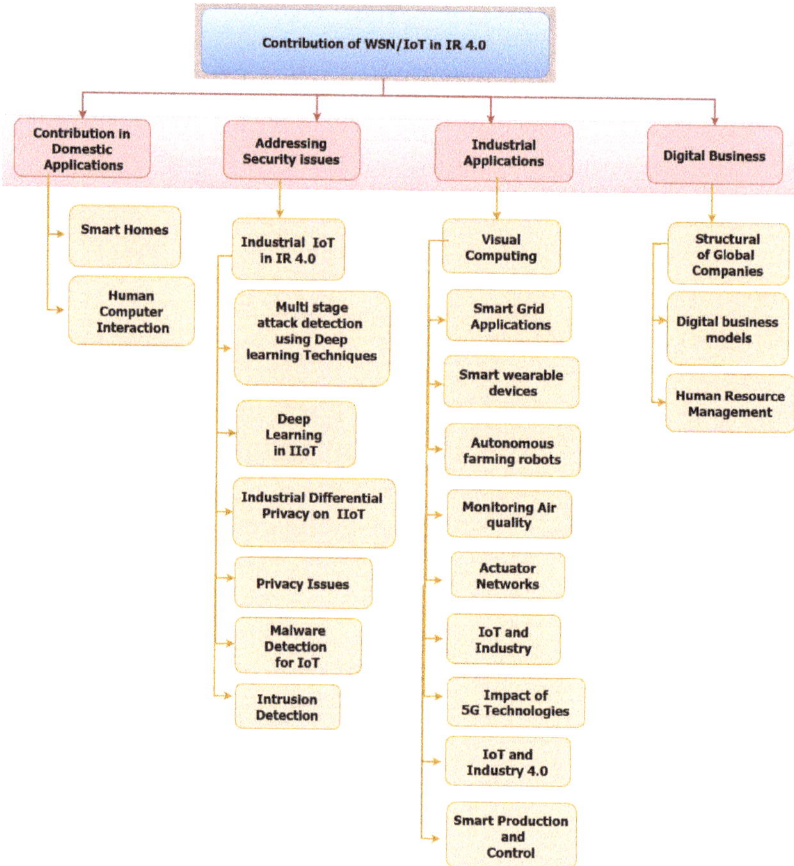

Figure 8. Taxonomy of existing studies.

The categories discussed in this paper and the contribution of WSN in IR 4.0 are listed in the form of a taxonomy presented in Figure 8. WSN is also used in health care management systems to monitor medically ill patients, periodically check their various measurements such as blood glucose levels and pulse, and wirelessly transmit this information to a central repository for further diagnosis [49]. WSN is also used to assist elderly and disabled people. Disabled people are informed of relevant information about real-time activities using smart devices, such as wristwatches [28,50]. In recent decades, WSN have been applied in many fields, including transportation, agriculture [51], automation, manufacturing process control, and supply chain management. In addition, WSN can be easily deployed, have low construction cost, no expenditure on wiring, and lower complexity [52,53].

WSN can be used in various manufacturing applications, such as industrial control, process automation, rescue, and defense. WSN is also used to control and automate industrial processes known as actuators. They can operate independently of a physical environment defined by predefined dimensions [54]. WSN is used to collect, track, and record data in smart factories. Data acquisition is usually done by product information in smart factories. After data collection, processing is done by intelligent machines and manufacturing systems. Nowadays, these factories are self-sufficient, cost-effective, and automated by integrating wireless communications with existing private networks and reducing labor [30].

In software, WSN takes maximum advantage of wireless technologies used to build industrial network infrastructure [55]. On the other side, Industry 4.0 is integrating big data analytics and cloud services [56], 3D printing, computer security, autonomous robotics, the Internet of Things (IoT), 5G, Augmented Reality (AR), and modeling [57,58].

4.2. RQ2: Contributions of IoT in IR 4.0

An integrated digital system would introduce a new intelligent and economical manufacturing process using cutting-edge technology for a variety of existing items and processes [59]. The data collected from production process warehouses and consumer information can be critically analyzed to make a decision in real time under Industry 4.0. The real-time decision-making capability of each small and medium organization enables them to efficiently accept new technologies [60,61]. Industrial IoT delivers solutions and services that provide insights into an organization's ability to monitor and control its operations and assets. IIoT software and tools provide important solutions for better process, layout scheduling, organization, and administration.

In addition, IIoT enables real-time and decision-making features among numerous networked devices that can communicate and interact with each other [62]. Because of the rapid communication and data transfer, attackers can attack data and cause harm to an organization, resulting in cyber attacks. Cyber attacks have become a major challenge for the industrial Internet of Things (IIoT). Therefore, integrating IoT with Industry 4.0 plays a critical role in securing IoT devices from attacks. Unique security objectives and challenges of IIoT have been introduced to overcome industrial-level issues. IIoT challenges and objectives relate to IoT being used by consumers and its capabilities leading to longer life of IoT devices and sensor nodes. In [63], the authors analyzed security challenges and attacks at three levels of the network (perception, network, and application). They considered cryptographic challenges, authentication, network monitoring, and access control mechanisms. The IIoT also addresses local network connectivity and protection from attackers inside. Cyber attacks have become a serious challenge for the IIoT. Hackers attack infrastructure/devices through intrusion and hiding, resulting in poor performance. A bidirectional long and short term memory network with a multi-feature layer has been developed to avoid temporal attacks. Machine learning-based networks that learn temporal attacks from historical data and make associations with test data can effectively identify and detect different attacks within different intervals [64].

DL-IIoT has enormous potential to improve data processing and contribute to IR 4.0. Similarly, machine learning algorithms, such as logistic regression, are widely used for malware detection and security threat protection [65]. Deep learning algorithms are also used for intelligent analysis and processing. Deep learning [46]algorithms such as CNN, auto-encoders, and recurrent neural networks have applications such as intelligent assembly and manufacturing, network monitoring, and accident prevention. The application of deep learning algorithms in IIoT has also enabled various smart applications such as manufacturing, active attack detection and prevention systems, smart meters, and smart agriculture [66]. DL-IIoT relies heavily on data collection, which affects the privacy of the organization's data. Therefore, differentiated privacy is used to protect user privacy, reduce privacy risk, and achieve high performance in IIoT.

On the other hand, IoT and IIoT must provide "differentiated privacy" for individuals and industrial data [67–69]. The contribution of IoT in Industry 4.0 has improved the average availability and sustainability of the enterprise by knowing market trends and decreasing unanticipated downturns [70]. The taxonomy of existing studies and the contribution of IoT in IR 4.0 is shown in Figure 8.

4.3. RQ3: Type of WSN Coverage Area for IR 4.0

WSN coverage is an important factor in sensor quality. Sensing and connectivity are key features of WSN. The former indicate how well a particular sensor behaves and monitors a particular area of interest in which it is deployed. Connectivity shows how well different nodes communicate with each other. The types of wireless sensor network coverage are as follows.

4.3.1. Area Coverage

Sensors usually perform well in area coverage and monitors the field of interest (FoI). This is also called "blanket coverage" because each node communicates with others. Each sensor is placed so that the coverage of the other WSN sensors covers each other [32].

4.3.2. Barrier Coverage

Barrier coverage of the sensor network comes into play when some intruders try to breach the security layer of the network. Sensors are easy to handle and deploy; therefore, their wireless nature makes them vulnerable to malicious security attacks [71]. The sensor nodes are primarily distributed throughout the network and are deployed in chains to detect interruptions.

4.3.3. Point Coverage

Point coverage aims to find a target within range using nearby nodes. It is also known as target coverage. It is characterized by consuming less energy in a given zone than in the entire region of the FoI. Only a few targets are covered by individual nodes, while other targets can be detected under other sensor scopes. The primary goal is to select a specific target within the FoI to reduce energy consumption [71].

4.4. RQ4: Classification of Network Intruders

There are two main types of intruders: internal and external. Internal intruders are people inside the organization; they can be either a customer or a legitimate user, such as an employee of the organization's network. External intruders are people outside the organization, whether external or internal. Each intruder can be involved in numerous illegal activities, working alone, as a group, or with agencies. These entities are described in detail below.

4.4.1. Solo Entities

Solo entities are those that work alone with minimal safety. They are usually experts in their domains by employing a single piece of code as equipment, such as viruses, worms, and sniffers to misuse frameworks. They usually gain access to the organization's framework through hardware damage and web loopholes. Their targets are usually little, and attacks are slightly less critical. Moreover, large and complex systems that may contain flaws are more vulnerable to attacks. Monetary institutions are also more exposed to attacks as they exchange sensitive information [72].

4.4.2. Organized Groups

Solo entities are those who work alone with minimal security. They are usually experts in their field, using a single piece of code as equipment, such as viruses, worms, and sniffers to abuse frameworks. They generally gain access to the organization's framework through hardware damage and web loopholes. Their targets are usually small, and attacks are somewhat less critical. In addition, large and complex systems that may contain flaws are more vulnerable to attack. Monetary institutions are also more exposed to attacks because they exchange sensitive information [72].

4.4.3. Intelligence Agencies

Intelligent agencies from other countries are involved in this type of attack. These agencies constantly seek to test the military architecture of other nations, including contemporary monitoring and covert political and military activities. To do so, they require many resources, from software to hardware, research, development, personnel, and finances. Because they have all these resources at their disposal, some agencies now pose a serious threat to economic and military espionage. Such organizations pose the greatest threat to networks and must be closely monitored to protect the nation's important assets [73].

4.5. RQ5: Network Security Attack in IoT and WSN Layers

Threats become more attractive and dangerous as technology increases. Although new security mechanisms are being developed, intruders can easily find other ways to attack systems. Table 8 explains the network security attacks in the IoT and WSN domains. The attacks are categorized according to the open system interconnection (OSI) layered point of view [74].

Table 8. Network Security Attacks on IoT and WSN Layers.

Sr. No.	Layer Name	Attacks
1	Physical layer	Interception, radio interference, jamming, tempering, Sybil attack.
2	Data link layer	Replay attack, Spoofing, altering routing attack, Sybil Attack, collision, traffic analysis, and monitoring, exhaustion.
3	Network layer	Black hole attack, wormhole attack, sinkhole attack, grey hole attack, selective forwarding attack, hello flood attack, misdirection attack, internet smurf attack, spoofing attack.
4	Transport layer	De-synchronization, transport layer flooding attack.
5	Application layer	Spoofing, alter routing attack, false data ejection, path-based DoS.

4.5.1. Denial of Service Attacks (DOS)

A Denial-of-service (DoS) attack is a malicious attack in which attackers make the victim's system unresponsive and difficult to reach for the legitimate user by making many requests to the expected URL than the server can handle [75,76]. DoS attacks typically occur when authenticated clients have not been granted access to the information or service [77]. A distributed denial of service (DDoS) attack is a type of DoS attack that uses multiple users or infected systems to attack a victim's system or to attack a website so that it becomes unresponsive. This attack also prevents the website from functioning properly and disrupts regular traffic. In WSN, a DDoS attack changes the routing protocol information DSR, resulting in a huge amount of unauthorized traffic and making the network/website unavailable. On the other hand, a low-rate denial-of-service (LDoS) attack is another type of DoS that penetrates the WSN's routing protocol, thus compromising the security and trust mechanisms. An LDoS attack is difficult to detect due to its non-stationary nature and low signal strength52. In this attack, illegitimate traffic affects the operational capability of a network. It causes severe outages and monetary losses.

4.5.2. Replay Attacks

When an attacker replays a flood of messages between the sender and the receiver, updating the line table (DT) to steal information, it is called a replay attack [33]. It is a network layer attack in which a third party intercepts the data during transmission. The attacker retransmits this data by either modifying or delaying it, spoofing the sender's IP address to the attacker's IP address, and impersonating the legitimate sender.

4.5.3. Trojan Worms, Viruses, and Malware

An attacker can use malicious software to manipulate data, steal information, or even launch a denial-of-service attack on a device. A worm, such as a Trojan horse, can infect one's computer when one downloads a file or receives an update. The worm then multiplies and attacks other machines on the network. Unlike a virus, many Trojan horses usually reside on one's own computer. A virus can infect the host's file when sent via email and then spread to other users [78]. Malware is malicious content that can interfere with a computer's operation and slow its performance. When data from IoT devices is compromised, malware can infest the cloud or data centers. These attacks breach the primary security mechanisms of any OS/server, such as a firewall and window defender.

4.5.4. Black Hole Attacks

This is a network layer attack known as a packet drop attack. In this attack, a node sends an RREQ packet to all its neighbors in the network, and the router is supposed to forward the packet instead of discarding it. The nature of this attack is similar to a DDoS attack. Attackers are used to attack routers by sending many false requests to prevent legitimate routers from forwarding packets. This is also called a first-come, first-served attack because the attacker can also use a malicious router or reprogram it to block packets instead of sending correct information [78]. These attacks reduce the average throughput. When combined with a sinkhole attack, this attack affects performance and stops all traffic around the black hole. When combined with the sinkhole attack, this attack severely degrades traffic and modifies or discards content during transmission.

4.5.5. Sink Hole Attacks

This is a network layer attack where attackers attract all network traffic from nearby nodes to a compromised node and appear as attractive and trusted nodes. This attack is also used to initiate other attacks such as spoofing attacks, DoS, and modification of routing information in WSN [34]. When combined with selective routing and worm attacks, sinkhole attacks become even more dangerous. A sinkhole attack is initiated in two ways, either by hacking a node within the network or by a malicious node impersonating itself as the shortest path to the base station [35,36]. A sinkhole attack impacts the routing configuration/protocols of the forwarding node. Due to this behavior, it is considered as an error or malicious node by the neighboring nodes, which affects the network performance. This leads to mis-routing and incorrect displays of the routing protocol.

4.5.6. Wormhole Attacks

A wormhole attack is a network layer attack in which an invader attacks the WSN through two or more compromised nodes. The invaders forward the data from one malicious node to another node at the end of the network through the tunnel. The wormhole appears to other nodes as a fictitious neighbor. Wormhole nodes usually transmit data directly from one node to the destination without including other nodes in the path. Due to this nature, other nodes in WSN easily trust those closest to other nodes, which causes many routing problems. Moreover, they can build better communication channels for long-range communication [37]. Wormhole attacks affect the performance of many network services, such as time synchronization, localization, and data fusion.

4.5.7. Selective Forwarding (Gray Hole)

A selective forwarding attack (SFA) is a special type of black hole attack in which the compromised node drops some selective packets instead of all packets. Invaders drop packets containing critical information, such as military information, without noticing them or allowing others that may contain non-critical information to pass. This can lead to worse effects and a decrease in network efficiency [34–37]. Selective forwarding attacks impact network performance and consume limited energy resources.

4.6. RQ6: Issues in WSN and IoT Frameworks

In this section, we mention various WSN and IoT frameworks that highlight the importance of WSN and IoT in different aspects of life. Although many advances have been made in IoT, there are still many problems, as shown in Figure 9, that need to be reduced and solved efficiently to avoid any damage [79,80].

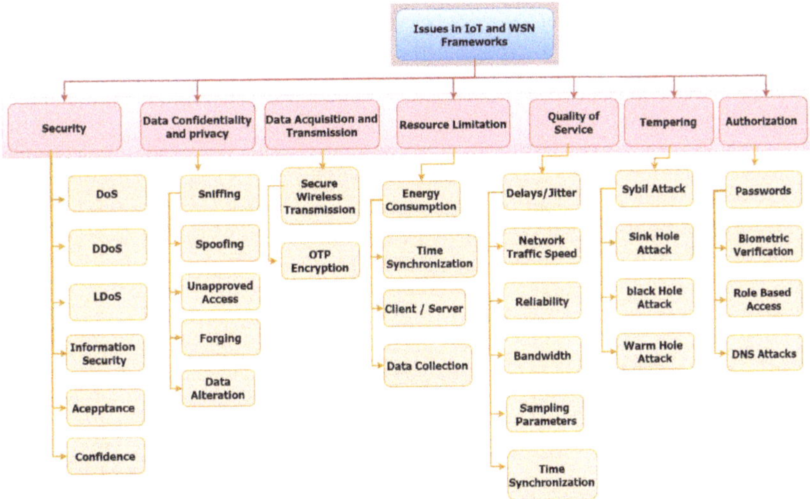

Figure 9. Issues in WSN and IoT Framework.

4.6.1. Security

Security is essential for any organization to protect its environment, systems, devices and applications from outside attacks. Data and communication technologies are increasing every day. Therefore, data and information security are necessary tasks [81]. In addition to data, its transmission over the network should also be protected. Although technology has evolved and security mechanisms have improved, attackers have still found many ways to breach the security level [38,39]. With the increasing number of IoT devices, new security issues have emerged. For real-time applications, the most important thing is to keep the WSN secure. The network and its associated router or hub should enforce an access control mechanism to prevent unauthorized users. Each node connected to another node is security relevant, whether it is a restricted device or a smart device. Acceptance, confirmation, categorization, trust, and information security are the most important security requirements to be considered in IoT networks and WSN. It is challenging to provide security measures for flexible detection devices. Therefore, protecting information from dictatorial forces or illegal access is called security [82,83].

4.6.2. Data Confidentiality and Privacy

Data confidentiality is a significant issue in IoT and network security. In IoT frameworks, the client gains access to the information and system management in an unintended environment due to issues such as the use of sensor nodes. Attackers can physically capture them and extract data using an energy analysis attack [84]. Refurbished devices made from these captured devices can launch new attacks and violate security. Therefore, the IoT device should verify whether or not the user or device has been granted permission to access the system. The practice of controlling access to data by granting or denying permission based on a set of laws. Many devices/clients must be authenticated by management to access the system. Data confidentiality and access are the main issues in the Internet of Things (IoT). Researchers are trying to figure out how to handle the personalities of customers, items/articles, and devices in a secure manner. Due to the ubiquitous nature of IoT and WSN systems, privacy and confidentiality are major concerns in IoT devices and frameworks. Some issues, such as sniffing and spoofing, unauthorized access, data changing, forging, and unapproved alteration of IoT and WSN nodes, pose significant uncertainties in IoT. An attacker can use various IoT devices and applications to capture sensitive and personal data that is visible to outsiders.

4.6.3. Data Acquisition and Transmission

The primary goal of IoT is to collect data and transmit it to where it is needed in a network. Sensors are the devices used to collect data from the environment to transmit it to the base station. After the raw data is collected, it is sent to the Sink Hub for processing. Data collection and transmission are other problems in IoT and WSN because data is exposed and modified during transmission. Data acquisition is an energy-consuming process, so extra care must be taken during gathering and transmission. Intruders can steal the data during transmission if it is not encrypted or transmitted over a secure channel. The intruder can take over a node and reprogram it with a malicious code, damaging the entire network. Therefore, security is required for this process. Sometimes intruders attack the databases of organizations to violate the confidentiality of the data [85]. Also, the intruder may destroy the node or collect important or unusual information that could be used against the system. For this reason, researchers present many security mechanisms. They protect end-to-end communication links using one-time-pad (OTP) encryption method and also identify the vulnerabilities in the DBMS application using SEPTIC method.

4.6.4. Resource Limitations

If necessary resources in WSN and IoT are abandoned or not handled efficiently, it may affect the performance of the network. The network consists of many nodes and sensors that require energy to operate well [86]. Various MAC layer protocols have been developed to reduce the energy consumption of sensors or nodes. These energy-efficient algorithms work primarily by regulating the synchronization of network traffic over time and the time period during which a node becomes active in a network [87,88]. In contrast, the communication medium is another basic requirement, since nodes rely on the Internet for data transmission. There is a constant need for energy, otherwise the network will fail. The nodes have limited resources because battery capacity, correspondence capacity, and computing power are low. Again, security is the main problem, because the security measurement expenses require more resources to maintain the speed of the network, which is not affordable. As a result of low regulated security, attacks can subvert software execution and protocols used in the network [89,90].

4.6.5. Quality of Service

Quality of Service (QoS) manages networks and resources to strengthen IoT connectivity. QoS manages delay, jitter, reliability, and bandwidth by classifying network traffic. It plays an important role in optimizing systems. Quality of Service means that energy efficiency, reliability, bit error rate, and latency should be good enough to capture data over a network. Therefore, it is classified in two ways: program-specific and network-specific. The QoS perspective of the network refers to the effective management of network resources and transmission performance, while the perspective of the program refers to mobility, time synchronization, and sampling parameters. Similarly, many algorithms have been developed to distribute heavy traffic evenly, and the energy consumption load in a network uses a cluster-head approach to achieve high performance and efficiency [91].

4.6.6. Tampering

Sensors can be placed either indoors or outdoors. Indoor sensors can be easily managed and protected, while outdoor sensors are more vulnerable to attackers due to remote locations with poor security, harsh climates, etc 78. The probability that these sensors will be physically attacked is much higher; therefore, physical protection cannot be guaranteed. A DoS attack manipulates the network by breaking the connection or changing the current network. The attacker can also replace the original node with a fake or malicious node, causing a severe attack on the network [92]. In a Sybil attack, a malicious node penetrates each cluster head of the network and affects the operation of the routing protocol. Compromised nodes can be used to launch new attacks without exposing themselves [41,93]. These nodes are difficult to detect and isolate, allowing an attacker to alter data or transmit malware throughout the network that causes significant damage [94]. Constant monitoring of the network is necessary to ensure that WSN nodes cannot be tampered with and that network performance remains stable [95].

4.6.7. Authorization and Authentication

Nodes are the building blocks of the Internet of Things that must be defined in the network. Transmission between devices and access to the entire network span a wide range in IoT and WSN. IoT devices perform role-based access control, and their devices are allowed to do only what is required [96,97]. Devices and their data must be protected from physical and logical attacks on the network. Attacks on the DNS cache could affect the overall performance of the network. Authentication is the process by which each node on the network can access data based on a fixed connection to a server or cloud-based server. If the authentication process is not administered properly, it will lead to security issues and questions. An attacker can easily access the network and make it fail temporarily by making too many wrong attempts.

Authentication is complicated due to the massive proliferation of wireless media and the nature of sensor networks. Authentication is usually done using the credentials of a legitimate user [98,99]. However, this technique is not secure enough. Therefore, passwords should be changed regularly and computers should not be left unattended to make this technique robust. Both the sender and the recipient should perform authentication to verify the origin of the communication [100,101].

4.7. RQ7: Limitations of the Literature Review

In this section, Table 9 explains the proposed solutions of the work conducted by various authors and the contributions with the limitations of their work are also described. The goal was to find research gaps in this area to help other researchers. The research gaps will allow researchers to develop solutions and new methods that could help fill the missing piece.

Table 9. Contributions and limitation of the literature.

Reference	Title of Article	Proposed Solution	Limitations and Future Work
Sharma et al. [9]	Enhancement of relay nodes communication approach in WSN-IoT for underground coal mine	They designed relay node structures for a wireless sensor network and load balancing to improve network lifetime parameters. They designed an IoT-based WSN to provide advance warning of any natural disaster in coal mines.	There were several analysis parameters to analyze the networks, such as network lifetime, communication and transmission cost, energy consumption, and coverage of the whole area.
Faheem et al. [49]	Bio-inspired routing protocol for WSN-based smart grid applications in the context of Industry 4.0	They designed a comprehensive, optimized, and QoS monitoring multi-hop network system for real-time data transmission in Industry 4.0. This self-optimized smart routing protocol (SIRP) was efficiently used for WSN-based SG applications.	In the future, they will attempt to enhance their developed SIRP routing scheme and communications architecture to collect QoS-aware data for different WSN-based smart grid applications with little data redundancy.
Arslan et al. [52]	IoT and wireless sensor network-based autonomous farming robot	They developed a computer vision-based algorithm used for the classification of weed and a non-image. Wireless sensor nodes detect weed images through image processing methods and gather light, temperature, humidity, and moisture data.	The limitation of this work is that they did not provide any GUI or mobile application control to work robot autonomously.
Chen et al. [53]	Efficient and secure three-party mutual authentication key agreement protocol for WSN in IoT environments	They proposed a practical and secure approach to merge IoT and WSN. Their scheme had high performance, low communication, and computational costs, low energy consumption, and provided effective authentication of the user in IoT.	The limitation of this study is that they did not provide a solution to the security threats in a heterogeneous IoT environment. In the future, they will evaluate the reliability and scalability of their systems of heterogeneous environments.
Rathee et al. [102]	A secure IoT sensors communication in Industry 4.0 using blockchain technology	Wireless sensor network security improved using blockchain and compared security metrics. &It ensured confidentiality and responsibility and tracked each sensor's operation. The blockchain was used to store IoT artifacts and sensors.	The developed IoT sensor takes time to test a single block before it is put to the blockchain.
Mellado et al. [103]	Prediction of satellite shadowing in smart cities with application to IoT	The technology had a minimal processing load. It was highly desirable to create a coverage map that can optimize network resources in satellites.	There is a lack of evaluation of requirements for satellite-based IoT and output connectivity protocols through simulations in actual situations.

Table 9. *Cont.*

Reference	Title of Article	Proposed Solution	Limitations and Future Work
Garg et al. [101]	Towards secure and provable authentication for the internet of things: realizing Industry 4.0	The effectiveness of the developed protocol was evaluated with frequently utilized AVISPA, PUFs, and ECC encryption algorithms. A proposed technique was developed to create a durable, stable, and efficient user architecture that promotes shared authentication for IoT and server nodes and is resistant to cyber threats.	This protocol is for academic and research purposes only, and its implementation has not yet been tested in the real world.
Behera et al. [104]	Residual energy-based cluster-head selection in WSN for IoT application	The method takes into account the intended value of initial energy, residual energy, and cluster heads to choose the specific set of cluster heads in the network that adapts IoT applications to maximize flow, durability, and residual energy.	They did not review existing path selection factors in a node mobility network that altered its role constantly.
Wan et al. [105]	Software-defined industrial Internet of Things in the context of industry 4.0	They proposed a new idea of information interaction in Industry 4.0 using software-defined IIoT. They enhanced the network size using IIoT. The IIoT architecture manages physical devices and information exchange methods via a customized networking protocol.	The limitation of the study is the effective coordination between IIoT where the network is heterogeneous for transmission of information.
Tan et al. [106]	Application of IoT-aided simulation to manufacturing systems in cyber-physical systems	They discussed the construction and implementation methods of digital twin (DT). In this study also explained the issues involved in developing DT with the help of IoT manufacturing devices. DT is the simulation tool that can gather and synchronize data for the real world to a real-time environment.	The absence of experimentation and optimization in predicting future locations or results are other essential aspects of DT.
Rahman et al. [107]	DistB-SDoIndustry: enhancing security in Industry 4.0 services based on the distributed blockchain through software-defined networking-IoT enabled architecture	In this work, the authors develop a distributed blockchain-based security system integrated with the help of IoT and SDN. Blockchain is used for data security and confidentiality, while SDN-IoT incorporates sensor networks and IoT devices to improve the security services in Industry 4.0.	Limitations of this study are that the developed model SDN-IoT was still in the initial stage, so it was not able to detect different types of risks, such as service denial (DoS) and flood attack and packet filtering. The developed system had no proper GUI, so the throughput, packet arrival time, and response time were rarely challenging to analyze.

Table 9. Cont.

Reference	Title of Article	Proposed Solution	Limitations and Future Work
Haque et al. [108]	Comparative study of IoT-based topology maintenance protocol in a wireless sensor network for structural health monitoring	They developed a computer-based monitoring system to analyze the vibration or earthquake measurement. WSN are used to sense structural damages and identify their pinpoint location. They also proposed a topology-based maintenance system to analyze network architecture. Their system was an energy-efficient system that automatically turned off nodes where no traffic was detected.	The limitation of this study is that WSN nodes are not capable enough to provide scalability for large coverage areas.
Wan et al. [109]	Toward dynamic resources management for IoT-based manufacturing	To build a fully interactive environment and dynamic management of resources, an ontology-based technology, SDN, communication technology device to device combined with ontology modeling and multi-agency technology were used to accomplish sophisticated administration of resources. They solved load secluding problems using Jena logic reasoning and contract-net protocol-based technology in Industry 4.0.	The limitation of this work was the high time complexity of the load balancing algorithm to complete the task efficiently. It was challenging to refine the process due to the complex nature of multi-agent technology, and referencing rules were much more complex.
Bulaghi et al. [110]	SENET: a novel architecture for IoT-based body sensor networks	Multiple algorithms, such as particle swarm optimization (PSO), ant colony optimization (ACO), and genetic algorithms (GA) were used to save energy of WSN. They evaluated WSN energy consumption using optimization algorithms and calculated the total number of uncovered points, their stability, and dependability.	The design meets some disadvantages and does not work in real-time data.
Thiago et al. [111]	A comprehensive dependability model for QoM-aware Industrial WSN	When performing visual area coverage in occluded scenarios. They proposed a mathematical model named quality of monitoring parameter (QoM) to assess the dependability of WSN, their availability, and reliability considering hardware, networking, and visual coverage failures.	Their developed method was inefficient at analyzing the system's dependability in real-time applications due to failures or repairs happening as soon.

Table 9. *Cont.*

Reference	Title of Article	Proposed Solution	Limitations and Future Work
Patricia et al. [112]	Wireless sensor network combined with cloud computing for air quality monitoring	They designed a small size, low cost, and efficient system to monitor the air quality using wireless sensor nodes. They performed multiple algorithms such as multilayer perceptron, SVM, and PCA to discriminate and quantify the volatile organic compounds.	The limitation of this study is that sensor nodes are less efficient at covering a large area to monitor and cannot do real-time testing and the field measurements of sensors.
Li et al. [113]	Edge computing-enabled wireless sensor networks for multiple data collection tasks in smart agriculture	They designed a data collection algorithm considering data quality factors in smart agriculture. Then modeled the data collection process by merging WSN and IoT.	The developed edge computing driven framework [47] and data collection algorithm were not capable of collecting data in a real agriculture environment.
Kumar et al. [114]	Cluster centroid-based energy-efficient routing protocol for WSN-assisted IoT	They developed a system that was capable of self-organization of local nodes to save energy. Their system adopted new algorithms to rotate head clusters based on centroid locations in IoT using WSN. The technique exceeds conventional protocols for efficiency criteria, such as the consumption of energy by the network, intermediate sensor node, packet distribution ratio, packet failure percentage, and network output. Their work was best for the base station located in the network.	The routing protocol was not optimal, routing strategies were lacking, and packet loss was caused if the base stations were even in the network. In the future, they will enhance this work by using a multi-hop path strategy to the base station. In this technique, the cluster head will transmit data to the base station, even outside the network.
Haseeb et al. [115]	An energy-efficient and secure IoT-based WSN framework: an application to smart agriculture	They proposed an IoT-based WSN framework that collected data from agriculture and transmitted it to the nearest base station. They enhanced network throughput, low latency rate, energy consumption, and packet drop ratio. They also provided security to the data transmission channel using the recurrence of the linear generator.	The limitation of this work is that they did not assess the device consistency in a mobile IoT. Therefore, they will analyze the performance and reliability of developed frameworks in the transportation system and mobile-based IoT network.

5. Challenges and Open Issues

Intelligent systems can address various problems faced by industry, but there have been some challenges in integrating IoT and WSN into Industry 4.0. Technological improvements in IoT and WSN have increased concerns about security and data management [96]. As more and more data is generated, it is difficult for factories and industries to manage it properly. Artificial intelligence algorithms have been implemented to manage Big Data and make systems and devices act more intelligently. The algorithms are used to process the data in different time periods. For education, the data must be shared in a central repository, while enterprises are mainly reluctant to share their private data due to poor and insufficient organizational support for data in Industry 4.0. There are also safety management issues in Industry 4.0 [116].

Big data: The emergence of various technologies and the explosion of their use have led to the outstanding development of Big Data technology and processing. Every device produces a huge amount of data. Due to the growing amount of big data, the improvements of Big Data packages encounter limitations and demand situations that need to be "overcome" in order to manage the amount of data used efficiently.

Adapting to 6G: 6G is another challenge for wireless sensor networks and the Internet of Things. Processing power is a major challenge in developing low-power and low-cost 6G devices. In addition, 6G brings privacy and security challenges for WSN and IoT.

Updates:system components could not be upgraded due to interoperability between protocols, systems, and their components. Therefore, systems are more vulnerable to attack if any part of a single system from a network is infected in intelligent factories.

Environment: security is also a critical challenge in WSN [97,100,106]. WSN nodes are not secure when deployed in a prone environment due to the wireless transmission of data. An attacker can access them from anywhere in the world and manipulate them easily. Internet attacks can also affect the vulnerability of sensor nodes.

Supply chain management systems: IoT devices are spreading erery day, posing new challenges to the integrity and scalability of supply chain management systems [117,118]. Simultaneously connecting IoT devices to the cloud or the Internet requires a lot of access control, fault tolerance, data management, privacy, and security.

Limited resources: are other challenges in WSN domain that affect the energy of sensor nodes used in the network. Sensor nodes usually change their mode from sleep to active and vice versa. Therefore, sleep mode is considered as "outside the network" while active mode brings some other issues such as energy consumption [119]. Due to the high energy consumption, they also became dead. Sensor nodes usually have limited power, processing, and memory. In addition, sensor mobility is another problem that hinders the integration of mobile sensor nodes with the Internet.

6. Future Directions

Industry 4.0 leads to the merging of people and technology to complement human activities with intelligent machines. Industry 4.0 will lead to customized human fashion that will minimize the oversupply and unavailability of supplies or items. Human-machine interaction will increase productivity and customer satisfaction with customized products.

The next version of Industry 4.0 is Industry 5.0, which is expected to be more user-friendly and better integrate technology with society and the environment. It depends mainly on robots. Robots are already the backbone of manufacturing, and Industry 4.0 technologies [120,121] provide flexibility in manufacturing. Industry 5.0 combines human creativity and craftsmanship with the speed, productivity (e.g., CPS) [122], and consistency of robots. In this version, robots can be programmed to work alongside humans.

Soft computing: can be used to reduce the dimensions of these large dimensional data sets [123]. Good features are essential to make efficient decisions. This is the reason why soft computing techniques are used to obtain useful features.

Explainable artificial intelligence (XAI): can be used for interpretability of the decision made by the classification model. Classification models make decisions in a black box where the user does not know how the decision is made. XAI converts this black box into a white box and interprets the decision made by a model. XAI increases user confidence to take further action [124].

Federated learning (FL): is an optimal choice for privacy preservation. FL works by training global and local models on the edge device. The model on the edge device does not share the data with the global, thus keeping the data private at each edge device. Only parameters are shared globally to retrain the global model and optimize the inference results [125,126].

Secure devices: sensor nodes are designed to consume less energy and become active when they are needed or an event occurs [59]. Further improvements are also needed to prevent attacks from the Internet. While the IoT has no limitations in terms of processing or energy. Due to the tremendous proliferation of IoT devices, this paradigm is now being shifted from the IoT to the Internet of Everything.

Sustainability: IoT systems are now moving toward the idea of self-organization, and systems are becoming capable of responding in an automated and adaptive manner and dealing with changes and uncertainties in the environment [118].

Education 5.0: in this digital era, education must also change from traditional to integrating hardware and software with co-bots to develop new skills and a smart society. Educational institutes are now using pedagogical tools to provide a better experience. Even though IoT-based education is still not widespread, there is still room for further improvement, such as sensor node coverage and efficiency, wireless data transmission of data [127], battery life, and high-cost nodes.

General directions: there are many challenges. Future directions may address heterogeneous interoperability of systems, self-organization protocols, routing schemes for managing IoT networks, data management [79], cross-platform optimization, and the development of network security algorithms to secure wireless transmission from data manipulation, stealing, or hacking activities. On the hardware side, researchers are developing energy-efficient sensor nodes [91,115], with net-zero power to reduce maximum power consumption.

7. Conclusions

In this digital and modern era, technology is evolving every day. Due to the massive proliferation of technology, IoT and WSN play an important role in Industry 4.0 to develop smart applications, design networked data centers, and build autonomous smart industries. Data networks have been created and improved with the help of new and smart devices. In this systematic literature review, WSN and IoT network threats were analyzed and a descriptive comparative study was conducted. These networks are the main attack surfaces for attackers to draw meaningful patterns from system and user data. Wireless sensor networks (WSN) and the Internet of Things (IoT) have rapidly (and widely) evolved to meet the increasing demand for conventional application scenarios, such as plant automation and remote process control systems. These smart devices are also being used to improve the efficiency of existing networks and create new opportunities for automating and securing industrial processes. In this article, we explore seven research questions: (i) What are the contributions of WSN in IR4. 0? (ii) What are the contributions of IoT in IR 4.0? (iii) What are the types of WSN coverage areas for IR 4.0? (iv) What are the main types of network intrusions in WSN and IoT systems? (v) What are the main network security attacks in WSN and IoT? (vi) What are the major issues in IoT and WSN systems? (vii) What are the limitations and research gaps in the current work? The main purpose of the fourth industrial revolution with WSN and IoT explicitly shows that the evolutionary transition needs to be intensified and extended to include emerging research areas and intimidating technological challenges. This article covers all elements of WSN, from the design phase to the security requirements, from the implementation phase to the classification of the network, and from the difficulties and challenges of WSN. Future studies will address the problems in the coverage regions of wireless sensor networks and provide effective solutions to the existing problems and challenges in this area. The use and application of WSN and IoT in Industry 4.0 involves the processing of extracted data and the efficient and secure transmission of this data to a remote location.

Author Contributions: Conceptualization, M.M., M.R., and S.H.; data curation, S.H.; formal analysis, A.R.J.; funding acquisition, T.R.G.; investigation, A.R.J.; methodology, A.R.J.; project administration, J.C.-W.L. and T.R.G.; resources, G.S. and T.R.G.; software, A.R.J.; supervision, A.R.J. and J.C.-W.L.; validation, S.H. and M.M.; visualization, S.H. and M.R.; writing—review and editing, G.S., T.R.G., and M.R. All authors have read and agreed to the published version of the manuscript.

Funding: This paper is partially supported by the Western Norway University of Applied Sciences, Bergen, Norway.

Institutional Review Board Statement: Not applicable.

Informed Consent Statement: Not applicable.

Data Availability Statement: The data presented in this study are available on request from the corresponding author.

Conflicts of Interest: The authors declare no conflict of interest.

References

1. Khorov, E.; Lyakhov, A.; Andrey, A.K.; Guschin, A. A survey on IEEE 802.11ah: An enabling networking technology for smart cities. *J. Comput. Commun.* **2015**, *58*, 53–69. [CrossRef]
2. Ashibani, Y.; Mahmoud, Q.H. Cyber physical systems security: Analysis, challenges and solutions. *Comput. Secur.* **2017**, *68*, 81–97. [CrossRef]
3. Ahmad, W.; Rasool, A.; Javed; Baker, A.R.T.; Jalil, Z. Cyber Security in IoT-Based Cloud Computing: A Comprehensive Survey. *Electronics* **2022**, *11*, 16. [CrossRef]
4. Landaluce, H.; Arjona, L.; Perallos, A.; Falcone, F.; Angulo, I.; Muralter, F. A review of iot sensing applications and challenges using RFID and wireless sensor networks. *J. Sens.* **2020**, *20*, 1–18. [CrossRef] [PubMed]
5. Dalipi, F.; Yayilgan, S.Y. Security and privacy considerations for IoT application on smart grids: Survey and research challenges. In Proceedings of the Future Internet of Things and Cloud Workshops (FiCloudW), Vienna, Austria, 22–24 August 2016; pp. 63–68.
6. Tuna, G.; Gungor, V.C. A survey on deployment techniques, localization algorithms, and research challenges for underwater acoustic sensor networks. *J. Commun. Syst.* **2017**, *30*, e3350. [CrossRef]
7. Faheem, M.; Tuna, G.; Gungor, V.C. Lrp: Link quality-aware queue-based spectral clustering routing protocol for underwater acoustic sensor networks. *J. Commun. Syst.* **2017**, *30*, e3257. [CrossRef]
8. Parra, L.; Sendra, S.; Lloret, J.; Rodrigues, J.J. Design and deployment of a smart system for data gathering in aquaculture tanks using wireless sensor networks. *J. Commun. Syst.* **2017**, *30*, e3335. [CrossRef]
9. Sharma, R.; Prakash, S. Enhancement of relay nodes communication approach in WSN-IoT for underground coal mine. *J. Inf. Optim. Sci.* **2020**, *41*, 521–531. [CrossRef]
10. Rashid, B.; Rehmani, M.H. Applications of wireless sensor networks for urban areas: A survey. *J. Netw. Comput. Appl.* **2016**, *60*, 192–219. [CrossRef]
11. Ravi, C.; Tigga, A.; Reddy, G.T.; Hakak, S.; Alazab, M. Driver Identification Using Optimized Deep Learning Model in Smart Transportation. *ACM Trans. Internet Technol.* **2020** . [CrossRef]
12. Tandale, U.; Momin, B.; Seetharam, D.P. An empirical study of application layer protocols for IoT. In Proceedings of the International Conference on Energy, Communication, Data Analytics and Soft Computing (ICECDS), Chennai, India, 1–2 August 2017; pp. 2447–2451.
13. Kumar, R.; Kumar, P.; Srivastava, G.; Gupta, G.P.; Tripathi, R.; Gadekallu, T.R.; Xiong, N.N. PPSF: A privacy-preserving and secure framework using blockchain-based machine-learning for IoT-driven smart cities. *IEEE Trans. Netw. Sci. Eng.* **2021**, *8*, 2326–2341. [CrossRef]
14. Farsi, M.; Daneshkhah, A.; Hosseinian-Far, A.; Jahankhani, H. (Eds.) *Digital Twin Technologies and Smart Cities*; Springer: Berlin/Heidelberg, Germany, 2020.
15. Majeed, M.A.A.; Rupasinghe, T.D. Internet of things (IoT) embedded future supply chains for Industry 4.0: An assessment from an ERP-based fashion apparel and footwear industry. *Int. J. Supply Chain. Manag.* **2017**, *6*, 25–40.
16. Oliff, H.; Liu, Y. Towards Industry 4.0 Utilizing Data-Mining Techniques: A Case Study on Quality Improvement. *Procedia CIRP* **2017**, *63*, 167–172. [CrossRef]
17. Ovsthus, A.A.K.S.K.; Kristensen, L.M. An Industrial Perspective on Wireless Sensor Networks—A Survey of Requirements, Protocols, and Challenges. *IEEE Commun. Surv. Tutor.* **2014**, *16*, 1391–1412.
18. Muneeba, N.; Javed, A.R.; Tariq, M.A.; Asim, M.; Baker, T. Feature engineering and deep learning-based intrusion detection framework for securing edge IoT. *J. Super Comput.* **2022**, 1–15.
19. Osterrieder, P.; Budde, L.; Friedli, T. The smart factory as a key construct of Industry 4.0: A systematic literature review. *Int. J. Prod. Econ.* **2020**, *221*, 107476. [CrossRef]
20. Devesh, M.; Kant, A.K.; Suchit, Y.R.; Tanuja, P.; Kumar, S.N. Fruition of CPS and IoT in Context of Industry 4.0. In *Intelligent Communication, Control and Devices*; Springer: Singapore, 2020; pp. 367–375.
21. Saniuk, S.; Grabowska, S.; Gajdzik, B. Social expectations and market changes in the context of developing the Industry 4.0 concept. *J. Sustain.* **2020**, *12*, 1362. [CrossRef]
22. Deepa, N.; Pham, Q.V.; Nguyen, D.C.; Bhattacharya, S.; Prabadevi, B.; Gadekallu, T.R.; Pathirana, P.N. A survey on blockchain for big data: Approaches, opportunities, and future directions. *Future Gener. Comput. Syst.* **2022**, *131*, 209–226. [CrossRef]
23. Number of Internet of Things (IoT) Connected Devices Worldwide from 2019 to 2030. 2021. Available online: https://www.statista.com/statistics/1183457/iot-connected-devices-worldwide/ (accessed on 21 January 2022).

24. Jovanović, B. Key IoT Statistics. 2021. Available online: Https://dataprot.net/statistics/iot-statistics/ (accessed on 20 December 2021).
25. AI Multiple. *30 Internet of Things—IoT Stats from Reputable Sources in 2021*. Available online: https://research.aimultiple.com/iot-stats/ (accessed on 21 February 2022).
26. O'Dea, S. Wide-Area and Short-Range IoT Devices Installed Base Worldwide 2014–2026. 2021. Available online: Https://www.statista.com/statistics/1016276/wide-area-and-short-range-iot-device-installed-base-worldwide/ (accessed on 14 December 2021).
27. Pundir, S.; Wazid, M.; Singh, D.P.; Das, A.K.; Rodrigues, J.J.P.C.; Park, Y. Intrusion Detection Protocols in Wireless Sensor Networks Integrated to the Internet of Things Deployment: Survey and Future Challenges. *IEEE Access* **2020**, *8*, 3343–3363. [CrossRef]
28. Kumar, S.; Tiwari, P.; Zymbler, M. Internet of Things is a revolutionary approach for future technology enhancement: A review. *J. Big Data* **2019**, *6*, 1–21. [CrossRef]
29. Musaab Zaidan, A.; Bahaa, A.; Talal, B.; Kiah, M.M.; Laiha, M. A Review of Smart Home Applications based on Internet of Things. *J. Netw. Comput. Appl.* **2017**, *97*, 48–65.
30. Raza, S.; Faheem, M.; Genes, M. Industrial wireless sensor and actuator networks in Industry 4.0: Exploring requirements, protocols, and challenges—A MAC survey. *Int. J. Commun. Syst.* **2019**, *32*, 1–32. [CrossRef]
31. Misra, G.; Kumar, V.; Agarwal, A.; Agarwal, K. Internet of things (IoT): A technological analysis and survey on vision, concepts, challenges, innovation directions, technologies, and applications. *Am. J. Electr. Electron. Eng.* **2016**, *4*, 2332.
32. Singh, A.; Sharma, T.P. A survey on area coverage in wireless sensor networks. In Proceedings of the International Conference on Control, Instrumentation, Communication and Computational Technologies (ICCICCT), Kanyakumari District, India, 10–11 July 2014.
33. Gendreau, A.A.; Moorman, M. Survey of intrusion detection system towards an end-to-end secure internet of things. In Proceedings of the IEEE 4th International Conference on Future Internet of Things and Cloud, Rome, Italy, 23–25 August 2016; pp. 84–90.
34. Kibirige, W.G. *A Survey on Detection of Sinkhole Attack in Wireless Sensor Network*; Department of Informatics Sokoine University of Agriculture: SUA Morogoro, Tanzania, 2015.
35. Butun, I.; Österberg, P.; Song, H. Security of the Internet of Things: Vulnerabilities, Attacks, and Countermeasures. *IEEE Commun. Surv. Tutor.* **2020**, *22*, 616–644. [CrossRef]
36. Pamarthi, S.; Narmadha, R. Literature review on network security in Wireless Mobile Ad-hoc Network for IoT applications: Network attacks and detection mechanisms. *Int. J. Intell. Unmanned Syst.* 2021, ahead-of-print.
37. Parvathy, K. Wormhole Attacks in Wireless Sensor Networks (WSN) & Internet of Things (IoT): A Review. *Int. J. Recent Technol. Eng. (IJRTE)* **2021**, *10*.
38. Ayotunde, A.F. Internet of Things security: A survey. *J. Netw. Comput. Appl.* **2017**, *8*, 10–28.
39. Granjal, J.; Monteiro, E.; Silva, J. Security for the Internet of things: A survey of existing protocols and open research issues. *IEEE Commun. Surv. Tutor.* **2015**, *17*, 1294–1312. [CrossRef]
40. Sharmaa, V.; Patelb, R.B.; Bhadauriaa, H.S.; Prasadc, D. Deployment schemes in wireless sensor network to achieve blanket coverage in large-scale open area: A review. *Egypt. Inform. J.* **2016**, *17*, 45–56. [CrossRef]
41. Trotta, D.; Garengo, P. Industry 4.0 key research topics: A bibliometric review. In Proceedings of the 2018 7th International Conference on Industrial Technology and Management (ICITM), Oxford, UK, 7–9 May 2018; pp. 113–117.
42. Iwendi, C.; Jalil, Z.; Javed, A.R.; Reddy, T.; Kaluri, R.; Srivastava, G.; Jo, O. Keysplitwatermark: Zero watermarking algorithm for software protection against cyber-attacks. *IEEE Access* **2020**, *8*, 72650–72660. [CrossRef]
43. Wang, W.; Qiu, C.; Yin, Z.; Srivastava, G.; Gadekallu, T.R.; Alsolami, F.; Su, C. Blockchain and PUF-based Lightweight Authentication Protocol for Wireless Medical Sensor Networks. *IEEE Internet Things* **2021**. [CrossRef]
44. Hina, M.; Ali, M.; Javed, A.R.; Srivastava, G.; Gadekallu, T.R.; Jalil, Z. Email Classification and Forensics Analysis using Machine Learning. In Proceedings of the IEEE SmartWorld, Ubiquitous Intelligence and Computing, Advanced & Trusted Computing, Scalable Computing and Communications, Internet of People and Smart City Innovation, Atlanta, GA, USA, 18–21 October 2021; pp. 630–635.
45. Javed, A.R.; Beg, M.O.; Asim, M.; Baker, T.; Al-Bayatti, A.H. Alphalogger: Detecting motion-based side-channel attack using smartphone keystrokes. *J. Ambient. Intell. Humaniz. Comput.* **2020**, 1–14. [CrossRef]
46. Azmoodeh, A.; Dehghantanha; Choo, K.K.R. Robust malware detection for Internet of (battlefield) Things devices using deep eigenspace learning. *IEEE Trans. Sustain. Comput.* **2019**, *4*, 88–95. [CrossRef]
47. Wu, Y.; Dai, H.N.; Wang, H. Convergence of Blockchain and Edge Computing for Secure and Scalable IIoT Critical Infrastructures in Industry 4.0. *IEEE Internet Things J.* **2020**, *8*, 2300–2317. [CrossRef]
48. Ghayvat, H.; Mukhopadhyay, S.; Gui, X.; Suryadevara, N. WSN-and IoT-based smart homes and their extension to intelligent building. *Sjpurnal Sens.* **2015**, *15*, 10350–10379. [CrossRef] [PubMed]
49. Faheem, M. Bio-inspired routing protocol for WSN-based smart grid applications in the context of Industry 4.0. *Trans. Emerg. Telecommun. Technol.* **2019**, *30*, 1–24. [CrossRef]
50. Periša, M.; Kuljanić, T.M.; Cvitić, I.; Kolarovszki, P. Conceptual model for informing user with an innovative smart wearable device in Industry 4.0. *J. Wireess Netw.* **2019**, *9*, 1615–1626. [CrossRef]

51. Kumar, R.; Kumar, P.; Tripathi, R.; Gupta, G.P.; Gadekallu, T.R.; Srivastava, G. Sp2f: A secured privacy-preserving framework for smart agricultural unmanned aerial vehicles. *J. Comput. Netw.* **2020**, *187*, 10781. [CrossRef]
52. Khan, A.; Aziz, S.; Bashir, M.; Khan, M.U. IoT and Wireless Sensor Network based Autonomous Farming Robot. In Proceedings of the International Conference on Emerging Trends in Smart Technologies (ICETST), Karachi, Pakistan, 26–27 March 2020; pp. 1–5.
53. Chen, C.T.; Lee, C.C.; Lin, I.C. Efficient and secure three-party mutual authentication key agreement protocol for WSN in IoT environments. *PLoS ONE* **2020**, *15*, e0232277.
54. Arroyo, P.; Lozano, J.; Suárez, J. Evolution of Wireless Sensor Network for Air Quality Measurements. *J. Electron.* **2018**, *7*, 342. [CrossRef]
55. Wittenberg, C. Cause the Trend Industry 4.0 in the Automated Industry to New Requirements on User Interfaces? *Int. Conf.-Hum.-Comput. Interact.* **2015**, *9171*, 238–245.
56. Henze, M.; Hermerschmidt, L.; Kerpen, D.; Häubling, R.; Rumpe, B.; Wehrle, K. A comprehensive approach to privacy in the cloud-based Internet of Things. *J. Future Gener. Comput. Syst.* **2016**, *56*, 701–718. [CrossRef]
57. Stverkova, H.; Pohludka, M. Business Organisational Structures of Global Companies: Use of the Territorial Model to Ensure Long-Term Growth. *Soc. Sci.* **2018**, *7*, 98. [CrossRef]
58. Rao, G.S.; Prasad, R. Impact of 5G Technologies on Industry 4.0. *Wirel. Pers. Commun.* **2018**, *100*, 145–159. [CrossRef]
59. Lange, S. Digitalization, and energy consumption. Does ICT reduce energy demand? *J. Ecol. Econ.* **2020**, *176*, 106760. [CrossRef]
60. Müller, J.M. Fortune favors the prepared: How SMEs approach business model innovations in Industry 4.0. *Technol. Forecast. Soc. Change* **2018**, *132*, 2–17. [CrossRef]
61. Masood, T.; Sonntag, P. Industry 4.0: Adoption challenges and benefits for SMEs. *J. Comput. Ind.* **2020**, *121*, 103261. [CrossRef]
62. Zhu, J. Measurement and analysis of corporate operating vitality in the age of digital business models. *Appl. Econ. Lett.* **2020**, *27*, 511–517. [CrossRef]
63. Serror, M.; Hack, S.; Henze, M.; Schuba, M.; Wehrle, K. Challenges and Opportunities in Securing the Industrial Internet of Things. *IEEE Trans. Ind. Inform.* **2021**, *17*, 2985–2996. [CrossRef]
64. Li, X.; Xu, M.; Vijayakumar, P.; Kumar, N.; Liu, X. Detection of low-frequency and multi-stage attacks in industrial internet of things. *IEEE Trans. Veh. Technol.* **2020**, *69*, 8820–8831. [CrossRef]
65. Akram, Z.; Majid, M.; Habib, S. A Systematic Literature Review: Usage of Logistic Regression for Malware Detection. In Proceedings of the International Conference on Innovative Computing (ICIC), Seoul, Korea, 9–10 November 2021; pp. 1–8.
66. Khalil, R.A.; Saeed, N.; Masood, M.; Fard, Y.M.; Alouini, M.-S.; Al-Naffouri, T.Y. Deep learning in the industrial internet of things: Potentials, challenges, and emerging applications. *IEEE Internet Things J.* **2021**, *8*, 11016–11040. [CrossRef]
67. Jiang, B.; Li, J.; Yue, G.; Song, H. Differential privacy for Industrial Internet of Things: Opportunities, applications and challenges. *IEEE Internet Things J.* **2021**, *8*, 10430–10451. [CrossRef]
68. Ahmed, A.; Javed, A.R.; Jalil, Z.; Srivastava, G.; Gadekallu, T.R. Privacy of Web Browsers: A Challenge in Digital Forensics. In Proceedings of the International Conference on Genetic and Evolutionary Computing, Jilin, China, 21–21 October **2021**; Springer: Singapore, 2021; pp. 493–504.
69. Vangala, A.; Das, A.K.; Kumar, N.; Alazab, M. Smart secure sensing for IoT-based agriculture: Blockchain perspective. *IEEE Sensors J.* **2021**, *21*, 17591–17607. [CrossRef]
70. Maddikunta, P.K.R.; Pham, Q.V.; Prabadevi, B.; Deepa, N.; Dev, K.; Gadekallu, T.R.; Liyanage, M. Industry 5.0: A survey on enabling technologies and potential applications. *J. Ind. Inf. Integr.* **2021**, 100257. [CrossRef]
71. Liu, B.; Dousse; Wang, J.; Saipulla, A. Strong barrier coverage of wireless sensor networks. In Proceedings of the 9th ACM International Symposium on Mobile Ad Hoc Networking and Computing, Hong Kong, China, 26–30 May 2008.
72. Abomhara, M.; Køien, G.M. Cyber security and the internet of things: Vulnerabilities, threats, intruders and attacks. *J. Cyber Secur. Mobil.* **2015**, *4*, 65–88. [CrossRef]
73. Benazzouz, Y.; Munilla, C.; Gunalp, O.; Gallissot, M.; Gurgen, L. Sharing user iot devices in the cloud. In Proceedings of the IEEE World Forum Internet Things (WF-IoT), Seoul, Korea, 6–8 March 2014; pp. 373–374.
74. Kumar, S.A.; Vealey, T.; Srivastava, H. Security in internet of things: Challenges, solutions and future directions. In Proceedings of the 49th Hawaii International Conference on System Sciences (HICSS), Koloa, HI, USA, 5–8 January 2016; pp. 5772–5781.
75. Rehman, S.U.; Khaliq, M.; Imtiaz, S.I.; Rasool, A.; Shafiq, M.; Javed, A.R.; Jalil, Z.; Bashir, A.K. Diddos: An approach for detection and identification of distributed denial of service (ddos) cyberattacks using gated recurrent units (gru). *Future Gener. Comput. Syst.* **2021**, *118*, 453–466. [CrossRef]
76. Iwendi, C.; Rehman, S.U.; Javed, A.R.; Khan, S.; Srivastava, G. Sustainable security for the internet of things using artificial intelligence architectures. *Acm Trans. Internet Technol. (TOIT)* **2021**, *21*, 1–22. [CrossRef]
77. Anwar, R.W.; Bakhtiari, M.; Zainal, A.; Abdullah, A.H.; Qureshi, K.N. Security issues and attacks in wireless sensor network. *World Appl. Sci. J.* **2014**, *30*, 1224–1227.
78. Chandan, R.; Mishra, P.K. A Review of Security Challenges in Ad-Hoc Network. *Int. J. Appl. Eng. Res.* **2018**, *13*, 16117–16126.
79. Lee, J. Recent advances and trends in predictive manufacturing systems in a big data environment. *J. Manuf. Lett.* **2015**, *1*, 38–41. [CrossRef]
80. Wollschlaeger, M.; Sauter, T.; Jasperneite, J. The Future of Industrial Communication: Automation Networks in the Era of the Internet of Things and Industry 4.0. *IEEE Ind. Electron. Mag.* **2017**, *11*, 17–27. [CrossRef]

81. Elsadig, M.A.; Altigani, A.; Baraka, M.A.A. Security issues and challenges on wireless sensor networks. *Int. J. Adv. Trends Comput. Sci. Eng.* **2019**, *8*, 1551–1559. [CrossRef]
82. Hossain, A.J.; Fotouhi, M.; Hasan, R. Towards an Analysis of Security Issues, Challenges, and Open Problems in the Internet of Things. In Proceedings of the IEEE World Congress on Services, New York, NY, USA, 27 June–2 July 2015; pp. 21–28.
83. Ling, Z.; Liu, K.; Xu, Y.; XinwenFu, Y. An End-toEnd View of IoT Security and Privacy. In Proceedings of the IEEE Global Communications Conference, Singapore, 4–8 December 2017; pp. 1–7.
84. Wazid, M. Design and Analysis of Intrusion Detection Protocols for Hierarchical Wireless Sensor Networks. Ph.D. Dissertation, Center Secure, Theory Algorithmic Res, International Institute of Information Technology, Hyderabad, India, 2017.
85. Haseeb, K.; Islam, N.; Saba, T.; Rehman, A.; Mehmood, Z. LSDAR: A light-weight structure-based data aggregation routing protocol with secure Internet of things integrated next-generation sensor networks. *J. Sustain. Cities Soc.* **2020**, *54*, 101995. [CrossRef]
86. Capriglione, D.; Casinelli, D.; Ferrigno, L. Analysis of quantities influencing the performance of time synchronization based on linear regression in low-cost WSN. *J. Meas.* **2016**, *77*, 105–116. [CrossRef]
87. Ng, K.P.; Tsimenidis, C.; Woo, W.L. C–Sync: Counter-based synchronization for duty-cycled wireless sensor networks. *J. Hoc Netw.* **2017**, *61*, 51–64. [CrossRef]
88. Tian, Y.P. Time synchronization in WSN with random bounded communication delays. *Trans. Autom. Control* **2017**, *62*, 5445–5450. [CrossRef]
89. Drushti, D.; Upadhyay, H. Security and Privacy Consideration for Internet of Things in Smart Home Environments. *Int. J. Eng. Res. Dev.* **2015**, *10*, 73–83.
90. Fink, G.A.; Zarzhitsky, D.V.; Carroll, T.E.; Farquhar, E.D. Security and privacy grand challenges for the Internet of Things. In Proceedings of the International Conference on Collaboration Technologies and Systems (CTS), Atlanta, GA, USA, 1–5 June 2015; IEEE: Piscataway, NJ, USA, 2015; pp. 27–34.
91. Faheem, M.; Abbas, M.Z.; Tuna, G.; Gungor, C. EDHRP: Energy efficient event driven hybrid routing protocol for densely deployed wireless sensor network. *J. Netw. Comput. Appl.* **2015**, *58*, 309–326. [CrossRef]
92. Jamshidi, M.; Zangeneh, E.; Esnaashari, M.; Darwesh, A.M.; Meybodi, A.J. A Novel Model of Sybil Attack in Cluster-Based Wireless Sensor Networks and Propose a Distributed Algorithm to Defend It. *J. Wirel. Pers. Commun.* **2019**, *105*, 145–173. [CrossRef]
93. Yang, G.; Dai, L.; Wei, Z. Challenges, threats, security issues, and new trends of underwater wireless sensor networks. *J. Sens.* **2018**, *18*, 3907. [CrossRef]
94. Ziegeldorf, J.H.; Morchon, O.G.; Wehrle, K. Privacy in the Internet of Things: Threats and challenges. *Secur. Commun. Netw.* **2015**, *12*, 2728–2742. [CrossRef]
95. Vasilomanolakis, E.; Daubert, J.; Luthra, M.; Gazis, V.; Wiesmaier, A.; Kikiras, P. On the Security and Privacy of Internet of Things Architectures and Systems. In Proceedings of the International Workshop onSecure Internet of Things (SIoT), Vienna, Austria, 21–25 September 2015.
96. Mehta, R.; Sahnib, J.; Khannac, K. Internet of Things: Vision, Applications and Challenges. In Proceedings of the International Conference on Computational Intelligence and Data Science (ICCIDS 2018), Procedia Computer Science, Gurugram, India, 7–8 April 2018; Volume 132, pp. 1263–1269.
97. Sfar, A.R.; Natalizio, E.; Challal, Y.; Chtourou, Z. A roadmap for security challenges in the Internet of Things. *J. Digit. Commun. Netw.* **2018**, *4*, 118–137. [CrossRef]
98. Challa, S.; Das, A.K.; Odelu, V.; Kumar, N.; Khan, S.; Kumari, M.K. An efficient ECC-based provably secure three-factor user authentication and key agreement protocol for wireless healthcare sensor networks. *J. Comput. Electr. Eng.* **2018**, *69*, 534–554. [CrossRef]
99. Ranjan, A.K.; Somani, G. *Access Control and Authentication in the Internet of Things Environment*; Springer International Publishing: Cham, Germany, 2016; pp. 283–305.
100. Sicari, S.; Rizzardi, A.; Grieco, L.A.; Coen-Porisini, A. Security, privacy and trust in Internet of things: The road ahead. *J. Comput. Netw.* **2015**, *76*, 146–164. [CrossRef]
101. Garg, S.; Kaur, K.; Kaddoum, G.; Choo, K.K.R. Toward Secure and Provable Authentication for Internet of Things: Realizing Industry 4.0. *J. IEEE Internet Things* **2020**, *7*, 4598–4606. [CrossRef]
102. Rathee, G.; Balasaraswathi, M.; Chandran, K.P. A secure IoT sensors communication in Industry 4.0 using blockchain technology. *J. Ambient. Intell. Humaniz. Comput.* **2021**, *12*, 533–545. [CrossRef]
103. Mellado, S.H.; Clemente, R.M.; Lecuyer, V.B. Prediction of satellite shadowing in smart cities with application to IoT. *J. Sens.* **2020**, *20*, 1–19.
104. Behera, T.M.; Mohapatra, G.S.; Samal, U.C.; Han, M.G.S.; Daneshmand, M.; Gandomi, A.H. Residual energy-based cluster-head selection in WSN for IoT application. *J. IEEE Internet Things* **2019**, *6*, 5132–5139. [CrossRef]
105. Wan, J.; Tang, S.; Shu, Z.; Li, D.; Wang, S.; Imran, M.; Vasilakos, A.V. Software-Defined Industrial Internet of Things in the Context of Industry 4.0. *J. Sens.* **2016**, *16*, 7373–7380. [CrossRef]
106. Tan, Y.; Yang, W.; Yoshida, K.; Takakuwa, S. Application of IoT-aided simulation to manufacturing systems in the cyber-physical system. *J. Mach.* **2019**, *7*, 2. [CrossRef]

107. Rahman, A.; Sara, U. DistB-SDoIndustry: Enhancing security in Industry 4.0 services based on distributed blockchain through software defined networking-IoT enabled architecture. *Int. J. Adv. Comput. Sci. Appl. (IJACSA)* **2020**, *11*, 674–682. [CrossRef]
108. Haque, M.E.; Asikuzzaman, M.; Khan, I.U.; Ra, I.H.; Hossain, M.S.; Shah, S.B.H. Comparative study of IoT-based topology maintenance protocol in a wireless sensor network for structural health monitoring. *J. Remote Sens.* **2020**, *12*, 2538. [CrossRef]
109. Wan, J.; Chen, B.; Imran, M.; Tao, F.; Li, D.; Liu, C.; Ahmad, S. Toward Dynamic Resources Management for IoT-Based Manufacturing. *IEEE Commun. Mag.* **2018**, *56*, 52–59. [CrossRef]
110. Bulaghi, Z.A.; Navin, A.H.Z.; Hosseinzadeh, M.; Rezaee, A. SENET: A novel architecture for IoT-based body sensor networks. *Inform. Med. Unlocked* **2020**, *20*, 100365. [CrossRef]
111. Jesus, T.C.; Portugal, P.; Costa, D.G.; Vasques, F. A comprehensive dependability model for QOM-aware industrial wsn when performing visual area coverage in occluded scenarios. *J. Sens.* **2020**, *20*, 1–22. [CrossRef]
112. Arroyo, P.; Herrero, J.L.; Suárez, J.I.; Lozano, J. Wireless sensor network combined with cloud computing for air quality monitoring. *J. Sens.* **2019**, *19*, 691. [CrossRef]
113. Li, X.; Zhu, L.; Chu, X.; Fu, H. Edge Computing-Enabled Wireless Sensor Networks for Multiple Data Collection Tasks in Smart Agriculture. *J. Sens.* **2020**, *2020*. [CrossRef]
114. Prophess, N.; Kumar, R.; Gnanadhas, J.B. Cluster Centroid-Based Energy Efficient Routing Protocol for WSN-Assisted IoT, in journal of Advances in Science. *Technol. Eng. Syst.* **2020**, *5*, 296–313.
115. Haseeb, K.; Din, I.U.; Almogren, A.; Islam, N. An energy-efficient and secure IoT-based WSN framework: An application to smart agriculture. *J. Sens.* **2020**, *20*, 2081. [CrossRef]
116. Forcina, A.; Falcone, D. The role of Industry 4.0 enabling technologies for safety management: A systematic literature review. *Int. Conf. Ind. 4.0 Smart Manuf.* **2021**, *180*, 436–445. [CrossRef]
117. Birkela, H.; Müllerb, J.M. Potentials of Industry 4.0 for supply chain management within the triple bottom line of sustainability—A systematic literature review. *J. Clean. Prod.* **2021**, *289*, 125612. [CrossRef]
118. Mrugalska, B.; Ahmed, J. Organizational Agility in Industry 4.0: A Systematic Literature Review. *J. Sustain.* **2021**, *13*, 8272. [CrossRef]
119. Atif, S.; Ahmed, S.; Wasim, M.; Zeb, B.; Pervez, Z.; Quinn, L. Towards a Conceptual Development of Industry 4.0, Servitisation, and Circular Economy: A Systematic Literature Review-2021. *J. Sustain.* **2021**, *13*, 6501. [CrossRef]
120. Jesús, H.O.; William, G.M.; Leonardo, Z.C. Industry 4.0: Current Status and Future Trends. In *Industry 4.0* ; InTechOpen: London, UK, 2020.
121. Wei, Q.; Siqi, C.; Mugen, P. Recent advances in Industrial Internet: Insights and challenges. *J. Digit. Commun. Netw.* **2020**, *6*, 1–13.
122. Taigang, K.V.L.; Lifeng, Z. Industry 4.0: Towards future industrial opportunities and challenges. In Proceedings of the 12th International Conference on Fuzzy Systems and Knowledge Discovery (FSKD), Zhangjiajie, China, 15–17 August 2015; pp. 2147–2152.
123. Zhang, W.; Zhang, Y.; Gu, X.; Wu, C.; Han, L. Soft Computing. In *Application of Soft Computing, Machine Learning, Deep Learning and Optimizations in Geoengineering and Geoscience*; Springer: Singapore, 2022; pp. 7–19.
124. Wang, S.; Qureshi, M.A.; Miralles-Pechuaán, L.; Huynh-The, T.; Gadekallu, T.R.; Liyanage, M. Explainable AI for B5G/6G: Technical Aspects, Use Cases, and Research Challenges. *arXiv* **2021**, arXiv:2112.04698.
125. Ramu, S.P.; Boopalan, P.; Pham, Q.V.; Maddikunta, P.K.R.; The, T.H.; Alazab, M.; Gadekallu, T.R. Federated Learning enabled Digital Twins for smart cities: Concepts, recent advances, and future directions. *Sustain. Cities Soc.* **2022**, 103663. [CrossRef]
126. Połap, D.; Srivastava, G.; Lin, J.C.W.; Woźniak, M. Federated Learning Model with Augmentation and Samples Exchange Mechanism. In Proceedings of the International Conference on Artificial Intelligence and Soft Computing, Zakopane, Poland, 12–14 October 2021; Springer: Cham, Germany, 2021; pp. 214–223.
127. Saibabu, G.; Jain, A.; Sharma, V.K. Security Issues and Challenges in IoT Routing over Wireless Communication. *Int. J. Innov. Technol. Explor. Eng. (IJITEE)* **2020**, *9*, 1572–1580.

MDPI
St. Alban-Anlage 66
4052 Basel
Switzerland
Tel. +41 61 683 77 34
Fax +41 61 302 89 18
www.mdpi.com

Sensors Editorial Office
E-mail: sensors@mdpi.com
www.mdpi.com/journal/sensors

www.ingramcontent.com/pod-product-compliance
Lightning Source LLC
Chambersburg PA
CBHW061758210926
51850CB00006B/592
9783036574684